Handbook of Experimental Pharmacology

Volume 165

Editor-in-Chief

K. Starke, Freiburg i. Br.

Editorial Board

G.V.R. Born, London
M. Eichelbaum, Stuttgart
D. Ganten, Berlin
F. Hofmann, München
B. Kobilka, Stanford, CA
W. Rosenthal, Berlin
G. Rubanyi, Richmond, CA

Springer
*Berlin
Heidelberg
New York
Hong Kong
London
Milan
Paris
Tokyo*

Cell Adhesion

Contributors
G. Berx, L. Borradori, T. Brabletz, P.F. Bradfield, H.-J. Choi,
S. Getsios, L.M. Godsel, K.J. Green, A.K. Horst, A.C. Huen,
W. Ikeda, B.A. Imhof, K. Irie, R.L. Juliano, J. Koster,
Y. Miyamoto, U. Müller, W.J. Nelson, B. Nieswandt,
S. Offermanns, T.D. Perez, P. Reddig, T. Sakisaka, H. Semb,
K. Shimizu, A. Sonnenberg, E.T. Stoeckli, K. Strumane,
Y. Takai, F. Van Roy, C. Wagener, W.I. Weis

Editors
Jürgen Behrens and W. James Nelson

Springer

Professor
Dr. Jürgen Behrens
Nikolaus-Fiebiger-Zentrum
für Molekulare Medizin
Universität Erlangen-Nürnberg
Glückstrasse 6
91054 Erlangen, Germany
e-mail: jbehrens@molmed.uni-erlangen.de

W. James Nelson, Ph.D.
Department of Molecular & Cellular Physiology
Beckman Center B121
Stanford University School of Medicine
Stanford, CA 94305-5345
USA
e-mail: wjnelson@stanford.edu

With 59 Figures and 10 Tables

ISSN 0171-2004
ISBN 3-540-20941-7 Springer-Verlag Berlin Heidelberg New York

Library of Congress Cataloging-in-Publication Data
Cell adhesion / contributors, G Berx ... [et al.] ; editors, Jürgen Behrens and W. James Nelson.
  p. m. – (Handbook of experimental pharmacology ; v. 165)
Includes bibliographical references and index.
ISBN 3-540-20941-7 (alk. paper)
1. Cell adhesion molecules. 2. Cell adhesion. I. Berx, G. II. Behrens, Jürgen, Dr. III. Nelson, W. J. (W. James) IV. Series.
QP905.H3 vol. 165   [QP552.C42]   615'.1s–dc22   [571.6]   2004042951

This work is subject to copyright. All rights are reserved, whether the whole or part of the material is concerned, specifically the rights of translation, reprinting, re-use of illustrations, recitation, broadcasting, reproduction on microfilm or in any other way, and storage in data banks. Duplication of this publication or parts thereof is permitted only under the provisions of the German Copyright Law of September 9, 1965, in its current version, and permission for use must always be obtained from Springer-Verlag. Violations are liable to Prosecution under the German Copyright Law.

Springer-Verlag is a part of Springer Science+Business Media
springeronline.com

© Springer-Verlag Berlin Heidelberg 2004
Printed in Germany

The use of general descriptive names, registered names, etc. in this publication does not imply, even in the absence of a specific statement, that such names are exempt from the relevant protective laws and regulations and free for general use.

Product liability: The publishers cannot guarantee the accuracy of any information about dosage and application contained in this book. In every individual case the user must check such information by consulting the relevant literature.

Editor: Dr. R. Lange
Desk Editor: S. Dathe
Cover design: design & production GmbH, Heidelberg
Typesetting: Stürtz AG, 97080 Würzburg

Printed on acid-free paper   27/3150 hs – 5 4 3 2 1 0

Preface

As with other areas of biological research, the progress that has been made over the past 25 years in the field of cell adhesion is impressive. In the late 1970s, the search for specific cell surface receptors for adhesion processes was initiated using mainly biochemical and immunological approaches. Since then, the introduction of novel methods of cellular and molecular biology and powerful techniques for manipulating gene expression in transgenic and knock-out mice have greatly advanced the field. Not only do we now know the precise molecular structure of many of the cell adhesion receptors that were postulated to exist in the early days, but we have a clear picture of their function, and evidence of their involvement in signal transduction. We have also found clues to the role of cell adhesion in normal embryonal development and adult physiology, and we have evidence that disturbance in cell adhesion can cause disease.

In this volume, our goal was to provide an overview of the main topics of current cell adhesion research, including structural analyses of cell adhesion molecules and studies of their functional role in vitro and in vivo. We have focussed mainly on the four major families of cell-adhesion receptors, i.e. the cadherins, the integrins, the Ig superfamily and the selectin-based adhesion system. The chapters by Perez and Nelson and Choi and Weis describe the structural basis of cadherin function, focussing on the extracellular domain of cadherins and the cytoplasmic tail interactions with catenins, respectively. Semb reviews the in vivo functions of cadherins with an emphasis on normal embryonal development, and Strumane et al. describe the alterations of cadherins in cancer development and metastasis. A chapter by Brabletz is devoted to the in vivo role of the catenins, highlighting the signal transduction function of β-catenin in the Wnt pathway. The molecular composition and function of desmosomes are the topic of the chapter by Godsel and colleagues, introducing desmosomal cadherins and their link to the cytoskeleton. Integrins are dealt with in three reviews: an overview of signal transduction triggered by integrin-mediated adhesion is given by Miyamoto et al., the in vivo function of integrins as revealed by mouse models is described by Müller, and the role of integrins and associated components in the hemidesmosome is reviewed by Koster and co-authors. In the section on Ig superfamily adhesion receptors, Horst and Wagener and Irie and colleagues describe the structure and cellular function of two major subfamilies, the CEACAMs and nectins, respectively. Stoeckli has contributed a chapter on the role of the Ig superfamily in brain development where the diversity of these adhesion receptors is particularly prominent. Two chapters deal with the physiological function of adhesion receptors in the adult organism. The chapter by Bradfield and Imhof focusses on leukocyte–endothelium interactions and introduces the selectins and their interplay with Ig receptors and integrins, and the chapter by Nieswandt and Offermanns deals with platelet aggregation in hemostasis and thrombosis, where detailed molecular analysis has already reached the level of pharmacological application.

What will be the future trends in cell adhesion research and the potential application of this knowledge for human disease, and in particular for pharmacological applications (which is the underlying interest of this handbook series)? It can be expected that our knowledge of the in vivo functions of cell adhesion receptors and mechanisms will continue to increase as more mouse models are generated. Naturally, this will have an impact on our understanding of disease processes that are related to defective cell adhesion. Paradigms for this, to name only two, are the implications of E-cadherin being a metastasis suppressor in human cancer, or the detailed knowledge of blood clotting being based on specific cell adhesion processes. Moreover, the perception of cell adhesion molecules as signalling receptors and their detailed structural analysis will hopefully set the stage for new pharmacological approaches. Thus, cell adhesion research will stay as exciting as it has been.

I hope that this book will serve as a helpful guide for students and for researchers from other disciplines, and also be a valuable source of reference for our colleagues in the field.

Jürgen Behrens, Erlangen

List of Contributors

(Addresses stated at the beginning of respective chapters)

Berx, G.   69
Borradori, L.   243
Brabletz, T.   105
Bradfield, P.F.   405

Choi, H.-J.   23

Getsios, S.   137
Godsel, L.M.   137
Green, K.J.   137

Horst, A.K.   283
Huen, A.C.   137

Ikeda, W.   343
Imhof, B.A.   405
Irie, K.   343

Juliano, R.L.   197

Koster, J.   243

Miyamoto, Y.   197
Müller, U.   217

Nelson, W.J.   3
Nieswandt, B.   437

Offermanns, S.   437

Perez, T.D.   3

Reddig, P.   197

Sakisaka, T.   343
Semb, H.   53
Shimizu, K.   343
Sonnenberg, A.   243
Stoeckli, E.T.   373
Strumane, K.   69

Takai, Y.   343

Van Roy, F.   69

Wagener, C.   283
Weis, W.I.   23

List of Contents

Part I. Cadherins

Cadherin Adhesion: Mechanisms and Molecular Interactions 3
    T.D. Perez, W.J. Nelson

Structural Aspects of Adherens Junctions and Desmosomes 23
    H.-J. Choi, W.I. Weis

Cadherins in Development 53
    H. Semb

Cadherins in Cancer. 69
    K. Strumane, G. Berx, F. Van Roy

In Vivo Functions of Catenins 105
    T. Brabletz

The Molecular Composition and Function of Desmosomes 137
    L.M. Godsel, S. Getsios, A.C. Huen, K.J. Green

Part II. Integrins and Extracellular Matrix

Regulation of Signal Transduction by Integrins 197
    Y. Miyamoto, P. Reddig, R.L. Juliano

Integrins and Extracellular Matrix in Animal Models 217
    U. Müller

Hemidesmosomes: Molecular Organization and Their Importance
for Cell Adhesion and Disease. 243
    J. Koster, L. Borradori, A. Sonnenberg

Part III. Immunoglobulin Superfamily

CEA-Related CAMs 283
    A.K. Horst, C. Wagener

Roles of Nectins in Cell Adhesion, Signaling and Polarization 343
    K. Irie, K. Shimizu, T. Sakisaka, W. Ikeda, Y. Takai

Ig Superfamily Cell Adhesion Molecules in the Brain 373
    E.T. Stoeckli

Part IV. Physiology and Pathology of Cell Adhesion

Adhesion Mechanisms of Endothelial Cells 405
    P.F. Bradfield, B.A. Imhof

Pharmacology of Platelet Adhesion and Aggregation 437
    B. Nieswandt, S. Offermanns

Subject Index 473

Part I
Cadherins

Cadherin Adhesion: Mechanisms and Molecular Interactions

T. D. Perez · W. J. Nelson (✉)

Department of Molecular and Cellular Physiology, Stanford University,
School of Medicine, 279 Campus Dr, Beckman Center B121, Stanford,
CA 94305-5435, USA
wjnelson@stanford.edu

1	Introduction	3
2	Ca^{++} and Cadherin Adhesion	6
3	**Cadherin Adhesion Structure: *Trans*-Dimer**.	7
3.1	Early Studies	7
3.2	Recent Studies, Evolving Models	11
3.3	Further Functional Studies of EC Domains in Adhesion	13
4	**Cadherin Adhesion Structure: *Cis*-Dimer**	15
	References	19

Abstract Cadherins constitute a superfamily of cell–cell adhesion molecules expressed in many different cell types that are required for proper cellular function and maintenance of tissue architecture. Classical cadherins are the best understood class of cadherins. They are single membrane spanning proteins with a divergent extracellular domain of five repeats and a conserved cytoplasmic domain. Binding between cadherin extracellular domains is weak, but strong cell–cell adhesion develops during lateral clustering of cadherins by proteins that link the cadherin cytoplasmic domain to the actin cytoskeleton. Understanding how different regions of cadherins regulate cell–cell adhesion has been a major focus of study. Here, we examine evidence of the structure and function of the extracellular domain of classical cadherins in regard to the control of recognition and adhesive contacts between cadherins on opposing cell surfaces. Early experiments that focused on understanding the homotypic, Ca^{++}-dependent characteristics of cadherin adhesion are discussed, and data supporting the widely accepted *cis*- and *trans*-dimer models of cadherins are analyzed.

Keywords Cadherin · Cell–cell adhesion · Calcium · Structure · Dimer

1
Introduction

Cadherins are an important superfamily of cell–cell adhesion proteins comprising over 40 members (Takeichi 1990), and perhaps even more when the proto-cadherin family is included (Frank and Kemler 2002). Cadherins are expressed in many cells and tissues, and are evolutionarily conserved in vertebrates and invertebrates (Kemler 1992; Takeichi 1995; Gumbiner 1996). Here we focus on a subset of cadherins, the classical cadherins, that are the best described and understood. We focus on the structure and function of the extracellular domain that controls recognition and adhesive contacts between cadherins on opposing cell surfaces.

Classical cadherins, of which E-, P-, N-, and R-cadherin are members, were shown to mediate segregation of different cell populations in early studies. Segregation was based on homophilic adhesion in which cells expressing one cadherin subtype (E-cadherin for example) segregated in suspension from cells expressing a different cadherin subtype (P-cadherin). Cadherin-mediated homotypic adhesion appears to depend on binding specificity, Ca^{++}-dependence, and molecular contacts of cadherins and cadherin–cadherin adhesion. Sequence analysis of classical cadherins reveals that they have five tandemly repeated domains in the extracellular domain, termed extracellular cadherin repeats 1–5 (EC1–5) (Fig. 1) with EC1 at the amino terminus (Hatta et al. 1988). When synthesized, cadherins contain signal and precursor peptides that are cleaved during processing and maturation of the protein in the endoplasmic reticulum (ER) and Golgi. The precursor peptide is cleaved before cell surface presentation of the cadherin, and cleavage is required for adhesive function (Ozawa and Kemler 1990).

How do cadherin extracellular domains interact to form cell–cell adhesions? The most prevalent models describe two cadherins within the same membrane forming a lateral or *cis*-dimer and that this dimer promotes ad-

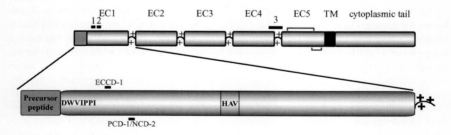

Fig. 1. Schematic of a classical cadherin. The schematic depicts the domain organization of a classical cadherin. Five extracellular repeats (EC1–5) are preceded by a precursor peptide which is cleaved during maturation. Ca^{++}-binding sites (+ *marks*) are located at each EC junction. EC5 is followed by a single transmembrane segment and a highly conserved cytosolic domain which associates with members of the catenin protein family. Monoclonal antibody binding regions are marked with *numbered black bars* above the repeats and correspond to binding sites of (*1*) ECCD-1, (*2*) PCD1 and NCD1, and (*3*) DECMA. Disulfide bonds are shown in EC5 as *gray brackets*. A close view of EC1 also shows the histidine-alanine-valine (HAV) tripeptide as well as the highly conserved first seven residues of the mature protein including Trp2

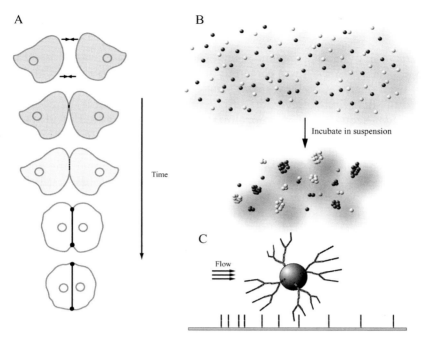

Fig. 2A–C. General methods to examine cadherin adhesion functions. **A** Schematic of cadherin distribution during epithelial cell–cell adhesion as imaged by live cell microscopy of Madin-Darby canine kidney (MDCK) cells expressing green fluorescent protein (GFP)-E-cadherin (Adams et al. 1998). **B** Cell segregation assay. Cells expressing one subtype of cadherin (*dark spheres*) are mixed with cells expressing a second subtype of cadherin (*light spheres*). The cells are incubated in suspension. Cadherins mediate segregation of cells expressing different cadherin subtypes (*no dark and light mixed aggregates*). **C** Adhesion flow assay. Recombinantly expressed fragments of the cadherin extracellular domain are attached to a substrate. Beads coated with the same cadherin fragment are flowed over the surface and their migration is monitored. Events in which the beads stop moving are interpreted as adhesion events

hesive dimerization (*trans*-dimer) of cadherins on adjacent cells (Fig. 2). The models differ in their description of the organization and mechanism of the extracellular domains in the *trans*-dimers. Two models describe *trans*-dimerization through EC1 alone while a third describes complete intercalation of cadherin extracellular domains. Each model further depicts that both *cis*- and *trans*-dimerization depend on Ca^{++} to bind between each EC repeat to stabilize and order extracellular domains. We will analyze the data supporting these conclusions.

2
Ca^{++} and Cadherin Adhesion

Understanding the Ca^{++}-dependence of cadherin adhesion was addressed in early experiments. Ca^{++} was shown to protect the cadherin extracellular domain from degradation by trypsin, suggesting a role in structural organization of the domain. However, the mechanism remains unclear. After solving the protein sequence of uvomorulin (E-cadherin), Ringwald et al. (1987) identified three internal repeats in the extracellular domain. Within these domains they identified two distinct putative Ca^{++}-binding loops based on sequence homology to known Ca^{++}-binding regions. Although the analysis of the extracellular domain was not exactly correct, the authors first demonstrated the basic structure of the cadherin extracellular domain, a functional domain composed of distinct, repeated units each of which are likely able to bind Ca^{++}. A year later with the sequencing of N-cadherin and use of a different alignment program, the five internal cadherin repeats were identified (Hatta et al. 1988). The authors, however, did not comment on Ca^{++} binding sites. Nevertheless, sequence alignment of the known cadherins shows conservation in each of the five internal repeats of the putative Ca^{++}-binding sites proposed by Ringwald et al. (1987).

Evidence of the functional role of Ca^{++} in cadherin adhesion came initially from studies on the recombinant extracellular domain of E-cadherin (Pokutta et al. 1994). Electron microscopy of the recombinant domain directly showed dependence on Ca^{++} for elongation and maintenance of a rigid bent rod-like structure, but the curvature was not commented upon. In the absence of Ca^{++}, the domain was apparently disordered and not elongated, and resembled a globular structure. Ca^{++} binding was reversible. These conformational changes were confirmed by circular dichroism spectroscopy. A large change in the ellipticity of the CD spectrum was observed between 202–215 nm after removal of Ca^{++}. The change in CD spectra was used to examine conformational changes at different Ca^{++} concentrations, and an average K_d of 42–45 μM was determined. As there were multiple Ca^{++} ions known to bind to the cadherin extracellular domain, it was not possible to determine whether all sites have the same K_d or the value obtained was an average. The authors also noticed an increase in the fluorescence of tryptophan upon Ca^{++} binding. They used this characteristic to titrate Ca^{++} as well and determine two different K_d values, one at 130 μM and the second at 210 μM. The results suggested that the K_d obtained from titration based on CD spectrum changes was an average, and the change in tryptophan fluorescence measured a Ca^{++} binding site of low affinity. Finally, Pokutta et al. (1994) measured Ca^{++} affinities by protection against tryptic digestion. From these experiments they observed a K_d of 24 μM, but there appeared to be cooperative binding here as well. Though unable to satisfactorily measure exact dissociation constants of the different Ca^{++} binding sites, the authors were able to prove the existence of high and low affinity binding sites and first observed and modeled the mechanism of Ca^{++}-dependent adhesion whereby Ca^{++} acted to rigidify the extracellular domain.

The dissociation constants of Ca^{++} at the various binding sites were also determined using a fragment of the first two EC repeats of E-cadherin (ECAD12) (Koch et al. 1997). By monitoring the CD spectra change of ECAD12, an average K_d of 360 μM for the Ca^{++} binding sites between EC1 and EC2 was recorded. ECAD12 bound three Ca^{++} ions, and using equilibrium dialysis two K_d values of 330 μM and one K_d value of 2 mM were determined. Interestingly, when the Ca^{++} binding sites of the full E-cadherin extracellular domain were analyzed by equilibrium dialysis, an average K_d of only 30 μM was calculated for a total of nine Ca^{++} ions bound to the protein fragment.

3
Cadherin Adhesion Structure: *Trans*-Dimer

3.1
Early Studies

Initial studies of cadherin extracellular domain adhesion focused on determining which EC repeats are involved in adhesion and specificity. Nose et al. (1990) used a domain swapping strategy to determine the location of sites controlling cadherin subtype specificity (i.e., which domains controlled E↔E, P↔P specificity). When expressed in cadherin-deficient fibroblast L cells, wildtype E- and P-cadherins mediated sorting and segregation of cells expressing the same cadherin (Fig. 3). However, when E-cadherin EC1 was switched with EC1 of P-cadherin to generate a chimera of P-cadherin EC1 and E-cadherin EC2–5, the chimeric protein now mediated adhesion with L cells expressing P-cadherin. The authors concluded that EC1 contained sites that determine cadherin adhesion specificity. They further narrowed the region to between residues 61 and 113. However, it is noteworthy that swapping sub-regions of EC1 (residues 1–31 and 1–67) did not provide a complete switch in adhesion specificity. Rather, there was a decrease in specificity; chimeric proteins were able to mediate some adhesion between either E-cadherin- or P-cadherin-expressing cells. Thus, adhesion specificity is located in the amino terminal EC1 repeat, but it cannot be excluded that other sites outside EC1 are also involved. Nevertheless, a role for EC1 was also supported by early studies using monoclonal antibodies directed against EC1 that block the adhesion function of cadherins (Yoshida-Noro et al. 1984; Behrens et al. 1985; Gumbiner and Simons 1986; Hatta and Takeichi 1986; Nose and Takeichi 1986). However, the residues recognized by these antibodies and involved in adhesion remain unknown.

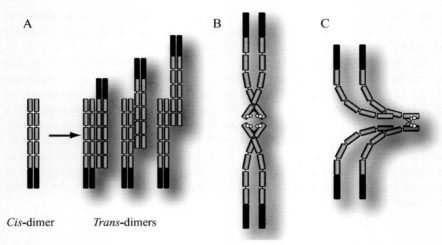

Fig. 3A–C. Cadherin adhesion models. **A** *Trans*-dimerization model of interdigitated cadherin domains by Sivasankar et al. (2001). *Cis*-dimerized cadherins in opposing membranes are able to bind through several different mechanisms. One mechanism involves complete interdigitation of cadherin extracellular domains. A second binding complex occurs through overlap of EC1–4, and a third occurs between EC1 and 2 alone. **B** *Cis*- and *trans*-dimerization model proposed by Pertz et al. (1999). W2 acts as an allosteric activator to allow EC1 domains of opposing cadherins to interact through an unknown mechanism. Ca^{++}-coordination first allows cadherins to *cis*-dimerize, but only at high Ca^{++}-concentrations (>1 mM) will W2 dock in the hydrophobic pocket. **C** *Cis*- and *trans*-dimer model proposed by Bogon et al. (2002). *Cis*-dimers form through an interaction of EC1 in one cadherin with EC2 and some of EC3 of a second cadherin. These *cis*-dimers would organize into a semi-crystalline array forming *trans*-dimers through exchange of W2 via the strand dimer. Of note is the fact that only one cadherin of a *cis*-dimer interacts with an opposing *cis*-dimer on a neighboring cell. In all schematics, only cadherins involved in *trans*-adhesion show W2 docking

Tripeptides based on protein–protein interfaces have been useful in examining adhesion of integrins to extracellular matrix, and a similar strategy has been used with cadherins. Blashuck et al. (1990) hypothesized a tripeptide adhesion sequence was present in cadherins. Sequence analysis identified several potential tripeptide adhesion sequences, and they tested whether these peptides blocked normal cadherin–cadherin adhesion, aggregation, and blastocyst compaction. The authors showed that only the histidine-alanine-valine (HAV) sequence from the EC1 repeat of all classical cadherins is likely important in adhesion. Mouse blastocysts incubated with a decapeptide derived from N-cadherin containing the HAV sequence failed to compact. Furthermore, rat dorsal root ganglia did not extend neurites over astrocytes in the presence of the HAV-containing peptide. Both blastocyst compaction and neurite extension are mediated by E-cadherin and N-cadherin, respectively. Because the HAV sequence is found in the EC1 of all classical cadherins, this evidence gives further support to the developing model of EC1 involvement in mediating both adhesion and specificity of cadherins.

The HAV sequence has continued to be studied as a potential mediator and recognition sequence of cadherin adhesion. Based on the inhibitory aspect of the HAV-containing decapeptide, Williams et al. (2000) showed that cyclic peptides containing HAV were better than linear peptides in inhibiting N-cadherin-mediated neurite extension of cerebellar neurons over N-cadherin-expressing 3T3 cells. In addition, they found that incorporation into the cyclic peptides of specific residues flanking the HAV sequence in N-cadherin increased their inhibitory activity. Interestingly, if they incorporated flanking residues from E-cadherin rather than N-cadherin, the cyclic peptides did not inhibit N-cadherin adhesion. It should be noted that several of the peptides derived from N-cadherin sequence failed to inhibit N-cadherin adhesion. It would have been more informative if the E-cadherin-derived cyclic peptides were shown to inhibit E-cadherin, but not N-cadherin adhesion. Further evidence for the involvement of EC1 in cadherin adhesion was revealed in a study from the same group on small peptide agonists of cadherin adhesion (Williams et al. 2002). The authors demonstrated that a recombinant N-cadherin EC1 domain was able to inhibit neurite outgrowth from cerebellar neurons over 3T3 cells expressing N-cadherin. Together, these results demonstrate that residues flanking the HAV sequence are potential mediators of cadherin adhesion specificity.

Despite strong evidence, using a variety of experimental approaches, for a role of EC1 in adhesion, some evidence points to a model in which additional EC repeats are required for adhesion. The monoclonal antibody DECMA-1 blocks E-cadherin adhesion (Vestweber and Kemler 1985). Mapping of the epitope showed that it is directed against the EC4/EC5 boundary rather than EC1, as shown for other inhibitory antibodies (see above and Fig. 1) (Ozawa et al. 1990). Ozawa et al. (1990) showed also that E-cadherin contains at least one disulfide bond, and that this bond is important, but not required, for cadherin–cadherin adhesion. Sequence analysis showed only EC5 to have potential disulfide bonds [confirmed by crystal structure (Boggon et al. 2002)] adding further support to some role of EC4–5 in cadherin adhesion.

Analysis of crystal structures of cadherin extracellular domains provided new information on molecular interactions between cadherin EC domains. Because EC1 was suspect in adhesive interactions, the initial focus of crystal structure studies was EC1. The first crystal of cadherin EC1 was from N-cadherin (Shapiro et al. 1995). It raised more questions than answers about past evidence of EC1. First, three different crystal forms were observed, one contained a single molecule in the asymmetric unit of the crystal lattice, and the other two contained two molecules per asymmetric unit in different orientations. The molecular interactions described (see below) were observed in all the crystal forms, but mostly only as crystal packing interactions. Second,

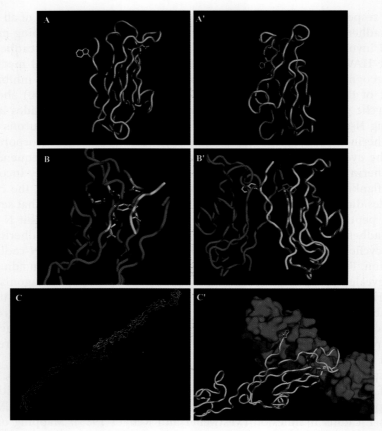

Fig. 4A–C. Structure of EC1 and strand dimer. **A** N-cadherin EC1 is shown in a ribbon with the HAV sequence and W2 side chains. **A′** Ninety degree rotation of the structure in A. **B** Close-up of W2 docking in the hydrophobic pocket in the strand dimer. W2 of one cadherin (*yellow*) is docked in a pocket of an opposing cadherin (*blue*). Side chains of residues composing the surface of the pocket are shown. **B′** EC1 domains participating in the strand dimer exchange of W2. **C** Strand dimer of the complete C-cadherin extracellular domain. **C′** A close-up of EC1 domains in the C-cadherin strand dimer. A surface model of cadherin (*blue*) is shown against a ribbon model (*yellow*). [A, A′, B, and B′ are from Protein Data Bank (PDB) accession 1NCI (Shapiro et al. 1995), and C and C′ from PDB accession 1L3 W (Boggon et al. 2002)]

the structure revealed that the HAV tripeptide, though on the surface, is partially buried and obscured, and a potential dimer could not be assigned to correlate with the HAV sequence (Fig. 4A and 4A′). The structure did help explain why a single linear sequence of the protein could not be identified as the determinant of cadherin specificity.

The shape of the cadherin repeat is a β-barrel structure, and residues in close proximity on the surface of the protein are not necessarily close in the primary structure of the protein. An intriguing interaction is between two domains with their long axes aligned in a roughly parallel, not antiparallel, orientation. The authors suggested that this interaction might be a putative lateral or *cis*-dimer (see below for discussion of *cis*-dimers). Additionally, tryptophan 2 (W2) of each domain was inserted into a hydrophobic pocket of the adjacent cadherin in what was termed a "strand dimer" (Fig. 4). The strand dimer was the major characteristic of lateral cadherin dimers. Interestingly, the hydrophobic pocket accepting the side chain of W2 is composed, in part, of the alanine from the HAV tripeptide.

3.2
Recent Studies, Evolving Models

The first crystal structure of the full cadherin extracellular domain showed that the strand dimer, inferred from early structures to represent a *cis*-dimer, may in fact be representative of the association of cadherins on neighboring cells (Boggon et al. 2002). The structure shows the characteristic strand dimer with the EC1 domains of the two molecules in a roughly parallel orientation; however, the rest of the extracellular domain adopts a curved structure rather than the assumed rigid straight structure (Fig. 4C). The curvature of the structure is such that the long axis of EC1 is roughly perpendicular to the long axis of EC5. Based on this new evidence, the authors proposed a model that the strand dimer mediated *trans*-dimerization and that *cis*-dimerization occurred through a previously undescribed interaction.

The key characteristic of the strand dimer is intercalation of W2 into the hydrophobic pocket of an opposing cadherin. Experiments focusing on W2 and the hydrophobic pocket demonstrate their importance (Tamura et al. 1998; Pertz et al. 1999; Ahrens et al. 2002; Perret et al. 2002). Mutation of W2, A78, or A80 (the alanines comprising parts of the hydrophobic pocket) inhibits cell aggregation, bead aggregation, and cell-bead binding. Of particular note are the studies by Pertz et al. (1999) which used a chimeric protein of the E-cadherin extracellular domain fused to the coiled-coil pentamerization domain of cartilage oligomatrix protein (ECADCOMP). Using electron microscopy to examine structure and interactions, Pertz et al. (1999) showed that ECADCOMP forms a pentamer in solution. In the presence of Ca^{++}, the E-cadherin extracellular domain adopts a bent rod structure, two of which in a pentamer can form a ring-like structure, inferred as a putative *cis*-interaction (Fig. 5). There are instances in which two ring-like structures are in contact in a putative *trans*-interaction. An ECADCOMP carrying the W2A mutation, while still able to adopt a ring structure, is never seen in association with a second ring. While all other data cannot distinguish between a role in *cis*- or *trans*-dimerization, these electron microscopy studies

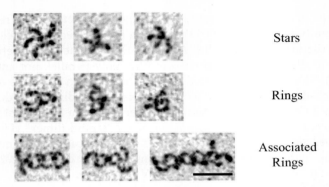

Fig. 5. ECADCOMP conformations. ECADCOMP visualized by rotary shadowing electron microscopy. The *top row* shows non-dimerized pentamers in a star-like pattern. The *middle row* shows pentamers in which two or four E-cadherin extracellular domains have formed a ring-like structure. The *bottom row* shows multimerized pentamers with a concentric ring orientation. The ring structures are proposed *cis*-dimers, and the concentric rings are proposed *trans*-dimers of *cis*-dimerized cadherins. Used by permission from Pertz et al. (1999)

support a model in which W2 docking in the hydrophobic pocket is required for *trans*-dimerization.

A second model has been proposed in which the strand dimer is not directly involved in *trans*-dimerization (Koch et al. 1999; Pertz et al. 1999). Crystal studies of EC1 and 2 of N-cadherin and E-cadherin showed an apparent *cis*-dimer lacking the strand dimer. In two of the crystals, the amino terminus was disordered and could not be resolved (Nagar et al. 1996; Tamura et al. 1998). The third crystal showed W2 docking into the hydrophobic pocket of its own protein (Pertz et al. 1999). Pertz et al. (1999) proposed that W2 is not involved in strand exchange and direct intermolecular interactions, but rather is required as an allosteric activator for *trans*-dimerization. The conclusion was based on the assumption that W2 is required for *cis*-dimerization, an assumption which the authors proved wrong. Note that ECADCOMP bearing a W2A mutation was seen by electron microscopy to form ring-like structures in an apparent *cis*-dimerization but was unable to oligomerize into concentric ring structures.

Other studies continue to support a model of W2 as an allosteric effector rather than a direct mediator of *trans*-adhesion. Nuclear magnetic resonance (NMR) analysis of Ca^{++}-dependent dimers and oligomers of E-cadherin EC1 and 2 [the same construct used by Pertz et al. (1999) for crystallographic studies] suggested that W2 is buried in the hydrophobic pocket of its own molecule (Haussinger et al. 2002). Analysis of $\{^{1}H\}$ (^{15}N) shifts of W2 indole group revealed no significant change in the orientation of W2 during Ca^{++}-dependent dimerization and oligomerization of E-cadherin EC1 and 2. Additionally, the {^{1}H}-^{15}N nuclear Overhauser enhancements (NOEs) of the W2 indole group in the presence and absence of 600 µM Ca^{++} indicate considerable flexibility of the residue in both monomeric and aggregated states that would not be expected if it were fixed in the hydrophobic pocket of a neighboring molecule as depicted in the strand dimer model. One caveat of this study, though, is the presence of an N-terminal methionine as a cloning and protein expression artifact. Previous studies showed that precise cleavage of the precursor peptide of cadherins is required for full adhesive functionality (Ozawa and Kemler 1990).

3.3
Further Functional Studies of EC Domains in Adhesion

EC1 was first modeled as the adhesive domain because it was found to be involved in specificity, contained the HAV adhesion sequence, and N-cadherin EC1 packed in an antiparallel orientation as if in a *trans*-dimer in one crystal form (Blaschuk et al. 1990; Nose et al. 1990; Shapiro et al. 1995). However, the precise orientation of EC1 in the *trans*-dimer remains unclear. Nevertheless, adhesion through EC1 remains the model best fitting all data, and additional studies have been presented which strengthen this view.

First, further work on ECADCOMP showed that adhesive pentamers oligomerized through the N-terminal regions of E-cadherin (Tomschy et al. 1996; Koch et al. 1999; Pertz et al. 1999). It is unclear, however, exactly which parts of the N-terminus are involved in adhesion. Second, the studies of Perret et al. (2002) demonstrated adhesive events between cadherin fragments of just EC1 and 2. To investigate kinetics of cadherin *trans* adhesion, they constructed an E-cadherin extracellular fragment consisting of EC1 and 2 with a C-terminal His tag (E-cad1/2). Beads were coated with an antibody against the His tag, and E-cad1/2 was added to them. A mica surface was prepared by adsorbing Ni^{++} to the surface and then chelating the E-cad1/2 His tag directly. The authors visualized the cadherin-coated beads as they rolled across the cadherin-coated surface and measured the duration and frequency of stop events interpreted as cadherin adhesion between the bead and surface (Fig. 3). An approximately fivefold decrease in the frequency of binding events was observed when the tryptophan analog I3A or an E-cad1/2 with the W2A mutation was used, thereby further supporting a role of EC1 and W2 in adhesion.

These results, however, are different from those of Chappuis-Flament et al. (2001). In similar adhesion flow experiments, the authors used various C-cadherin fragments, missing one or several of the EC repeats, fused to the F_c domain of IgG. The cadherin fragments were bound directly to protein A-coated beads through the F_c domain, and their function was tested by Ca^{++}-dependent bead aggregation. In these experiments, at least three cadherin EC repeats were required for strong Ca^{++}-dependent adhesion. The authors were unable to measure adhesion by bead aggregation or in a flow assay of a cadherin fragment of EC1 and 2 alone. However, EC1 and 2 were shown to be required in all the assays performed, while the additional repeat could be either EC3, EC4, or EC5. The third domain had to be an EC repeat, though, as fibronectin repeats fused to only EC1 and 2 failed to mediate bead aggregation. The differences in experimental design and aims between these experiments and those of Perret et al. (2002), specifically attachment of the cadherin fragment to the beads and analysis of single adhesion events versus bulk adhesion properties, could explain the different conclusions arrived by the two sets of experiments. However, it is clear from both studies that EC1 and 2 are required for adhesion.

A different experimental design has been used to examine a role of additional EC repeats, i.e., other than just EC1 in adhesion. Sivasankar and colleagues, using a surface force apparatus, demonstrated that the strongest adhesive interaction between cadherins on two surfaces occurred at a minimum distance of about 25 nm (Sivasankar et al. 1999; Leckband and Sivasankar 2000; Sivasankar et al. 2001). Interestingly, this is the approximate length of a cadherin extracellular domain if all the repeats would be in a straight line, one after the other. However, as noted earlier, electron microscopy of the extracellular domain of E-cadherin and the crystal structure of the full-length extracellular domain of C-cadherin depict the cadherin extracellular domain as a bent rod (Pokutta et al. 1994; Tomschy et al. 1996; Ahrens et al. 2002; Boggon et al. 2002). Due to the curve of the extracellular domain, the *trans*-adhesion dimer proposed by Boggon et al. (2002) would occur between surfaces approximately 25 nm apart. Additionally, analysis of oligomerized dimers of E-cadherin, using the c-Jun/c-Fos dimerization system, reveals a measurement of adhesive dimerization between EC1 of approximately 25 nm. If EC5 of each extracellular domain is oriented perpendicular to the surface and each cadherin is interacting with the cadherin on the opposite side (left–right, right–left), then the calculated distance between two theoretical surfaces bearing the E-cadherin c-Jun/c-Fos dimers is approximately 28 nm. In each of these models, however, a single cadherin itself would stand only 10–15 nm from the plasma membrane surface. This conclusion conflicts with measurements of the cadherin extracellular domain height analyzed with the surface force apparatus (Sivasankar et al. 2001). These measurements showed that at least some cadherins can extend up to 20 nm from a surface. Furthermore, adhesive interactions were detected with the surface force apparatus even if the cadherin surfaces were only allowed to approach to 30 and 40 nm of each other. This result suggests that adhesive interactions can occur between cadherins that are not in a bent configuration. Since the structure of the cadherin extracellular domain is potentially a bent rod, measurements with the surface force apparatus do support a model in which the N-terminal EC1 repeat is solely involved in *trans*-adhesion if the cadherin extracellular domain is able to adopt multiple rigid orientations, but they also support a model in which multiple EC repeats intercalate during cadherin adhesion.

Taken together, we know that EC1 is required for adhesion and at least partly is responsible for homotypic specificity. Multiple EC repeats are likely involved in adhesion as well, but it is unclear whether they directly form adhesive contacts with an opposing cadherin or simply correctly present EC1 to an opposing cadherin. All evidence supports a model in which EC1 alone interacts in adhesion, but further experiments are required: functional adhesion must be verified while identifying or knowing the orientation of the cadherin extracellular domains. The W2 residue plays a role in adhesion, but the exact mechanism remains to be determined. Experiments to date have not been able to distinguish between W2 as an allosteric activator of *trans*-adhesion or as a direct mediator of *trans*-adhesion through the strand dimer observed in crystal structures. Finally, though the HAV sequence does appear to have a function in adhesion (mutation of the alanine and use of HAV-containing peptides), the mechanism is unknown. Mutation studies of the other residues in the tripeptide may help to explain the function of HAV. Additionally, structural studies with HAV peptides may help to explain their ability to inhibit adhesion. The precise determination of all residues involved and required for adhesion would help in determining the true molecular interactions of cadherin adhesion.

4
Cadherin Adhesion Structure: *Cis*-Dimer

Formation of lateral cadherin dimers, referred to as *cis*-dimers, was first proposed based on crystal structures of EC1 of N-cadherin (Shapiro et al. 1995). This conclusion was based on the observation that individual EC1 domains packed in a parallel orientation, representative of proteins that had originated from the same cell membrane.

The first functional evidence for *cis*-dimerization of cadherins came from studies using purified C-cadherin extracellular domain (Brieher et al. 1996). C-cadherin extracellular domain separated into two peaks through a gel filtration column. Crosslinking of protein fractions showed that the higher molecular weight peak corresponded to dimer and the lower molecular weight peak corresponded to monomer. The putative dimer from the high molecular weight fractions was shown to have a higher adhesive potential in a cell adhesion assay than the low molecular weight monomer fractions. The authors concluded that these results confirm lateral dimerization of the cadherin extracellular domain-promoted adhesive dimerization. However, whether the extracellular domains in the dimer fractions were in a parallel (*cis*-dimer) or antiparallel (*trans*-dimer) orientation was not determined.

A key experiment on lateral dimerization of cadherins was performed in vivo (Takeda et al. 1999). Using cadherin-deficient L cells, full-length, wild-type E-cadherin or a chimeric E-cadherin fused directly to the actin-binding domain of α-catenin (Eαcat) was ectopically expressed. When cadherin-expressing cells were treated with the crosslinking agent 3,3'-dithiobis[sulfosuccinimidylpropionate] (DTSSP) and solubilized, monomers and dimers of E-cadherin were identified. These dimers could arise from either lateral or adhesive *trans*-interactions. To test this, cells expressing E-cadherin and cells expressing Eαcat were cocultured. The cocultures were crosslinked and then analyzed by immunoblot for α-catenin. Only monomers and homodimers of Eαcat were observed. If the dimers were adhesive dimers, heterodimers of E-cadherin and Eαcat would also be expected. The authors additionally showed that these lateral dimers were only found in adherent cells; if cells were grown in the presence of ethyleneglycoltetraacetic acid (EGTA), low-Ca^{++} medium, or cadherin-inhibiting antibodies, lateral dimers could not be crosslinked. The conclusion drawn was that lateral cadherin dimers are a functional unit for cadherin adhesion. However, a second conclusion is also supported by the data, that cadherins are only able to be crosslinked into lateral dimers because of the increase in local concentration of cadherins on the cell surface during cadherin-mediated adhesion.

Cis-dimer formation has been further investigated by immunoprecipitation experiments without first crosslinking the proteins (Chitaev and Troyanovsky 1998; Klingelhofer et al. 2000; Shan et al. 2000). Immunoprecipitations showed complexes of cadherins believed to represent *cis*-dimers isolated from cell lysates. Chitaev and Troyanovsky (1998) provide evidence that E-cadherin forms lateral dimers through a Ca^{++}-independent/W2-dependent mechanism. However, in a following study from the same group, Klingelhofer et al. (2000) suggest E-cadherin and P-cadherin can form hetero *cis*-dimers through multiple mechanisms. They present data suggesting heterocomplex formation in the presence or absence of Ca^{++} and also dependent and independent of W2. The W2-independent mechanism required the absence of Ca^{++}. Additionally, Shan et al. (2000) found R-cadherin/N-cadherin heterocomplexes could be coimmunoprecipitated from cells expressing both cadherin subtypes. They showed also that the R-cadherin/N-cadherin interaction appears to be real because E-cadherin, if coexpressed with R-cadherin, is not coimmunoprecipitated with R-cadherin. The fact that these lateral cadherin interactions are Ca^{++}-independent is in direct conflict with the data of Takeda et al. (1999) in which the lateral cadherin dimers they could crosslink were Ca^{++}-dependent. The data from these immunoprecipitation experiments are very difficult to interpret in light of the various models of W2 involvement in *trans*- and *cis*-dimerization, and different methods are required to fully address the formation of lateral cadherin complexes in vivo.

Cis-dimerization has also been examined using peptides (Williams et al. 2002). Short peptide HAV-containing antagonists were dimerized so that two adhesion sites were on the same molecule. These peptides were found to act as agonists for neurite outgrowth, a cadherin-mediated adhesion response, on 3T3 cells lacking N-cadherin expression. Dimeric peptides of a second putative adhesion site containing INPISG also activated neurite outgrowth in the cerebellar neuron system. Cyclic monomeric peptides were able to block the activation by these dimeric peptides. It can be concluded that lateral dimerization or clustering of cadherins is able to mediate cellular response. It is not clear, however, if the dimeric peptides promoted adhesion (no assays were performed to test this), and they could act independently of adhesion since N-cadherin binds and activates fibroblast growth factor (FGF) receptors in the neurons (Williams et al. 1994; Saffell et al. 1997; Williams et al. 2001).

Crystal structures of the first two EC repeats show possible lateral dimer organization. Two different crystals have been presented of E-cadherin EC1 and EC2 (Nagar et al. 1996; Pertz et al. 1999). Each shows the two domain fragments aligned lengthwise in a parallel orientation. The region of closest contact is around the Ca^{++} binding sites between EC1 and 2 such that the proteins together adopt a sort of twisted "X" configuration (Fig. 6). The first

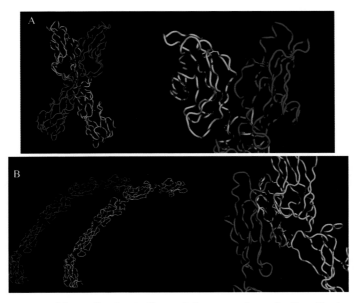

Fig. 6. Structures of the cadherin *cis*-dimer. **A** Structure from the E-cadherin crystal by Pertz et al (1999). The overall alignment of the two Ecad12 fragments is a slightly twisted "X". A close view of W2 in each EC1 reveals it is docking in the hydrophobic pocket of its own EC1. **B** Structure of the C-cadherin extracellular domain in a proposed *cis*-dimer. A groove in EC1 of one cadherin binds over a bulge in EC2 of a lateral cadherin. [**A** created from PDB accession 1FF5 (Pertz et al. 1999), and **B** created from PDB accession 1L3 W (Boggon et al. 2002)]

crystal structure showed a less likely dimerization interaction as it was mediated by several water molecules and the interface was relatively small (Nagar et al. 1996). The second published structure was able to resolve more of the protein to show W2 docking into the hydrophobic pocket of its own molecule (Pertz et al. 1999). Additionally, the authors observed a lateral dimer they believed to be more stable also in the shape of an intertwisted "X". There have been no biological assays done to conclusively confirm either interaction. The N-cadherin EC1/2 structure showed what appeared to be a lateral dimer, but no strand dimer was detected as previously described for the N-cadherin EC1 domain alone (Tamura et al. 1998). Tamura et al. (1998) mention a crystal packing interface possibly involved in *trans* adhesion, but they allude to results suggesting it is not a real interface and the one described in earlier work of N-cadherin EC1 (Shapiro et al. 1995) is more likely correct.

With the crystallization of the complete cadherin extracellular domain and the discovery that the strand dimer possibly mediated *trans*-, not *cis*-, dimerization, Boggon et al. (2002) had to formulate a new molecular model for the *cis*-dimer. The authors described a putative *cis*-dimer interaction based on the crystal packing interactions between a groove in EC1 of one cadherin and a bulge on EC2 of a second cadherin (Fig. 6). The proposed interaction showed two adjacent cadherins aligned with their N-termini pointed in the same direction. This interaction would allow cadherins within the same membrane to organize in a continuous linear array. Interestingly, the surface of the groove in EC1 is composed in small part by the histidine and valine of the HAV tripeptide. Functional significance has not been verified.

Electron microscopy analysis of ECADCOMP supports a different *cis*-dimer model. In their studies on Ca^{++}-dependent adhesion, Tomschy et al. (1996) and Pertz et al. (1999) describe association of E-cadherin extracellular domains within a single pentamer before extracellular domains in different pentamers adhere. They base this on the observation that ring-like structures are found in some isolated pentamers at high Ca^{++} concentrations, but they are always seen as the adherent cadherins when two or more pentamers associate. In contrast to the model proposed by Boggon et al. (2002), the cadherins in this *cis*-dimer would bend toward each other and the N-termini of the extracellular domains would point in opposite directions. The conclusion that these data represent a real *cis* interaction and that this interaction is required before *trans* adhesion is valid. However, there is another equally possible explanation: Because the extracellular domain is a curved rod and the extracellular domains are constrained by the pentamerization domain, they simply lay on the surface in an orientation that appears like a *cis* interaction. When two pentamers associate through a single extracellular domain, a second is brought into the structure quickly because the effective concentration of extracellular domains is very high, and the two adherent pairs simply lay flat to appear as two adjacent ring-like structures, again because of the constraint imposed by the pentamerization domain.

The *cis*-dimer model requires more conclusive validation. The fact that lateral complexes of cadherin can form in vivo is certain, but the question remains concerning whether these complexes serve a specific role in adhesion or if they are simply the result of cadherin clustering by its anchor to the actin cytoskeleton. Data should be obtained that show a distinct difference in the adhesive properties of known cadherin *cis*-dimers versus single cadherin extracellular domains. All experiments to date have either not shown a difference in adhesion due to *cis*-dimerization or have not determined the specific orientation of the dimers. In addition, the molecular interactions of cadherin *cis*-dimers must be determined. Models based on the packing in crystal structures continue to change and offer conflicting views. As remains with the *trans*-dimer model, residues involved in *cis*-dimerization must be determined.

Acknowledgements Work from the Nelson Laboratory is supported by NIH GM55227, and T.D.P. is also supported by a Howard Hughes Medical Institute Predoctoral Fellowship.

References

Adams CL, Chen YT, Smith SJ, Nelson WJ (1998) Mechanisms of epithelial cell–cell adhesion and cell compaction revealed by high-resolution tracking of E-cadherin-green fluorescent protein. J Cell Biol 142:1105–19

Ahrens T, Pertz O, Haussinger D, Fauser C, Schulthess T, Engel J (2002) Analysis of heterophilic and homophilic interactions of cadherins using the c-Jun/c-Fos dimerization domains. J Biol Chem 277:19455–60

Behrens J, Birchmeier W, Goodman SL, Imhof BA (1985) Dissociation of Madin-Darby canine kidney epithelial cells by the monoclonal antibody anti-arc-1: mechanistic aspects and identification of the antigen as a component related to uvomorulin. J Cell Biol 101:1307-15

Blaschuk OW, Sullivan R, David S, Pouliot Y (1990) Identification of a cadherin cell adhesion recognition sequence. Dev Biol 139:227–9

Boggon TJ, Murray J, Chappuis-Flament S, Wong E, Gumbiner BM, Shapiro L (2002) C-cadherin ectodomain structure and implications for cell adhesion mechanisms. Science 296:1308–13

Brieher WM, Yap AS, Gumbiner BM (1996) Lateral dimerization is required for the homophilic binding activity of C-cadherin. J Cell Biol 135:487–96

Chitaev NA, Troyanovsky SM (1998) Adhesive but not lateral E-cadherin complexes require calcium and catenins for their formation. J Cell Biol 142:837–46

Frank M, Kemler R (2002) Protocadherins. Curr Opin Cell Biol 14:557–62

Gumbiner B, Simons K (1986) A functional assay for proteins involved in establishing an epithelial occluding barrier: identification of a uvomorulin-like polypeptide. J Cell Biol 102:457–68

Gumbiner BM (1996) Cell adhesion: the molecular basis of tissue architecture and morphogenesis. Cell 84:345–57

Hatta K, Takeichi M (1986) Expression of N-cadherin adhesion molecules associated with early morphogenetic events in chick development. Nature 320:447–9

Hatta K, Nose A, Nagafuchi A, Takeichi M (1988) Cloning and expression of cDNA encoding a neural calcium-dependent cell adhesion molecule: its identity in the cadherin gene family. J Cell Biol 106:873–81

Haussinger D, Ahrens T, Sass HJ, Pertz O, Engel J, Grzesiek S (2002) Calcium-dependent homoassociation of E-cadherin by NMR spectroscopy: changes in mobility, conformation and mapping of contact regions. J Mol Biol 324:823–39

Kemler R (1992) Classical cadherins. Semin Cell Biol 3:149–55

Klingelhofer J, Troyanovsky RB, Laur OY, Troyanovsky S (2000) Amino-terminal domain of classic cadherins determines the specificity of the adhesive interactions. J Cell Sci 113 (Pt 16):2829–36

Koch AW, Pokutta S, Lustig A, Engel J (1997) Calcium binding and homoassociation of E-cadherin domains. Biochemistry 36:7697–705

Koch AW, Bozic D, Pertz O, Engel J (1999) Homophilic adhesion by cadherins. Curr Opin Struct Biol 9:275–81

Leckband D, Sivasankar S (2000) Mechanism of homophilic cadherin adhesion. Curr Opin Cell Biol 12:587–92

Nagar B, Overduin M, Ikura M, Rini JM (1996) Structural basis of calcium-induced E-cadherin rigidification and dimerization. Nature 380:360–4

Nose A, Takeichi M (1986) A novel cadherin cell adhesion molecule: its expression patterns associated with implantation and organogenesis of mouse embryos. J Cell Biol 103:2649–58

Nose A, Tsuji K, Takeichi M (1990) Localization of specificity determining sites in cadherin cell adhesion molecules. Cell 61:147–55

Ozawa M, Kemler R (1990) Correct proteolytic cleavage is required for the cell adhesive function of uvomorulin. J Cell Biol 111:1645–50

Ozawa M, Hoschutzky H, Herrenknecht K, Kemler R (1990) A possible new adhesive site in the cell-adhesion molecule uvomorulin. Mech Dev 33:49–56

Perret E, Benoliel AM, Nassoy P, Pierres A, Delmas V, Thiery JP, Bongrand P, Feracci H (2002) Fast dissociation kinetics between individual E-cadherin fragments revealed by flow chamber analysis. EMBO J 21:2537–46

Pertz O, Bozic D, Koch AW, Fauser C, Brancaccio A, Engel J (1999) A new crystal structure, Ca2+ dependence and mutational analysis reveal molecular details of E-cadherin homoassociation. EMBO J 18:1738–47

Pokutta S, Herrenknecht K, Kemler R, Engel J (1994) Conformational changes of the recombinant extracellular domain of E-cadherin upon calcium binding. Eur J Biochem 223:1019–26

Ringwald M, Schuh R, Vestweber D, Eistetter H, Lottspeich F, Engel J, Dolz R, Jahnig F, Epplen J, Mayer S, et al (1987) The structure of cell adhesion molecule uvomorulin. Insights into the molecular mechanism of Ca2+-dependent cell adhesion. EMBO J 6:3647–53

Saffell JL, Williams EJ, Mason IJ, Walsh FS, Doherty P (1997) Expression of a dominant negative FGF receptor inhibits axonal growth and FGF receptor phosphorylation stimulated by CAMs. Neuron 18:231–42

Shan WS, Tanaka H, Phillips GR, Arndt K, Yoshida M, Colman DR, Shapiro L (2000) Functional cis-heterodimers of N- and R-cadherins. J Cell Biol 148:579–90

Shapiro L, Fannon AM, Kwong PD, Thompson A, Lehmann MS, Grubel G, Legrand JF, Als-Nielsen J, Colman DR, Hendrickson WA (1995) Structural basis of cell–cell adhesion by cadherins. Nature 374:327–37

Sivasankar S, Brieher W, Lavrik N, Gumbiner B, Leckband D (1999) Direct molecular force measurements of multiple adhesive interactions between cadherin ectodomains. Proc Natl Acad Sci U S A 96:11820–4

Sivasankar S, Gumbiner B, Leckband D (2001) Direct measurements of multiple adhesive alignments and unbinding trajectories between cadherin extracellular domains. Biophys J 80:1758–68

Takeda H, Shimoyama Y, Nagafuchi A, Hirohashi S (1999) E-cadherin functions as a cis-dimer at the cell–cell adhesive interface in vivo. Nat Struct Biol 6:310–2

Takeichi M (1990) Cadherins: a molecular family important in selective cell–cell adhesion. Annu Rev Biochem 59:237–52

Takeichi M (1995) Morphogenetic roles of classic cadherins. Curr Opin Cell Biol 7:619–27

Tamura K, Shan WS, Hendrickson WA, Colman DR, Shapiro L (1998) Structure-function analysis of cell adhesion by neural (N-) cadherin. Neuron 20:1153–63

Tomschy A, Fauser C, Landwehr R, Engel J (1996) Homophilic adhesion of E-cadherin occurs by a co-operative two-step interaction of N-terminal domains. EMBO J 15:3507–14

Vestweber D, Kemler R (1985) Identification of a putative cell adhesion domain of uvomorulin. EMBO J 4:3393–8

Williams EJ, Furness J, Walsh FS, Doherty P (1994) Activation of the FGF receptor underlies neurite outgrowth stimulated by L1, N-CAM, and N-cadherin. Neuron 13:583–94

Williams E, Williams G, Gour BJ, Blaschuk OW, Doherty P (2000) A novel family of cyclic peptide antagonists suggests that N-cadherin specificity is determined by amino acids that flank the HAV motif. J Biol Chem 275:4007–12

Williams EJ, Williams G, Howell FV, Skaper SD, Walsh FS, Doherty P (2001) Identification of an N-cadherin motif that can interact with the fibroblast growth factor receptor and is required for axonal growth. J Biol Chem 276:43879–86

Williams G, Williams EJ, Doherty P (2002) Dimeric versions of two short N-cadherin binding motifs (HAVDI and INPISG) function as N-cadherin agonists. J Biol Chem 277:4361–7

Yoshida-Noro C, Suzuki N, Takeichi M (1984) Molecular nature of the calcium-dependent cell–cell adhesion system in mouse teratocarcinoma and embryonic cells studied with a monoclonal antibody. Dev Biol 101:19–27

Structural Aspects of Adherens Junctions and Desmosomes

H.-J. Choi · W. I. Weis (✉)

Department of Structural Biology, Stanford University School of Medicine,
299 Campus Drive, West Stanford, CA 94305-5126, USA
bill.weis@stanford.edu

1	Overview	24
2	Cadherins.	25
2.1	Classical Cadherin Cytoplasmic Domain.	25
2.2	Desmosomal Cadherins	26
3	β-Catenin and Plakoglobin	27
3.1	Structure of the β-Catenin Arm Repeat Domain and Implications for Other Arm Proteins	28
3.2	The β-Catenin/E-Cadherin Complex	30
3.3	Role of Non-arm Repeat Regions	33
3.4	Desmosomal Cadherin–Plakoglobin Interactions	34
4	α-Catenin.	37
4.1	α-Catenin Homodimerization and β-Catenin Binding Region	37
4.2	α-Catenin M Region	40
4.3	Actin Binding by α-Catenin	42
4.4	Regulation of α-Catenin Interactions	43
5	Intermediate Filament Binding by Desmoplakin	44
References		46

Abstract The cadherin-containing intercellular junctions, adherens junctions and desmosomes share an overall logical organization in which the extracellular regions of the cadherins on opposing cells interact, while their cytoplasmic domains are linked to the cytoskeleton through protein assemblies. In adherens junctions, β-catenin binds to the cytoplasmic domain of cadherins and to α-catenin, which links the cadherin/β-catenin complex to the actin cytoskeleton. In desmosomes, the β-catenin homolog plakoglobin binds to desmosomal cadherins. The desmosomal cadherin/plakoglobin complex is linked to the intermediate filament system by the protein desmoplakin. In the past decade, components of these systems have been purified to homogeneity and studied biochemically and structurally, providing the beginnings of a mechanistic description of junction architecture and dynamics.

Keywords Adherens junction · Desmosome · Cadherin · Catenin · Desmoplakin

1
Overview

Intercellular junctions are integral components of epithelial tissues. They provide the mechanical linkage between cells, which enables morphogenetic changes during development and imparts strength to the tissue (Yap et al. 1997; Kowalczyk et al. 1999; Perez-Moreno et al. 2003). During the development and maintenance of epithelial cell and tissue polarity, these junctions serve as landmarks for delivery of various membrane components to these surfaces, and thereby define the apical and basolateral membranes (Nelson 2003). Early electron microscopy studies of epithelial tissues defined several kinds of junctions, including tight junctions, adherens junctions, and desmosomes (Farquhar and Palade 1963). Adherens junctions and desmosomes contain cadherins, which are single-pass transmembrane proteins that mediate Ca^{2+}-dependent adhesion. The extracellular regions of cadherins present on opposing cell surfaces interact with one another, while the cytoplasmic regions of cadherins are linked to the cytoskeleton, thereby mechanically coupling the cytoskeletons of adjacent cells.

In adherens junctions, β-catenin or its homolog plakoglobin binds to the cytoplasmic domain of cadherins and to α-catenin (Fig. 1). In turn, α-catenin binds to F-actin as well as a number of actin cytoskeleton-associated proteins. Another β-catenin relative, p120^{ctn}, binds to the cytoplasmic domain of cadherins. In desmosomes, plakoglobin binds to the desmosomal cadherins, desmocollins, and desmogleins. The desmosomal cadherin/plakoglobin complex interacts with desmoplakin, which binds to intermediate

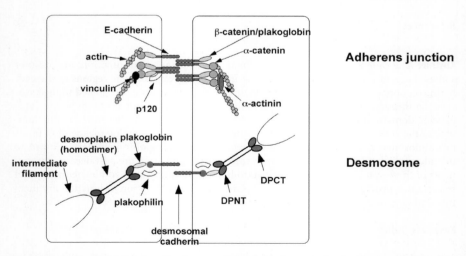

Fig. 1. Overall structure of cadherin-containing junctions. Only those components discussed in this chapter are shown

filaments (Fig. 1). Desmosomes also contain plakophilins, which are related to p120^{ctn}.

Progress towards a molecular understanding of junctional assemblies has been facilitated by the purification, biochemical characterization, and determination of the atomic structures of several adherens junction and desmosomal proteins. The molecular basis of the extracellular domain interactions is discussed in the chapter by T.D. Perez and W.J. Nelson (this volume). Here, we discuss data for interactions among the intracellular components of cadherin-containing junctions. Because relatively few direct structural data exist for desmosomal proteins, we consider most of them in the context of their homology to their adherens junction counterparts. There are many types of classical cadherins, as well as several isoforms of desmosomal cadherins, plakophilins, and α-catenins, but we have generally ignored these differences and focused on features common to each group of proteins.

2
Cadherins

2.1
Classical Cadherin Cytoplasmic Domain

The cytoplasmic domains of classical cadherins, typically ~150 amino acids in length, are the most highly conserved in the family (Nollet et al. 2000). The β-catenin/plakoglobin binding site is found in the last 100 amino acids (Ozawa et al. 1990; Stappert and Kemler 1994; Huber and Weis 2001), whereas the juxtamembrane region of 50 amino acids contains the binding site for p120^{ctn} (Ozawa and Kemler 1998; Thoreson et al. 2000). Cadherins and β-catenin form a 1:1 complex in vitro and in vivo (Ozawa and Kemler 1992; Aberle et al. 1994; Hinck et al. 1994), a result confirmed by the crystal structure of the E-cadherin/β-catenin complex (Huber and Weis 2001; see below). Circular dichroism spectroscopy, fluorescence anisotropy, and ^{1}H nuclear magnetic resonance (NMR), as well as an aberrantly large hydrodynamic radius and extreme susceptibility to proteases, demonstrated that the cytoplasmic domain of murine E-cadherin is unstructured in the absence of β-catenin (Huber et al. 2001). The same is true of the *Drosophila* E-cadherin, which is only 33% identical in sequence, implying that this is a general property of classical cadherin cytoplasmic domains (Huber et al. 2001). These studies were carried out with bacterially expressed material, but it was shown that the binding properties of β-catenin to E-cadherin is the same regardless of whether the bacterially expressed or endogenous cadherin isolated from Madin-Darby canine kidney (MDCK) cells is used, confirming that the domain is unstructured. This property arises from the unusual sequence and composition of the domain, which is highly enriched in acidic amino acids and deficient in hydrophobic amino acids, which is a hallmark of natively unstructured proteins.

The lack of structure of the uncomplexed cadherin cytoplasmic domain may be important for cadherin turnover. E-cadherin has a relatively short half-life of about 5 h in cultured MDCK cells (Shore and Nelson 1991). β-Catenin associates with E-cadherin shortly after cadherin synthesis (Hinck et al. 1994), and E-cadherin mutants deficient for β-catenin binding are retained within the endoplasmic reticulum and are rapidly degraded (Chen et al. 1999). PEST motifs, which are sequences that target proteins for proteolytic degradation (Rechsteiner and Rogers 1996), are highly conserved in the β-catenin binding region of cadherins (Huber et al. 2001). The unstructured nature of the cytoplasmic domain likely makes it a good substrate for the cellular protein degradation machinery. We have proposed that binding to β-catenin may sequester the conserved degradation motifs that are exposed in the unstructured, isolated cadherin cytoplasmic domain, ensuring that only those cadherin molecules associated with β-catenin reach the cell surface to participate in cell adhesion (Huber et al. 2001).

Kemler and colleagues noted that consensus sites for both casein kinase II (CKII) and glycogen synthase 3β (GSK-3β) are highly conserved in the cadherin cytoplasmic domains (Lickert et al. 2000). Mutagenesis of these sites to alanine or acids to mimic phosphorylation indicates that phosphorylation is associated with a quantitative strengthening of adhesion (Lickert et al. 2000). Phosphorylation of the bacterially expressed E-cadherin domain with these enzymes produces no change in its migration on a gel filtration column, indicating that phosphorylation does not cause a structuring of the isolated domain (A. Huber and W. Weis, unpublished observations). It does, however, produce a large increase in the affinity of E-cadherin for β-catenin (Huber et al. 2001).

2.2
Desmosomal Cadherins

Desmosomal cadherins mediate calcium-dependent cell–cell adhesion by homophilic or heterophilic interaction between the extracellular domains of cadherins on adjacent cells. There are two subfamilies of desmosomal cadherins, desmogleins (Dsg) and desmocollins (Dsc). Each of these exists as three isoforms, termed Dsg1, 2, and 3 and Dsc1, 2, and 3, which are expressed in cell type-specific and temporally specific ways. The different isoforms appear to function in cell positioning and epithelial tissue morphogenesis (Runswick et al. 2001).

Desmosomal cadherins have an overall structure similar to that of classical cadherins, with an extracellular domain comprising five cadherin repeats, a transmembrane anchor, and a cytoplasmic domain. The extracellular domains of Dscs and Dsgs show ~50% sequence similarity to those of classical cadherins (Nollet et al. 2000). A recent tomographic reconstruction of epidermal desmosomes indicated that the overall extracellular domain structure is similar to that seen in the crystal structure of the classical C-cadherin extracellular domain (He et al. 2003).

The cytoplasmic domains of desmosomal cadherins have been reported to bind to several desmosomal proteins, including plakoglobin (Mathur et al. 1994; Chitaev et al. 1996; Wahl et al. 1996; Witcher et al. 1996), plakophilins (Hatzfeld et al. 2000; Chen et al. 2002; Bonne et al. 2003), and desmoplakin (Smith and Fuchs 1998), but direct binding by purified proteins has been shown only for plakoglobin. The cytoplasmic domain of desmogleins is much larger than those of other cadherins, roughly 470 amino acids. The primary structure of this region can be divided into several subdomains. The membrane-proximal region is a cysteine-rich sequence known as the "intraceullular anchoring" (IA) domain. The IA domain is followed by a sequence homologous to the β-catenin/plakoglobin-binding region of classical cadherins. This region mediates binding to plakoglobin, and has been called the cadherin segment (CS) domain or catenin-binding (C) domain. Following the catenin-binding domain, desmogleins feature three additional subdomains: a proline-rich region, a 30 residue-repeat domain, and a C-terminal glycine-rich sequence. The functions of these desmoglein-specific subdomains are not known at present.

Each desmocollin isoform has two alternative splice variants that change the size of the cytoplasmic domain. The functional significance of these variants is not known. The larger variants, denoted with an "a", feature a Dsc-specific IA domain that is followed by a catenin-binding domain. Dsca variants associate with desmosomal proteins such as plakoglobin and plakophilin. The smaller Dscb variants lack most of the catenin-binding domain, and instead have an additional 11 amino acid sequence. The only reported interaction of a Dscb variant is that between plakophilin-3 and Dsc3b (as well as Dsc3a) (Bonne et al. 2003).

The strong homology between the catenin-binding domains of desmosomal cadherins and their classical counterparts makes it likely that this region is not structured in the absence of plakoglobin. At present, there are no biophysical data available for the other portions of the large Dsg cytoplasmic domain.

3
β-Catenin and Plakoglobin

β-Catenin binds to the cytoplasmic domain of classical cadherins and to α-catenin. In addition to its role in adherens junctions, β-catenin serves as a transcriptional coactivator in the Wnt signaling pathway that controls cell fate determination during embryogenesis and in the renewal of certain tissues in the adult. In this role, β-catenin binds directly to Tcf-family transcription factors, the tumor suppressor proteins Adenomatous Polyposis Coli (APC) and Axin, and several Wnt-pathway specific and general transcription factors (Bienz and Clevers 2000; Polakis 2000). Tcfs, the transcriptional inhibitor ICAT (inhibitor of β-catenin and Tcf), and APC can compete directly with cadherin for binding to β-catenin (Hülsken et al. 1994; Rubinfeld et al. 1995; von Kries et al. 2000; Graham et al. 2002).

The primary structure of β-catenin comprises an amino-terminal domain of approximately 150 amino acids, followed by a 520 residue domain containing 12 sequence repeats of 42 amino acids known as armadillo (arm) repeats, and a C-terminal region of about 100 amino acids. The N- and C-terminal regions appear to be unstructured, as assessed by circular dichroism of the first 80 residues (S. Pokutta and W.I. Weis, unpublished observations), and the aberrantly large migration of the full-length or N-terminally truncated molecule on gel filtration columns (Daniels and Weis 2002). Limited proteolysis experiments reveal that these regions are flexibly linked to the arm repeat domain (Huber et al. 1997). These experiments also demonstrated that the arm repeats form a single, protease-resistant domain.

Plakoglobin, also known as γ-catenin, is highly homologous to β-catenin, particularly in the arm repeat domains, which are ~80% identical in sequence. Like β-catenin, plakoglobin is found in adherens junctions, where it binds to the cytoplasmic domain of cadherins and to α-catenin. Plakoglobin is unique, however, in that it is also an integral component of desmosomes, where it binds to the desmosomal cadherins (Mathur et al. 1994; Chitaev et al. 1996; Wahl et al. 1996; Witcher et al. 1996) and to desmoplakin (Kowalczyk et al. 1997).

3.1
Structure of the β-Catenin Arm Repeat Domain and Implications for Other Arm Proteins

The crystal structure of the β-catenin arm repeat domain reveals that each arm repeat consists of three α-helices, designated H1, H2, and H3 (Huber et al. 1997) (Fig. 2a). The H2 and H3 helices are antiparallel to one another, and H1 is orthogonal to the others, forming the turn that links H3 of one repeat to H2 of the following repeat. The repeats pack next to one another in a regular fashion, effectively creating a series of four-helix bundles. The typical knobs-into-holes packing of helices creates a fairly uniform twist of about 30° between repeats. In this manner, the repeats form an elongated, right-handed superhelix of helices. The structure is very regular, except for a prominent kink of about 50° that occurs around repeat 9 (Fig. 2a). The superhelical structure features a groove created by the surfaces of the H3 helices. For the first 10 repeats, this groove is highly basic in character. Subsequent crystal structures and mutagenesis analyses has shown that portions

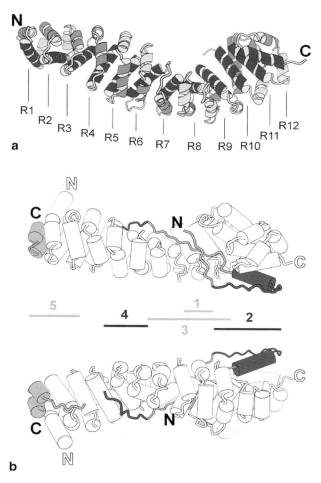

Fig. 2a, b. The structure of β-catenin and its complex with the E-cadherin cytoplasmic domain. **a** Ribbon diagram of the armadillo domain of β-catenin. Each arm repeat is composed of three helices H1, H2, and H3, shown in *light*, *medium*, and *dark grey*, respectively. Repeat 7 lacks H1, and there is a 13-residue insert between H2 and H3 of repeat 10 that is disordered in the crystal structures. **b** Composite model of the E-cadherin cytoplasmic domain/β-catenin complex. The two views are related by a 180° rotation about the horizontal axis. The five primary structure regions of E-cadherin are indicated by alternating *light* and *dark grey*

of this groove interact with several β-catenin ligands, including cadherins, Tcfs, APC, and Axin (Graham et al. 2000, 2001; von Kries et al. 2000; Huber and Weis 2001; Poy et al. 2001; Spink et al. 2001; Daniels and Weis 2002; Graham et al. 2002; Xing et al. 2002) (Fig. 2b and below).

The strong sequence identity between the β-catenin and plakoglobin arm repeat domains implies that the latter has a structure very similar to that of β-catenin. Curiously, we have found that the plakoglobin arm domain is much more sensitive to trypsin digestion than that of β-catenin, implying a degree of flexibility not present in the latter (S. Pokutta, H.-J. Choi, and W.I. Weis, unpublished observations). It is unclear if this property is related to the ability to interact with desmosomal as well as classical cadherins.

Arm repeat domains are also present in the adherens junction protein $p120^{ctn}$ and its desmosomal relatives, the plakophilins. The $p120^{ctn}$ and plakophilin arm repeat domains are somewhat different from those of β-catenin and plakoglobin: they have only 10 repeats, and there are substantial inserts in repeats 4 and 6 (Anastasiadis and Reynonds 2000). The precise functions of these proteins are not clear at present. $p120^{ctn}$ binds to the juxtamembrane region of the classical cadherin cytoplasmic domain, and the arm repeat region is responsible for this interaction (Ozawa and Kemler 1998; Thoreson et al. 2000). Recent data indicate that $p120^{ctn}$ has a role in maintaining normal levels of cadherins, possibly by influencing the intracellular trafficking of cadherins (Chen et al. 2003; Davis et al. 2003; Miranda et al. 2003; Xiao et al. 2003). This protein may also influence cadherin–actin cytoskeleton connections by virtue of its ability to interact with Rho guanosine triphosphatases (GTPases; reviewed in Anastasiadis and Reynolds 2001). Plakophilins are thought to have a role in targeting and clustering of other desmosomal components, and they have been reported to interact with plakoglobin, desmosomal cadherins, and desmoplakin (Kowalczyk et al. 1997; Smith and Fuchs 1998; Hatzfeld et al. 2000; Chen et al. 2002). All of these interactions map to the N-terminal "head" region. There is no binding partner currently ascribed to the plakophilin arm repeats, although it has been suggested that this domain may be involved in regulating cytoskeletal dynamics (Hatzfeld et al. 2000).

3.2
The β-Catenin/E-Cadherin Complex

The arm repeat domain of β-catenin and plakoglobin mediates binding to cadherins (Hülsken et al. 1994; Troyanovsky et al. 1994a; Pai et al. 1996; Witcher et al. 1996; Chitaev et al. 1998). Cocrystal structures of both unphosphorylated and phosphorylated E-cadherin bound to the β-catenin arm repeat domain have been determined (Huber and Weis 2001). In both cases, the crystals contained two copies in the asymmetric unit, providing a total of four independent views of the complex. The structures reveal that the last 100 amino acids of the 151 amino acid domain are ordered (Fig. 2b). Several interaction regions of the cadherin appear to be dynamic, however, as they are not visible in all four copies. A composite view of the complex is shown in Fig 2b. With all regions bound, the cadherin is seen to cover the entire length of the β-catenin arm domain, burying ~6100 Å^2 of surface area. It is important to note that the β-catenin residues that form the cadherin interface are highly conserved in plakoglobin, which explains the ability of plakoglobin to bind to classical cadherins in adherens junctions.

For convenience, we divide the visible cadherin sequence into five regions. These regions are visible to different extents in the four copies: region 1 is visible in two copies, regions 2 and 3 in all copies, region 4 only in the phosphorylated cadherin, and region 5 in three of the four copies (Huber and Weis 2001). Thus, some of the interaction regions appear to be more dynamic than others, but these observations, along with several lines of biochemical evidence, indicate that regions 2–5 all contribute to the affinity of the complex. Region 2 (residues 639–666) contains an amphipathic α helix that packs against a hydrophobic patch at the end of the arm repeat domain. In addition, Asp665 forms a hydrogen bond with Tyr654 of β-catenin. This tyrosine is a substrate of the *src* kinase, and phosphorylation of Tyr654 reduces the affinity of E-cadherin/β-catenin interaction about six- to tenfold (Roura et al. 1999; Piedra et al. 2001). This is consistent with earlier mutagenesis data indicating that deletion of region 2 reduces, but does not abolish, the interaction between E-cadherin and β-catenin (Stappert and Kemler 1994).

Region 3 (residues 653–683) is essentially an extended peptide that binds in the groove between repeats 5 and 10. This region is essential for the interaction of cadherins with β-catenin. There are five key side chains of E-cadherin involved in the interaction, as well as a number of hydrogen bonds formed between the cadherin backbone and side chains of β-catenin. Asp674 and Glu682 of cadherin form salt bridges with Lys435 and Lys312 of β-catenin, and the side chains of cadherin residues Leu676, Leu677, and Phe679 pack into small hydrophobic pockets present on the surface of β-catenin. The importance of this region is highlighted by the observation that changing Lys435 or Lys312 to glutamate residues, which would introduce a repulsive electrostatic force, abolishes binding to cadherin as well as Tcfs and APC (Graham et al. 2000). Crystal structures of other β-catenin ligands bound to the arm repeats, including several Tcf co-crystals (Graham et al. 2000, 2001; Poy et al. 2001), APC (Spink et al. 2001), and ICAT (Daniels and Weis 2002; Graham et al. 2002), reveal that all of these ligands bind in this region using a similar extended peptide structure. From these structures, a consensus β-catenin-binding motif is apparent: $Dx\theta\theta x\varphi x_{2-7}E$, where θ represents an aliphatic hydrophobic and φ represents an aromatic amino acid (Daniels and Weis 2002). It should be noted that despite the charged nature of the groove, binding of cadherin is not especially sensitive to salt [50% loss of binding at 1 M NaCl (Huber et al. 2001)], consistent with the mixed hydrophobic/polar interactions observed in the interfaces.

Region 4 (residues 684–694) is only seen in the phospho-cadherin structure, where it binds to repeats 3–5. Mass spectrometry revealed that species modified with 5–6 phosphates bind selectively to β-catenin, but only three phosphoserines were visible in these structures (Huber and Weis 2001). Ser684, a consensus CKII site, is phosphorylated but forms only a water-mediated contact with β-catenin. In contrast, serines 686 and 692 are consensus sites for GSK-3β, and both are phosphorylated and make direct contact with β-catenin. These interactions stabilize the conformation of a 10 amino acid stretch of the sequence, which features several hydrophobic contacts in addition to the phosphoserine interactions (Fig. 2b).

The final region 5 comprises two α helices that bind to the N-terminus of the arm repeats, thereby serving as a "cap" that buries a relatively hydrophobic surface. These helices are seen in three of the four copies of the complex; in the fourth they are displaced by a crystal lattice interaction. This observation suggests that the interaction is somewhat dynamic. Deletion of this region reduces, but does not abolish, the interaction (Stappert and Kemler 1994), indicating that it contributes to the affinity of the cadherin/β-catenin interaction but is not essential.

The dynamic nature of the interaction and the lack of structure of E-cadherin when not bound to β-catenin likely has important functional consequences. Mutagenesis data indicate that only region 3 is absolutely required for binding, whereas regions 2, 4, and 5 contribute to the overall affinity of the interaction. For example, phosphorylation of Tyr654, which would disrupt the region 2 helix interaction, slightly reduces affinity (Roura et al. 1999), whereas phosphorylation of serines in region 4 greatly enhances the affinity (Huber and Weis 2001) (A.H. Huber and W.I. Weis, in preparation). The lack of structure of these regions when not bound to β-catenin would also be expected to make them readily accessible to kinases and other modification enzymes, allowing graded modulation of affinity under different conditions. (It was noted above that uncomplexed cadherin is likely a good substrate for the protein degradation machinery.) It has been shown that phosphorylation of region 4 is correlated with an increased adhesiveness (Lickert et al. 2000), suggesting that the increased affinity of the cadherin/β-catenin interaction underlies this change. Similarly, it is tempting to relate decreased adhesiveness in cells expressing oncogenic tyrosine kinases like *src*, where phosphorylation of the cadherin/catenin complex is observed (Matsuyoshi et al. 1992; Behrens et al. 1993; Hamaguchi et al. 1993; Takeda et al. 1995), with a decreased affinity. Such causal relationships have not been established, however, and in some cases the opposite effect seems to hold (Daniel and Reynolds 1997; Calautti et al. 1998).

As discussed in the chapter by Perez and Nelson (this volume), there is evidence that cadherins can form lateral dimers on the cell membrane. In some models of cadherin-based adhesion, these so-called *cis* dimers are essential for the formation of properly adhesive interactions between cadherins on opposing cells. If so, dimerization mediated by the cytoplasmic interactions of the cadherins could be important for regulating adhesion, akin to "inside-out" signaling of integrins. Given the unstructured nature of the cadherin cytoplasmic domain, however, it is highly unlikely that this domain can mediate specific homodimerization. Moreover, β-catenin binds to cadherins while the latter are still in the endoplasmic reticulum, and the two proteins move together to the cell surface (Hinck et al. 1994; Näthke et al. 1994; Chen et al. 1999). Therefore, dimerization of the cadherin domain would have to be in the context of its complex with β-catenin. Gel filtration of β-catenin alone or β-catenin bound to E-cadherin cytoplasmic domain indicates that the complex is monomeric (Huber et al. 2001), a result confirmed by two crystal forms of uncomplexed β-catenin (Huber et al. 1997) and two crystal forms of the E-cadherin/β-catenin complex (Huber and Weis 2001). These data clearly demonstrate that there is no dimerization mediated by the attachment to β-catenin. It cannot be ruled out that binding of p120^{ctn} to the juxtamembrane region of cadherins mediates dimerization, but evidence that this protein more likely controls cadherin trafficking or other cytoskeletal functions makes this less likely. As discussed below, the interaction of the cadherin/β-catenin complex with α-catenin could lead to dimerization.

3.3
Role of Non-arm Repeat Regions

The region of β-catenin just N-terminal to the arm repeat domain houses the α-catenin binding site (Aberle et al. 1996; Pokutta and Weis 2000). The sequences N-terminal to the α-catenin binding site have no known function in the adherens junction, although they may be important for regulating some of the interactions (see below). The first 45 amino acids contain the phosphorylation sites required for β-catenin destruction in the Wnt pathway (Polakis 2000). Similarly, the C-terminal region does not appear to be required for adhesive interactions, and there are no known adherens junction proteins that bind in this region: *Drosophila* embryos containing a C-terminally deleted β-catenin show normal adherens junctions but are defective in Wnt signaling (Cox et al. 1996). The C-terminal region is required, however, for interactions with several general transcription factors in its role in the Wnt pathway (reviewed in Bienz and Clevers 2000; Daniels and Weis 2002).

There are currently no structural data available for the non-arm repeat regions of β-catenin. We have found that purified β-catenin constructs containing either the N- or C-terminal domains run larger than their expected size on gel filtration columns, as opposed to the arm repeat domain alone, which runs at its predicted molecular mass. The β-catenin C-terminal domain is required for binding to the histone acetyltransferase p300 (Hecht et al. 2000; Miyagishi et al. 2000; Takemaru and Moon 2000). A complex composed of a β-catenin construct spanning the arm repeats and the C-terminal region, the relevant domain of p300, and the β-catenin binding domain of

Lef-1 or Tcf-4, sizes at its predicted mass (Daniels and Weis 2002). These observations suggest that the N- and C-terminal regions are intrinsically unstructured, but become structured upon complex formation with partner proteins. The details of these interactions are unknown at present.

Several lines of evidence have suggested that the N- and the C-terminal regions of β-catenin can interact with the arm repeat domain and affect its interactions with other partners (Cox et al. 1999; Wahl et al. 2000; Zhurinsky et al. 2000; Piedra et al. 2001; Castaño et al. 2002). Both the N- and C-terminal regions expressed as recombinant proteins can interact with the arm repeats (Piedra et al. 2001; Castaño et al. 2002). Phosphorylation of Tyr654, in the last arm repeat, disrupts the interaction with, and causes increased proteolytic sensitivity of, the C-terminal region, suggesting that the latter folds back to interact with the arm repeats (Piedra et al. 2001). The interaction appears to be mediated largely by the last 22 amino acids of β-catenin (Castaño et al. 2002). It was suggested that this sequence is homologous to the region 2 helix and the beginning of region 3 of the cadherin cytoplasmic domain (Figs. 2b, 3), but the β-catenin sequence contains two consecutive proline residues that would be expected to prevent helix formation. The recombinant β-catenin C-terminal region appears to compete with E-cadherin, but not Tcf-4, for binding to the arm repeats, and both of these ligands interact better with the arm repeat domain alone than with full-length β-catenin. Likewise, a small enhancement in α-catenin binding is observed in the absence of the C-terminal region. Finally, protease sensitivity experiments indicate that the C-terminal epitope of β-catenin is lost more quickly in the presence of either ligand (Castaño et al. 2002). These data suggest that the arm repeats interact with the β-catenin C-terminus, and that binding to ligands could involve displacement of the C-terminus (Piedra et al. 2001).

Some data suggest a more positive role of the non-arm repeat β-catenin sequences in the binding to cadherins. In proteolytic sensitivity assays, full-length β-catenin protects the E-cadherin cytoplasmic domain from degradation more effectively than the armadillo repeats alone, suggesting that these regions of β-catenin might interact with cadherin (Huber et al. 2001). Binding of the N-terminal region is enhanced by the presence of the C-terminal region. When the C-terminal region is present, binding to E-cadherin appears to be enhanced by the N-terminal region, indicating that the N-terminus can modulate the effect of the C-terminus (Castaño et al. 2002). While all of these results are intriguing, a molecular explanation of the effects of the non-arm regions will require direct structural data.

3.4
Desmosomal Cadherin–Plakoglobin Interactions

The structure of the E-cadherin/β-catenin complex, combined with the high homology between the plakoglobin and β-catenin arm repeat domains and

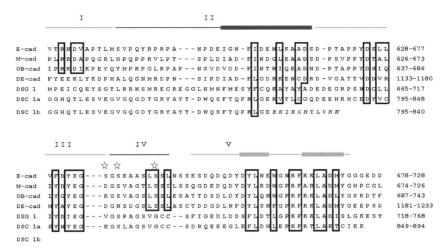

Fig. 3. Structure-based sequence alignment of the catenin-binding regions of classical and desmosomal of cadherin cytoplasmic domains. *Rectangles* above the sequences indicate α-helices seen in E-cadherin bound to β-catenin (Huber and Weis 2001). The *stars* indicate the phosphoserines visible in region 4 of the E-cadherin/β-catenin complex. Also shown is the Dsc1b sequence; the additional 11 amino acids that are not homologous to the catenin-binding region are shown in *italics*

between the classical and desmosomal cadherin catenin-binding domains, can be used to understand a great deal of the biochemical data on desmosomal cadherin/plakoglobin interactions (Huber and Weis 2001). The ability of plakoglobin to replace β-catenin in adherens junctions is explained by the observation that all β-catenin residues which interact with E-cadherin are conserved in plakoglobin. Conversely, although β-catenin is not normally found in desmosomes, it has been found to associate with desmosomal cadherins under some circumstances (Ruiz et al. 1996; Norvell and Green 1998; Wahl et al. 2000).

A structure-based alignment of the classical and desmosomal cadherin sequences reveals that interaction regions 3 and 5 are especially well conserved, indicating that both Dsc and Dsg are likely to form interactions with plakoglobin that are structurally similar to those formed between E-cadherin and β-catenin (Huber and Weis 2001) (Fig. 3). The "hydrophobic cap," or region 5, seen in the E-cadherin structure (Fig. 2b) may have a particularly important role in desmosomal cadherin/plakoglobin interactions. Deletion of plakoglobin arm repeats 1–3, the site of the region 5 interaction in the E-cadherin/β-catenin complex, abolishes binding to both Dsg and Dsc (Witcher et al. 1996; Chitaev et al. 1998), whereas a similar deletion in β-catenin diminishes, but does not eliminate, E-cadherin binding (Stappert and Kemler 1994). Several residues that form the outer rim of the E-cadherin cap are replaced with amino acids of increased hydrophobicity in the desmosomal cadherins, which could favor burial of the cap in the interface with plakoglobin (Huber and Weis 2001). In addition to regions 3 and 5, the interaction between the α helix of E-cadherin region 2 and the C-terminal arm repeats appears to be conserved in the desmocollins, but not desmogleins. This may explain the observation that plakoglobin arm repeats 11 and 12 are required for the interaction with Dsc but not Dsg (Witcher et al. 1996). These data indicate that the catenin-binding domains of desmosomal cadherins, which were originally defined by exon boundaries, should be extended N-terminally by 7 amino acids to include the complete region 2 helix (Fig. 3).

With the exception of the E-cadherin/β-catenin interaction (A.H. Huber and W.I. Weis, in preparation), no quantitative affinity measurements have been reported for cadherin/catenin interactions. Plakoglobin binds best to Dsg-1 but less well to Dsc-1a and E-cadherin (H.-J. Choi and W.I. Weis, unpublished observations). The equivalent rank order of binding to β-catenin has not been tested, but there is some evidence that the sequences flanking the armadillo repeat regions of β-catenin and plakoglobin determines their relative affinities for the different kinds of cadherins. β-Catenin does not normally associate with desmoglein in vivo, but the arm repeat domain of β-catenin, which has 76% sequence identity to that of plakoglobin, can associate with desmoglein 2 (Wahl et al. 2000). However, a chimera composed of the N-terminal and C-terminal domains of β-catenin and the arm repeats of plakoglobin does not associate with desmoglein 2, suggesting that the N-terminal and C-terminal regions of β-catenin, which have 41% and 29% sequence identity, respectively, to those of plakoglobin prevent the association with desmoglein and thus play a role in determining the specificity. In the case of desmocollin, although the catenin binding site is required for binding to plakoglobin, the 12 amino acids in the preceding IA domain are also involved in binding to plakoglobin and desmoplakin, suggesting a possible role in the specific inclusion of desmocollin/plakoglobin complex in desmosome (Troyanovsky et al. 1994b).

Plakoglobin is unique in that it serves both in adherens junctions and desmosomes. This raises the question of how α-catenin, which interacts with either β-catenin or plakoglobin in adherens junctions, is excluded from desmosomes. Mutagenesis data indicate that the α-catenin- and desmosomal cadherin-binding sites on plakoglobin overlap, suggesting that α-catenin is prevented from binding to plakoglobin by desmosomal cadherins (Troyanovsky et al. 1996; Wahl et al. 1996; Witcher et al. 1996). Mutating hydrophobic residues in the α-catenin binding helix of plakoglobin (see below) eliminates both α-catenin and desmoglein binding (Chitaev et al. 1998). These data indicate that desmosomal cadherin/plakoglobin complexes feature interactions beyond those observed in the E-cadherin/β-catenin complex.

4
α-Catenin

α-Catenin links the classical cadherin/β-catenin complex to the actin cytoskeleton. This 906 amino acid protein binds directly to F-actin (Rimm et al. 1995; Pokutta et al. 2002) and to a number of actin-binding proteins, including vinculin (Watabe-Uchida et al. 1998; Weiss et al. 1998), α-actinin (Knudsen et al. 1995; Nieset et al. 1997), l-afadin (Tachibana et al. 2000; Pokutta et al. 2002), and ZO-1 and ZO-2 (Itoh et al. 1997; Itoh et al. 1999). It is thought that the multiple partners of this protein serve to integrate signals for adherens junction assembly and crosstalk between adherens junctions and other adhesion systems (Perez-Moreno et al. 2003).

α-Catenin is homologous to the actin-binding protein vinculin in three discrete regions of primary structure (Nagafuchi et al. 1991). Proteolytic fragmentation of the full-length protein yields two stable fragments of 21 and 29 kDa, corresponding roughly to the first two vinculin homology regions (Pokutta and Weis 2000; Yang et al. 2001). Thus, α-catenin is divided into structurally distinct domains that are likely connected by flexible linkers.

4.1
α-Catenin Homodimerization and β-Catenin Binding Region

Although cadherins, β-catenin, and α-catenin form a 1:1:1 complex in solution (Aberle et al. 1994) and in cells (Ozawa and Kemler 1992; Hinck et al. 1994), α-catenin in isolation forms a homodimer at micromolar concentrations (Rimm et al. 1995; Koslov et al. 1997). The amino-terminal 21-kDa domain, residues 82–264 of the murine sequence, mediates homodimerization of α-catenin (Pokutta and Weis 2000). The crystal structure of this region shows that each protomer consists of five α-helices (Fig. 4a). The first α helix (residues 82–113) is followed by a very long, 50-residue helix whose N-terminal half, residues 117–145, pairs in an antiparallel manner with the first helix. The C-terminal half of this helix is the first helix of an antiparallel four-helix bundle. The dimer forms by pairing of the two N-terminal helices with their equivalents in another protomer, which forms a non-covalent four-helix bundle (Fig. 4b).

The domain formed by residues 82–278 of α-catenin (Fig. 4a) does not bind to β-catenin, but a construct comprising residues 57–264 binds to β-catenin comparably to the full-length α-catenin (Pokutta and Weis 2000). Plotting residues 57–81 on a helical wheel diagram indicates that this sequence has the potential to form an amphipathic α helix. The α-catenin binding site on β-catenin lies in the region between residues 118–149 (Aberle et al. 1996). This sequence also can form an amphipathic helix, and mutations in residues occupying the *a* and *d* positions of this putative helix, i.e., those that form

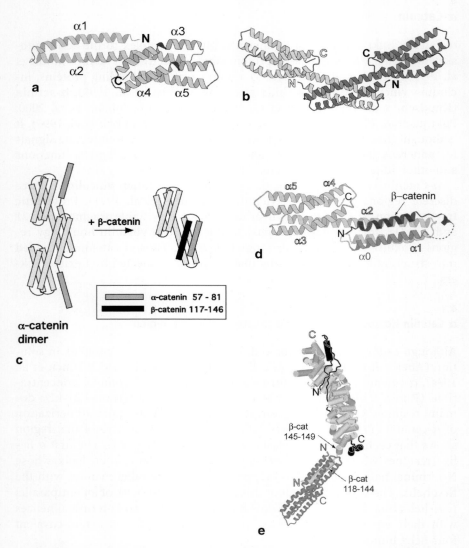

Fig. 4a–e. The α-catenin homodimerization and β-catenin binding domain. **a** Ribbon diagram of the structure of the protomer. The positions of two conserved prolines that kink helices α3 and α4 are shown in *black*. **b** Structure of the dimer. **c** Model for the interaction of α-catenin 57–264 with β-catenin 118–149. The α-catenin homodimerization domain protomer 82–264 is shown in *white*. α-Catenin 57–81 is shown in *light grey*. In the absence of β-catenin, this sequence is flexibly linked to the rest of the domain, as implied by the trypsin sensitivity at residue 81 (Pokutta and Weis 2000). In the presence of β-catenin (*black*), α-catenin 57–81 (α0) contributes one helix to the four-helix bundle that defines the interface of the two proteins. **d** Structure of βα-cat. Color scheme as in **c**. **e** Model of the E-cadherin/β-catenin/α-catenin complex

the hydrophobic face, destroyed binding to α-catenin (Aberle et al. 1996). Based on these observations, it was proposed that in the presence of β-catenin, the homodimer of α-catenin dissociates, and the two helices α1 and α2 contributed by one protomer are replaced by the 57–81 helix (α0) and the β-catenin helix (Pokutta and Weis 2000) (Fig. 4c). To test this model, a chimeric protein was constructed in which β-catenin residues 118–149 were fused via a 5-glycine linker to the N-terminus of the β-catenin binding (residues 57–264) region of α-catenin. This construct sizes as a monomer on gel filtration columns, whereas 57–264 runs as a dimer, confirming that the homodimer is disrupted in the presence of the β-catenin peptide (Pokutta and Weis 2000). The crystal structure of the β-catenin/α-catenin chimera, denoted $\beta\alpha$-cat, confirmed the model (Pokutta and Weis 2000) (Fig. 4d). The structure of α-catenin 82–264 in the chimera is very close to that found in the homodimer.

The binding of β-catenin to α-catenin can be described as "helix exchange," in which the α-catenin α0 helix and β-catenin 118–149 sequence replace the α1 and α2 helices from the partner protomer of the homodimer (Fig. 4c). Both the α-catenin homodimer and the β-catenin/α-catenin heterodimer interfaces are part of a four-helix bundle involving α1 and α2, but the different sequences give rise to structural differences. The most notable difference is in the packing angle of the helices. In the homodimer, the α1 and α2 helices of one protomer lie at an angle of approximately 40° with respect to the α1 and α2 helices of the partner. In the heterodimer, the α0 and β-catenin helices lie at an angle of about 10° with respect to α1 and α2, producing a somewhat larger buried surface area. In addition to the helix packing angles, some of the difference in interface area is due to the fact that β-catenin residues 142–145 adopt an extended, rather than a helical, conformation, and residues 146–149 form a single turn of helix that lies in a different direction than the helix formed by residues 121–141. Tyr142 packs into the core of the four-helix bundle, where its aromatic ring makes a number of critical contacts with α-catenin. This and other differences create a larger amount of buried surface area than would be possible if the helix continued straight through. The larger interface found in the β-catenin heterodimer is consistent with the observation that the heterodimer is favored over the homodimer in solution (Koslov et al. 1997).

An important future goal is to obtain direct experimental data for the higher order structure of the cadherin/β-catenin/α-catenin assembly. Some insight into this structure comes from sequence alignments of the β-catenin arm repeats, which indicate that β-catenin residues 146–149, which form a single turn of helix in the β/α chimera, could correspond to the first helix of the first armadillo repeat (Huber et al. 1997). In several crystal structures of β-catenin bound to ligands (Graham et al. 2000; Huber and Weis 2001), this region is part of a kinked but continuous α helix between residues 134–160 that is stabilized by lattice contacts. It was proposed that this could be part of the α-catenin binding site (Graham et al. 2000). Inspection of the chimera structure, however, shows that the N-terminal β-catenin helix cannot continue past residue 142, because doing so would introduce steric clashes with several residues on both β-catenin and α-catenin. This explains why residues 142–145 are non-helical when bound to α-catenin. Superposition of residues 146–149 from the βα chimera onto the corresponding residues in the E-cadherin or XTcf-3 complex structures gives a model of α-catenin bound to β-catenin that is free of steric clashes (Huber and Weis 2001) (Fig. 4e). Comparison of the E-cadherin and XTcf-3 structures reveals that the helical regions of β-catenin preceding Thr150 are rotated with respect to one another by 100° about the helix axis. This probably reflects hinge motion due to the presence of proline at residue 154, which cannot form a hydrogen bond with the backbone of Thr150. In the models of βα-cat superimposed on the β-catenin complexes, the 100° rotation gives rise to substantially different relative positions of α- and β-catenin. In the case of E-cadherin, the superposition gives rise to a steric clash with the cadherin region 5 helix, leading to the proposal that the dynamics of the interaction between cadherin region 5 and β-catenin could influence the relative positions of α- and β-catenin in the adherens junction complex (Huber and Weis 2001).

4.2
α-Catenin M Region

In two independent studies, trypsin digestion of full-length α-catenin produced somewhat different fragments overlapping the central vinculin homology region: one group obtained the fragment 385–632 (Yang et al. 2001), while the other found a larger fragment spanning residues 385–651 (Pokutta and Weis 2000). Crystals of both fragments of this central or middle (M) domain were obtained, and the structure consists of two independent antiparallel four-helix bundles connected by a short linker (Yang et al. 2001; Pokutta et al. 2002) (Fig. 5a). In the crystals of the longer construct, residues 633–651 were never observed, consistent with this region being flexibly linked to the rest of the protein. The crystals of the shorter M domain construct (Yang et al. 2001) contain a dimer in the asymmetric unit, whereas three molecules are present in the asymmetric unit of the longer construct (Pokutta et al. 2002). In both cases, there are small (10–14°) differences in the relative positions of the two four-helix bundles in the different copies present in the asymmetric unit. More dramatic, however, is the difference between the two crystal forms. In the shorter construct, the two four-helix bundle subdomains lie roughly perpendicular to one another, whereas in the crystals of the longer construct the two subdomains are separated by about 40° to produce a more "closed" conformation (Fig. 5b). The interfaces and the number of contacts between the two bundles in both crystal forms is relatively small, which

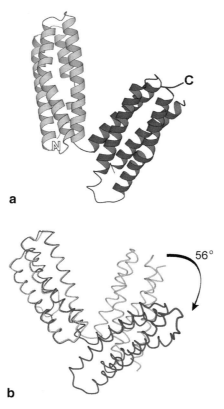

Fig. 5a, b. The α-catenin M domain. **a** Ribbon diagram. The two four-helix bundles are shown in *light* and *dark grey*. **b** Comparison of the structure in the two crystal forms. The first four-helix bundle seen in the crystals of the shorter construct (Yang et al. 2001) (*dark grey*) is superimposed on the same region from the crystal structure of the longer construct (Pokutta et al. 2002) (*light grey*) to reveal the difference in the relative orientations of the second bundle

presumably allows for the observed flexibility in the relative positions of the two subdomains.

The M domain does not have as clearly a defined role as the N-terminal dimerization/β-catenin binding and the C-terminal actin binding domains. The putative vinculin and α-actinin binding sites map to the very N-terminus of the M-domain but include upstream sequences removed by trypsin. Recently, the binding site for l-afadin, a protein that links the immunoglobulin superfamily adhesion protein nectin to the actin cytoskeleton (Mandai et al. 1997; Tachibana et al. 2000), was mapped to the M domain, and it was shown that the recombinant M domain can bind to l-afadin present in MDCK cell lysates (Pokutta et al. 2002). The role of the nectin system is not yet clear, but the discovery that l-afadin can bind to α-catenin suggests physical interactions with adherens junctions, consistent with its cellular localization (Tachibana et al. 2000). Expression of each M fragment subdomain separately demonstrated that both domains are required for full binding to l-afadin (Pokutta et al. 2002). Interestingly, l-afadin bound to the M domain more strongly than to full-length α-catenin, suggesting that there is a cryptic binding site that presumably is modulated by other factors (Pokutta et al. 2002). The crystallographically-observed flexibility in interdomain orientation could be related to conformational modulation of the α-catenin structure needed to achieve full binding to l-afadin and perhaps other cellular partners.

Tsukita and colleagues have shown that fusion of α-catenin to the cytoplasmic domain of E-cadherin can restore strong cell adhesion to cell lines deficient in either cadherins or α-catenin, presumably by directly linking cadherins to the actin cytoskeleton (Nagafuchi et al. 1994). Further studies using this system demonstrated that a fusion of α-catenin residues 509–643, spanning the second helical subdomain of the M fragment, to E-cadherin supports cell aggregation and weak adhesion (Imamura et al. 1999). This fusion protein is not linked to the actin cytoskeleton, so the result suggests that this portion of α-catenin can support lateral clustering of cadherins. The authors denote this region as the "adhesion modulation domain" of α-catenin. Based on the non-crystallographic dimer of the shorter M domain fragment, it was proposed that dimerization could modulate adhesion (Yang et al. 2001), consistent with the notion that *cis*-dimers of cadherins have an important role in modulating adhesion. As discussed above and by Perez and Nelson in this volume, the role of such dimerization is still not clear. Moreover, the M fragment crystallizes with different oligomer interfaces, and it was found that both fragments can be crosslinked to oligomers larger than dimers, making unclear the relevance of the dimerization of the shorter M fragment (Pokutta et al. 2002). Nonetheless, the various oligomerization modes of this region could contribute to the lateral clustering of cadherins or have other modulatory roles.

4.3
Actin Binding by α-Catenin

α-Catenin binds to filamentous (F) actin, an activity associated with the C-terminal portion of the protein (Rimm et al. 1995; Pokutta et al. 2002). Near the C-terminus of α-catenin, residues 678–864 have the highest level of homology to vinculin (27% identity). The equivalent vinculin residues are sufficient to bind actin. The structure of this vinculin fragment region is an antiparallel five-helix bundle (Bakolitsa et al. 1999) (Fig. 6). The residues responsible for packing the five helices together are well conserved in α-catenin, implying that it too has this structure. Surprisingly, an α-catenin con-

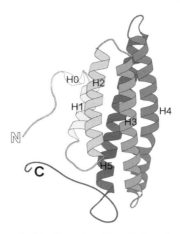

Fig. 6. The structure of the actin-binding vinculin tail domain (Bakolitsa et al. 1999)

struct spanning only residues 678–864 does not bind to actin, whereas residues 671–906 bind robustly in an actin cosedimentation assay, implying that the C-terminal 42 amino acids unique to α-catenin are required for actin binding (Pokutta et al. 2002). This difference remains a puzzle, although it is important to note that α-catenin binds to actin constitutively, whereas vinculin requires activation in order to relieve an autoinhibitory interaction between its actin-binding and N-terminal domains.

The mechanism of F-actin binding by α-catenin or vinculin is not known. The actin-binding five-helix bundle is structurally similar to apolipoprotein E, whose helices are thought to "unfurl" from one another in order to interact with lipoprotein surfaces (Narayanaswami et al. 1999; Weers et al. 1999). Liddington and colleagues found that the vinculin domain becomes more susceptible to proteases in the presence of actin, and proposed that activation of vinculin in the presence of actin leads to a similar structural change and the interaction with the actin filament (Bakolitsa et al. 1999).

4.4
Regulation of α-Catenin Interactions

The structures of three portions of α-catenin representing roughly 75% of the sequence are known. Both the dimerization/β-catenin binding and M-domain regions are composed of a series of four-helix bundles, and the actin-binding domain has a five-helix bundle. The tryptic sensitivity of full-length α-catenin (Yang et al. 2001; Pokutta et al. 2002) implies that the linkers between the dimerization and M domains, and between the M and actin-binding domains, are flexible. Crystallographic data likewise indicate significant flexibility between the two M subdomains (Yang et al. 2001; Pokutta et al. 2002) (Fig. 5b). It seems likely that intramolecular flexibility, combined with changes in oligomeric interactions α-catenin promoters, is likely to have an important role in the action of regulatory factors on the interactions of α-catenin. It remains to be seen whether there are direct intramolecular interactions between different portions of α-catenin under certain conditions.

The homologies between three regions of α-catenin and vinculin indicate that they are closely related. Nonetheless, α-catenin does not appear to have the activation barrier to actin binding shown by vinculin. On the other hand, we have found that binding to β-catenin is not especially strong, due to the need to disrupt the homodimer interface. It is clear that there must be regulatory factors in the cell that control the ability to form stable complexes with β-catenin and not revert to homodimers. Whether this is due to post-translational modifications such as phosphorylation, or allosteric effectors perhaps working through other regions of the protein, is not known. It will be important to find such factors in order to understand the regulated assembly and disassembly of adherens junctions.

5
Intermediate Filament Binding by Desmoplakin

Desmoplakin is a large, roughly 3,000 amino acid, protein that links the desmosomal cadherin/plakoglobin/plakophilin complex to intermediate filaments (IFs). In this sense, desmoplakin is the functional equivalent of α-catenin in adherens junctions, but unlike the cadherins and arm-family proteins present in both kinds of junctions, it is unrelated in sequence to α-catenin. The primary structure (Fig. 7a) consists of a 1,056 residue N-terminal domain (DPNT) responsible for the interactions with plakoglobin and possibly desmosomal cadherins and plakophilin, an 899 amino acid coiled-coil region that dimerizes the protein, and a C-terminal 956 amino acid domain (DPCT) that binds to IFs. There are at present no structural data available for DPNT or the dimerization domain.

DPCT contains three modules, termed A, B, and C, composed of 4.5 copies of a 38 amino acid repeat (Green et al. 1990). One or more of these modules are found in other IF binding proteins, termed the plakin family, that are found in desmosomes and hemidesmosomes. Chymotryptic fragmentation of bacterially expressed DPCT generated three stable fragments corresponding to the A, B, and C modules (Choi et al. 2002). Each module was shown to bind to vimentin weakly; stronger binding is obtained using the full DPCT, a fragment spanning the three modules, or a fragment spanning fragments B and C. The structures of B and C revealed that, unlike arm repeats, the 4.5 copies of the 38 amino acid motif form a globular structure (Choi et al. 2002) (Fig. 7b). Sequence analysis reveals unique packing inter-

Structural Aspects of Adherens Junctions and Desmosomes

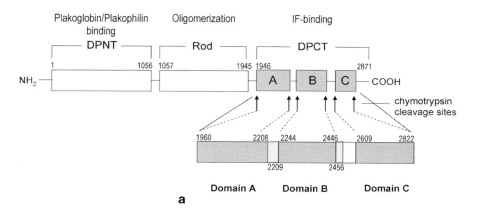

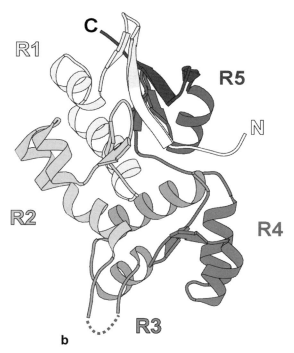

Fig. 7a, b. Desmoplakin structure. **a** Primary structure. The sites of chymotryptic fragmentation which produce the three IF-binding domains A, B, and C, are indicated. **b** Structure of DPCT module C. The four complete 38 residue repeats are noted *R1–R4*, and the fifth half repeat is denoted *R5*

actions conserved in the individual repeats, making it likely that all plakins will have this basic globular subdomain. Thus, plakin family members containing more than one module are predicted to be arranged more like the "beads on a string" found, for example, in immunoglobulin superfamily proteins. Indeed, the structure of the DPCT AB fragment shows two independently folded domains (H.-J. Choi and W.I. Weis, unpublished).

The inability to cocrystallize filamentous proteins with their binding partners has hindered molecular understanding of the DPCT–IF interaction, as well as the α-catenin/actin interaction. It was noted, however, that a positively charged groove is a highly conserved surface feature of the plakin module, and the groove is wide enough to accommodate an α helix such as those found in IFs (Choi et al. 2002). A combination of site-directed mutagenesis and other methods will be required to confirm this hypothesis.

As in the case of adherens junctions, there are at present no data that address the higher order structure of the cytoplasmic desmosome assembly. Progress in this area will depend on resolving ambiguities as to which components actually bind to one another. Interactions between desmosomal cadherins and plakophilin, desmosomal cadherins and desmoplakin, and between desmoplakin and both plakoglobin and plakophilin, have been reported, giving the tentative model shown in Fig. 1. It must be cautioned, however, that apart from desmosomal cadherin/plakoglobin and DPCT/IF interactions, none of the reported interactions have been assessed using purified, native proteins. These reports have relied upon colocalization by immunofluorescence microscopy, immunoprecipitation, yeast two-hybrid, and blot overlay assays in which at least one protein partner is denatured (Bornslaeger et al. 1996; Kowalczyk et al. 1997; Smith and Fuchs 1998). In the absence of direct binding data with purified, natively folded proteins, the reported interactions must be considered tentative. As more components are purified and assembled in vitro, a clearer picture of the desmosome should emerge.

Acknowledgements We thank S. Halfon for comments on the manuscript. Work on cadherins and catenins in our laboratory is supported by grant GM56169 from the National Institutes of Health to W.I.W. H.-J.C. was supported by postdoctoral fellowships from the Korean Organization for Science and Engineering and the American Heart Association.

References

Aberle H, Butz S, Stappert J, Weissig H, Kemler R, Hoschuetzky H (1994) Assembly of the cadherin-catenin complex in vitro with recombinant proteins. J Cell Sci 107:3655–3663

Aberle H, Schwartz H, Hoschuetzky H, Kemler R (1996) Single amino acid substitutions in proteins of the *armadillo* gene family abolish their binding to α-catenin. J Biol Chem 271:1520–1526

Anastasiadis PZ, Reynolds AB (2001) Regulation of Rho ATPases by p120-catenin. Curr Opin Cell Biol 13:604–610

Anastasiadis PZ, Reynonds AB (2000) The p120 catenin family: complex roles in adhesion, signaling and cancer. J Cell Sci 113:1319–1334

Bakolitsa C, de Pereda JM, Bagshaw CR, Critchley DR, Liddington RC (1999) Crystal structure of the vinculin tail suggests a pathway for activation. Cell 99:603–613

Behrens J, Vakaet L, Friis R, Winterhager E, van Roy F, Mareel MM, Birchmeier W (1993) Loss of epithelial differentiation and gain of invasiveness correlates with tyrosine phosphorylation of the E-cadherin/β-catenin complex in cells transformed with a temperature-sensitive v-*src* gene. J Cell Biol 120:757–766

Bienz M, Clevers H (2000) Linking colorectal cancer to Wnt signaling. Cell 103:311–320

Bonne S, Gilbert B, Hatzfeld M, Chen X, Green KJ, van Roy F (2003) Defining desmosomal plakophilin-3 interactions. J Cell Biol 161:403–416

Bornslaeger EA, Corcoran CM, Stappenbeck TS, Green KJ (1996) Breaking the connection: displacement of the desmosomal plaque protein desmoplakin from cell–cell interfaces disrupts anchorage of intermediate filament bundles and alters intercellular junction assembly. J Cell Biol 134:985–1001

Calautti E, Cabodi S, Stein PL, Hatzfeld M, Kedersha N, Dotto GP (1998) Tyrosine phosphorylation and src family kinases control keratinocyte cell–cell adhesion. J Cell Biol 141:1449–1465

Castaño J, Raurell I, Piedra JA, Miravet S, Duñach M, García de Herreros A (2002) β-Catenin N- and C-terminal tails modulate the coordinated binding of adherens junction proteins to β-catenin. J Biol Chem 277:31541–31550

Chen X, Bonné S, Hatzfeld M, van Roy F, Green KJ (2002) Protein binding and functional characterization of plakophilin 2. J Biol Chem 277:10512–10522

Chen X, Kojima S, Borisy GG, Green KJ (2003) p120 catenin associates with kinesin and facilitates the transport of cadherin-catenin complexes to intercellular junctions. J Cell Biol 163:547–557

Chen Y-T, Stewart DB, Nelson WJ (1999) Coupling assembly of the E-cadherin/b-catenin complex to efficient endoplasmic reticulum exit and basal-lateral membrane targeting of E-cadherin in polarized MDCK cells. J Cell Biol 144:687–699

Chitaev NA, Leube RE, Troyanovsky RB, Eshkind LG, Franke WW, Troyanovsky SM (1996) The binding of plakoglobin to desmosomal cadherins: patterns of binding sites and topogenic potential. J Cell Biol 133:359–369

Chitaev NA, Averbakh AZ, Troyanovsky RB, Troyanovsky SM (1998) Molecular organization of the desmoglein-plakoglobin complex. J Cell Sci 111:1941–1949

Choi H-J, Park-Snyder S, Pascoe LT, Green KJ, Weis WI (2002) Structures of two intermediate filament-binding fragments of desmoplakin reveal a unique repeat motif structure. Nat Struct Biol 9:612–620

Cox RT, Kirkpatrick C, Peifer M (1996) Armadillo is required for adherens junction assembly, cell polarity, and morphogenesis during *Drosophila* embryogenesis. J Cell Biol 134:133–148

Cox RT, Pai LM, Kirkpatrick C, Stein J, Peifer M (1999) Roles of the C-terminus of armadillo in Wingless signaling in Drosophila. Genetics 153:319–332

Daniel JM, Reynolds AB (1997) Tyrosine phosphorylation and cadherin/catenin function. BioEssays 19:883–891

Daniels DL, Weis WI (2002) ICAT inhibits β-catenin binding to Tcf/Lef-family transcription factors and the general coactivator p300 using independent structural modules. Mol Cell 10:573–584

Davis MA, Ireton RC, Reynolds AB (2003) A core function for p120-catenin in cadherin turnover. J Cell Biol 163:525–534
Farquhar MG, Palade GE (1963) Junctional complexes in various epithelia. J Cell Biol 17:375–412
Graham TA, Weaver C, Mao F, Kimmelman D, Xu W (2000) Crystal structure of a β-catenin/Tcf complex. Cell 103:885–896
Graham TA, Ferkey DM, Mao F, Kimelman D, Xu W (2001) Tcf4 can specifically recognize β-catenin using alternative conformations. Nat Struct Biol 8:1048–1052
Graham TA, Clements WK, Kimelman D, Xu W (2002) The crystal structure of the β-catenin/ICAT complex reveals the inhibitory mechanism of ICAT. Mol Cell 10:563–571
Green KJ, Parry DAD, Stenert PM, Virata MLA, Wagner RM, Angst DB, Nilles LA (1990) Structure of the human desmoplakins. Implications for function in the desmosomal plaque. J Biol Chem 265:2603–2612
Hamaguchi M, Matsuyoshi N, Ohnishi Y, Gotoh B, Takeichi M, Nagai Y (1993) p60v-src causes tyrosine phosphorylation and inactivation of the N-cadherin–catenin cell adhesion system. EMBO J 12:307–314
Hatzfeld M, Haffner C, Schulze K, Vinzens U (2000) The function of plakophilin 1 in desmosome assembly and actin filament organization. J Cell Biol 149:209–222
He W, Cowin P, Stokes DL (2003) Untangling desmosomal knots with electron tomography. Science 302:109–113
Hecht A, Vleminckx K, Stemmler MP, van Roy F, Kemler R (2000) The p300/CBP acetyltransferases function as transcriptional coactivators of β-catenin in vertebrates. EMBO J 19:1839–1850
Hinck L, Nathke IS, Papkoff J, Nelson WJ (1994) Dynamics of cadherin/catenin complex formation: novel protein interactions and pathways of complex assembly. J Cell Biol 125:1327–1340
Huber AH, Weis WI (2001) The structure of the β-catenin/E-cadherin complex and the molecular basis of diverse ligand recognition by β-catenin. Cell 105:391–402
Huber AH, Nelson WJ, Weis WI (1997) Three-dimensional structure of the armadillo repeat region of β-catenin. Cell 90:871–882
Huber AH, Stewart DB, Laurents DV, Nelson WJ, Weis WI (2001) The cadherin cytoplasmic domain is unstructured in the absence of β-catenin: a possible mechanism for regulating cadherin turnover. J Biol Chem 276:12301–12309
Hülsken J, Birchmeier W, Behrens J (1994) E-cadherin and APC compete for the interaction with β-catenin and the cytoskeleton. J Cell Biol 127:2061–2069
Imamura Y, Itoh M, Maeno Y, Tsukita S, Nagafuchi A (1999) Functional domains of α-catenin required for the strong state of cadherin-based cell adhesion. J Cell Biol 144:1311–1322
Itoh M, Nagafuchi A, Moroi S, Tsukita S (1997) Involvement of ZO-1 in cadherin-based cell adhesion through its direct binding to α catenin and actin filaments. J Cell Biol 138:181–192
Itoh M, Morita K, Tsukukita S (1999) Characterization of ZO-2 as a MAGUK family member associated with tight as well as adherens junctions with a binding affinity to occludin and α-catenin. J Biol Chem 274:5981–5986
Knudsen KA, Soler AP, Johnson KR, Wheelock MJ (1995) Interaction of α-actinin with the cadherin/catenin cell–cell adhesion complex via α-catenin. J Cell Biol 130:67–77
Koslov ER, Maupin P, Pradhan D, Morrow JS, Rimm DL (1997) α-catenin can form asymmetric homodimeric complexes and/or heterodimeric complexes with β-catenin. J Biol Chem 272:27301–27306

Kowalczyk AP, Bornslaeger EA, Borgwardt JE, Palka HL, Bhaliwal AS, Corcoran CM, Denning MF, Green KJ (1997) The amino-terminal domain of desmoplakin binds to plakoglobin and clusters desmosomal cadherin-plakoglobin complexes. J Cell Biol 139:773–784

Kowalczyk AP, Bornslaeger EA, Norvell SM, Palka HL, Green KJ (1999) Desmosomes: intercellular adhesive junctions specialized for attachment of intermediate filaments. Int Rev Cytol 185:237–302

Lickert H, Bauer A, Kemler R, Stappert J (2000) Casein kinase II phosphorylation of E-cadherin Increases E-cadherin/β-catenin interaction and strengthens cell–cell adhesion. J Biol Chem 275:5090–5095

Mandai K, Nakanishi H, Satoh A, Obaishi H, Wada M, Nishioka H, Itoh M, Mizoguchi A, Aoki T, Fujimoto T, Matsuda Y, Tskukita S, Takai Y (1997) Afadin: a novel actin filament-binding protein with one PDZ domain localized at cadherin-based cell-to-cell adherens junction. J Cell Biol 139:517–528

Mathur M, Goodwin L, Cowin P (1994) Interactions of the cytoplasmic domain of the desmosomal cadherin Dsg1 with plakoglobin. J Biol Chem 269:14075–14080

Matsuyoshi N, Hamaguchi M, Taniguchi S, Nagafuchi A, Tsukita S, Takeichi M (1992) Cadherin-mediated cell–cell adhesion is perturbed by v-*src* tyrosine phosphorylation in metastatic fibroblasts. J Cell Biol 118:703–714

Miranda KC, Joseph SR, Yap AS, Teasdale RD, Stow JL (2003) Contextual binding of p120^{ctn} to E-cadherin at the basolateral plasma membrane in polarized epithelia. J Biol Chem 278:43480–43488

Miyagishi M, Fujii R, Hatta M, Yoshida E, Araya N, Nagafuchi A, Ishihara S, Nakajima T, Fukamizu A (2000) Regulation of Lef-mediated transcription and p53-dependent pathway by associating β-catenin with CPB/p300. J Biol Chem 275:35170–35175

Nagafuchi A, Takeichi M, Tsukita S (1991) The 102 kD cadherin-associated protein: similarity to vinculin and posttranscriptional regulation of expression. Cell 65:849–857

Nagafuchi A, Ishihara S, Tsukita S (1994) The roles of catenins in the cadherin-mediated cell adhesion: functional analysis of E-cadherin-α-catenin fusion molecules. J Cell Biol 127:235–245

Narayanaswami V, Wang J, Schieve D, Kay CM, Ryan RO (1999) A molecular trigger of lipid binding-induced opening of a helix bundle exchangeable apolipoprotein. Proc Natl Acad Sci U S A 96:4366–4371

Näthke IS, Hinck L, Swedlow JR, Papkoff J, Nelson WJ (1994) Defining interactions and distributions of cadherin and catenin complexes in polarized epithelial cells. J Cell Biol 125:1341–1352

Nelson WJ (2003) Adaptation of core mechanisms to generate cell polarity. Nature 422:766–774

Nieset JE, Redfield AR, Jin F, Knudsen KA, Johnson KR, Wheelock MJ (1997) Characterization of the interactions of α-catenin with α-actinin and β-catenin/plakoglobin. J Cell Sci 110:1013–1022

Nollet F, Kools P, van Roy F (2000) Phylogenetic analysis of the cadherin superfamily allows identification of six major subfamilies besides several solitary members. J Mol Biol 299:551–572

Norvell SM, Green KJ (1998) Contributions of extracellular and intracellular domains of full length and chimeric cadherin molecules to junction assembly in epithelial cells. J Cell Sci 111:1305–1318

Ozawa M, Kemler R (1992) Molecular organization of the uvomorulin-catenin complex. J Cell Biol 116:989–996

Ozawa M, Kemler R (1998) The membrane-proximal region of the E-cadherin cytoplasmic domain prevents dimerization and negatively regulates adhesion activity. J Cell Biol 142:1605–1613

Ozawa M, Ringwald M, Kemler R (1990) Uvomorulin-catenin complex formation is regulated by a specific domain in the cytoplasmic region of the cell adhesion molecule. Proc Natl Acad Sci USA 87:4246–4250

Pai L-M, Kirkpatrick C, Blanton J, Oda H, Takeichi M, Peifer M (1996) *Drosophila* α-catenin and E-cadherin bind to distinct regions of *Drosophila* Armadillo. J Biol Chem 271:32411–32420

Perez-Moreno M, Jamora C, Fuchs E (2003) Sticky business: orchestrating cellular signals at adherens junctions. Cell 112:535–548

Piedra J, Martínez D, Castaño J, Miravet S, Duñach M, García de Herreros A (2001) Regulation of β-catenin structure and activity by tyrosine phosphorylation. J Biol Chem 276:20436–20443

Pokutta S, Weis WI (2000) Structure of the dimerization and β-catenin binding region of α-catenin. Mol Cell 5:533–543

Pokutta S, Drees F, Takai Y, Nelson WJ, Weis WI (2002) Biochemical and structural definition of the l-afadin- and actin-binding sites of α-catenin. J Biol Chem 277:18868–18874

Polakis P (2000) Wnt signaling and cancer. Genes Dev 14:1837–1851

Poy F, Lepourcelet M, Shivdasani RA, Eck MJ (2001) Structure of a human Tcf4-β-catenin complex. Nat Struct Biol 8:1053–1057

Rechsteiner M, Rogers SW (1996) PEST sequences and regulation by proteolysis. Trends Biochem Sci 21:267–271

Rimm DL, Koslov ER, Kebriaei P, Cianci CD, Morrow JS (1995) α_1(E)-catenin is an actin-binding and -bundling protein mediating the attachment of F-actin to the membrane adhesion complex. Proc Natl Acad Sci USA 92:8813–8817

Roura S, Miravet S, Piedra J, García de Herreros A, Dunach M (1999) Regulation of E-cadherin/catenin association by tyrosine phosphorylation. J Biol Chem 274:36734–36740

Rubinfeld B, Souza B, Albert I, Munemitsu S, Polakis P (1995) The APC protein and E-cadherin form similar but independent complexes with α-catenin, β-catenin, and plakoglobin. J Biol Chem 270:5549–5555

Ruiz P, Brinkmann V, Ledermann B, Behrend M, Grund C, Thalhammer C, Vogel F, Birchmeir C, Günthert U, Franke WW, Birchmeier W (1996) Targeted mutation of plakoglobin in mice reveals essential functions of desmosomes in the embryonic heart. J Cell Biol 135:215–225

Runswick SK, O'Hare MJ, Jones L, Streuli CH, Garrod DR (2001) Desmosomal adhesion regulates epithelial morphogenesis and cell positioning. Nat Cell Biol 3:823–830

Shore EM, Nelson WJ (1991) Biosynthesis of the cell adhesion molecule uvomorulin (E-cadherin) in Madin-Darby canine kidney epithelial cells. J Biol Chem 266:19672–19680

Smith EA, Fuchs E (1998) Defining the interactions between intermediate filaments and desmosomes. J Cell Biol 141:1229–1241

Spink KE, Fridman SG, Weis WI (2001) Molecular mechanisms of β-catenin recognition by Adenomatous Polyposis Coli revealed by the structure of an APC/β-catenin complex. EMBO J 20:6203–6212

Stappert J, Kemler R (1994) A short core region of E-cadherin is essential for catenin binding and is highly phosphorylated. Cell Adhes Commun 2:319–327

Tachibana K, Nakanishi H, Mandai K, Ozaki K, Ikeda W, Yamamoto Y, Nagafuchi A, Tsukita S, Takai Y (2000) Two cell adhesion molecules, nectin and cadherin, interact through their cytoplasmic domain-associated proteins. J Cell Biol 150:1161–1175

Takeda H, Nagafuchi A, Yonemura S, Tsukita S, Behrens J, Birchmeier W, Tsukita S (1995) V-src kinase shifts the cadherin-based cell adhesion from the strong to the weak state and β catenin is not required for the shift. J Cell Biol 131:1839–1847

Takemaru K, Moon RT (2000) The transcriptional coactivator CBP interacts with β-catenin to activate gene expression. J Cell Biol 149:249–254

Thoreson MA, Anastasiadis PZ, Daniel JM, Ireton RC, Wheelock MJ, Johnson KR, Hummingbird DK, Reynolds AB (2000) Selective uncoupling of p120^{ctn} from E-cadherin disrupts strong adhesion. J Cell Biol 148:189–201

Troyanovsky RB, Chitaev NA, Troyanovsky SM (1996) Cadherin binding sites of plakoglobin: localization, specificity and role in targeting to adhering junctions. J Cell Sci 109:3069–3078

Troyanovsky SM, Troyanovsky RB, Eshkind LG, Krutovskikh VA, Leube RE, Franke WW (1994a) Identification of the plakoglobin-binding domain in desmoglein and its role in plaque assembly and intermediate filament anchorage. J Cell Biol 127:151–160

Troyanovsky SM, Troyanovsky RB, Eshkind LG, Leube RE, Franke WW (1994b) Identification of amino acid sequence motifs in desmocollin, a desmosomal glycoprotein, that are required for plakoglobin binding and plaque formation. Proc Natl Acad Sci USA 91:10790–10794

von Kries JP, Winbeck G, Asbrand C, Schwarz-Romond T, Sochnikova N, Dell'Oro A, Behrens J, Birchmeier W (2000) Hot spots in β-catenin for interactions with LEF-1, conductin, and APC. Nat Struct Biol 7:800–807

Wahl JD, Sacco PA, McGranahan-Sadler TM, Sauppe LM, Wheelock MJ, Johnson KR (1996) Plakoglobin domains that define its association with the desmosomal cadherins and the classical cadherins: identification of unique and shared domains. J Cell Sci 109:1143–1154

Wahl JKI, Nieset JE, Sacco-Bubulya PA, Sadler TM, Johnson KR, Wheelock MJ (2000) The amino- and carboxyl-terminal tails of β-catenin reduce its affinity for desmoglein 2. J Cell Sci 113:1737–1745

Watabe-Uchida M, Uchida N, Imamura Y, Nagafuchi A, Fujimoto K, Uemura T, Vermeulen S, van Roy F, Adamson ED, Takeichi M (1998) α-catenin-vinculin interaction functions to organize the apical junctional complex in epithelial cells. J Cell Biol 142:847–857

Weers PMM, Narayanaswami V, Kay CM, Ryan RO (1999) Interaction of exchangeable apolipoprotein with phospholipid vesicles and lipoprotein particles. J Biol Chem 274:21804–21810

Weiss EE, Kroemker M, Rüdiger A-H, Jockusch BM, Rüdiger M (1998) Vinculin is part of the cadherin-catenin junctional complex: complex formation between α-catenin and vinculin. J Cell Biol 141:755–764

Witcher LL, Collins R, Puttagunta S, Mechanic SE, Munson M, Gumbiner B, Cowin P (1996) Desmosomal cadherin binding domains of plakoglobin. J Biol Chem 271:10904–10909

Xiao K, Allison DF, Buckley KM, Kottke MD, Vincent PA, Faundez V, Kowalczyk AP (2003) Cellular levels of p120 catenin function as a set point for cadherin expression levels in microvascular endothelial cells. J Cell Biol 163:535–545

Xing Y, Clemens WK, Kimmelman D, Xu W (2002) Crystal structure of a β-catenin/Axin complex suggests a mechanism for the β-catenin destruction complex. Genes Dev 17:2753–2764

Yang J, Dokurno P, Tonks NK, Barford D (2001) Crystal structure of the M-fragment of α-catenin: implications for modulation of cell adhesion. EMBO J 20:3645–3656

Yap AS, Brieher WM, Gumbiner BM (1997) Molecular and functional analysis of cadherin-based adherens junctions. Annu Rev Cell Dev Biol 13:119–146

Zhurinsky J, Shtutman M, Ben-Ze'ev A (2000) Differential mechanisms of LEF/TCF family-dependent transcriptional activation by beta-catenin and plakoglobin. Mol Cell Biol 20:4238–4252

Cadherins in Development

H. Semb

Section of Endocrinology, Lund University, BMC, B10, Klinikgatan 26, 22184 Lund, Sweden
henrik.semb@endo.mas.lu.se

1	Introduction	53
2	Cell Adhesion	55
3	Cell and Tissue Polarity	57
4	Cell Sorting	58
5	Cell Migration	59
6	Cell Survival	61
7	Tube Formation	62
8	Future	64
References		65

Abstract The cadherin family of cell adhesion molecules has emerged as a key regulator of embryonic morphogenesis. Although we are beginning to learn more about the developmental functions of non-classic cadherins, most of our current knowledge of the involvement of cadherins in various cellular processes that guide morphogenesis, such as adhesion, migration, cell shape changes, proliferation, and survival are based on the analysis of classic cadherins. Key issues for future studies include deeper knowledge of how the regulation of cadherin activity contributes to specific aspects of morphogenesis, and whether all cadherin-mediated morphogenetic activities can be directly or indirectly attributed to its role in cell–cell adhesion or whether they are executed via adhesion-independent mechanisms.

Keywords Cadherin · Adhesion · Migration · Development · Morphogenesis

1
Introduction

Cell differentiation and morphogenesis constitute the two most important developmental processes by regulating the specialization and arrangement of cells into functional tissues and organs. Morphogenesis includes a plethora of events such as cell sorting, delamination, condensation, migration, invagination, cavitation, mesenchymal-to-epithelial and epithelial-to-mesenchymal conversion, tube formation, convergent extension, radial intercalation, and

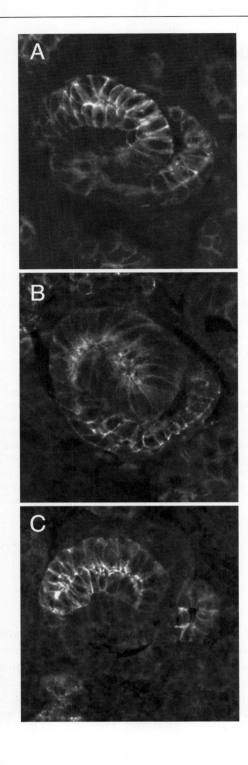

epiboly. The cellular processes that regulate these events include cell adhesion, cell shape changes, cell motility, proliferation, and apoptosis. Pioneering work by Malcolm Steinberg resulted in the "differential adhesion hypothesis," which proposed that the segregation or "sorting out" of different embryonic cell types could be explained by qualitative or quantitative differences in cell adhesion (Steinberg 1962, 1963, 1970). This hypothesis was presented before any cell adhesion molecule (CAM) had been identified. Soon after the first CAMs were identified it became clear that several different subfamilies of CAMs existed; some that mediate adhesion between cells (e.g., cadherins) and some that mediate adhesion between cells and extracellular matrix (ECM; e.g., integrins). Cadherins have been suggested to regulate most, if not all major morphogenetic processes, including cell sorting, delamination, condensation, migration, invagination, cavitation, mesenchymal-to-epithelial and epithelial-to-mesenchymal conversion, tube formation, and convergent extension. However, in several cases the implications are based on circumstantial evidence, such as changes in the expression pattern of cadherins (Fig. 1), and it is only in some of these cases that the evidence is direct. Here I review the current status of the functional role of cadherins in development, with special emphasis on morphogenesis. The review will focus on examples where cadherins have genetically been shown to play a role. Due to space limitations and recent excellent reviews on the topic (i.e., Tepass et al. 2000; Yagi and Takeichi 2000; Hirano et al. 2003), cadherin function during development of the nervous system is not discussed. A wide variety of model organisms, including *Caenorhabditis elegans*, *Drosophila*, zebrafish, *Xenopus*, chicken and mouse, have been used for studies of cadherin function during development. Most of the mechanistic insights, however, have been obtained from the organisms that offer greatest experimental advantages in terms of studying gene function, such as *Drosophila* and *Xenopus*.

2
Cell Adhesion

The functional role of cadherins in cell–cell adhesion has been extensively characterized both in vitro and in vivo. In "Cadherin Adhesion: Mechanisms and Molecular Interactions" (by T.D. Perez and W.J. Nelson, this volume),

◄─────────────────────────────

Fig. 1A–C. Expression of cadherins in S-shaped bodies during kidney tubulogenesis. The image shows the partially overlapping and partially differential expression of cadherins in S-shaped bodies during tubulogenesis as visualized by immunofluorescence staining of R-cadherin (**A**), N-cadherin (**B**), and P-cadherin (**C**)

the role of cadherin in cell–cell adhesion is extensively discussed. In mammalian embryogenesis the first epithelium, the trophectoderm, forms as the outer cell layer of the blastocyst, and E-cadherin is the only classic cadherin that is expressed in adherens junctions of this epithelium (Hyafil et al. 1980). In the absence of E-cadherin this epithelium fails to form, demonstrating that E-cadherin is required for epithelial cell–cell adhesion and that this adhesion is required for the formation of an epithelium (Larue et al. 1994; Riethmacher et al. 1995). Similar to vertebrates, the *Drosophila* E-cadherin, DE-cadherin, is required for formation and maintenance of epithelia during embryogenesis (Tepass et al. 1996; Uemura et al. 1996). In *C. elegans*, the homologs of the adherens junction proteins, namely classic cadherin, β-catenin, and α-catenin, are encoded by *hmr-1*, *hmp-2*, and *hmp-1*, respectively. Surprisingly, mutations of these genes result in less severe phenotypes than anticipated. Whereas hmr-1, hmp-1, and hmp-2 are required for both cell migration during body enclosure and cell shape changes during body elongation, epithelial integrity and formation of adherens junctions are unaffected in their absence (Costa et al. 1998). A possible explanation for the deviation from the classic cadherin/catenin function in vertebrates and *Drosophila* is that early embryonic adhesion in *C. elegans* may be carried out by related cadherin molecules or by non-cadherin-based adhesion systems.

During heart development, the only classic cadherin known to be expressed in the myocardium is N-cadherin (Hatta et al. 1987; Duband et al. 1988). Mouse embryos lacking N-cadherin exhibit several developmental abnormalities, including malformed somites and yolk sac, undulated neural tube, and a severe cardiovascular defect, which is the apparent cause of death around embryonic day (E) 9.5–10. Detailed analysis of the heart phenotype in N-cadherin-deficient embryos demonstrated that N-cadherin is required for cell–cell adhesion within the myocardium, and in the absence of N-cadherin this cell adhesion defect causes the collapse of the outflow tract leading to severe pericardial effusion and ballooning of the pericardial sac (Radice et al. 1997). Elegant transgenic complementation experiments addressed two important questions: (1) Could N-cadherin function within the myocardium be replaced by another classic cadherin, such as E-cadherin; and (2) which of the phenotypes observed in the N-cadherin-deficient embryos were caused by a cell autonomous function of N-cadherin and which were secondary to the cardiovascular defects? Surprisingly, the cardiac phenotype was restored independent of whether N-cadherin or E-cadherin was expressed in the N-cadherin-null background, demonstrating that myocyte adherence requires classic cadherin-mediated adhesion but that this adhesion is subtype-unspecific. Furthermore, in the yolk sac, but not the neural tube or somite, the phenotype was corrected in the rescued embryos, demonstrating that the yolk sac defect was secondary to the cardiovascular defects (Luo et al. 2001).

Another cell adhesion defect that was apparent in N-cadherin-deficient mice was a fragmentation of somites (Radice et al. 1997). In addition to N-cadherin, cadherin-11 is expressed in somites. However, cadherin11-deficient mice exhibit no anomaly within the somites. Interestingly, in the absence of both N-cadherin and cadherin-11, the somites become fragmented into smaller pieces than in N-cadherin-deficient embryos, suggesting that both cadherins cooperate in maintaining adhesion within the somite (Horikawa et al. 1999).

Recently, mapping and sequence analysis of the zebrafish *parachute* (*pac*) mutations demonstrated that *pac* corresponds to the zebrafish *N-cadherin* homolog (Lele et al. 2002). Analysis of several *N-cadherin* mutant alleles revealed cell adhesion and convergent cell movement defects during neurulation (see below). In *pac* embryos, the pseudostratified neuroepithelium in the dorsal midbrain and hindbrain is lost. Instead, the cells delaminate and form cavitated cell aggregates, resulting in defects in neural tube integrity. Transplantation experiments demonstrated that *pac* neuroepithelial cells formed aggregates that segregated from wild-type neighbors within the neural tube (Lele et al. 2002). These results are in agreement with recent findings in *N-cadherin*$^{-/-}$ chimeric mice (Kostetskii et al. 2001). Altogether these findings indicate that N-cadherin is directly involved in neuroepithelial cell-cell adhesion in vivo.

In addition to gene targeting, other experimental strategies, such as expression of dominant negative cadherin mutants, and RNA inhibition experiments have been used to demonstrate that cadherins regulate cellular cohesion during development (Horsfield et al. 2002; reviewed in Takeichi 1995; Gumbiner 1996; Marrs and Nelson 1996; Vleminckx and Kemler 1999; Horsfield et al. 2002).

3
Cell and Tissue Polarity

Cell polarity is not only a fundamental attribute of most cells, in the form of apical–basal polarity, but it also includes the polarity of cells within the plane of the tissue layer to which they belong. Furthermore, during patterning of the early embryo, three axes are formed: the anteroposterior, the dorsoventral, and the left–right axis. Although the instructive role of cadherins in apical–basal cell polarity (see the chapter by T.D. Perez and W.J. Nelson, this volume) has been known for some time, recent evidence indicates that cadherins also play critical roles in planar polarity and left–right axis formation.

In *Drosophila*, planar polarity is regulated by several cadherins, including flamingo (fmi), a seven-transmembrane protocadherin cadherin, and fat-like cadherins, including fat (ft) and dachsous (ds) (Adler et al. 1998; Usui et al. 1999). As for apical–basal polarity, the planar polarity is initiated in a similar way by cadherins, which are initially ubiquitously distributed throughout the membrane but subsequently become localized to specific compartments. For planar polarity this is controlled by the frizzled (fz) signaling pathway (Adler and Lee 2001). For example, within the *Drosophila* hair-forming cells in the developing wings, signals from the asymmetrically distributed fz lead to the accumulation of fmi at the proximal/distal cell boundaries. If the polarized distribution of fmi is perturbed, the wing hairs loose their correct orientation (Usui et al. 1999). In order for fmi to retain its polarized distribution it requires fmi to be present in the membrane of the adjacent cells, suggesting a homophilic binding mechanism. The polarized distribution of fmi at proximal/distal boundaries is regulated by diego, an ankyrin-repeat protein, which promotes clustering of fmi (Feiguin et al. 2001).

Planar polarity exists in vertebrates as well, where it is necessary for convergent extension movements during zebrafish and *Xenopus* gastrulation (Heisenberg et al. 2000; Tada and Smith 2000; Wallingford et al. 2000; Yagi and Takeichi 2000; Yamanaka et al. 2002). During convergent extension, cells on the lateral sides of the embryo intercalate and migrate towards the dorsal midline. However, although cadherin-mediated adhesion is required for planar polarity during convergent extension movement in vertebrates, there is no evidence that planar polarity is regulated by unequally distributed cadherins as in *Drosophila*.

Vertebrates develop a left–right asymmetry, which becomes obvious during formation of the heart and gut. Recently, N-cadherin was shown to be asymmetrically distributed in the chicken node and primitive streak during gastrulation. Blocking N-cadherin function by functional-blocking antibodies resulted in inverted heart looping and random expression of downstream components of the left–right pathway, such as *snail* and *Pitx2* (Garcia-Castro et al. 2000). The underlying mechanism for how N-cadherin affects left–right asymmetry is unclear, but it certainly is an interesting topic for future studies. Of note is that left–right asymmetry is unaffected in N-cadherin-deficient mice (Radice et al. 1997).

4
Cell Sorting

To test whether the original "differential adhesion hypothesis" also applied to the segregation of cells in vivo, *Drosophila* genetics was used to study the mechanism for oocyte localization within the egg chamber—a process that is important for anterior–posterior axis formation in the fly embryo (Gonzalez-Reyes and St Johnston 1994). The oocyte and the posterior follicle cells express higher levels of DE-cadherin and catenins than the other follicle cells and nurse cells (Godt and Tepass 1998; Gonzalez-Reyes and St Johnston 1998). In flies lacking DE-cadherin the oocyte becomes misplaced within the chamber. Importantly, when DE-cadherin is removed from a subpopulation of the follicle cells, the oocyte preferentially associates with DE-cadherin-expressing follicle cells independent of their localization within the egg chamber (Godt and Tepass 1998; Gonzalez-Reyes and St Johnston 1998). By demonstrating that cadherin expression is required for oocyte localization within the egg chamber, these experiments provided evidence that the "differential adhesion hypothesis" also applies to the segregation of cells in vivo.

Additional evidence in support of differential adhesion being a key element of patterning has recently been put forward. For example, in the developing spinal cord, motor neurons become segregated into distinct functional units termed motor pools. Each pool expresses a unique combination of cadherins, for example eF and A motor neurons differ only in the expression of one cadherin, MN-cadherin. By ectopically expressing MN-cadherin in the chicken spinal cord, the differential expression of MN-cadherin in these motor neurons was perturbed, resulting in abnormal intermingling and positioning of eF and A motor neurons (Price et al. 2002).

5
Cell Migration

During development it is common that tissue rearrangements occur while cells are still in contact with each other, suggesting dynamic changes in cell adhesion. It has been difficult to discriminate between a permissive and a more direct role of CAMs during cell motility, since in order for cells to move relative to one another cell adhesive interactions must be highly regulated. However, cell–cell adhesion is most likely necessary for generating the tractional forces needed for cells to rearrange within solid tissues.

When neural crest cells delaminate from the neural tube they change their expression pattern of cadherins. Neural crest cells express N-cadherin and cadherin-6B within the neural tube, but when they migrate from the tube, they down-regulate these two cadherins and induce expression of cadherin-7 (Nakagawa and Takeichi 1995; Duband et al. 1988, 1998). Overexpression of cadherins that are normally expressed in the neuroepithelium or in the migrating neural crest cells resulted in deficient emigration of neural crest cells from the neural tube, indicating that the regulation of cadherin expression or its activity is required for neural crest delamination from the neural tube (Nakagawa and Takeichi 1998; Dufour et al. 1999).

During *Xenopus* gastrulation, cells undergo convergent extension movements, which involve migration of cells towards the dorsal midline via intercalation of cells, leading to an extension of the embryo along the anterior–posterior axis (Shih and Keller 1992). Whereas a reduction in C-cadherin activity is necessary for these movements to occur (Zhong et al. 1999), inhibiting C-cadherin results in defects both in cell migration and tissue integrity (Lee and Gumbiner 1995). Thus, whether the disruption in cell movement is due to a direct role of C-cadherin in cell motility or secondary to the disrupted tissue architecture remains unclear.

Paraxial protocadherin (PAPC) is expressed in the *Xenopus* and zebrafish mesoderm during gastrulation. PAPC mediates homotypic cell adhesion, and loss-of-function experiments showed that it is involved in the convergence and cell movements of the mesoderm during gastrulation (Kim et al. 1998; Yamamoto et al. 1998). Importantly, however, the fact that overexpression of PAPC stimulated convergent extension suggests that PAPC directly promotes cell motility by acting as a CAM. An alternative interpretation for PAPC's role in cell movement includes a role as a signaling receptor that e.g., could regulate tissue polarity, which is observed during convergent extension.

In zebrafish, unlike in mammals, birds, and amphibia, neurulation does not involve invagination of an epithelial sheet. Instead, neuroectodermal cells converge towards the midline, followed by cavitation to form the neurocoel (Kimmel et al. 1995). The convergent extension cell movement is driven by medial cell intercalation, which presumably requires efficient coordination of cell adhesiveness and motility (Warga and Kimmel 1990). During neurulation, N-cadherin-deficient cells fail to undergo convergent extension, and instead sort out to more lateral positions compared with wild-type cells during neurulation (Lele et al. 2002). Whether N-cadherin is needed for either of these processes or both has yet to be demonstrated.

Perhaps the best example of a direct involvement of cadherins in cell migration comes from studies of *Drosophila* oogenesis. In *Drosophila*, follicles' border cells migrate on the surface of germline cells. Both these cells express DE-cadherin, and removing DE-cadherin from either cell type disrupted migration of the border cells. Surprisingly, border cells were capable of aggregating in the absence of DE-cadherin, indicating that their adhesion is independent of DE-cadherin function. When DE activity was reduced the border cells migrated more slowly, suggesting that the speed of migration is dependent on DE-cadherin expression or activity (Niewiadomska et al. 1999).

In order for cells to migrate to a particular location, local adhesive interactions do not suffice. For this purpose the cell must integrate both attractive and repulsive cues. This has been extensively studied during formation of axon trajectories. In *Drosophila*, N-cadherin-mediated adhesion is causally involved in axon patterning by regulating axon–axon interactions (Iwai et al. 1997). Repulsive cues mediated by the receptor/ligand complex Roundabout (Robo) and Slit are also required for axon guidance (Seeger et al. 1993; Kidd et al. 1998). Recent experiments performed with cell lines demonstrated that cadherin-mediated adhesion and Robo/Slit-mediated repulsion are functionally integrated. Slit-mediated activation of Robo inhibits

N-cadherin activity by inducing a complex between Robo, Abelson kinase, and N-cadherin, which results in phosphorylation of β-catenin and concomitant prevention of the critical association of actin with N-cadherin (Rhee et al. 2002). Although the local formation of a such receptor complex seems ideal for steering the growth cone while still allowing adhesion and growth in other directions, it will be interesting to find out if similar mechanisms apply to directional cell migration outside of the nervous system and particularly whether these molecular interactions perform similar functions in the intact organism.

6
Cell Survival

Genetic ablation of various cadherins revealed an essential role of these molecules as survival factors. For example, when E-cadherin was specifically ablated from the differentiating alveolar epithelial cells of the mammary gland, already at parturition the mutant epithelium exhibited characteristic features of an involuting mammary gland after weaning. Normally the epithelium undergoes apoptosis during involution, but when E-cadherin was ablated, massive apoptosis occurred already at parturition (Boussadia et al. 2002).

R-cadherin is transiently expressed within the developing proximal ureteric bud epithelium, and in the absence of R-cadherin this epithelium exhibits altered morphology and branching behavior. These findings correlated with increased apoptosis within the epithelium (Dahl et al. 2002). Whether the decreased cell survival within the ureteric bud epithelium of R-cadherin-deficient embryos is caused by a direct role of R-cadherin in cell survival or whether it is secondary to the observed changes in morphogenetic behavior of the epithelium remains unanswered.

N-cadherin-deficient embryos are smaller and developmentally delayed compared with their wild-type littermates. Whereas no change in proliferation rate was found in the mutant embryos, apoptosis was increased, especially in the collapsing neural folds and somites (Radice et al. 1997). When the cardiovascular defects in N-cadherin-deficient embryos were corrected by cardiac-specific expression of N-cadherin or E-cadherin, the embryos survived longer but were nevertheless smaller compared with their wild-type littermates. The fact that the increased apoptosis seen in N-cadherin-deficient embryos was unaffected in cardiac-specific rescued embryos, demonstrates that N-cadherin's involvement in cell survival is not secondary to cardiovascular defects (Luo et al. 2001).

The analysis of VE-cadherin's functional role in endothelial cell survival has been most instructive in understanding by which mechanism cadherins regulate cell survival. VE-cadherin is only found at endothelial adherens junctions (Lampugnani et al. 1992). To study the role of VE-cadherin and its binding to β-catenin in intracellular signaling, three strains of mice were generated: (1) VE-cadherin-deficient mice, (2) mice that expressed a mutant VE-cadherin lacking the β-catenin-binding cytoplasmic tail, and (3) mice with undetectable levels of VE-cadherin due to the insertion of an intronic *neomycin* gene. All three strains died at E9.5 due to vascular insufficiency (Carmeliet et al. 1999). Normally, blood vessels form via vasculogenesis and angiogenesis. The former involves differentiation of endothelial cells, which subsequently become assembled into vessels, whereas the latter implicates growth of new vessels from preexisting ones. The initial endothelial cell differentiation and cell assembly into primitive vessels occurred normally in the VE-cadherin mutant animals, indicating that initial vasculogenesis was unaffected. However, vascular defects were more severe at later stages, when the primitive vascular networks normally expand via sprouting angiogenesis and remodel in a branched network of large and small vessels. Throughout the entire vasculature, endothelial cells became disconnected from each other and detached from the basement membrane. Many of these endothelial cells were apoptotic. Thus, VE-cadherin is not required for endothelial cell differentiation or proliferation but for endothelial cell survival. In vitro studies showed that the endothelial cell survival factor vascular endothelial growth factor (VEGF) rescued survival of wild-type but not mutant cells. Furthermore, a multicomponent complex comprising VE-cadherin, β-catenin, VEGFR-2, and PI3-kinase is necessary for the endothelial survival function of VEGF through activation of Akt. Disruption of this complex by ablation of VE-cadherin or by deleting the β-catenin-binding domain of VE-cadherin renders endothelial cells refractory to the VEGF survival signal. Interestingly, the role of VE-cadherin is specific for VEGF and not for basic fibroblast growth factor (bFGF), which rescued survival regardless of the presence or absence of VE-cadherin (Carmeliet et al. 1999).

7
Tube Formation

Tubular networks, such as the vasculature and within epithelial organs, e.g., the lung, kidney, pancreas and trachea, come in many forms and are full of complexities. However, the control of growth, branching, and fusion of all these tubes involves similar dynamic cell–cell rearrangements, which are guided by the coordination of cell migration and cell adhesion. Analogous to convergent extension, tubular networks form via the reposition of cells, implicating dynamic changes in the activity of cell–cell adhesion. For discussion of the general mechanisms in tube formation several recent excellent reviews are recommended (Affolter et al. 2003; Lubarsky and Krasnow 2003).

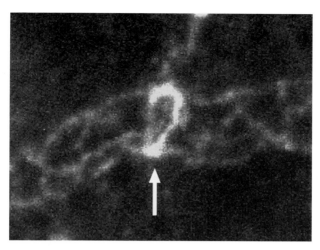

Fig. 2. Lumenal fusion between tracheal metameres connects neighboring branches and is mediated by two specialized fusion cells, one at the tip of each connecting branch. The image shows part of the tracheal dorsal trunk, including the fusion point (*arrow*), after branch fusion is completed. The apical cell junctions are visualized by antibody labeling for DE-cadherin, and demonstrate that DE-cadherin is highly up-regulated in the fusion cells, where it becomes deposited as a ring-like structure between the two connecting branches and surrounding the lumen circumference. (Photo courtesy of Johanna Hemphälä and Christos Samakovlis)

Most of our present understanding of the role of cadherins in tube formation has been obtained by studying complete or partial loss-of-function mutants of the *DE-cadherin* gene. Whereas zygotic expression is required for processes of epithelial cell rearrangements (Fig. 2), including the extension of the Malpighian tubules and extension and fusion of terminal tracheal branches, maternal DE-cadherin is sufficient for maintaining the general epithelial architecture (Tepass et al. 1996; Uemura et al. 1996). Whether cadherins perform similar functions in branched tubular organs of higher organisms seem plausible but remains to be shown. These findings contributed significantly to the concept of how cadherin-mediated cell–cell adhesion may contribute to cell movement, i.e., not by a permissive block in cell–cell adhesion, but rather more directly by actively switching cell–cell adhesion on and off. For the latter, several possible mechanisms can be envisioned, including regulation of the expression, turnover, or activity of cadherins (for details see the chapter by T.D. Perez and W.J. Nelson, this volume).

Various in vitro and cell culture studies on the small GTPase Rac have suggested that it is a key regulator of cadherin-mediated cell adhesion (Fukata and Kaibuchi 2001; Braga 2002). However, these findings have not until recently been verified in the context of embryonic development. Reduced Rac activity prevented many aspects of cell rearrangement during tracheal development. This phenotype was associated with an increase in the level of DE-cadherin and its associated molecules, and expansion of DE-cadherin localization to the basolateral membrane. In contrast, hyperactivation of Rac prevented incorporation of newly synthesized DE-cadherin into cell junction and reduced cell adhesiveness, transforming the tracheal epithelium into mesenchyme. Based on these findings, it was proposed that switching of Rac between active and inactive states promotes turnover of the DE-cadherin complex at the cell junctions, a prerequisite for maintaining the plasticity of the tracheal epithelium during branching morphogenesis (Chihara et al. 2003).

8
Future

By determining cell and tissue architecture, cadherins act as the predominant CAMs involved in morphogenesis during development. Nevertheless, there is much yet to be understood regarding by which mechanism cadherins execute each of the distinct morphogenetic events they participate in.

In many organisms most cells express more than one cadherin. Whereas distinct subcellular distribution of different cadherin subtypes within the same cell suggests different functions, mutational and gene-targeting analysis of individual cadherins has not conclusively demonstrated that they perform different functions within the same cell. In fact, analysis of mice that either lack one or more than one type of cadherin expressed in the same cell suggests that the function of a mutated cadherin can be surrogated by a closely related cadherin (Horikawa et al. 1999; Dahl et al. 2002; Hollnagel et al. 2002). This issue has been raised previously, but no answer has yet been put forward. The opposite problem of embryonic lethality (Larue et al. 1994; Riethmacher et al. 1995; Radice et al. 1997) can currently be overcome by spatiotemporal knockout of the cadherin.

Another question is whether the function of cadherins in various aspects of morphogenesis, including cell adhesion, cell motility, cell survival, and cell rearrangement, can all be directly or indirectly attributed to its role in cell–cell adhesion, or whether some of these processes may be regulated by pathways independent of physical cell–cell adhesion. There is no clear answer to this question yet, but with further studies of the signaling function of cadherin, how this activity relates to its role in cell–cell adhesion, and the regulation of these activities in the context of embryonic development, this question may soon be resolved.

References

Adler PN, Lee H (2001) Frizzled signaling and cell–cell interactions in planar polarity. Curr Opin Cell Biol 13:635–40

Adler PN, Charlton J, Liu J (1998) Mutations in the cadherin superfamily member gene dachsous cause a tissue polarity phenotype by altering frizzled signaling. Development 125:959–68

Affolter M, Bellusci S, Itoh N, Shilo B, Thiery JP, Werb Z (2003) Tube or not tube: remodeling epithelial tissues by branching morphogenesis. Dev Cell 4:11–8

Boussadia O, Kutsch S, Hierholzer A, Delmas V, Kemler R (2002) E-cadherin is a survival factor for the lactating mouse mammary gland. Mech Dev 115:53–62

Braga VM (2002) Cell–cell adhesion and signalling. Curr Opin Cell Biol 14:546–56

Carmeliet P, Lampugnani MG, Moons L, Breviario F, Compernolle V, Bono F, Balconi G, Spagnuolo R, Oostuyse B, Dewerchin M, Zanetti A, Angellilo A, Mattot V, Nuyens D, Lutgens E, Clotman F, de Ruiter MC, Gittenberger-de Groot A, Poelmann R, Lupu F, Herbert JM, Collen D, Dejana E (1999) Targeted deficiency or cytosolic truncation of the VE-cadherin gene in mice impairs VEGF-mediated endothelial survival and angiogenesis. Cell 98:147–57

Chihara T, Kato K, Taniguchi M, Ng J, Hayashi S (2003) Rac promotes epithelial cell rearrangement during tracheal tubulogenesis in Drosophila. Development 130:1419–1428

Costa M, Raich W, Agbunag C, Leung B, Hardin J, Priess JR (1998) A putative catenin-cadherin system mediates morphogenesis of the Caenorhabditis elegans embryo. J Cell Biol 141:297–308

Dahl U, Sjodin A, Larue L, Radice GL, Cajander S, Takeichi M, Kemler R, Semb H (2002) Genetic dissection of cadherin function during nephrogenesis. Mol Cell Biol 22:1474–87

Duband JL, Volberg T, Sabanay I, Thiery JP, Geiger B (1988) Spatial and temporal distribution of the adherens-junction-associated adhesion molecule A-CAM during avian embryogenesis. Development 103:325–44

Dufour S, Beauvais-Jouneau A, Delouvee A, Thiery JP (1999) Differential function of N-cadherin and cadherin-7 in the control of embryonic cell motility. J Cell Biol 146:501–16

Feiguin F, Hannus M, Mlodzik M, Eaton S (2001) The ankyrin repeat protein Diego mediates Frizzled-dependent planar polarization. Dev Cell 1:93–101

Fukata M, Kaibuchi K (2001) Rho-family GTPases in cadherin-mediated cell–cell adhesion. Nat Rev Mol Cell Biol 2:887–97

Garcia-Castro MI, Vielmetter E, Bronner-Fraser M (2000) N-Cadherin, a cell adhesion molecule involved in establishment of embryonic left-right asymmetry. Science 288:1047–51

Godt D, Tepass U (1998) Drosophila oocyte localization is mediated by differential cadherin-based adhesion. Nature 395:387–91

Gonzalez-Reyes A, St Johnston D (1994) Role of oocyte position in establishment of anterior-posterior polarity in Drosophila. Science 266:639–42

Gonzalez-Reyes A, St Johnston D (1998) The Drosophila AP axis is polarised by the cadherin-mediated positioning of the oocyte. Development 125:3635–44

Gumbiner BM (1996) Cell adhesion: the molecular basis of tissue architecture and morphogenesis. Cell 84:345–57

Hatta K, Takagi S, Fujisawa H, Takeichi M (1987) Spatial and temporal expression pattern of N-cadherin cell adhesion molecules correlated with morphogenetic processes of chicken embryos. Dev Biol 120:215–27

Heisenberg CP, Tada M, Rauch GJ, Saude L, Concha ML, Geisler R, Stemple DL, Smith JC, Wilson SW (2000) Silberblick/Wnt11 mediates convergent extension movements during zebrafish gastrulation. Nature 405:76–81

Hirano S, Suzuki ST, Redies CM (2003) The cadherin superfamily in neural development: diversity, function and interaction with other molecules. Front Biosci 8: D306–55

Hollnagel A, Grund C, Franke WW, Arnold HH (2002) The cell adhesion molecule M-cadherin is not essential for muscle development and regeneration. Mol Cell Biol 22:4760–70

Horikawa K, Radice G, Takeichi M, Chisaka O (1999) Adhesive subdivisions intrinsic to the epithelial somites. Dev Biol 215:182–9

Horsfield J, Ramachandran A, Reuter K, LaVallie E, Collins-Racie L, Crosier K, Crosier P (2002) Cadherin-17 is required to maintain pronephric duct integrity during zebrafish development. Mech Dev 115:15–26

Hyafil F, Morello D, Babinet C, Jacob F (1980) A cell surface glycoprotein involved in the compaction of embryonal carcinoma cells and cleavage stage embryos. Cell 21:927–34

Iwai Y, Usui T, Hirano S, Steward R, Takeichi M, Uemura T (1997) Axon patterning requires DN-cadherin, a novel neuronal adhesion receptor, in the Drosophila embryonic CNS. Neuron 19:77–89

Kidd T, Brose K, Mitchell KJ, Fetter RD, Tessier-Lavigne M, Goodman CS, Tear G (1998) Roundabout controls axon crossing of the CNS midline and defines a novel subfamily of evolutionarily conserved guidance receptors. Cell 92:205–15

Kim SH, Yamamoto A, Bouwmeester T, Agius E, Robertis EM (1998) The role of paraxial protocadherin in selective adhesion and cell movements of the mesoderm during Xenopus gastrulation. Development 125:4681–90

Kimmel CB, Ballard WW, Kimmel SR, Ullmann B, Schilling TF (1995) Stages of embryonic development of the zebrafish. Dev Dyn 203:253–310

Kostetskii I, Moore R, Kemler R, Radice GL (2001) Differential adhesion leads to segregation and exclusion of N-cadherin-deficient cells in chimeric embryos. Dev Biol 234:72–9

Lampugnani MG, Resnati M, Raiteri M, Pigott R, Pisacane A, Houen G, Ruco LP, Dejana E (1992) A novel endothelial-specific membrane protein is a marker of cell–cell contacts. J Cell Biol 118:1511–22

Larue L, Ohsugi M, Hirchenhain J, Kemler R (1994) E-cadherin null mutant embryos fail to form a trophectoderm epithelium. Proc Natl Acad Sci U S A 91:8263–7

Lee CH, Gumbiner BM (1995) Disruption of gastrulation movements in Xenopus by a dominant-negative mutant for C-cadherin. Dev Biol 171:363–73

Lele Z, Folchert A, Concha M, Rauch GJ, Geisler R, Rosa F, Wilson SW, Hammerschmidt M, Bally-Cuif L (2002) parachute/N-cadherin is required for morphogenesis and maintained integrity of the zebrafish neural tube. Development 129:3281–94

Lubarsky B, Krasnow MA (2003) Tube morphogenesis: making and shaping biological tubes. Cell 112:19–28

Luo Y, Ferreira-Cornwell M, Baldwin H, Kostetskii I, Lenox J, Lieberman M, Radice G (2001) Rescuing the N-cadherin knockout by cardiac-specific expression of N- or E-cadherin. Development 128:459–69

Marrs JA, Nelson WJ (1996) Cadherin cell adhesion molecules in differentiation and embryogenesis. Int Rev Cytol 165:159–205

Nakagawa S, Takeichi M (1995) Neural crest cell–cell adhesion controlled by sequential and subpopulation-specific expression of novel cadherins. Development 121:1321–32

Nakagawa S, Takeichi M (1998) Neural crest emigration from the neural tube depends on regulated cadherin expression. Development 125:2963–71

Niewiadomska P, Godt D, Tepass U (1999) DE-Cadherin is required for intercellular motility during Drosophila oogenesis. J Cell Biol 144:533–47

Price SR, De Marco Garcia NV, Ranscht B, Jessell TM (2002) Regulation of motor neuron pool sorting by differential expression of type II cadherins. Cell 109:205–16

Radice GL, Rayburn H, Matsunami H, Knudsen KA, Takeichi M, Hynes RO (1997) Developmental defects in mouse embryos lacking N-cadherin. Dev Biol 181:64–78

Rhee J, Mahfooz NS, Arregui C, Lilien J, Balsamo J, VanBerkum MF (2002) Activation of the repulsive receptor Roundabout inhibits N-cadherin-mediated cell adhesion. Nat Cell Biol 4:798–805

Riethmacher D, Brinkmann V, Birchmeier C (1995) A targeted mutation in the mouse E-cadherin gene results in defective preimplantation development. Proc Natl Acad Sci U S A 92:855–9

Seeger M, Tear G, Ferres-Marco D, Goodman CS (1993) Mutations affecting growth cone guidance in Drosophila: genes necessary for guidance toward or away from the midline. Neuron 10:409–26

Shih J, Keller R (1992) Cell motility driving mediolateral intercalation in explants of Xenopus laevis. Development 116:901–14

Steinberg MS (1962) On the mechanism of tissue reconstruction by dissociated cells. I. Population kinetics, differential adhesiveness, and the absence of directed migration. Proc Natl Acad Sci USA 48:1577–1582

Steinberg MS (1963) Reconstruction of tissues by dissociated cells. Science 141:401–408

Steinberg MS (1970) Does differential adhesion govern self-assembly processes in histogenesis? Equilibrium configurations and the emergence of a hierarchy among populations of embryonic cells. J Exp Zool 173:395–433

Tada M, Smith JC (2000) Xwnt11 is a target of Xenopus Brachyury: regulation of gastrulation movements via Dishevelled, but not through the canonical Wnt pathway. Development 127:2227–38

Takeichi M (1995) Morphogenetic roles of classic cadherins. Curr Opin Cell Biol 7:619–27

Tepass U, Gruszynski-DeFeo E, Haag TA, Omatyar L, Torok T, Hartenstein V (1996) shotgun encodes Drosophila E-cadherin and is preferentially required during cell rearrangement in the neurectoderm and other morphogenetically active epithelia. Genes Dev 10:672–85

Tepass U, Truong K, Godt D, Ikura M, Peifer M (2000) Cadherins in embryonic and neural morphogenesis. Nat Rev Mol Cell Biol 1:91–100

Uemura T, Oda H, Kraut R, Hayashi S, Kotaoka Y, Takeichi M (1996) Zygotic Drosophila E-cadherin expression is required for processes of dynamic epithelial cell rearrangement in the Drosophila embryo. Genes Dev 10:659–71

Usui T, Shima Y, Shimada Y, Hirano S, Burgess RW, Schwarz TL, Takeichi M, Uemura T (1999) Flamingo, a seven-pass transmembrane cadherin, regulates planar cell polarity under the control of Frizzled. Cell 98:585–95

Vleminckx K, Kemler R (1999) Cadherins and tissue formation: integrating adhesion and signaling. Bioessays 21:211–20

Wallingford JB, Rowning BA, Vogeli KM, Rothbacher U, Fraser SE, Harland RM (2000) Dishevelled controls cell polarity during Xenopus gastrulation. Nature 405:81–5

Warga RM, Kimmel CB (1990) Cell movements during epiboly and gastrulation in zebrafish. Development 108:569–80

Yagi T, Takeichi M (2000) Cadherin superfamily genes: functions, genomic organization, and neurologic diversity. Genes Dev 14:1169–80

Yamamoto A, Amacher SL, Kim SH, Geissert D, Kimmel CB, De Robertis EM (1998) Zebrafish paraxial protocadherin is a downstream target of spadetail involved in morphogenesis of gastrula mesoderm. Development 125:3389–97

Yamanaka H, Moriguchi T, Masuyama N, Kusakabe M, Hanafusa H, Takada R, Takada S, Nishida E (2002) JNK functions in the non-canonical Wnt pathway to regulate convergent extension movements in vertebrates. EMBO Rep 3:69–75

Zhong Y, Brieher WM, Gumbiner BM (1999) Analysis of C-cadherin regulation during tissue morphogenesis with an activating antibody. J Cell Biol 144:351–9

Cadherins in Cancer

K. Strumane · G. Berx · F. Van Roy (✉)

Department for Molecular Biomedical Research,
Ghent University and Flanders Interuniversity Institute for Biotechnology (V.I.B),
Technologiepark 927, 9052 Zwijnaarde, Belgium
f.vanroy@dmbr.UGent.be

1	E-Cadherin Function in Cancer Cells	70
2	E-Cadherin Expression in Human Carcinomas	73
3	Genetic Alterations of the E-Cadherin Gene in Cancer	75
4	Transcriptional Downregulation of E-Cadherin	77
4.1	E-Cadherin Gene Silencing by Promoter Hypermethylation	78
4.2	Transacting Regulators	79
5	Post-translational E-Cadherin Inactivation	81
5.1	Tyrosine Phosphorylation	81
5.2	E-Cadherin Protein Truncations	82
5.3	Cell Surface Proteoglycans and Mucin-Like Glycoproteins	82
6	Other Cadherins in Cancer	83
7	Upregulation of E-Cadherin and Tumor Therapy	85
	References	86

Abstract The presence of a functional E-cadherin/catenin cell–cell adhesion complex is a prerequisite for normal development and maintenance of epithelial structures in the mammalian body. This implies that the acquisition of molecular abnormalities that disturb the expression or function of this complex is related to the development and progression of most, if not all, epithelial cell-derived tumors, i.e. carcinomas. E-cadherin downregulation is indeed correlated with malignancy parameters such as tumor progression, loss of differentiation, invasion and metastasis, and hence poor prognosis. Moreover, E-cadherin has been shown to be a potent invasion suppressor as well as a tumor suppressor. Disturbed expression profiles of the E-cadherin/catenin complex have been demonstrated in histological sections of many human tumor types. In different kinds of carcinomas, biallelic downregulation of the E-cadherin gene, resulting in tumor-restricted decrease or even complete loss of E-cadherin expression, appears to be caused by a variety of inactivation mechanisms. Gene deletion due to loss of heterozygosity of the *CDH1* locus on 16q22.1 frequently occurs in many carcinoma types. However, somatic inactivating mutations resulting in aberrant E-cadherin expression by loss of both wild-type alleles is rare and restricted to only a few cancer types. A majority of carcinomas thus seems to show deregulated E-cadherin expression by other mechanisms. The present evidence proposes transcriptional repression as a powerful and recurrent molecular mechanism for silencing E-cadherin expression. The predominant mechanisms emerging in most carcinomas are hypermethylation of the E-cadherin promoter and expression of transrepressor molecules such as SIP1, Snail, and Slug that bind sequence elements in the proximal E-cadherin promoter. Interestingly, complex differential expression of other cadherins seems to be associated with loss of E-cadherin and to reinforce effects of this loss on tumor progression. Multiple agents can upregulate and stabilize the E-cadherin/catenin complex. Especially for those tumors with transcriptional and thus reversible downregulation of E-cadherin expression, these drug agents offer important therapeutic opportunities.

Keywords E-cadherin · Downregulation · Carcinoma

1
E-Cadherin Function in Cancer Cells

During the multi-step process of tumor development, clonal accumulation of genetic and epigenetic events determines the transition from a normal to a malignant cell status. E-cadherin downregulation is a rate-limiting step for several stages of tumor progression through its effect on several important regulatory circuits, such as control of cell proliferation, balance between cell survival and apoptosis, and tumor cell migration, invasion, and metastatic dissemination.

E-cadherin is a cell surface molecule mediating homophilic cell–cell adhesion (Fig. 1). E-cadherin functionality depends on its linkage to the actin cytoskeleton via the cytoplasmic catenins. The best studied role of E-cadherin in tumor progression is its invasion suppressor function, demonstrated in vitro by the abrogation of the invasive phenotype in malignant epithelial and fibroblastic tumor cells following transfection with E-cadherin cDNA (Chen and Obrink 1991; Frixen et al. 1991; Vleminckx et al. 1991; Luo et al. 1999; Moersig et al. 2002; Nawrocki-Raby et al. 2003). Conversely, invasiveness of the transfected cells is restored by treatment with E-cadherin-blocking antibodies or by reducing E-cadherin expression with an E-cadherin antisense RNA. Interestingly, transduction of melanoma cells with E-cadherin expressing adenovirus leads to adhesion to keratinocytes resulting in growth control, induction of apoptosis, and invasion suppression in a skin reconstruction model (Hsu et al. 2000).

In vivo experiments with Rip1Tag2 transgenic mice, which express the SV40 T-antigen under control of the insulin promoter, showed that loss of E-cadherin expression coincides with the transition from a well-differentiated β-cell adenoma to an invasive β-cell carcinoma (Perl et al. 1998). This transgenic mouse strain was then intercrossed with transgenic mice expressing in their pancreatic β-cells either wild-type E-cadherin or a dominant-negative form of E-cadherin. These composite transgenic mice demonstrated that loss of E-cadherin-mediated cell–cell adhesion is required and sufficient for the progression from a benign adenoma to a malignant carcinoma.

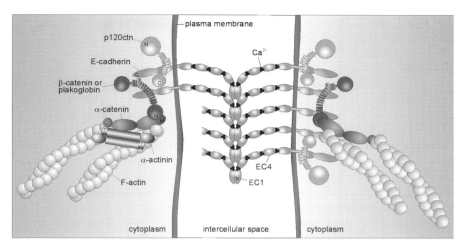

Fig. 1. Schematic representation of the E-cadherin/catenin adhesion complex at the zonula adherens of epithelial cells. The combination of the *cis* and *trans* interfaces, formed in the intercellular space of neighboring cells by the aminoterminal extracellular domains of classical cadherins, results in a one-layer lattice (Boggon et al. 2002). The carboxyterminal E-cadherin domain binds to cytoplasmic catenins. p120ctn binds to a more membrane-proximal domain of E-cadherin than β-catenin or plakoglobin. The aminoterminal end of β-catenin binds α-catenins. The latter form directly, or indirectly via α-actinin, the anchoring link to the filamentous actin (*F-actin*) cytoskeleton. *Arm*, armadillo; *C*, carboxyterminal; *EC*, extracellular cadherin repeat; *N*, aminoterminal; *PM*, plasma membrane

E-cadherin has also been shown to exert a metastasis-suppressive effect in vivo. E-cadherin-negative human breast cancer cells, which form osteolytic bone metastases when introduced directly into the arterial blood stream in nude mice, show suppressed metastatic potential upon expression of transfected E-cadherin (Mbalaviele et al. 1996). On the other hand, re-expression of E-cadherin on circulating tumor cells can also facilitate cell survival within metastatic deposits (Graff et al. 2000), and in experimental tumor systems, large metastases were indeed shown to express more E-cadherin than small ones (Mareel et al. 1991; Karube et al. 2002). Therefore, dynamic regulation of E-cadherin expression by a reversible epigenetic mechanism, such as promoter methylation, may drive metastatic progression (Graff et al. 2000).

In addition to its invasion and metastasis suppression functions, E-cadherin acts as a tumor suppressor. As pointed out in more detail below, the E-cadherin gene *CDH1* was defined as a tumor suppressor gene on the basis of its biallelic inactivation in lobular breast tumors and diffuse gastric tumors (Becker et al. 1994, 1995a, 1996, 1998), thus fulfilling Knudson's two-hit model for tumor suppressor genes (Knudson 1971). Moreover, transfection experiments demonstrated the suppressive effect of E-cadherin expression on tumor cell proliferation in vitro (Miyaki et al. 1995; St Croix et al. 1998; Vizirianakis et al. 2002) and in vivo (Meiners et al. 1998). However, the molecular mechanism of the tumor-suppressive function of E-cadherin is not yet known in full detail. Most of the characterized tumor suppressor gene products are involved in signal transduction pathways that lead either to cell-cycle arrest or apoptosis. Likewise, E-cadherin-dependent growth inhibition was shown in vitro to involve inhibition of mitogenic signaling pathways initiated from receptor tyrosine kinases such as epidermal growth factor receptor (EGFR; Takahashi and Suzuki 1996), which in turn results in the upregulation of the cyclin-dependent kinase inhibitor p27^{kip1} to levels sufficient for cell cycle arrest (St Croix et al. 1998). Protein kinase C-induced apoptosis of mammary and prostate epithelial cells was counteracted by E-cadherin-dependent cell aggregation (Day et al. 1999). This aggregation was shown to result in activation (i.e. dephosphorylation) of the retinoblastoma tumor suppressor protein and in cell cycle arrest. Another study suggested that E-cadherin-mediated cell growth arrest was based on negative interference with nuclear β-catenin transcriptional activity, a process in which β-catenin interaction with the cytoplasmic tail of E-cadherin was essential (Stockinger et al. 2001). Reduction of β-catenin transcriptional activity was also observed upon transfection of highly invasive bronchial tumor cells with E-cadherin (Nawrocki-Raby et al. 2003). These transfectants showed decreased invasiveness in various assays and were characterized by a striking decrease in the level or activity of several metalloproteinases.

Despite being a rate-limiting step in early tumor progression, E-cadherin downregulation on its own seems to be insufficient to initiate tumor development. For example, transgenic mice expressing a dominant-negative E-cadherin in pancreatic β-cells do not show pancreatic β-cell carcinogenesis (Dahl et al. 1996). In addition, E-cadherin inactivation targeted to the mammary glands of mice affects the terminal differentiation program of the lactating mammary gland but does not induce tumor formation (Boussadia et al. 2002). Although additional permissive factors appear to be needed for tumor initiation to occur, regulation of E-cadherin expression might be involved, as was shown in composite transgenic mice, in which a targeted mutation in the *APC* (adenomatous polyposis coli) gene is combined with a targeted mutation in the *CDH1* gene (Smits et al. 2000). These mice show enhanced APC-driven tumor initiation compared to the APC$^{+/1638\ N}$ mice, whereas E-cadherin haplo-insufficiency is not sufficient by itself to trigger tumorigenesis, as confirmed by the lack of tumors in the E-cad$^{+/-}$ mice. These findings support the suggestiong that there may be crosstalk between E-cadherin and the Wnt signaling pathway. It remains uncertain whether this link is physiologically relevant and whether the connection is direct or indirect. However, the involvement of E-cadherin in β-catenin-mediated neoplastic transformation is indicated by the observation that in a colorectal cancer cell line, a 3-bp deletion in the β-catenin gene, eliminating Ser-45, affects its interaction with E-cadherin (Chan et al. 2002).

Apparently, the consequences of E-cadherin downregulation can differ considerably according to the cell and tissue types involved. This is documented by E-cadherin germline mutations that predispose predominantly to diffuse-type gastric cancer and less so to lobular breast cancer, as will be elaborated below. All data taken together, E-cadherin downregulation may either have a causal role in tumorigenesis or else promote malignant progression of tumor cells, according to the cell/tissue types affected.

2
E-Cadherin Expression in Human Carcinomas

Tumor-restricted E-cadherin downregulation has been correlated with clinico-pathological parameters and has been shown to be a powerful diagnostic and prognostic tool, for instance for breast carcinomas (Acs et al. 2001; Elzagheid et al. 2002). The tumor suppressive functions of E-cadherin described above are also apparent from the extensive data on E-cadherin expression profiles that are available today for many different types of human carcinomas (Strumane et al. 2003) (Fig. 2). Therefore, in addition to the histological grading systems generally used for carcinomas, the use of E-cadherin expression as a molecular diagnostic and prognostic parameter may give important information on the behavior of individual tumor cases.

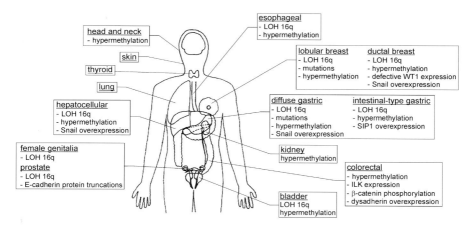

Fig. 2. Overview of carcinoma types in different tissues of the human male/female body, for which downregulation of the E-cadherin/catenin complex has been reported in the literature (reviewed by Strumane et al. 2003). The mechanisms of downregulation indicated for particular carcinomas are further detailed in the text

Interfering with the expression of catenins can also result in functional downregulation of E-cadherin. Therefore, it is important to integrate the analysis of catenins in E-cadherin expression studies of tumor samples. The extent of aberrant expression of E-cadherin, as well as of catenins, clearly correlates with the grade of malignancy in cervical, skin, and prostate tumors (Umbas et al. 1997; Carico et al. 2001; Papadavid et al. 2001, 2002). The reflection of the different stages of tumor progression in quantitative or qualitative differences in E-cadherin/catenin immunoreactivity has also been suggested for esophageal cancer (Bailey et al. 1998; Washington et al. 1998), colorectal tumors (Hao et al. 1997), and melanomas (Silye et al. 1998). In colon tumors, E-cadherin expression was found to be decreased in the central areas of the tumors, while it was completely absent at the invasive fronts (Brabletz et al. 2001).

Remarkably, E-cadherin, α-catenin and β-catenin are re-expressed in the metastatic lesions from patients with primary breast tumors that show reduced expression of all of these proteins (Bukholm et al. 2000). In contrast, plakoglobin is even more reduced in the metastases compared to the primary tumors, pointing to a different function of plakoglobin. Unlike these findings, metastases of primary colorectal, pancreatic, and lung tumors show E-cadherin and catenin expression patterns that are similar to those in the primary tumors (Gunji et al. 1998; Sulzer et al. 1998; Ghadimi et al. 1999). The regulatory mechanism by which E-cadherin may be re-expressed in metastases of cancer cells is still unclear. Probably, this depends on multiple factors, including the mechanisms of downregulation in the primary tumor and the particular tumor cell microenvironment.

Exceptionally, the E-cadherin/catenin complex is hyperactivated in tumors. A notable example is inflammatory breast carcinoma, where upregulated E-cadherin expression contributes to the formation of compact tumor cell clumps (Tomlinson et al. 2001; Colpaert et al. 2003). This process is associated with markedly decreased sialyl-Lewis (x/a) carbohydrate ligand-binding epitopes on overexpressed MUC1 and other surface molecules that bind endothelial E-selectin (Alpaugh et al. 2002a,b). This leads to lack of binding of the tumor embolus to the surrounding endothelium and mediates formation of lymphovascular emboli and passive, though efficient, metastatic spread. Upregulation of the E-cadherin/catenin complex has also been described for some types of hepatocellular carcinomas (Shimoyama and Hirohashi 1991; Ihara et al. 1996), metaplastic ovarian surface epithelium (OSE) cells (Soler et al. 1997; Sundfeldt et al. 1997; Auersperg et al. 1999), and testis tumor cells (Saito et al. 2000a). In bone and soft tissue sarcomas, expression of E-cadherin correlates with epithelial differentiation (Sato et al. 1999; Laskin and Miettinen 2002; Yoo et al. 2002). In synovial sarcoma, preserved expression of E-cadherin and α-catenin is associated with a better overall survival rate (Saito et al. 2000b).

3
Genetic Alterations of the E-Cadherin Gene in Cancer

The human E-cadherin gene (*CDH1*) is located on chomosome 16q22.1 (Mansouri et al. 1988; Natt et al. 1989, Berx 1995b). The *CDH1* gene bridges a region of 100 kbp and comprises 16 exons (Berx et al. 1995b). The long arm of chromosome 16 shows a high frequency of loss of heterozygosity (LOH) in various human tumor types (Fig. 2), such as breast carcinoma with 36%–67% (Cleton-Jansen et al. 2001), esophagus carcinoma with 65% (Wijnhoven et al. 1999), hepatocellular carcinoma with 49% (Wang et al. 2000a; Matsumura et al. 2001), prostate cancer with 42% (Carter et al. 1990; Suzuki et al. 1996; Latil et al. 1997; Pan et al. 1998), endometrial carcinoma with 40% (Kihana et al. 1996), advanced gastric carcinomas with 33% (Kimura et al. 1997), and bladder cancer with 30% LOH of the informative cases (Taddei et al. 2000). It has been suggested that high frequency of LOH for the *CDH1* locus is related to p53 alterations, allowing progression of genomic instability. A statistically significant correlation between p53 inactivation and LOH for 16q22.1 has indeed been shown in hepatocellular carcinomas (Wang et al. 2000a). A strong association between alterations of the p53 protein and downregulation of E-cadherin protein expression has also been found in lobular and ductal breast carcinomas (Bukholm et al. 1997) and in bladder carcinomas (del Muro et al. 2000), but not in soft tissue sarcomas (Yoo et al. 2002).

High incidence of LOH for a specific chromosomal region is considered to be a hallmark for the localization of a tumor suppressor gene. So far, clear evidence that *CDH1* is the tumor suppressor gene associated with LOH on 16q has only been found for lobular breast carcinomas, which show a high frequency of E-cadherin-inactivating mutations (Berx et al. 1995a). In infiltrative lobular breast cancer, E-cadherin gene mutations have been demonstrated to be scattered throughout the exons coding for the extracellular domain (Berx et al. 1998) (Fig. 3A). The majority of these mutations are out-of-frame mutations, predicted to yield secreted truncated E-cadherin fragments or no stable protein at all. If combined with LOH of the second allele, it is obvious that such *CDH1* mutations fully abrogate E-cadherin mediated cell–cell adhesion.

In addition, sporadic diffuse gastric tumors, but not intestinal-type gastric tumors, show a high incidence of E-cadherin inactivating mutations. Besides skipping of particular exons, which is the predominant defect causing in-frame deletions (Becker et al. 1994; Berx et al. 1998), truncation and missense mutations have also been reported for this histological tumor subtype (Caldas et al. 1999; Machado et al. 1999). In enigmatic contrast with lobular breast cancer, to date no 16q LOH in combination with E-cadherin mutations could be demonstrated for diffuse gastric cancer. Tumor-associated E-cadherin in-frame deletion mutations have been shown to alter cellular

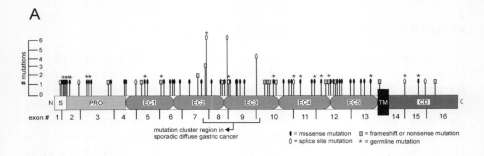

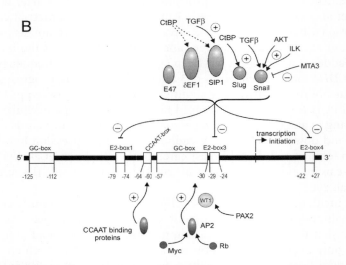

Fig. 3A, B. Different levels of regulation of E-cadherin expression. A Inactivating mutations in the E-cadherin gene are found scattered throughout the coding exons 1 to 16. *C*, carboxyterminal; *CD*, cytoplasmic domain; *EC*, extracellular cadherin repeat; *N*, aminoterminal; *PRO*, propeptide; *S*, signal sequence; *TM*, transmembrane region. Updated after (Guilford et al. 1999; Nollet et al. 1999). B Modular structure of the E-cadherin promoter. Transcriptional downregulation can occur by DNA hypermethylation of CpG sequences as well as by alterations in the expression of transacting factors that bind regulatory elements in the E-cadherin promoter, as depicted here and discussed in the text. There is no E2-box2 in the human E-cadherin promoter

morphology, to decrease cellular adhesion, and to increase cellular motility (Handschuh et al. 1999). Tumor-confined missense mutations in the conserved calcium binding motifs of E-cadherin also yield similar effects (Handschuh et al. 2001). These findings provide a neat explanation for the observed diffuse infiltrative growth pattern of the lobular breast and diffuse gastric cancers.

Although high frequencies of 16q22.1 LOH have been reported for primary ovarian, endometrial and ductal breast carcinomas, only rare inactivating tumor-restricted E-cadherin mutations have been reported for these tumor types (Risinger et al. 1994; Cheng et al. 2001; Lei et al. 2002). In addition, thyroid cancers rarely bear E-cadherin mutations (Soares et al. 1997; Rocha et al. 2001). It should be noted in this regard that most LOH studies in the past were performed using polymorphic microsatellite markers located outside the E-cadherin gene locus. Therefore, the association between loss of E-cadherin function and LOH on 16q22 must be further explored. Indeed, for prostate carcinomas, no allelic imbalance was detected based on a *CDH1*-specific single nucleotide polymorphism (SNP) (Li et al. 1999; Murant et al. 2000). This suggests that the recurrent loss of E-cadherin function in prostate cancer is not the result of allelic deletion.

A possible role for loss of E-cadherin expression in early tumor development is supported by the presence of somatic mutations in early onset gastric cancers (Saito et al. 1999) and in early-stage in situ gastric and breast cancers (Becker et al. 1996; Muta et al. 1996; de Leeuw et al. 1997; Vos et al. 1997). Moreover, E-cadherin germline mutations have been detected in patients with hereditary diffuse gastric cancer syndrome (Gayther et al. 1998; Guilford et al. 1998, 1999; Richards et al. 1999; Shinmura et al. 1999; Yoon et al. 1999; Chun et al. 2001; Lewis et al. 2001; Humar et al. 2002; Oliveira et al. 2002; Yabuta et al. 2002; for a review see Caldas et al. 1999) (Fig. 3A). In contrast to expectations, the occurrence of both diffuse gastric and lobular breast carcinomas in the same patient has been reported only once to date (Keller et al. 1999). On the other hand, germline E-cadherin mutations were found in a family with diffuse gastric cancer and colon cancer (Salahshor et al. 2001b), and in a family with prostate and gastric cancers (Ikonen et al. 2001). However, *CDH1* germline mutations that have been found in families with hereditary prostate cancer are thought not to be responsible for the association of prostate cancer, gastric, and/or breast cancer in these families (Jonsson et al. 2002). An extensive allelic association study demonstrated that an E-cadherin germline missense mutation, while conferring no predisposition to familial breast cancer, does correlate with the ductal comedotype of familial breast tumors (Salahshor et al. 2001a).

4
Transcriptional Downregulation of E-Cadherin

Besides irreversible genetic inactivation, epigenetic mechanisms at the transcriptional level have also been proposed to explain E-cadherin downregulation in human carcinomas. So far, most expression analyses in human tumor specimens have been performed by immunohistochemistry. However, E-cadherin downregulation at the mRNA level has been determined by

RT-PCR analyses in, for instance, prostate cancer (Wang et al. 2000b), ductal breast carcinomas (Guriec et al. 1996; Cheng et al. 2001) and colorectal cancer (Munro et al. 1995), by Northern hybridization in thyroid carcinomas (Brabant et al. 1993), and by RNA in situ hybridization in colorectal cancer (Dorudi et al. 1995).

4.1
E-Cadherin Gene Silencing by Promoter Hypermethylation

Gene hypermethylation has been postulated to result in loss of E-cadherin expression. Aberrant hypermethylation of the promoter and/or the 5' CpG-island (Berx et al. 1995b) of the E-cadherin gene has been correlated with decreased E-cadherin expression in different types of carcinoma cell lines (Graff et al. 1995; Yoshiura et al. 1995) (Fig. 2). Meanwhile, a significant correlation between CpG hypermethylation in the E-cadherin promoter region and decreased E-cadherin expression has been reported for primary invasive ductal breast carcinomas (Nass et al. 2000; Cheng et al. 2001; Hu et al. 2002), lobular breast carcinomas (Droufakou et al. 2001), diffuse and intestinal gastric carcinomas (Tamura et al. 2000; Leung et al. 2001; Mingchao et al. 2001; Tamura et al. 2001), hepatocellular carcinomas (Kanai et al. 1997), colorectal lesions (Kanazawa et al. 2002), oral squamous cell carcinomas (Saito et al. 1998; Nakayama et al. 2001), esophageal carcinomas (Corn et al. 2001; Si et al. 2001), renal cell carcinomas (Nojima et al. 2001), bladder neoplasms (Bornman et al. 2001; Ribeiro-Filho et al. 2002), and even in leukemia (Melki et al. 1999; Melki et al. 2000).

Partial restoration of E-cadherin expression in some cell lines after treatment with demethylating agents suggests direct involvement of hypermethylation in silencing E-cadherin expression (Graff et al. 1995; Yoshiura et al. 1995). Promoter methylation in combination with LOH has also been proposed to be an efficient inactivation mechanism for the *CDH1* gene in sporadic cancers. In gastric cancers of patients carrying germline *CDH1* mutations, promoter methylation was proved to be the second inactivating hit (Grady et al. 2000). Interestingly, E-cadherin promoter methylation seems to be dynamic and unstable, reflecting the heterogeneous loss of E-cadherin expression during malignant cancer progression. Such a dynamic increase in E-cadherin CpG methylation has been documented for malignant progression of breast carcinomas (Graff et al. 2000; Nass et al. 2000) and hepatocellular carcinomas (Kanai et al. 1997; Kanai et al. 2000). As pointed out above, it should be noted that acquisition of invasiveness often accompanies diminished cell–cell adhesion, whereas the successful re-establishment of such adhesion may be necessary for tumor cell survival during dissemination and for the subsequent growth of metastases in distal, ectopic organ sites (Mareel et al. 1991; Mareel et al. 1995; Bukholm et al. 2000).

4.2
Transacting Regulators

In addition to DNA methylation, transrepression of the E-cadherin gene overruling transactivation can be a potent mechanism for its silencing. Epithelial-specific E-cadherin expression is controlled by conserved regulatory sequences in the *CDH1* promoter (Fig. 3B). The upstream regulatory sequences contain putatively positive regulatory GC- and CCAAT-boxes, as well as different E2-boxes with a repressor role (Behrens et al. 1991; Bussemakers et al. 1994; Giroldi et al. 1997).

The transcription factor activating enhancer binding protein (AP)2 has been shown to bind the GC-box in the mouse E-cadherin promoter and to induce E-cadherin transcription (Hennig et al. 1996). In addition, the retinoblastoma protein (Rb) and the proto-oncogene product c-Myc specifically activate transcription of the E-cadherin promoter by acting as co-activators of AP2 in epithelial cells (Batsche et al. 1998; Decary et al. 2002). Another candidate transcription factor interacting with the GC-box of the E-cadherin promoter is the zinc finger protein WT1, encoded by the tumor suppressor gene *Wt1* affected in Wilms' tumor of the kidney (Hosono et al. 2000). GC-box binding by WT1 has been shown to activate the E-cadherin promoter and to induce epithelial differentiation of fibroblasts upon ectopic expression.

Loss of E-cadherin expression in dedifferentiated cancer cells has been suggested to be due to transacting pathways, in which regulatory transrepression factors silence E-cadherin transcription through proximal promoter elements (Hennig et al. 1995; Ji et al. 1997; Hajra et al. 1999). Meanwhile, several different transcriptional repressors have been shown to bind directly to E-box elements in the E-cadherin promoter (Fig. 3B). These include the zinc finger transcription factors Snail (Batlle et al. 2000; Cano et al. 2000), Slug (Hajra et al. 2002; Bolos et al. 2003), SIP1 (ZEB2) (Comijn et al. 2001), δEF1 (ZEB1) (Grooteclaes and Frisch 2000), and the basic helix–loop–helix factor E12/E47 (Perez-Moreno et al. 2001). Overexpression of these different transcription factors in epithelial cells results in loss of E-cadherin-mediated cell–cell adhesion and gain of invasive properties. In addition, Snail and E12/E47 have the potential to induce tumor growth (Cano et al. 2000; Perez-Moreno et al. 2001).

Indirectly acting regulatory factors, for which no intrinsic E-cadherin promoter-binding activity has been demonstrated, include numerous factors with a repression function: integrin linked-kinase (ILK) (Wu et al. 1998; Somasiri et al. 2001), the serine/threonine kinase AKT or protein kinase (PK)B (Grille et al. 2003), C-terminal-binding protein (CtBP) (Grooteclaes et al. 2003; Shi et al. 2003), ErbB2 (D'Souza and Taylor-Papadimitriou 1994), c-Fos (Reichmann et al. 1992), transforming growth factor (TGF)-β (Miettinen et al. 1994; Piek et al. 1999), cyclo-oxygenase (COX)-2 (Tsujii and Du-

Bois 1995), interleukin (IL)-6 (Asgeirsson et al. 1998), and tumor necrosis factor (TNF)-α (Perry et al. 1999). Factors with an activation effect comprise hepatocyte nuclear factor (HNF)-4 (Spath and Weiss 1998), metastasis-associated gene 3 (MTA3) (Fujita et al. 2003), and PAX2 (Torban and Goodyer 1998).

Interestingly, some of the above-mentioned indirectly acting factors impinge on the expression status of various transcription factors known to interact with the E-cadherin promoter (Fig. 3B). High expression of the serine/threonine kinase ILK in epithelial cells mediates transcriptional activation of Snail expression, which in turn results in decreased E-cadherin expression (Tan et al. 2001). Likewise, expression of constitutively active AKT promotes epithelial-mesenchymal transition (EMT) and invasiveness by induction of Snail transcription whereas E-cadherin transcription is silenced (Grille et al. 2003). MTA3 is an estrogen receptor (ER)-regulated component of the Mi-2/NuRD transcriptional corepressor complex that directly targets the Snail promoter. Loss of ER signaling in breast cancer cells downregulates MTA3 levels, resulting in Snail-mediated repression of E-cadherin and providing a mechanistic link between estrogen receptor status and EMT of breast cancers (Fujita et al. 2003). Activation of type I cell surface serine kinase receptors of TGF-β induces EMT, which correlates with decreased E-cadherin expression, growth inhibition, and invasiveness (Miettinen et al. 1994; Piek et al. 1999; Yi et al. 2002). TGF-β induces the expression of the Smad-interacting zinc finger protein SIP1, resulting in E-cadherin repression (Comijn et al. 2001). Moreover, TGF-β has been shown to upregulate Snail and downregulate HNF4 expression in murine hepatocytes (Spagnoli et al. 2000). Smad4, although known to be an intracellular transmitter of TGF-β signals, has been proposed to mediate tumor suppression in colon carcinoma cells by transcriptional induction of E-cadherin and P-cadherin in a TGF-β-independent way (Muller et al. 2002). The paired box transcription factor (PAX)2 is a determining factor for the conversion and differentiation of kidney mesenchyme to epithelium. Transfection of PAX2 cDNA into the HEK293 human fetal kidney cell line increases the expression of the Wilms tumor (WT)1 and E-cadherin genes (Torban and Goodyer 1998). Inhibition of CtBP expression in fibroblasts or osteosarcoma cells results in upregulation of E-cadherin expression (Grooteclaes et al. 2003; Shi et al. 2003). This effect may be explained by the fact that different transcriptional repressors of E-cadherin, such as SIP1 and δEF1, are part of a CtBP corepressor complex (Shi et al. 2003).

However, in contrast with these suggestive findings, SIP1 and δEF1 apparently do not require CtBP interaction for efficient E-cadherin repression (Van Grunsven et al. 2003), suggesting that other putative CtBP interacting proteins such as Slug are involved in the CtBP-dependent repression of E-cadherin (Hemavathy et al. 2000).

Statistical analysis of expression data from clinical human carcinoma specimens should establish whether the reported positively and negatively acting transcription factors contribute significantly to the observed downregulation of E-cadherin expression in specific tumor types (Fig. 2). Activity and expression of ILK are elevated in a high percentage of colon polyps from familial adenomatous polyposis (FAP) patients, as well as in colon carcinomas (Marotta et al. 2001). AKT is frequently activated in human cancer (Testa and Bellacosa 2001). SIP1 overexpression has been linked to downregulated E-cadherin in intestinal-type gastric carcinomas (Rosivatz et al. 2002). Increased Snail expression has been found to be associated with reduced E-cadherin expression in ductal breast carcinomas (Cheng et al. 2001; Blanco et al. 2002), hepatocellular carcinomas (Jiao et al. 2002), diffuse-type gastric cancer (Rosivatz et al. 2002), and fibroblastoid cells (Gotzmann et al. 2002). Undetectable levels of the positively acting transcription factor WT1 also characterized most of the ductal breast carcinomas analyzed by Cheng et al. (2001). Nevertheless, a significant proportion of these WT1-negative tumors still expressed E-cadherin.

In addition, in ductal breast carcinomas, a negative correlation has been found between expression of E-cadherin/catenin complexes and ErbB2 receptor overexpression (Schönborn et al. 1997). In a set of 210 breast cancers, p53 accumulation was associated with both ErbB2 expression and reduced E-cadherin expression (Bukholm et al. 1997). However, no statistically significant association was seen when E-cadherin expression and ErbB2 overexpression were compared in this analysis. The highly aggressive inflammatory breast cancer form, which is exceptional in its persistent E-cadherin expression, shows no association between E-cadherin and ErbB2 expression either (Kleer et al. 2001). In cholangiocarcinomas of the liver, E-cadherin downregulation does not correlate with ErbB2 expression, whereas β-catenin downregulation does (Ashida et al. 1998). Although IL-6 levels were found to be increased in serum of breast cancer patients, no significant correlation with E-cadherin expression could be demonstrated (Asgeirsson et al. 1998).

5
Post-translational E-Cadherin Inactivation

5.1
Tyrosine Phosphorylation

Changes in the phosphorylation pattern of cadherins and catenins mediated by specific protein tyrosine kinases and phosphatases may represent the predominant regulatory mechanism of the E-cadherin/catenin complex at the post-translational level. The major targets of tyrosine kinases within the cadherin/catenin complex are the armadillo proteins β-catenin, plakoglobin, and p120ctn (Daniel and Reynolds 1997) (Fig. 1). Increased tyrosine phosphorylation of β-catenin or plakoglobin induces disassembly of the cadherin/catenin complex and results in vitro in impaired cell compaction and induction of invasion (Matsuyoshi et al. 1992; Behrens et al. 1993; Hu et al. 2001). For human colon, strong tyrosine phosphorylation of β-catenin was detected in focally dedifferentiated invasive cancer cells, but not in other cancer cells or normal tissues (Hirohashi 1998). For human stomach, tyrosine phosphorylated β-catenin was detected in poorly differentiated carcinomas, for which no inactivating E-cadherin mutations were found (Mazaki et al. 1996). Activation of the Src oncoprotein induces tyrosine phosphorylation and ubiquitination of both E-cadherin and β-catenin (Fujita et al. 2002). Hakai, a novel E3 ubiquitin-ligase, was found to bind E-cadherin in a tyrosine phosphorylation-dependent way. The resulting ubiquitination triggers internalization of the E-cadherin/catenin complex followed by either recycling of the complex to the plasma membrane or its degradation in lysosomes (Fujita et al. 2002; Pece and Gutkind 2002).

5.2
E-Cadherin Protein Truncations

Extracellular cleavage of the 120-kDa transmembrane E-cadherin protein by plasmin, or by the metalloproteinases stromelysin-1 (MMP-3) and matrilysin (MMP-7), releases an 80-kDa soluble form of the E-cadherin extracellular domain, called sE-cad (Lochter et al. 1997; Davies et al. 2001; Noë et al. 2001; Ryniers et al. 2002). Paracrine inhibition of E-cadherin function by sE-cad consolidates the observation that administration of synthetic peptides, comprising the histidine–alanine–valine (HAV) sequence of the first extracellular domain of E-cadherin, abolishes the function of the E-cadherin/catenin complex (Noë et al. 1999; Makagiansar et al. 2001). On the other hand, cytosolic truncation of E-cadherin, which removes the p120ctn and β-catenin binding domains, yields a non-functional membrane-bound E-cadherin fragment, a process found to precede epithelial apoptosis during prostate and mammary involution (Vallorosi et al. 2000). A carboxyterminally truncated 100-kDa transmembrane protein, which can be generated by calpain cleavage, has been found to accumulate in prostate tumors where calpain upregulation is observed (Rashid et al. 2001; Rios-Doria et al. 2003) (Fig. 2). Also, presenilin-1/γ-secretase cleaves E-cadherin at the membrane-cytoplasm interface, releasing an intracellular E-cadherin fragment, increasing cytosolic pools of α- and β-catenin, and promoting disassembly of adherens junctions (Marambaud et al. 2002).

5.3
Cell Surface Proteoglycans and Mucin-Like Glycoproteins

Enlarged, negatively charged cell surface proteoglycans and related molecules may interfere, possibly by steric hindrance, with E-cadherin-mediated homophilic interactions, and thus promote the invasive phenotype associated with loss of the adhesion function of E-cadherin (Vleminckx et al. 1994). Expression of the sialylated, mucin-like glycoprotein episialin/MUC1 (DF3 antigen) is restricted to the apical borders of normal secretory epithelium, whereas carcinoma cells aberrantly express this transmembrane protein at high levels over the entire cell surface (Kufe et al. 1984). In human breast cancer cell lines, MUC1 expression suppresses the function of the E-cadherin/catenin complex (Kondo et al. 1998). Another cancer-associated transmembrane glycoprotein with a mucin-like domain, called dysadherin, was found to reduce cell–cell adhesiveness, to promote metastasis, and to downregulate E-cadherin protein expression by an unknown mechanism in liver cancer cells (Ino et al. 2002). Dysadherin has been detected in a wide variety of cancer cells, but only in a limited number of normal cells. Overexpression of dysadherin has been found to correlate with reduced E-cadherin expression and poor prognosis in advanced colorectal carcinomas (Aoki et al. 2003) and pancreatic ductal adenocarcinomas (Shimamura et al. 2003) (Fig. 2).

6
Other Cadherins in Cancer

Besides loss of E-cadherin, induction of expression of other cadherins has been observed during acquisition of invasive and/or metastatic properties of cancer cells. Complex cadherin expression patterns have been described for human prostate cancer cells, including expression of N- and P-cadherin, cadherin-4, -6, and -11 (Bussemakers et al. 2000). Renal cell carcinomas showed expression of N-cadherin, cadherin-4, -6, -8, -11, and -14, but not P-cadherin or cadherin-12 or -13 (Shimazui et al. 1996; Blaschke et al. 2002). A cadherin switch from epithelial E-cadherin to mesenchymal cadherins, such as N-cadherin or cadherin-11 (OB-cadherin), has been proposed to promote the interaction of cancer cells with the stroma, thereby facilitating invasion and metastasis (Simonneau et al. 1995; Birchmeier et al. 1996; Hazan et al. 1997; Thiery 2002). Cadherin-11 was shown to be differentially expressed in the most invasive breast cancer cell lines (Pishvaian et al. 1999). On the other hand, compared to its levels in normal astrocytes, cadherin-11 was consistently decreased in astrocytoma tissues and cell lines, a process mediated by autocrine/paracrine TGF-α (Zhou and Skalli 2000). A similar finding was done for N-cadherin and cadherin-11 in normal osteoblasts, showing high expression compared to osteosarcoma (Kashima et al. 1999). This is in line with the fact that forced expression of either N-cadherin or cadherin-11 in malignant osteosarcoma cells inhibits cell migration in vitro and metastasis in vivo (Kashima et al. 2003). Moreover, in SKBR3 cells, which have a homozygous deletion of the E-cadherin gene, cadherin-11 expression can reduce the invasive activity and induce seemingly normal adherens junctions (Feltes et al. 2002). Interestingly, expression of a natural, alternatively spliced cadherin-11 transcript, encoding a possibly secreted protein with a truncated cytoplasmic domain unable to associate with any of the catenins, was reported to be increased in more aggressive osteosarcoma cells (Kawaguchi et al. 1999), and to stimulate invasion by transfected mammary cancer cells (Feltes et al. 2002).

Upregulated N-cadherin has been found in highly invasive cancer cell lines and tumor biopsies of squamous cell carcinoma, melanoma, and carcinoma of breast prostate and bladder, and generally correlates inversely with E-cadherin expression (Herlyn et al. 2000; Van Aken et al. 2001 for review). In some cases, however, E- and N-cadherin are co-expressed. Expression of transfected N-cadherin cDNA in particular breast cancer cells promotes motility and invasion in vitro and metastasis in vivo, even in the presence of continued expression of E-cadherin (Nieman et al. 1999; Hazan et al. 2000). How the invasion promoter function of N-cadherin overrules the invasion suppressor function of E-cadherin is not known. Twist, a helix–loop–helix transcription factor, is possibly involved in the E- to N-cadherin switch during EMT, and was found to be upregulated in a subset of diffuse-type gastric carcinomas (Rosivatz et al. 2002). In melanoma cells, N-cadherin mediates dissociation from keratinocytes, homotypic aggregation among melanoma cells, heterotypic adhesion to and migration over dermal fibroblasts, and various anti-apoptotic processes (Li et al. 2001). In N-cadherin-expressing prostate cancer cells, resistance to apoptosis was acquired as Bcl-2 levels were increased (Tran et al. 2002). Moreover, phosphatidylinositol 3-kinase was recruited to the N-caderin/catenin complex, resulting in activation of the Akt kinase and Bad phosphorylation.

An inverse correlation with E-cadherin expression has also been found for P-cadherin in ductal breast carcinoma, thyroid mucoepidermoid carcinoma and gastrointestinal neoplasia (Sanders et al. 2000; Gamallo et al. 2001; Rocha et al. 2002). It has been proposed that in ductal breast carcinomas, induced or persistent P-cadherin expression is even more relevant than the altered expression of any other cadherin or catenin (Soler et al. 1999; Paredes et al. 2002).

T-cadherin (cadherin-13, H-cadherin) acts as a tumor suppressor protein in various cancers. It is a unique member of the cadherin superfamily in that it shares the ectodomain organization with classical cadherins but is anchored to the plasma membrane through a glycosyl-phosphatidylinositol moiety (Ranscht and Dours-Zimmermann 1991). Downregulation of T-cadherin seems to play a role in inducing malignant phenotypic changes and tumorigenicity in breast, lung, ovary, bladder, colon, and skin cancers (Lee 1996; Sato et al. 1998; Kawakami et al. 1999; Maruyama et al. 2001; Toyooka et al. 2001, 2002; Takeuchi et al. 2002a,b). The majority of these studies reported that T-cadherin downregulation occurred by a combination of aberrant DNA methylation with gene deletion (LOH of 16q24). Recently, re-expression of T-cadherin in glioma cells was demonstrated to result in G2 phase arrest via induction of p21$^{CIP1/WAF1}$ (Huang et al. 2003).

Desmosomal cadherins, comprising desmogleins (Dsg) and desmocollins (Dsc), have been proposed to have invasion and metastasis suppressor activity. Transfection of fibroblasts with multiple desmosomal components (Dsg1, Dsc1, plakoglobin) reduced invasiveness, and this inhibitory effect was abrogated by peptides directed against the cell adhesion recognition (CAR) sites of Dsg1 and Dsc1 (Tselepis et al. 1998). Use of these CAR-specific function-blocking peptides revealed a role for desmosomal adhesion in cell sorting processes and epithelial morphogenesis that is possibly as important as the role of E-cadherin in this respect (Runswick et al. 2001).

7
Upregulation of E-Cadherin and Tumor Therapy

Re-expression of E-cadherin in E-cadherin-negative cancer cells might be considered a new anti-tumor/invasion treatment strategy. The list of agents reported to upregulate and/or stabilize the functions of the E-cadherin/catenin complex is steadily growing, as exemplified in Table 1 (Fujioka et al. 1995; Debruyne et al. 1999; Meng et al. 2000; El-Hariry et al. 2001a; Jiang et al. 2001; Rong et al. 2001; Slaton et al. 2001; Nam et al. 2002). Often the mechanism of E-cadherin activation by these agents is poorly understood. Moreover, their effects may very well be cell-type or system-dependent. Further detailed investigation is warranted. Importantly, one must be careful when proposing activation of the E-cadherin/catenin complex as tumor therapy. For instance, by virtue of ISG15 secretion, melanoma cells can induce E-cadherin expression by dendritic cells, leading to impaired migration of the latter and possible immune escape (Padovan et al. 2002). Upregulation of E-cadherin function in three-dimensional cultures has been shown to induce cell cycle arrest by upregulation of the cyclin-dependent kinase inhibitor p27^{KIP1} and to confer in this way resistance to apoptosis-inducing anticancer agents (St Croix and Kerbel 1997; St Croix et al. 1998). This effect in multicellular tumor aggregates contrasts with the reported apoptotic effect of (high levels of) transfected E-cadherin complexes in monolayer cultures (Stockinger et al. 2001; Lowy et al. 2002). Anyhow, therapeutic upregulation of E-cadherin might be restricted to early tumor stages, whereas in later stages, in which tumor cells generally have gained several additional genetic

Table 1. Examples of reagents that upregulate and/or stabilize the E-cadherin/catenin complex

Agent	Reference(s)
Hormones:	
Estradiol	Maccalman et al. 1994
Androgens	Carruba et al. 1995
Relaxin	Bani et al. 1999
Thyrotropin	Brabant et al. 1995
Phytoestrogen	Rong et al. 2001
Pharmacological agents:	
Adriamycin	Yang et al. 1999
Aspirin	Jiang et al. 2001
Docetaxel	Eckert et al. 1997
Tamoxifen	Bracke et al. 1994
Tangeritin	Bracke et al. 2002
Ubenimex (bestatin)	Fujioka et al. 1995
Growth factors, cytokines and their receptors:	
FGF1–2	El-Hariry et al. 2001b
IFN-α	Slaton et al. 2001
IGF-I/IGF-IR	Guvakova and Surmacz 1997; Boterberg et al. 2000
IL-12	Hiscox and Jiang 1997
ISG15	Padovan et al. 2002
Others:	
Azacytidine	Graff et al. 1995
EFAs	Pasqualini et al. 2003
FTI-277	Nam et al. 2002
Indole-3-carbinol	Meng et al. 2000
Irradiation	Akimoto et al. 1998
All-*trans*-retinoic acid	Vermeulen et al. 1995; Goossens et al. 2002

EFA, essential fatty acids; FGF, fibroblastic growth factor; FTI, Farnesyl transferase inhibitor; IFN, interferon; IGF-I, insulin-like growth factor I; IGF-IR, IGF-I receptor; IL, interleukin; ISG15, interferon stimulated gene (product) 15.

aberrations, E-cadherin upregulation might fall short of counteracting growth promoting processes in the tumor, while still conferring drug resistance. Thus, although a lower level of E-cadherin is an unfavorable prognostic indicator, it may actually predict a better response to a therapy targeting rapidly dividing cells. Multifactorial in vivo studies in tumor models may reveal which is the better therapeutic approach.

References

Acs G, Lawton TJ, Rebbeck TR, Li Volsi VA, Zhang PJ (2001) Differential expression of E-cadherin in lobular and ductal neoplasms of the breast and its biologic and diagnostic implications. Am J Clin Pathol 115:85–98

Akimoto T, Mitsuhashi N, Saito Y, Ebara T, Niibe H (1998) Effect of radiation on the expression of E-cadherin and alpha-catenin and invasive capacity in human lung cancer cell line in vitro. Int J Radiat Oncol Biol Phys 41:1171–1176

Alpaugh ML, Tomlinson JS, Kasraeian S, Barsky SH (2002a) Cooperative role of E-cadherin and sialyl-Lewis X/A-deficient MUC1 in the passive dissemination of tumor emboli in inflammatory breast carcinoma. Oncogene 21:3631–3643

Alpaugh ML, Tomlinson JS, Ye Y, Barsky SH (2002b) Relationship of sialyl-Lewis(x/a) underexpression and E-cadherin overexpression in the lymphovascular embolus of inflammatory breast carcinoma. Am J Pathol 161:619–628

Aoki S, Shimamura T, Shibata T, Nakanishi Y, Moriya Y, Sato Y, Kitajima M, Sakamoto M, Hirohashi S (2003) Prognostic significance of dysadherin expression in advanced colorectal carcinoma. Br J Cancer 88:726–732

Asgeirsson KS, Olafsdottir K, Jonasson JG, Ogmundsdottir HM (1998) The effects of IL-6 on cell adhesion and E-cadherin expression in breast cancer. Cytokine 10:720–728

Ashida K, Terada T, Kitamura Y, Kaibara N (1998) Expression of E-cadherin, alpha-catenin, beta-catenin, and CD44 (standard and variant isoforms) in human cholangiocarcinoma: an immunohistochemical study. Hepatology 27:974–982

Auersperg N, Pan J, Grove BD, Peterson T, Fisher J, MainesBandiera S, Somasiri A, Roskelley CD (1999) E-cadherin induces mesenchymal-to-epithelial transition in human ovarian surface epithelium. Proc Natl Acad Sci USA 96:6249–6254

Bailey T, Biddlestone L, Shepherd N, Barr H, Warner P, Jankowski J (1998) Altered cadherin and catenin complexes in the Barrett's esophagus-dysplasia-adenocarcinoma sequence: correlation with disease progression and dedifferentiation. Am J Pathol 152:135–144

Bani D, Flagiello D, Poupon MF, Nistri S, Poirson-Bichat F, Bigazzi M, Bani Sacchi T (1999) Relaxin promotes differentiation of human breast cancer cells MCF-7 transplanted into nude mice. Virchows Arch 435:509–519

Batlle E, Sancho E, Franci C, Dominguez D, Monfar M, Baulida J, de Herreros AG (2000) The transcription factor Snail is a repressor of E-cadherin gene expression in epithelial tumour cells. Nat Cell Biol 2:84–89

Batsche E, Muchardt C, Behrens J, Hurst HC, Cremisi C (1998) RB and c-Myc activate expression of the E-cadherin gene in epithelial cells through interaction with transcription factor AP-2. Mol Cell Biol 18:1–12

Becker I, Becker KF, Röhrl MH, Minkus G, Schütze K, Höfler H (1996) Single-cell mutation analysis of tumors from stained histologic slides. Lab Invest 75:801–807

Becker KF, Atkinson MJ, Reich U, Becker I, Nekarda H, Siewert JR, Höfler H (1994) E-cadherin gene mutations provide clues to diffuse type gastric carcinomas. Cancer Res 54:3845–3852

Behrens J, Löwrick O, Klein-Hitpass L, Birchmeier W (1991) The E-cadherin promoter: functional analysis of a G.C-rich region and an epithelial cell-specific palindromic regulatory element. Proc Natl Acad Sci USA 88:11495–11499

Behrens J, Vakaet L, Friis R, Winterhager E, van Roy F, Mareel MM, Birchmeier W (1993) Loss of epithelial differentiation and gain of invasiveness correlates with tyrosine phosphorylation of the E-cadherin/β-catenin complex in cells transformed with a temperature-sensitive v-src gene. J Cell Biol 120:757–766

Berx G, Cleton-Jansen A-M, Nollet F, de Leeuw WJF, van de Vijver MJ, Cornelisse C, van Roy F (1995a) E-cadherin is a tumor/invasion suppressor gene mutated in human lobular breast cancers. EMBO J 14:6107–6115

Berx G, Staes K, van Hengel J, Molemans F, Bussemakers MJG, van Bokhoven A, van Roy F (1995b) Cloning and characterization of the human invasion suppressor gene E-cadherin (CDH1). Genomics 26:281–289

Berx G, Cleton-Jansen A-M, Strumane K, de Leeuw WJF, Nollet F, van Roy FM, Cornelisse C (1996) E-cadherin is inactivated in a majority of invasive human lobular breast cancers by truncation mutations throughout its extracellular domain. Oncogene 13:1919–1925

Berx G, Becker K-F, Höfler H, van Roy F (1998) Mutation update: mutations of the human E-cadherin (CDH1) gene. Hum Mutat 12:226–237

Birchmeier C, Birchmeier W, Brand-Saberi B (1996) Epithelial-mesenchymal transitions in cancer progression. Acta Anat 156:217–226

Blanco MJ, Moreno-Bueno G, Sarrio D, Locascio A, Cano A, Palacios J, Nieto MA (2002) Correlation of Snail expression with histological grade and lymph node status in breast carcinomas. Oncogene 21:3241–3246

Blaschke S, Mueller CA, Markovic-Lipkovski J, Puch S, Miosge N, Becker V, Mueller GA, Klein G (2002) Expression of cadherin-8 in renal cell carcinoma and fetal kidney. Int J Cancer 101:327–334

Boggon TJ, Murray J, Chappuis-Flament S, Wong E, Gumbiner BM, Shapiro L (2002) C-cadherin ectodomain structure and implications for cell adhesion mechanisms. Science 296:1308–1313

Bolos V, Peinado H, Perez-Moreno MA, Fraga MF, Esteller M, Cano A (2003) The transcription factor Slug represses E-cadherin expression and induces epithelial to mesenchymal transitions: a comparison with Snail and E47 repressors. J Cell Sci 116:499–511

Bornman DM, Mathew S, Alsruhe J, Herman JG, Gabrielson E (2001) Methylation of the E-cadherin gene in bladder neoplasia and in normal urothelial epithelium from elderly individuals. Am J Pathol 159:831–835

Boterberg T, Vennekens KM, Thienpont M, Mareel MM, Bracke ME (2000) Internalization of the E-cadherin/catenin complex and scattering of human mammary carcinoma cells: MCF-7/AZ after treatment with conditioned medium from human skin squamous carcinoma cells COLO 16. Cell Adhes Commun 7:299–310

Boussadia O, Kutsch S, Hierholzer A, Delmas V, Kemler R (2002) E-cadherin is a survival factor for the lactating mouse mammary gland. Mech Develop 115:53–62

Brabant G, Hoangvu C, Cetin Y, Dralle H, Scheumann G, Molne J, Hansson G, Jansson S, Ericson LE, Nilsson M (1993) E-cadherin—a differentiation marker in thyroid malignancies. Cancer Res 53:4987–4993

Brabant G, Hoangvu C, Behrends J, Cetin Y, Potter E, Dumont JE, Maenhaut C (1995) Regulation of the cell–cell adhesion protein, E-cadherin, in dog and human thyrocytes in vitro. Endocrinology 136:3113–3119

Brabletz T, Jung A, Reu S, Porzner M, Hlubek F, KunzSchughart LA, Knuechel R, Kirchner T (2001) Variable beta-catenin expression in colorectal cancers indicates tumor progression driven by the tumor environment. Proc Natl Acad Sci USA 98:10356–10361

Bracke ME, Charlier C, Bruyneel EA, Labit C, Mareel MM, Castronovo V (1994) Tamoxifen restores the E-cadherin function in human breast cancer MCF-7/6 cells and suppresses their invasive phenotype. Cancer Res 54:4607–4609

Bracke ME, Boterberg T, Depypere HT, Stove C, Leclercq G, Mareel MM (2002) The citrus methoxyflavone tangeretin affects human cell–cell interactions. Advances in Experimental and Medical Biology 505:135–139

Bukholm IK, Nesland JM, Karesen R, Jacobsen U, Borresen-Dale AL (1997) Expression of E-cadherin and its relation to the p53 protein status in human breast carcinomas. Virchows Arch 431:317–321

Bukholm IK, Nesland JM, Borresen-Dale AL (2000) Re-expression of E-cadherin, alpha-catenin and beta-catenin, but not of gamma-catenin, in metastatic tissue from breast cancer patients. J Pathol 190:15–19

Bussemakers MJG, Giroldi LA, Vanbokhoven A, Schalken JA (1994) Transcriptional regulation of the human E-cadherin gene in human prostate cancer cell lines: characterization of the human E-cadherin gene promoter. Biochem Biophys Res Commun 203:1284–1290

Bussemakers MJG, Van Bokhoven A, Tomita K, Jansen CFJ, Schalken JA (2000) Complex cadherin expression in human prostate cancer cells. Int J Cancer 85:446–450

Caldas C, Carneiro F, Lynch HT, Yokota J, Wiesner GL, Powell SM, Lewis FR, Huntsman DG, Pharoah PDP, Jankowski JA, MacLeod P, Vogelsang H, Keller G, Park KGM, Richards FM, Maher ER, Gayther SA, Oliveira C, Grehan N, Wight D, Seruca R, Roviello F, Ponder BAJ, Jackson CE (1999) Familial gastric cancer: overview and guidelines for management. J Med Genet 36:873–880

Cano A, Perez-Moreno MA, Rodrigo I, Locascio A, Blanco MJ, del Barrio MG, Portillo F, Nieto MA (2000) The transcription factor Snail controls epithelial-mesenchymal transitions by repressing E-cadherin expression. Nat Cell Biol 2:76–83

Carico E, Atlante M, Bucci B, Nofroni I, Vecchione A (2001) E-cadherin and alpha-catenin expression during tumor progression of cervical carcinoma. Gynecol Oncol 80:156–161

Carruba G, Miceli D, Damico D, Farruggio R, Comito L, Montesanti A, Polito L, Castagnetta LAM (1995) Sex steroids up-regulate E-cadherin expression in hormone-responsive LNCap human prostate cancer cells. Biochem Biophys Res Commun 212:624–631

Carter BS, Ewing CM, Ward WS, Treiger BF, Aalders TW, Schalken JA, Epstein JI, Isaacs WB (1990) Allelic loss of chromosomes 16q and 10q in human prostate cancer. Proc Natl Acad Sci USA 87:8751–8755

Chan TA, Wang ZH, Dang LH, Vogelstein B, Kinzler KW (2002) Targeted inactivation of CTNNB1 reveals unexpected effects of beta-catenin mutation. Proc Natl Acad Sci USA 99:8265–8270

Chen WC, Obrink B (1991) Cell–cell contacts mediated by E-cadherin (uvomorulin) restrict invasive behavior of L-cells. J Cell Biol 114:319–327

Cheng CW, Wu PE, Yu JC, Huang CS, Yue CT, Wu CW, Shen CY (2001) Mechanisms of inactivation of E-cadherin in breast carcinoma: modification of the two-hit hypothesis of tumor suppressor gene. Oncogene 20:3814–3823

Chun YS, Lindor NM, Smyrk TC, Petersen BT, Burgart LJ, Guilford PJ, Donohue JH (2001) Germline E-cadherin gene mutations: is prophylactic total gastrectomy indicated? Cancer 92:181–187

Cleton-Jansen AM, Callen DF, Seshadri R, Goldup S, McCallum B, Crawford J, Powell JA, Settasatian C, van Beerendonk H, Moerland EW, Smit VT, Harris WH, Millis R, Morgan NV, Barnes D, Mathew CG, Cornelisse CJ (2001) Loss of heterozygosity mapping at chromosome arm 16q in 712 breast tumors reveals factors that influence delineation of candidate regions. Cancer Res 61:1171–1177

Colpaert CG, Vermeulen PB, Benoy I, Soubry A, Van Roy F, Van Beest P, Goovaerts G, Dirix LY, Van Dam P, Fox SB, Harris AL, Van Marck EA (2003) Inflammatory breast cancer shows angiogenesis with high endothelial proliferation rate and strong E-cadherin expression. Br J Cancer 88:718–725

Comijn J, Berx G, Vermassen P, Verschueren K, van Grunsven L, Bruyneel E, Mareel M, Huylebroeck D, van Roy F (2001) The two-handed E-box-binding zinc finger protein SIP1 downregulates E-cadherin and induces invasion. Mol Cell 7:1267–1278

Corn PG, Heath EI, Heitmiller R, Fogt F, Forastiere AA, Herman JG, Wu TT (2001) Frequent hypermethylation of the 5' CpG island of E-cadherin in esophageal adenocarcinoma. Clin Cancer Res 7:2765–2769

D'Souza B, Taylor-Papadimitriou J (1994) Overexpression of erb-B2 in human mammary epithelial cells signals inhibition of transcription of the E-cadherin gene. Proc Natl Acad Sci USA 91:7202–7206

Dahl U, Sjodin A, Semb H (1996) Cadherins regulate aggregation of pancreatic beta-cells in vivo. Development 122:2895–2902

Daniel JM, Reynolds AB (1997) Tyrosine phosphorylation and cadherin/catenin function. Bioessays 19:883–891

Davies G, Jiang WG, Mason MD (2001) Matrilysin mediates extracellular cleavage of E-cadherin from prostate cancer cells: a key mechanism in hepatocyte growth factor/scatter factor-induced cell–cell dissociation and in vitro invasion. Clin Cancer Res 7:3289–3297

Day ML, Zhao X, Vallorosi CJ, Putzi M, Powell CT, Lin C, Day KC (1999) E-cadherin mediates aggregation-dependent survival of prostate and mammary epithelial cells through the retinoblastoma cell cycle control pathway. J Biol Chem 274:9656–9664

de Leeuw WJF, Berx G, Vos CBJ, Peterse JL, van de Vijver MJ, Litvinov S, van Roy FM, Cornelisse CJ, Cleton-Jansen A-M (1997) Simultaneous loss of E-cadherin and catenins in invasive lobular breast cancer and lobular carcinoma in situ. J Pathol 183:404–411

Debruyne P, Vermeulen S, Mareel M (1999) The role of the E-cadherin/catenin complex in gastrointestinal cancer. Acta Gastroenterol Belg 62:393–402

Decary S, Decesse JT, Ogryzko V, Reed JC, Naguibneva I, Harel-Bellan A, Cremisi CE (2002) The retinoblastoma protein binds the promoter of the survival gene bcl-2 and regulates its transcription in epithelial cells through transcription factor AP-2. Mol Cell Biol 22:7877–7888

del Muro XG, Torregrosa A, Munoz J, Castellsague X, Condom E, Vigues F, Arance A, Fabra A, Germa JR (2000) Prognostic value of the expression of E-cadherin and beta-catenin in bladder cancer. Eur J Cancer 36:357–362

Dorudi S, Hanby AM, Poulsom R, Northover J, Hart IR (1995) Level of expression of E-cadherin mRNA in colorectal cancer correlates with clinical outcome. Br J Cancer 71:614–616

Droufakou S, Deshmane V, Roylance R, Hanby A, Tomlinson I, Hart IR (2001) Multiple ways of silencing E-cadherin gene expression in lobular carcinoma of the breast. Int J Cancer 92:404–408

Eckert K, Fuhrmann-Selter T, Maurer HR (1997) Docetaxel enhances the expression of E-Cadherin and carcinoembryonic antigen (CEA) on human colon cancer cell lines in vitro. Anticancer Res 17:7–12

El-Hariry I, Pignatelli M, Lemoine NR (2001a) FGF-1 and FGF-2 modulate the E-cadherin/catenin system in pancreatic adenocarcinoma cell lines. Br J Cancer 84:1656–1663

El-Hariry I, Pignatelli M, Lemoine NR (2001b) FGF-1 and FGF-2 regulate the expression of E-cadherin and catenins in pancreatic adenocarcinoma. Int J Cancer 94:652–661

Elzagheid A, Kuopio T, Ilmen M, Collan Y (2002) Prognostication of invasive ductal breast cancer by quantification of E-cadherin immunostaining: the methodology and clinical relevance. Histopathol 41:127–133

Feltes CM, Kudo A, Blaschuk O, Byers SW (2002) An alternatively spliced cadherin-11 enhances human breast cancer cell invasion. Cancer Res 62:6688–6697

Frixen UH, Behrens J, Sachs M, Eberle G, Voss B, Warda A, Löchner D, Birchmeier W (1991) E-cadherin-mediated cell–cell adhesion prevents invasiveness of human carcinoma cells. J Cell Biol 113:173–185

Fujioka S, Kohno N, Hiwada K (1995) Ubenimex activates the E-cadherin-mediated adhesion of a breast cancer cell line YMB-S. Jpn J Cancer Res 86:368–373

Fujita N, Jaye DL, Kajita M, Geigerman C, Moreno CS, Wade PA (2003) MTA3, a Mi-2/NuRD complex subunit, regulates an invasive growth pathway in breast cancer. Cell 113:207–219

Fujita Y, Krause G, Scheffner M, Zechner D, Leddy HEM, Behrens J, Sommer T, Birchmeier W (2002) Hakai, a c-Cbl-like protein, ubiquitinates and induces endocytosis of the E-cadherin complex. Nat Cell Biol 4:222–231

Gamallo C, Moreno-Bueno G, Sarrio D, Calero F, Hardisson D, Palacios J (2001) The prognostic significance of P-cadherin in infiltrating ductal breast carcinoma. Modern Pathol 14:650–654

Gayther SA, Gorringe KL, Ramus SJ, Huntsman D, Roviello F, Grehan N, Machado JE, Pinto E (1998) Identification of germ-line E-cadherin mutations in gastric cancer families of European origin. Cancer Res 58:4086–4089

Ghadimi BM, Behrens J, Hoffmann I, Haensch W, Birchmeier W, Schlag PM (1999) Immunohistological analysis of E-cadherin, alpha-, beta- and gamma-catenin expression in colorectal cancer: implications for cell adhesion and signaling. Eur J Cancer 35:60–65

Giroldi LA, Bringuier PP, de Weijert M, Jansen C, van Bokhoven A, Schalken JA (1997) Role of E boxes in the repression of E-cadherin expression. Biochem Biophys Res Commun 241:453–458

Goossens K, Deboel L, Swinnen JV, Roskams T, Manin M, Rombauts W, Verhoeven G (2002) Both retinoids and androgens are required to maintain or promote functional differentiation in reaggregation cultures of human prostate epithelial cells. Prostate 53:34–49

Gotzmann J, Huber H, Thallinger C, Wolschek M, Jansen B, Schulte-Hermann R, Beug H, Mikulits W (2002) Hepatocytes convert to a fibroblastoid phenotype through the cooperation of TGF-beta1 and Ha-Ras: steps towards invasiveness. J Cell Sci 115:1189–1202

Grady WM, Willis J, Guilford PJ, Dunbier AK, Toro TT, Lynch H, Wiesner G, Ferguson K, Eng C, Park JG, Kim SJ, Markowitz S (2000) Methylation of the CDH1 promoter as the second genetic hit in hereditary diffuse gastric cancer. Nat Genet 26:16–17

Graff JR, Herman JG, Lapidus RG, Chopra H, Xu R, Jarrard DF, Isaacs WB, Pitha PM, Davidson NE, Baylin SB (1995) E-cadherin expression is silenced by DNA hypermethylation in human breast and prostate carcinomas. Cancer Res 55:5195–5199

Graff JR, Gabrielson E, Fujii H, Baylin SB, Herman JG (2000) Methylation patterns of the E-cadherin 5′ CpG island are unstable and reflect the dynamic, heterogeneous loss of E-cadherin expression during metastatic progression. J Biol Chem 275:2727–2732

Grille SJ, Bellacosa A, Upson J, Klein-Szanto AJ, Van Roy F, Lee-Kwon W, Donowitz M, Tsichlis PN, Larue L (2003) The protein kinase Akt induces epithelial mesenchymal transition and promotes enhanced motility and invasiveness of squamous cell carcinoma lines. Cancer Res 63:2172–2178

Grooteclaes M, Deveraux Q, Hildebrand J, Zhang Q, Goodman RH, Frisch SM (2003) C-terminal-binding protein corepresses epithelial and proapoptotic gene expression programs. Proc Natl Acad Sci USA 100:4568–4573

Grooteclaes ML, Frisch SM (2000) Evidence for a function of CtBP in epithelial gene regulation and anoikis. Oncogene 19:3823–3828

Guilford P, Hopkins J, Harraway J, McLeod M, McLeod N, Harawira P, Taite H, Scoular R, Miller A, Reeve AE (1998) E-cadherin germline mutations in familial gastric cancer. Nature 392:402–405

Guilford PJ, Hopkins JBW, Grady WM, Markowitz SD, Willis J, Lynch H, Rajput A, Wiesner GL, Lindor NM, Burgart LJ, Toro TT, Lee D, Limacher JM, Shaw DW, Findlay MPN, Reeve AE (1999) E-cadherin germline mutations define an inherited cancer syndrome dominated by diffuse gastric cancer. Hum Mutat 14:249–255

Gunji N, Oda T, Todoroki T, Kanazawa N, Kawamoto T, Yuzawa K, Scarpa A, Fukao K (1998) Pancreatic carcinoma—correlation between E-cadherin and alpha-catenin expression status and liver metastasis. Cancer 82:1649–1656

Guriec N, Marcellin L, Gairard B, Calderoli H, Wilk A, Renaud R, Bergerat JP, Oberling F (1996) E-cadherin mRNA expression in breast carcinomas correlates with overall and disease-free survival. Invas Metast 16:19–26

Guvakova MA, Surmacz E (1997) Overexpressed IGF-I receptors reduce estrogen growth requirements, enhance survival, and promote E-cadherin-mediated cell–cell adhesion in human breast cancer cells. Exp Cell Res 231:149–162

Hajra KM, Ji XD, Fearon ER (1999) Extinction of E-cadherin expression in breast cancer via a dominant repression pathway acting on proximal promoter elements. Oncogene 18:7274–7279

Hajra KM, Chen DY, Fearon ER (2002) The SLUG zinc-finger protein represses E-cadherin in breast cancer. Cancer Res 62:1613–1618

Handschuh G, Candidus S, Luber B, Reich U, Schott C, Oswald S, Becke H, Hutzler P, Birchmeier W, Hofler H, Becker KF (1999) Tumour-associated E-cadherin mutations alter cellular morphology, decrease cellular adhesion and increase cellular motility. Oncogene 18:4301–4312

Handschuh G, Luber B, Hutzler P, Hofler H, Becker KF (2001) Single amino acid substitutions in conserved extracellular domains of E-cadherin differ in their functional consequences. J Mol Biol 314:445–454

Hao XP, Palazzo JP, Ilyas M, Tomlinson I, Talbot IC (1997) Reduced expression of molecules of the cadherin/catenin complex in the transition from colorectal adenoma to carcinoma. Anticancer Res 17:2241–2247

Hazan RB, Kang L, Whooley BP, Borgen PI (1997) N-cadherin promotes adhesion between invasive breast cancer cells and the stroma. Cell Adhes Commun 4:399–411

Hazan RB, Phillips GR, Qiao RF, Norton L, Aaronson SA (2000) Exogenous expression of N-cadherin in breast cancer cells induces cell migration, invasion, and metastasis. J Cell Biol 148:779–790

Hemavathy K, Guru SC, Harris J, Chen JD, Ip YT (2000) Human slug is a repressor that localizes to sites of active transcription. Mol Cell Biol 20:5087–5095

Hennig G, Behrens J, Truss M, Frisch S, Reichmann E, Birchmeier W (1995) Progression of carcinoma cells is associated with alterations in chromatin structure and factor binding at the E-cadherin promoter in vivo. Oncogene 11:475–484

Hennig G, Lowrick O, Birchmeier W, Behrens J (1996) Mechanisms identified in the transcriptional control of epithelial gene expression. J Biol Chem 271:595–602

Herlyn M, Berking C, Li G, Satyamoorthy K (2000) Lessons from melanocyte development for understanding the biological events in naevus and melanoma formation. Melanoma Res 10:303–312

Hirohashi S (1998) Inactivation of the E-cadherin-mediated cell adhesion system in human cancers. Am J Pathol 153:333–339

Hiscox S, Jiang WG (1997) Interleukin-12, an emerging anti-tumour cytokine. In Vivo 11:125–132

Hosono S, Gross I, English MA, Hajra KM, Fearon ER, Licht JD (2000) E-cadherin is a WT1 target gene. J Biol Chem 275:10943–10953

Hsu MY, Meier FE, Nesbit M, Hsu JY, VanBelle P, Elder DE, Herlyn M (2000) E-cadherin expression in melanoma cells restores keratinocyte-mediated growth control and down-regulates expression of invasion-related adhesion receptors. Am J Pathol 156:1515–1525

Hu PQ, EJ OK, Rubenstein DS (2001) Tyrosine phosphorylation of human keratinocyte beta-catenin and plakoglobin reversibly regulates their binding to E- cadherin and alpha-catenin. J Invest Dermatol 117:1059–1067

Hu XC, Loo WTY, Chow LWC (2002) E-cadherin promoter methylation can regulate its expression in invasive ductal breast cancer tissue in Chinese woman. Life Sci 71:1397–1404

Huang ZY, Wu YL, Hedrick N, Gutmann DH (2003) T-cadherin-mediated cell growth regulation involves G(2) phase arrest and requires p21 (CIP1/WAF1) expression. Mol Cell Biol 23:566–578

Humar B, Toro T, Graziano F, Muller H, Dobbie Z, KwangYang H, Eng C, Hampel H, Gilbert D, Winship I, Parry S, Ward R, Findlay M, Christian A, Tucker M, Tucker K, Merriman T, Guilford P (2002) Novel germline CDH1 mutations in hereditary diffuse gastric cancer families. Hum Mutat 19:518–525

Ihara A, Koizumi H, Hashizume R, Uchikoshi T (1996) Expression of epithelial cadherin and alpha- and beta-catenins in nontumoral livers and hepatocellular carcinomas. Hepatology 23:1441–1447

Ikonen T, Matikainen M, Mononen N, Hyytinen ER, Helin HJ, Tommola S, Tammela TL, Pukkala E, Schleutker J, Kallioniemi OP, Koivisto PA (2001) Association of E-cadherin germ-line alterations with prostate cancer. Clin Cancer Res 7:3465–3471

Ino Y, Gotoh M, Sakamoto M, Tsukagoshi K, Hirohashi S (2002) Dysadherin, a cancer-associated cell membrane glycoprotein, down-regulates E-cadherin and promotes metastasis. Proc Natl Acad Sci USA 99:365–370

Ji XD, Woodard AS, Rimm DL, Fearon ER (1997) Transcriptional defects underlie loss of E-cadherin expression in breast cancer. Cell Growth Differ 8:773–778

Jiang MC, Liao CF, Lee PH (2001) Aspirin inhibits matrix metalloproteinase-2 activity, increases E-cadherin production, and inhibits in vitro invasion of tumor cells. Biochem Biophys Res Commun 282:671–677

Jiao W, Miyazaki K, Kitajima Y (2002) Inverse correlation between E-cadherin and Snail expression in hepatocellular carcinoma cell lines in vitro and in vivo. Br J Cancer 86:98–101

Jonsson BA, Bergh A, Stattin P, Emmanuelsson M, Gronberg H (2002) Germline mutations in E-cadherin do not explain association of hereditary prostate cancer, gastric cancer and breast cancer. Int J Cancer 98:838–843

Kanai Y, Ushijima S, Hui AM, Ochiai A, Tsuda H, Sakamoto M, Hirohashi S (1997) The E-cadherin gene is silenced by CpG methylation in human hepatocellular carcinomas. Int J Cancer 71:355–359

Kanai Y, Ushijima S, Tsuda H, Sakamoto M, Hirohashi S (2000) Aberrant DNA methylation precedes loss of heterozygosity on chromosome 16 in chronic hepatitis and liver cirrhosis. Cancer Lett 148:73–80

Kanazawa T, Watanabe T, Kazama S, Tada T, Koketsu S, Nagawa H (2002) Poorly differentiated adenocarcinoma and mucinous carcinoma of the colon and rectum show higher rates of loss of heterozygosity and loss of E-cadherin expression due to methylation of promoter region. Int J Cancer 102:225–229

Karube H, Masuda H, Ishii Y, Takayama T (2002) E-cadherin expression is inversely proportional to tumor size in experimental liver metastases. J Surg Res 106:173–178

Kashima T, Kawaguchi J, Takeshita S, Kuroda M, Takanashi M, Horiuchi H, Imamura T, Ishikawa Y, Ishida T, Mori S, Machinami R, Kudo A (1999) Anomalous cadherin expression in osteosarcoma—possible relationships to metastasis and morphogenesis. Am J Pathol 155:1549–1555

Kashima T, Nakamura K, Kawaguchi J, Takanashi M, Ishida T, Aburatani H, Kudo A, Fukayama M, Grigoriadis AE (2003) Overexpression of cadherins suppresses pulmonary metastasis of osteosarcoma in vivo. Int J Cancer 104:147–154

Kawaguchi J, Takeshita S, Kashima T, Imai T, Machinami R, Kudo A (1999) Expression and function of the splice variant of the human cadherin-11 gene in subordination to intact cadherin-11. J Bone Miner Res 14:764–775

Kawakami M, Staub J, Cliby W, Hartmann L, Smith DI, Shridhar V (1999) Involvement of H-cadherin (CDH13) on 16q in the region of frequent deletion in ovarian cancer. Int J Oncol 15:715–720

Keller G, Vogelsang H, Becker I, Hutter J, Ott K, Candidus S, Grundei T, Becker KF, Mueller J, Siewert JR, Hofler H (1999) Diffuse type gastric and lobular breast carcinoma in a familial gastric cancer patient with an E-cadherin germline mutation. Am J Pathol 155:337–342

Kihana T, Yano N, Murao S, Iketani H, Hamada K, Yano J, Matsuura S (1996) Allelic loss of chromosome 16q in endometrial cancer: Correlation with poor prognosis of patients and less differentiated histology. Jpn J Cancer Res 87:1184–1190

Kimura T, Sato H, Manabe R, Konishi H, Kushima R, Sugihara H, Hattori T, Kodama T, Kashima K (1997) Analysis of microsatellite regions and DNA ploidy pattern in signet ring cell carcinomas of the stomach. Gan To Kagaku Ryoho 24:273–278

Kleer CG, van Golen KL, Braun T, Merajver SD (2001) Persistent E-cadherin expression in inflammatory breast cancer. Modern Pathol 14:458–464

Knudson AG (1971) Mutation and cancer: statistical study of retinoblastoma. Proc Natl Acad Sci USA 68:820–823

Kondo K, Kohno N, Yokoyama A, Hiwada K (1998) Decreased MUC1 expression induces E-cadherin-mediated cell adhesion of breast cancer cell lines. Cancer Res 58:2014–2019

Kufe D, Inghirami G, Abe M, Hayes D, Justi-Wheeler H, Schlom J (1984) Differential reactivity of a novel monoclonal antibody (DF3) with human malignant versus benign breast tumors. Hybridoma 3:223–232

Laskin WB, Miettinen M (2002) Epithelial-type and neural-type cadherin expression in malignant noncarcinomatous neoplasms with epithelioid features that involve the soft tissues. Arch Pathol Labor Med 126:425–431

Latil A, Cussenot O, Fournier G, Driouch K, Lidereau R (1997) Loss of heterozygosity at chromosome 16q in prostate adenocarcinoma: Identification of three independent regions. Cancer Res 57:1058–1062

Lee SW (1996) H-cadherin, a novel cadherin with growth inhibitory functions and diminished expression in human breast cancer. Nat Med 2:776–782

Lei HX, Sjoberg-Margolin S, Salahshor S, Werelius B, Jandakova E, Hemminki K, Lindblom A, Vorechovsky I (2002) CDH1 mutations are present in both ductal and lobular breast cancer, but promoter allelic variants show no detectable breast cancer risk. Int J Cancer 98:199–204

Leung WK, Yu J, Ng EKW, To KF, Ma PK, Lee TL, Go MYY, Chung SCS, Sung JJY (2001) Concurrent hypermethylation of multiple tumor-related genes in gastric carcinoma and adjacent normal tissues. Cancer 91:2294–2301

Lewis FR, Mellinger JD, Hayashi A, Lorelli D, Monaghan KG, Carneiro F, Huntsman DG, Jackson CE, Caldas C (2001) Prophylactic total gastrectomy for familial gastric cancer. Surgery 130:612–617

Li CD, Berx G, Larsson C, Auer G, Aspenblad U, Pan Y, Sundelin B, Ekman M, Nordenskjöld M, van Roy F, Bergerheim USR (1999) Distinct deleted regions on chromosome segment 16q23-24 associated with metastases in prostate cancer. Gene Chromos Cancer 24:175–182

Li G, Satyamoorthy K, Herlyn M (2001) N-cadherin-mediated intercellular interactions promote survival and migration of melanoma cells. Cancer Res 61:3819–3825

Lochter A, Galosy S, Muschler J, Freedman N, Werb Z, Bissell MJ (1997) Matrix metalloproteinase stromelysin-1 triggers a cascade of molecular alterations that leads to stable epithelial-to-mesenchymal conversion and a premalignant phenotype in mammary epithelial cells. J Cell Biol 139:1861–1872

Lowy AM, Knight J, Groden J (2002) Restoration of E-cadherin/beta-catenin expression in pancreatic cancer cells inhibits growth by induction of apoptosis. Surgery 132:141–148

Luo J, Lubaroff DM, Hendrix MJC (1999) Suppression of prostate cancer invasive potential and matrix metalloproteinase activity by E-cadherin transfection. Cancer Res 59:3552–3556

Maccalman CD, Farookhi R, Blaschuk OW (1994) Estradiol regulates E-cadherin mRNA levels in the surface epithelium of the mouse ovary. Clin Exp Metastas 12:276–282

Machado JC, Soares P, Carneiro F, Rocha A, Beck S, Blin N, Berx G, Sobrinho-Simoes M (1999) E-cadherin gene mutations provide a genetic basis for the phenotypic divergence of mixed gastric carcinomas. Lab Invest 79:459–465

Makagiansar IT, Avery M, Hu Y, Audus KL, Siahaan TJ (2001) Improving the selectivity of HAV-peptides in modulating E- cadherin-E-cadherin interactions in the intercellular junction of MDCK cell monolayers. Pharmaceut Res 18:446–453

Mansouri A, Spurr N, Goodfellow PN, Kemler R (1988) Characterization and chromosomal localization of the gene encoding the human cell adhesion molecule uvomorulin. Differentiation 38:67–71

Marambaud P, Shioi J, Serban G, Georgakopoulos A, Sarner S, Nagy V, Baki L, Wen P, Efthimiopoulos S, Shao Z, Wisniewski T, Robakis NK (2002) A presenilin-1/gamma-secretase cleavage releases the E-cadherin intracellular domain and regulates disassembly of adherens junctions. EMBO J 21:1948–1956

Mareel M, Bracke M, Van Roy F (1995) Cancer metastasis: negative regulation by an invasion suppressor gene. Cancer Detect Prev 19:451–464

Mareel MM, Behrens J, Birchmeier W, De Bruyne GK, Vleminckx K, Hoogewijs A, Fiers WC, van Roy FM (1991) Down-regulation of E-cadherin expression in Madin Darby canine kidney (MDCK) cells inside tumors of nude mice. Int J Cancer 47:922–928

Marotta A, Tan C, Gray V, Malik S, Gallinger S, Sanghera J, Dupuis B, Owen D, Dedhar S, Salh B (2001) Dysregulation of integrin-linked kinase (ILK) signaling in colonic polyposis. Oncogene 20:6250–6257

Maruyama R, Toyooka S, Toyooka KO, Harada K, Virmani AK, Zochbauer-Muller S, Farinas AJ, Vakar-Lopez F, Minna JD, Sagalowsky A, Czerniak B, Gazdar AF (2001) Aberrant promoter methylation profile of bladder cancer and its relationship to clinicopathological features. Cancer Res 61:8659–8663

Matsumura T, Makino R, Mitamura K (2001) Frequent down-regulation of E-cadherin by genetic and epigenetic changes in the malignant progression of hepatocellular carcinomas. Clin Cancer Res 7:594–599

Matsuyoshi N, Hamaguchi M, Taniguchi S, Nagafuchi A, Tsukita S, Takeichi M (1992) Cadherin-mediated cell–cell adhesion is perturbed by v-src tyrosine phosphorylation in metastatic fibroblasts. J Cell Biol 118:703–714

Mazaki T, Ishii Y, Fujii M, Iwai S, Ishikawa K (1996) Mutations of p53, E-cadherin, alpha- and beta-catenin genes and tyrosine phosphorylation of beta-catenin in human gastric carcinomas. Int J Oncol 9:579–583

Mbalaviele G, Dunstan CR, Sasaki A, Williams PJ, Mundy GR, Yoneda T (1996) E-cadherin expression in human breast cancer cells suppresses the development of osteolytic bone metastases in an experimental metastasis model. Cancer Res 56:4063–4070

Meiners S, Brinkmann V, Naundorf H, Birchmeier W (1998) Role of morphogenetic factors in metastasis of mammary carcinoma cells. Oncogene 16:9–20

Melki JR, Vincent PC, Clark SJ (1999) Concurrent DNA hypermethylation of multiple genes in acute myeloid leukemia. Cancer Res 59:3730–3740

Melki JR, Vincent PC, Brown RD, Clark SJ (2000) Hypermethylation of E-cadherin in leukemia. Blood 95:3208–3213

Meng QH, Qi M, Chen DZ, Yuan RQ, Goldberg ID, Rosen EM, Auborn K, Fan SJ (2000) Suppression of breast cancer invasion and migration by indole-3-carbinol: associated with up-regulation of BRCA1 and E-cadherin/catenin complexes. J Mol Med 78:155–165

Miettinen PJ, Ebner R, Lopez AR, Derynck R (1994) TGF-beta induced transdifferentiation of mammary epithelial cells to mesenchymal cells: involvement of type I receptors. J Cell Biol 127:2021–2036

Mingchao, Devereux TR, Stockton P, Sun K, Sills RC, Clayton N, Portier M, Flake G (2001) Loss of E-cadherin expression in gastric intestinal metaplasia and later stage p53 altered expression in gastric carcinogenesis. Exp Toxicol Pathol 53:237–246

Miyaki M, Tanaka K, Kikuchi-Yanoshita R, Muraoka M, Konishi M, Takeichi M (1995) Increased cell-substratum adhesion, and decreased gelatinase secretion and cell growth, induced by E-cadherin transfection of human colon carcinoma cells. Oncogene 11:2547–2552

Moersig W, Horn S, Hilker M, Mayer E, Oelert H (2002) Transfection of E-cadherin cDNA in human lung tumor cells reduces invasive potential of tumors. Thorac Cardiovasc Surg 50:45–48

Muller N, Reinacher-Schick A, Baldus S, Van Hengel J, Berx G, Baar A, Van Roy F, Schmiegel W, Schwarte-Waldhoff I (2002) Smad4 induces the tumor suppressor E-cadherin and P-cadherin in colon carcinoma cells. Oncogene 21:6049–6058

Munro SB, Turner IM, Farookhi R, Blaschuk OW, Jothy S (1995) E-cadherin and OB-cadherin mRNA levels in normal human colon and colon carcinoma. Exp Mol Pathol 62:118–122

Murant SJ, Rolley N, Phillips SMA, Stower M, Maitland NJ (2000) Allelic imbalance within the E-cadherin gene is an infrequent event in prostate carcinogenesis. Gene Chromos Cancer 27:104–109

Muta H, Noguchi M, Kanai Y, Ochiai A, Nawata H, Hirohashi S (1996) E-cadherin gene mutations in signet ring cell carcinoma of the stomach. Jpn J Cancer Res 87:843–848

Nakayama S, Sasaki A, Mese H, Alcalde RE, Tsuji T, Matsumura T (2001) The E-cadherin gene is silenced by CpG methylation in human oral squamous cell carcinomas. Int J Cancer 93:667–673

Nam JS, Ino Y, Sakamoto M, Hirohashi S (2002) Ras farnesylation inhibitor FTI-277 restores the E-cadherin/catenin cell adhesion system in human cancer cells and reduces cancer metastasis. Jpn J Cancer Res 93:1020-1028

Nass SJ, Herman JG, Gabrielson E, Iversen PW, Parl FF, Davidson NE, Graff JR (2000) Aberrant methylation of the estrogen receptor and E-cadherin 5' CpG islands increases with malignant progression in human breast cancer. Cancer Res 60:4346-4348

Natt E, Magenis RE, Zimmer J, Mansouri A, Scherer G (1989) Regional assignment of the human loci for uvomorulin (UVO) and chymotrypsinogen B (CTRB) with the help of two overlapping deletions on the long arm of chromosome 16. Cytogenet Cell Genet 50:145-148

Nawrocki-Raby B, Gilles C, Polette M, Martinella-Catusse C, Bonnet N, Puchelle E, Foidart J-M, van Roy F, Birembaut P (2003) E-cadherin mediates MMP down-regulation in highly invasive bronchial tumor cells. Am J Pathol 163:653-661

Nieman MT, Prudoff RS, Johnson KR, Wheelock MJ (1999) N-cadherin promotes motility in human breast cancer cells regardless of their E-cadherin expression. J Cell Biol 147:631-643

Noë V, Willems J, Van de Kerckhove J, Van Roy F, Bruyneel E, Mareel M (1999) Inhibition of adhesion and induction of epithelial cell invasion by HAV-containing E-cadherin-specific peptides. J Cell Sci 112:127-135

Noë V, Fingleton B, Jacobs K, Crawford HC, Vermeulen S, Steelant W, Bruyneel E, Matrisian LM, Mareel M (2001) Release of an invasion promoter E-cadherin fragment by matrilysin and stromelysin-1. J Cell Sci 114:111-118

Nojima D, Nakajima K, Li LC, Franks J, Ribeiro L, Ishii N, Dahiya R (2001) CpG methylation of promoter region inactivates E-cadherin gene in renal cell carcinoma. Mol Carcinogen 32:19-27

Nollet F, Berx G, van Roy F (1999) The role of the E-cadherin/catenin adhesion complex in the development and progression of cancer. Mol Cell Biol Res Commun 2:77-85

Oliveira C, Bordin MC, Grehan N, Huntsman D, Suriano G, Machado JC, Kiviluoto T, Aaltonen L, Jackson CE, Seruca R, Caldas C (2002) Screening E-cadherin in gastric cancer families reveals germline mutations only in hereditary diffuse gastric cancer kindred. Hum Mutat 19:510-517

Padovan E, Terracciano L, Certa U, Jacobs B, Reschner A, Bolli M, Spagnoli GC, Borden EC, Heberer M (2002) Interferon stimulated gene 15 constitutively produced by melanoma cells induces E-cadherin expression on human dendritic cells. Cancer Res 62:3453-3458

Pan Y, Matsuyama H, Wang NN, Yoshihiro S, Haggarth L, Li C, Tribukait B, Ekman P, Bergerheim USR (1998) Chromosome 16q24 deletion and decreased E-cadherin expression: possible association with metastatic potential in prostate cancer. Prostate 36:31-38

Papadavid E, Pignatelli M, Zakynthinos S, Krausz T, Chu AC (2001) The potential role of abnormal E-cadherin and alpha-, beta- and gamma-catenin immunoreactivity in the determination of the biological behaviour of keratoacanthoma. Br J Dermatol 145:582-589

Papadavid E, Pignatelli M, Zakynthinos S, Krausz T, Chu AC (2002) Abnormal immunoreactivity of the E-cadherin/catenin (alpha-, beta-, and gamma-) complex in premalignant and malignant non-melanocytic skin tumours. J Pathol 196:154-162

Paredes J, Milanezi F, Viegas L, Amendoeira I, Schmitt F (2002) P-cadherin expression is associated with high-grade ductal carcinoma in situ of the breast. Virchows Arch 440:16-21

Pasqualini ME, Heyd VL, Manzo P, Eynard AR (2003) Association between E-cadherin expression by human colon, bladder and breast cancer cells and the 13-HODE:15-HETE ratio. A possible role of their metastatic potential. Prostaglandins Leukot Essent Fatty Acids 68:9–16

Pece S, Gutkind JS (2002) E-cadherin and Hakai: signalling, remodeling or destruction? Nat Cell Biol 4:E72–74

Perez-Moreno MA, Locascio A, Rodrigo I, Dhondt G, Portillo F, Nieto MA, Cano A (2001) A new role for E12/E47 in the repression of E-cadherin expression and epithelial-mesenchymal transitions. J Biol Chem 276:27424–27431

Perl AK, Wilgenbus P, Dahl U, Semb H, Christofori G (1998) A causal role for E-cadherin in the transition from adenoma to carcinoma. Nature 392:190–193

Perry I, Tselepis C, Hoyland J, Iqbal TH, Sanders DSA, Cooper BT, Jankowski JAZ (1999) Reduced cadherin/catenin complex expression in celiac disease can be reproduced in vitro by cytokine stimulation. Lab Invest 79:1489–1499

Piek E, Moustakas A, Heldin CH, Ten Dijke P (1999) TGF-β type I receptor/ALK-5 and Smad proteins mediate epithelial to mesenchymal transdifferentiation in NMuMG breast epithelial cells. J Cell Sci 112:4557–4568

Pishvaian MJ, Feltes CM, Thompson P, Bussemakers MJ, Schalken JA, Byers SW (1999) Cadherin-11 is expressed in invasive breast cancer cell lines. Cancer Res 59:947–952

Ranscht B, Dours-Zimmermann MT (1991) T-cadherin, a novel cadherin cell adhesion molecule in the nervous system lacks the conserved cytoplasmic region. Neuron 7:391–402

Rashid MG, Sanda MG, Vallorosi CJ, RiosDoria J, Rubin MA, Day ML (2001) Posttranslational truncation and inactivation of human E-cadherin distinguishes prostate cancer from matched normal prostate. Cancer Res 61:489–492

Reichmann E, Schwarz H, Deiner EM, Leitner I, Eilers M, Berger J, Busslinger M, Beug H (1992) Activation of an inducible c-FosER fusion protein causes loss of epithelial polarity and triggers epithelial-fibroblastoid cell conversion. Cell 71:1103–1116

Ribeiro-Filho LA, Franks J, Sasaki M, Shiina H, Li LC, Nojima D, Arap S, Carroll P, Enokida H, Nakagawa M, Yonezawa S, Dahiya R (2002) CpG hypermethylation of promoter region and inactivation of E-cadherin gene in human bladder cancer. Mol Carcinogen 34:187–198

Richards FM, McKee SA, Rajpar MH, Cole TRP, Evans DGR, Jankowski JA, McKeown C, Sanders DSA, Maher ER (1999) Germline E-cadherin gene (CDH1) mutations predispose to familial gastric cancer and colorectal cancer. Hum Mol Genet 8:607–610

Rios-Doria J, Day KC, Kuefer R, Rashid MG, Chinnaiyan AM, Rubin MA, Day ML (2003) The role of calpain in the proteolytic cleavage of E-cadherin in prostate and mammary epithelial cells. J Biol Chem 278:1372–1379

Risinger JI, Berchuck A, Kohler MF, Boyd J (1994) Mutations of the E-cadherin gene in human gynecologic cancers. Nat Genet 7:98–102

Rocha AS, Soares P, Seruca R, Maximo V, Matias-Guiu X, Cameselle-Teijeiro J, Sobrinho-Simoes M (2001) Abnormalities of the E-cadherin/catenin adhesion complex in classical papillary thyroid carcinoma and in its diffuse sclerosing variant. J Pathol 194:358–366

Rocha AS, Soares P, Machado JC, Maximo V, Fonseca E, Franssila K, Sobrinho-Simoes M (2002) Mucoepidermoid carcinoma of the thyroid: a tumour histotype characterised by P-cadherin neoexpression and marked abnormalities of E-cadherin/catenins complex. Virchows Arch 440:498–504

Rong HJ, Boterberg T, Maubach J, Stove C, Depypere H, Van Slambrouck S, Serreyn R, De Keuketeire D, Mareel M, Bracke M (2001) 8-Prenylnaringenin, the phytoestrogen in hops and beer, upregulates the function of the E-cadherin/catenin complex in human mammary carcinoma cells. Eur J Cell Biol 80:580–585

Rosivatz E, Becker I, Specht K, Fricke E, Luber B, Busch R, Hofler H, Becker KF (2002) Differential expression of the epithelial-mesenchymal transition regulators Snail, SIP1, and Twist in gastric cancer. Am J Pathol 161:1881–1891

Runswick SK, O'Hare MJ, Jones L, Streuli CH, Garrod DR (2001) Desmosomal adhesion regulates epithelial morphogenesis and cell positioning. Nat Cell Biol 3:823–830

Ryniers F, Stove C, Goethals M, Brackenier L, Noë V, Bracke M, Vandekerckhove J, Mareel M, Bruyneel E (2002) Plasmin produces an E-cadherin fragment that stimulates cancer cell invasion. Biol Chem 383:159–165

Saito A, Kanai Y, Maesawa C, Ochiai A, Torii A, Hirohashi S (1999) Disruption of E-cadherin-mediated cell adhesion systems in gastric cancers in young patients. Jpn J Cancer Res 90:993–999

Saito T, Katagiri A, Watanabe R, Tanikawa T, Kawasaki T, Tomita Y, Takahashi K (2000a) Expression of E-cadherin and catenins on testis tumor. Urol Int 65:140–143

Saito T, Oda Y, Sakamoto A, Tamiya S, Kinukawa N, Hayashi K, Iwamoto Y, Tsuneyoshi M (2000b) Prognostic value of the preserved expression of the E-cadherin and catenin families of adhesion molecules and of beta-catenin mutations in synovial sarcoma. J Pathol 192:342–350

Saito Y, Takazawa H, Uzawa K, Tanzawa H, Sato K (1998) Reduced expression of E-cadherin in oral squamous cell carcinoma: relationship with DNA methylation of 5' CpG island. Int J Oncol 12:293–298

Salahshor S, Haixin L, Huo H, Kristensen VN, Loman N, Sjoberg-Margolin S, Borg A, Borresen-Dale AL, Vorechovsky I, Lindblom A (2001a) Low frequency of E-cadherin alterations in familial breast cancer. Breast Cancer Res 3:199–207

Salahshor S, Hou H, Diep CB, Loukola A, Zhang H, Liu T, Chen J, Iselius L, Rubio C, Lothe RA, Aaltonen L, Sun XF, Lindmark G, Lindblom A (2001b) A germline E-cadherin mutation in a family with gastric and colon cancer. Int J Mol Med 8:439–443

Sanders DSA, Perry I, Hardy R, Jankowski J (2000) Aberrant P-cadherin expression is a feature of clonal expansion in the gastrointestinal tract associated with repair and neoplasia. J Pathol 190:526–530

Sato H, Hasegawa T, Abe Y, Sakai H, Hirohashi S (1999) Expression of E-cadherin in bone and soft tissue sarcomas: A possible role in epithelial differentiation. Hum Pathol 30:1344–1349

Sato M, Mori Y, Sakurada A, Fujimura S, Horii A (1998) The H-cadherin (CDH13) gene is inactivated in human lung cancer. Hum Genet 103:96–101

Schönborn I, Zschiesche W, Behrens J, Herrenknecht K, Birchmeier W (1997) Expression of E-cadherin/catenin complexes in breast cancer: correlation with favourable prognostic factors and survival. Int J Oncol 11:1327–1334

Shi Y, Sawada JI, Sui G, Affar EB, Whetstine JR, Lan F, Ogawa H, Luke MPS, Nakatani Y, Shi Y (2003) Coordinated histone modifications mediated by a CtBP co-repressor complex. Nature 422:735–738

Shimamura T, Sakamoto M, Ino Y, Sato Y, Shimada K, Kosuge T, Sekihara H, Hirohashi S (2003) Dysadherin overexpression in pancreatic ductal adenocarcinoma reflects tumor aggressiveness: relationship to E-cadherin expression. J Clin Oncol 21:659–667

Shimazui T, Giroldi LA, Bringuier PP, Oosterwijk E, Schalken JA (1996) Complex cadherin expression in renal cell carcinoma. Cancer Res 56:3234–3237

Shimoyama Y, Hirohashi S (1991) Cadherin intercellular adhesion molecule in hepatocellular carcinomas: loss of E-cadherin expression in an undifferentiated carcinoma. Cancer Lett 57:131–135

Shinmura K, Kohno T, Takahashi M, Sasaki A, Ochiai A, Guilford P, Hunter A, Reeve AE, Sugimura H, Yamaguchi N, Yokota J (1999) Familial gastric cancer: clinicopathological characteristics, RER phenotype and germline p53 and E-cadherin mutations. Carcinogenesis 20:1127–1131

Si HX, Tsao SW, Lam KY, Srivastava G, Liu Y, Wong YC, Shen ZY, Cheung ALM (2001) E-cadherin expression is commonly downregulated by CpG island hypermethylation in esophageal carcinoma cells. Cancer Lett 173:71–78

Silye R, Karayiannakis AJ, Syrigos KN, Poole S, VanNoorden S, Batchelor W, Regele H, Sega W, Boesmueller H, Krausz T, Pignatelli M (1998) E-cadherin/catenin complex in benign and malignant melanocytic lesions. J Pathol 186:350–355

Simonneau L, Kitagawa M, Suzuki S, Thiery JP (1995) Cadherin 11 expression marks the mesenchymal phenotype: towards new functions for cadherins? Cell Adhes Commun 3:115–130

Slaton JW, Karashima T, Perrotte P, Inoue K, Kim SJ, Izawa J, Kedar D, McConkey DJ, Millikan R, Sweeney P, Yoshikawa C, Shuin T, Dinney CP (2001) Treatment with low-dose interferon-alpha restores the balance between matrix metalloproteinase-9 and E-cadherin expression in human transitional cell carcinoma of the bladder. Clin Cancer Res 7:2840–2853

Smits R, Ruiz P, DiazCano S, Luz A, JagmohanChangur S, Breukel C, Birchmeier C, Birchmeier W, Fodde R (2000) E-cadherin and adenomatous polyposis coli mutations are synergic in intestinal tumor initiation in mice. Gastroenterology 119:1045–1053

Soares P, Berx G, van Roy F, Sobrinho-Simões M (1997) E-cadherin gene alterations are rare events in thyroid tumors. Int J Cancer 70:32–38

Soler AP, Knudsen KA, Tecson-Miguel A, McBrearty FX, Han AC, Salazar H (1997) Expression of E-cadherin and N-cadherin in surface epithelial-stromal tumors of the ovary distinguishes mucinous from serous and endometrioid tumors. Hum Pathol 28:734–739

Soler AP, Knudsen KA, Salazar H, Han AC, Keshgegian AA (1999) P-cadherin expression in breast carcinoma indicates poor survival. Cancer 86:1263–1272

Somasiri A, Howarth A, Goswami D, Dedhar S, Roskelley CD (2001) Overexpression of the integrin-linked kinase mesenchymally transforms mammary epithelial cells. J Cell Sci 114:1125–1136

Spagnoli FM, Cicchini C, Tripodi M, Weiss MC (2000) Inhibition of MMH (Met murine hepatocyte) cell differentiation by TGF(beta) is abrogated by pre-treatment with the heritable differentiation effector FGF1. J Cell Sci 113:3639–3647

Spath GF, Weiss MC (1998) Hepatocyte nuclear factor 4 provokes expression of epithelial marker genes, acting as a morphogen in dedifferentiated hepatoma cells. J Cell Biol 140:935–946

St Croix B, Sheehan C, Rak JW, Florenes VA, Slingerland JM, Kerbel RS (1998) E-cadherin-dependent growth suppression is mediated by the cyclin-dependent kinase inhibitor p27(KIP1). J Cell Biol 142:557–571

St Croix BS, Kerbel RS (1997) Cell adhesion and drug resistance in cancer. Curr Opin Oncol 9:549–556

Stockinger A, Eger A, Wolf J, Beug H, Foisner R (2001) E-cadherin regulates cell growth by modulating proliferation- dependent beta-catenin transcriptional activity. J Cell Biol 154:1185–1196

Strumane K, van Roy F, Berx G (2003) The role of E-cadherin in epithelial differentiation and cancer progression. In: Pandalai SG (ed) Recent research development in cellular biochemistry. Transworld Research Network, Kerala, pp 33–77

Sulzer MA, Leers MPG, vanNoord JA, Bollen ECM, Theunissen PHMH (1998) Reduced E-cadherin expression is associated with increased lymph node metastasis and unfavorable prognosis in non-small cell lung cancer. Amer J Respir Crit Care Med 157:1319–1323

Sundfeldt K, Piontkewitz Y, Ivarsson K, Nilsson O, Hellberg P, Brannstrom M, Janson PO, Enerback S, Hedin L (1997) E-cadherin expression in human epithelial ovarian cancer and normal ovary. Int J Cancer 74:275–280

Suzuki H, Komiya A, Emi M, Kuramochi H, Shiraishi T, Yatani R, Shimazaki J (1996) Three distinct commonly deleted regions of chromosome arm 16q in human primary and metastatic prostate cancers. Gene Chromos Cancer 17:225–233

Taddei I, Piazzini M, Bartoletti R, Dal Canto M, Sardi I (2000) Molecular alterations of E-cadherin gene: possible role in human bladder carcinogenesis. Int J Mol Med 6:201–208

Takahashi K, Suzuki K (1996) Density-dependent inhibition of growth involves prevention of EGF receptor activation by E-cadherin-mediated cell–cell adhesion. Exp Cell Res 226:214–222

Takeuchi T, Liang SB, Ohtsuki Y (2002a) Downregulation of expression of a novel cadherin molecule, T- cadherin, in basal cell carcinoma of the skin. Mol Carcinogen 35:173–179

Takeuchi T, Liang SB, Matsuyoshi N, Zhou SX, Miyachi Y, Sonobe H, Ohtsuki Y (2002b) Loss of T-cadherin (CDH13, H-cadherin) expression in cutaneous squamous cell carcinoma. Lab Invest 82:1023–1029

Tamura G, Yin J, Wang S, Fleisher AS, Zou TT, Abraham JM, Kong DH, Smolinski KN, Wilson KT, James SP, Silverberg SG, Nishizuka S, Terashima M, Motoyama T, Meltzer SJ (2000) E-cadherin gene promoter hypermethylation in primary human gastric carcinomas. J Natl Cancer Inst 92:569–573

Tamura G, Sato K, Akiyama S, Tsuchiya T, Endoh Y, Usuba O, Kimura W, Nishizuka S, Motoyama T (2001) Molecular characterization of undifferentiated-type gastric carcinoma. Lab Invest 81:593–598

Tan C, Costello P, Sanghera J, Dominguez D, Baulida J, de Herreros AG, Dedhar S (2001) Inhibition of integrin linked kinase (ILK) suppresses beta- catenin-Lef/Tcf-dependent transcription and expression of the E-cadherin repressor, Snail, in APC−/− human colon carcinoma cells. Oncogene 20:133–140

Testa JR, Bellacosa A (2001) AKT plays a central role in tumorigenesis. Proc Natl Acad Sci USA 98:10983–10985

Thiery JP (2002) Epithelial-mesenchymal transitions in tumour progression. Nat Rev Cancer 2:442–554

Tomlinson JS, Alpaugh ML, Barsky SH (2001) An intact overexpressed E-cadherin/alpha,beta-catenin axis characterizes the lymphovascular emboli of inflammatory breast carcinoma. Cancer Res 61:5231–5241

Torban E, Goodyer PR (1998) Effects of PAX2 expression in a human fetal kidney (HEK293) cell line. Biochim Biophys Acta 1401:53–62

Toyooka KO, Toyooka S, Virmani AK, Sathyanarayana UG, Euhus DM, Gilcrease M, Minna JD, Gazdar AF (2001) Loss of expression and aberrant methylation of the CDH13 (H- cadherin) gene in breast and lung carcinomas. Cancer Res 61:4556–4560

Toyooka S, Toyooka KO, Harada K, Miyajima K, Makarla P, Sathyanarayana UG, Yin J, Sato F, Shivapurkar N, Meltzer SJ, Gazdar AF (2002) Aberrant methylation of the CDH13 (H-cadherin) promoter region in colorectal cancers and adenomas. Cancer Res 62:3382–3386

Tran NL, Adams DG, Vaillancourt RR, Heimark RL (2002) Signal transduction from N-cadherin increases Bcl-2. Regulation of the phosphatidylinositol 3-kinase/Akt pathway by homophilic adhesion and actin cytoskeletal organization. J Biol Chem 277:32905–32914

Tselepis C, Chidgey M, North A, Garrod D (1998) Desmosomal adhesion inhibits invasive behavior. Proc Natl Acad Sci USA 95:8064–8069

Tsujii M, DuBois RN (1995) Alterations in cellular adhesion and apoptosis in epithelial cells overexpressing prostaglandin endoperoxide synthase 2. Cell 83:493–501

Umbas R, Isaacs WB, Bringuier PP, Xue Y, Debruyne FMJ, Schalken JA (1997) Relation between aberrant alpha-catenin expression and loss of E-cadherin function in prostate cancer. Int J Cancer 74:374–377

Vallorosi CJ, Day KC, Zhao X, Rashid MG, Rubin MA, Johnson KR, Wheelock MJ, Day ML (2000) Truncation of the beta-catenin binding domain of E-cadherin precedes epithelial apoptosis during prostate and mammary involution. J Biol Chem 275:3328–3334

Van Aken E, De Wever O, Correia da Rocha AS, Mareel M (2001) Defective E-cadherin/catenin complexes in human cancer. Virchows Arch 439:725–751

Van Grunsven LA, Michiels C, Van De Putte T, Nelles L, Wuytens G, Verschueren K, Huylebroeck D (2003) Interaction between Smad-interacting protein-1 and the corepressor C-terminal binding protein is dispensable for transcriptional repression of E-cadherin. J Biol Chem 278:26135–26145

Vermeulen S, Bruyneel E, van Roy F, Mareel M, Bracke M (1995) Activation of the E-cadherin/catenin complex in human MCF-7 breast cancer cells by all-*trans*-retinoic acid. Br J Cancer 72:1447–1453

Vizirianakis IS, Chen YQ, Kantak SS, Tsiftsoglou AS, Kramer RH (2002) Dominant-negative E-cadherin alters adhesion and reverses contact inhibition of growth in breast carcinoma cells. Int J Oncol 21:135–144

Vleminckx K, Vakaet Jr L, Mareel M, Fiers W, van Roy F (1991) Genetic manipulation of E-cadherin expression by epithelial tumor cells reveals an invasion suppressor role. Cell 66:107–119

Vleminckx KL, Deman JJ, Bruyneel EA, Van den Bossche GMR, Keirsebilck AA, Mareel MM, van Roy FM (1994) Enlarged cell-associated proteoglycans abolish E-cadherin functionality in invasive tumor cells. Cancer Res 54:873–877

Vos CBJ, Cleton-Jansen A-M, Berx G, de Leeuw WJF, ter Haar NT, van Roy F, Cornelisse CJ, Peterse JL, van de Vijver MJ (1997) E-cadherin inactivation in lobular carcinoma in situ of the breast: an early event in tumorigenesis. Br J Cancer 76:1131–1133

Wang G, Huang CH, Zhao Y, Cai L, Wang Y, Xiu SJ, Jiang ZW, Yang S, Zhao T, Huang W, Gu JR (2000a) Genetic aberration in primary hepatocellular carcinoma: correlation between p53 gene mutation and loss-of-heterozygosity on chromosome 16q21-q23 and 9p21-p23. Cell Res 10:311–323

Wang JZ, Krill D, Torbenson M, Wang Q, Bisceglia M, Stoner J, Thomas A, De Flavia P, Dhir R, Becich MJ (2000b) Expression of cadherins and catenins in paired tumor and non-neoplastic primary prostate cultures and corresponding prostatectomy specimens. Urol Res 28:308–315

Washington K, Chiappori A, Hamilton K, Shyr Y, Blanke C, Johnson D, Sawyers J, Beauchamp D (1998) Expression of beta-catenin, alpha-catenin, and E-cadherin in Barrett's esophagus and esophageal adenocarcinomas. Modern Pathol 11:805–813

Wijnhoven BPL, de Both NJ, van Dekken H, Tilanus HW, Dinjens WNM (1999) E-cadherin gene mutations are rare in adenocarcinomas of the oesophagus. Br J Cancer 80:1652–1657

Wu CY, Keightley SY, Leung-Hagesteijn C, Radeva G, Coppolino M, Goicoechea S, McDonald JA, Dedhar S (1998) Integrin-linked protein kinase regulates fibronectin matrix assembly, E-cadherin expression, and tumorigenicity. J Biol Chem 273:528–536

Yabuta T, Shinmura K, Tani M, Yamaguchi S, Yoshimura K, Katai H, Nakajima T, Mochiki E, Tsujinaka T, Takami M, Hirose K, Yamaguchi A, Takenoshita S, Yokota J (2002) E-cadherin gene variants in gastric cancer families whose probands are diagnosed with diffuse gastric cancer. Int J Cancer 101:434–441

Yang SZ, Kohno N, Kondo K, Yokoyama A, Hamada H, Hiwada K, Miyake M (1999) Adriamycin activates E-cadherin-mediated cell–cell adhesion in human breast cancer cells. Int J Oncol 15:1109–1115

Yi JY, Hur KC, Lee E, Jin YJ, Arteaga CL, Son YS (2002) TGF-beta1-mediated epithelial to mesenchymal transition is accompanied by invasion in the SiHa cell line. Eur J Cell Biol 81:457–468

Yoo J, Park S, Kang CS, Kang SJ, Kim YK (2002) Expression of E-cadherin and p53 proteins in human soft tissue sarcomas. Arch Pathol Labor Med 126:33–38

Yoon KA, Ku JL, Yang HK, Kim WH, Park SY, Park JG (1999) Germline mutations of E-cadherin gene in Korean familial gastric cancer patients. J Hum Genet 44:177–180

Yoshiura K, Kanai Y, Ochiai A, Shimoyama Y, Sugimura T, Hirohashi S (1995) Silencing of the E-cadherin invasion-suppressor gene by CpG methylation in human carcinomas. Proc Natl Acad Sci USA 92:7416–7419

Zhou RX, Skalli O (2000) Identification of cadherin-11 down-regulation as a common response of astrocytoma cells to transforming growth factor-alpha. Differentiation 66:165–172

In Vivo Functions of Catenins

T. Brabletz

Institute of Pathology, University of Erlangen-Nürnberg, Krankenhausstr. 8-10, 91054 Erlangen, Germany
thomas.brabletz@patho.imed.uni-erlangen.de

1	The E-Cadherin/Catenin Complex.	106
2	β-Catenin	107
2.1	The Wnt/Wingless Pathway.	109
2.2	Physiological Roles of β-Catenin and Wnt Signaling: Embryonic Development and Adult Tissue Homeostasis	109
2.3	Roles of β-Catenin in Disease	111
2.3.1	β-Catenin and Cancer.	111
2.3.1.1	Colorectal Cancer	112
2.3.1.2	Other Cancers Associated with β-Catenin or Wnt Pathway Components.	121
2.3.2	Other Diseases Associated with β-Catenin	122
2.3.2.1	Neurodegenerative Diseases	122
2.3.2.2	Therapeutical Options	122
3	Physiological and Pathological Roles of Other Catenins	123
3.1	α-Catenin	123
3.2	γ-Catenin.	124
3.3	p120 Catenin.	126
4	Conclusions	127
References		128

Abstract The adhesion of cells to neighbor cells determines cellular and tissue morphogenesis and regulates major cellular processes including motility, growth, survival, and differentiation. Regions of cell–cell adhesion are adherens junctions, desmosomes, and tight junctions. Cadherins are transmembrane molecules whose extracellular domains transmit the direct interaction of two cells. The intracellular cadherin domains bind directly or indirectly to the submembranous catenins, which are linked to the cytoskeleton. Four types of catenins, α-catenin, β-catenin, γ-catenin, and p120 catenin are known. Three of them, β-, γ-, and p120 catenin, are structurally related and possess similar protein interaction domains, the so-called armadillo repeats. These catenins are also parts of signal transduction pathways and play a role in phenotypical changes of cells, e.g., during switches from adherent to migratory cells. The function of catenins in such basic cellular processes also determines a role of catenins in embryogenesis, adult tissue homeostasis, and disease. In particular, β-catenin is known to be an important oncoprotein in human cancer development.

Keywords Catenin · Wnt signaling · APC · Colorectal cancer · Tumor progression · E-cadherin

1
The E-Cadherin/Catenin Complex

Cell–cell contacts are made by adherens junctions, desmosomes, and tight junctions. Determinants of an epithelial phenotype are homophilic cell adhesions and cellular polarity, defining basal and apical orientation. Both are

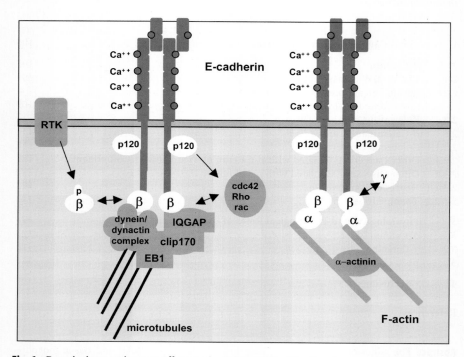

Fig. 1. Protein interactions at adherens junctions (AJ). AJs allow homophilic cell–cell adhesion through direct, Ca^{++}-dependent interaction of E-cadherin molecules. E-cadherin is indirectly linked to the actin and microtubule cytoskeletons, through associated proteins, which is essential for cell–cell adhesion. E-cadherin's direct interaction partner β-catenin (β) binds to α-catenin (α) and links cadherin/catenin complexes to the actin cytoskeleton. Thereby α-catenin is the central player in the linkage to F-actin, a process critical for coordinating actin dynamics in the cell. β-Catenin binds to the E-cadherin cytoplasmic tail via 12 so-called armadillo repeats. The affinity for this key interaction is increased by phosphorylation of several serine residues in the cadherin tail and reduced by phosphorylation (p) of β-catenin Y654, a known site of action for activated growth factor receptor tyrosine kinases (RTK). Thus, through posttranslational modifications, the strength of the AJ complex can be adapted to the particular needs of the epithelial cell within the context of its tissue. β-Catenin and γ-catenin (γ) compete for binding to E-cadherin and α-catenin. β-Catenin also interacts with microtubule-associated proteins such as IQGAP and the dynein/dynactin complex. β-Catenin and IQGAP also crossreact with the Rho/rac-family of small GTPases. p120 Catenin (*p120*) binds independently to E-cadherin and promotes E-cadherin clustering; p120 catenin has also been suggested to regulate the guanosine triphosphatase (GTPase) Rho

mediated by cell surface expression of the adherens junction molecule E-cadherin. Loss of these epithelial characteristics may be due to loss of E-cadherin function and can indicate a switch towards a dedifferentiated, mesenchyme-like phenotype (Hirohashi 1998; Perl et al. 1998). The E-cadherin/catenin complex is the main molecular complex of adherens junctions (Fig. 1). It not only mediates homophilic interaction between epithelial cells, but some components are both active parts of a signal transduction pathways (see Sects. 4 and 22) and themselves exposed to regulatory signals (Van Aken et al. 2001; Perez-Moreno et al. 2003). E-cadherin binds directly to β-catenin, γ-catenin (plakoglobin), and p120 catenin, and is indirectly linked to actin and the cytoskeleton via α-catenin. β-Catenin participates in the correct positioning and function of E-cadherin (Barth et al. 1997); thus, membranous expression of both proteins determines the epithelial phenotype. E-cadherin's direct interacting partner, β-catenin, also binds to several proteins. However, intercellular adhesion in β-catenin-deficient mouse embryos is maintained, since γ-catenin can substitute this function of β-catenin (Huelsken et al. 2000). β-Catenin associates with α-catenin and links cadherin/catenin complexes to the actin cytoskeleton. Its ability to interact with some microtubule-associated proteins such as IQGAP, adenomatous polyposis coli (APC), and the dynein/dynactin complex may link E-cadherin to the microtubule network (Perez-Moreno et al. 2003).

Catenins are not only components of adherens junctions but also of desmosomes, where γ-catenin mediates the interaction between desmosomal cadherins and the intermediate filament system (Fuchs et al. 1998).

2
β-Catenin

Among the catenins, β-catenin plays a particular role, since it is fundamentally involved in both developmental processes and pathomechanisms of various diseases, in particular cancer. Therefore, this section about the regulation of the different β-catenin functions is intended to help place into context the following discussions about its physiological and pathological roles. Besides its function in cell–cell adhesion as an interacting partner of E-cadherin, a second, completely different function of β-catenin was defined: In a nuclear pool, β-catenin interacts with DNA-binding proteins of the Tcf/LEF (T cell factor/lymphocyte enhancer factor)-family and acts as a transcriptional activator (Behrens et al. 1996; Molenaar et al. 1996) in the so-called canonical Wnt/wingless pathway. This pathway is highly conserved between Drosphila and vertebrates and a potent regulator of early embryonic and organ development (Cadigan et al. 1997). Thus, the intracellular distribution of β-catenin is of great importance for the different functions of β-catenin and the subsequent behavior of differentiating epithelial cells or tumor cells.

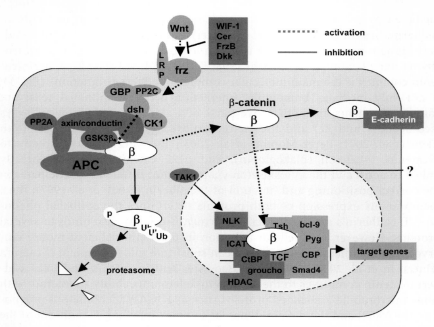

Fig. 2. The Wnt signaling pathway. Wnts are secreted glycoproteins that bind to and activate frizzled (*frz*) seven-transmembrane-span receptors. Low-density lipoprotein receptor-related proteins LRP5/6 act as an essential co-receptor. Various secreted factors, like WIF-1, Cerberus (*cer*), FrzB and Dickkopf (*Dkk*), antagonize this interaction. Wnt signaling leads to a stabilization of cytoplasmic β-catenin, the main effector of the Wnt pathway. In the absence of Wnts, β-catenin is phosphorylated (*p*) at the N-terminal serine and threonine residues 33, 37, 41, and 45 by glycogen synthase kinase 3β (*GSK3β*), which triggers ubiquitination and subsequent degradation in proteasomes. This is only possible in a multiprotein complex consisting of APC and the scaffolding protein axin/conductin. In the presence of Wnts, dishevelled (*dsh*) blocks β-catenin degradation, possibly by recruiting the GSK3β inhibitor GBP. β-Catenin degradation is modulated by the casein kinase CK1 and the protein phosphatases PP2A and PP2C. Stabilized β-catenin either is recruited to adherens junctions or, induced by unknown signals (?), accumulates in the nucleus. There it exerts its oncogenic function as a transcriptional activator after association with DNA-binding proteins of the Tcf/LEF-family of transcription factors. Coactivators, such as CBP, pygopus (*Pyg*), bcl-9, teashirt (*Tsh*) support the activation of target genes (see Table 1). Phosphorylation of Tcfs by NF-κB essential modulator (NEMO)-like kinase (*NLK*), a target of the MAP kinase kinase kinase TAK1, as well as interaction with inhibitor of β-catenin (ICAT) negatively regulate β-catenin transcriptional activity. In the absence of of β-catenin, certain Tcfs suppress target gene transcription by interacting with the corepressors C-terminal-binding protein (*CtBP*) and histone deacetylase (*HDAC*) bound to groucho. Interaction with Smad4 might connect the Wnt and transforming growth factor (TGF)-β pathways. (Modified from Huelsken and Behrens 2002)

One of the fundamental results in recent colon cancer research was the demonstration that the APC tumor suppressor protein interacts with β-catenin (Rubinfeld et al. 1993) and is a negative regulator of the Wnt pathway (Fig. 2) (Huelsken et al. 2002).

2.1
The Wnt/Wingless Pathway

Nuclear β-catenin is the main downstream effector molecule of the Wnt pathway. It has no DNA-binding domain but builds up a transcriptional activator complex together with Tcf/LEF family members and various cofactors (see reviews by J. Behrens 1999, Bienz et al. 2000, and Sieber et al. 2000). Tcf/LEF molecules are DNA-binding proteins which recognize a consensus promoter sequence (TCF-binding sites: WWCAAAG), but do not have a strong intrinsic transcriptional activator function (Clevers et al. 1997). In the absence of Wnt signaling, Tcfs are bound by transcriptional repressors like groucho, which keep the target gene promoter inactive (Cavallo et al. 1998). Nuclear accumulation of β-catenin displaces groucho from Tcf-binding. β-Catenin/Tcf plays a role in chromatin remodeling and leads to changes in promoter architecture, which makes the promoter accessible for other transcription factors. Thus, TCFs and the corresponding promoter binding sites play a key role in transcriptional regulation by integrating simultaneous signaling by various pathways (Riese et al. 1997). In the absence of Wnt signaling, β-catenin is degraded after phosphorylation by the glycogen synthase kinase3β (GSK3β) in a multiprotein complex. The main components of this complex are the tumor suppressor APC and the scaffolding protein axin/conductin. Only if these molecules are functional, the amount of β-catenin can be efficiently regulated. Loss of function of either APC or axin/conductin leads to a reduced degradation and subsequent overexpression of free cytoplasmic β-catenin, which can exert its nuclear function without efficient control. Since many target genes of β-catenin are themselves oncogenes (see Sect. 8), both APC and axin/conductin function as tumor suppressor. Physiologically Wnt signaling is induced by binding of the soluble Wnt-factors to their membrane receptors of the frizzled-(frz) family, which leads to an inactivation of GSK3β and subsequent accumulation of β-catenin. However, although active Wnt signaling or APC mutation in tumor cells leads to accumulation of the cytoplasmic, free pool of β-catenin, it is not fully understood what activates the decisive nuclear translocation of β-catenin. A complex interaction with other pathways might be responsible (see Sect. 8).

2.2
Physiological Roles of β-Catenin and Wnt Signaling:
Embryonic Development and Adult Tissue Homeostasis

Wnt signaling, characterized by nuclear accumulation and function of β-catenin, regulates fundamental embryonic processes. It was shown to induce epithelial-mesenchymal transitions (EMT) during the gastrulation in sea urchin (Logan et al. 1999) and also in human cell culture systems (Eger et al. 2000; Morali et al. 2001). Targeted disruption of β-catenin in mice leads to very early embryonic lethality by affecting development at gastrulation (Haegel et al. 1995). Overactivation of the Wnt pathway by injection of β-catenin in susceptible cells leads to an axis duplication in Xenopus embryos (Molenaar et al. 1996), indicating its regulatory role in axis formation.

Moreover Wnt signaling regulates differentiation and development of various organs in fetal development. In skin development, β-catenin is involved in hair follicle morphogenesis and control of the epidermal stem cell compartment. The findings suggest that transient β-catenin stabilization may be a key player in the long-sought epidermal signal leading to hair development and implicate aberrant β-catenin activation in hair tumors (Gat et al. 1998; Huelsken et al. 2001). Wnt signaling is required for thymocyte development and activates Tcf-1-mediated transcription (Staal et al. 2001). Also, β-catenin plays an important role in neural development: Inactivation of the β-catenin gene in mice by Wnt1–Cre-mediated deletion results in dramatic brain malformation and failure of craniofacial development (Brault et al. 2001). The correct development of various other tissues and organs, like in angiogenesis (Ishikawa et al. 2001) and adipogenesis (Ross et al. 2000), also depends on Wnt signaling and β-catenin. During bone and joint development (Enomoto-Iwamoto et al. 2002), β-catenin is activated through the lipoprotein receptor related protein (LRP5) as Wnt coreceptor (Gong et al. 2001).

Of particular interest for tumor formation (see Sect. 8) is, that the Wnt pathway is also essential for intestinal development. Thereby, β-catenin binds an intestine- and mammary epithelial-specific Tcf-family member, termed Tcf-4 (Barker et al. 1999). Tcf-4 expression characterizes the intestinal stem cell compartment, and targeted disruption of Tcf-4 in mice leads to a severe disturbance of gut development (Korinek et al. 1998).

The molecular role of β-catenin in the development of the above organs and tissues can be deduced by the functions of its defined target genes. Many of these target genes have the ability to support stem cell formation, and in a current model, β-catenin defines the stem-cell compartment (Taipale et al. 2001). This is also supported by a lack of stem cells after targeted depletion of β-catenin in skin (Huelsken et al. 2001) and gut (Korinek et al. 1998).

The function of β-catenin in characterizing the stem cell compartment is also important in the adult organism. Thereby β-catenin maintains tissue homeostasis, in particular in strong-proliferative, self-renewing tissues, like skin, hair, and intestine.

Compiling all data, it is becoming evident that Wnt signaling and nuclear β-catenin as its main effector do not simply control singular events, but regulate the complex process of morphogenesis, which needs a temporal and spatial coordination of individual processes such as cell–cell attachment, migration, proliferation, and differentiation.

2.3
Roles of β-Catenin in Disease

2.3.1
β-Catenin and Cancer

The many functions of β-catenin as the main effector of the Wnt pathway indicate a potential danger of dysregulated nuclear accumulation and activation of β-catenin. In particular its possible role as a determinator of stem

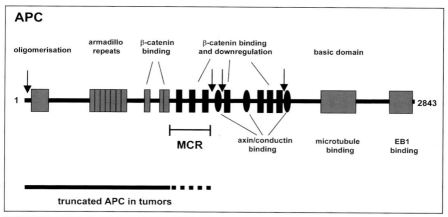

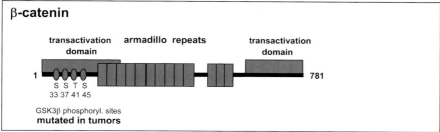

Fig. 3. Domaine structure of APC and β-catenin. APC binds β-catenin with its 15 aa repeats (*gray*) and 20 aa repeats (*dark*). The armadillo repeat domain is used for binding to other proteins, such as the Rac-specific guanine nucleotide exchange factor Asef. Shown are also nuclear export signals (*arrows*). In most tumors, point mutations are found in a small area called the mutation cluster region (MCR), leaving a truncated protein. The selective pressure might be directed against the presence of axin/conductin binding sites, and therefore against an efficient downregulation of β-catenin. Of note, truncated APC proteins in tumors also lost the most efficient nuclear export signal (NES), but retain the armadillo repeats, which might be important for a dominant, oncogenic function of the mutated APC protein. The second APC allele is either point mutated or completely lost. β-Catenin binds E-cadherin, APC, TCF/LEF, and different transcriptional co-factors at its armadillo repeats. Oncogenic mutations in tumors (with normal APC) are found at one of the GSK3β phosphorylation sites, resulting in a reduced degradation and overexpression of β-catenin

cell characteristics and as a mediator between an epithelial phenotype (β-catenin in adherens junctions) and a de-differentiated, mesenchyme-like phenotype (nuclear β-catenin), predisposes β-catenin as a potential oncoprotein.

Two main types of mutations can lead to an overexpression and oncogenic activation of β-catenin, and are in fact detectable in human cancers, in particular colorectal cancer: loss of function mutations of the APC gene and activating, oncogenic mutations in the β-catenin gene (Bienz et al. 2000; Polakis 2000; Bienz et al. 2003) (Fig. 3).

2.3.1.1
Colorectal Cancer

Colorectal cancer is one of the most common cancers in developed countries. The decisive genetic alteration, which is found in more than 80% of sporadic colorectal carcinomas (Kinzler et al. 1996) and was initially discovered as the germline mutation in the hereditary colorectal cancer syndrome called "familial adenomatous polyposis (FAP)", is the loss of function mutation of the APC tumor suppressor gene (Groden et al. 1991; Kinzler et al. 1991). Its causal nature in colorectal carcinogenesis was demonstrated in different mouse models (Kucherlapati et al. 2001). A natural mutant, which was called Min-mouse and has a truncating mutation at APC aa 850, develops multiple intestinal neoplasia (Su et al. 1992). Homozygous deletions of APC are embryonic lethal but the heterozygous phenotypes, such as targeted deletions at aa 1638, (Fodde et al. 1994) and aa 716 (Oshima et al. 1995), often develop intestinal tumors. A rapid colorectal adenoma formation is initiated by conditional targeting of the APC gene in the intestine (aa 580) (Shibata et al. 1997). One of the fundamental results in colon cancer research was the demonstration that the APC protein interacts with β-catenin (Rubinfeld et al. 1993) and is essential for its degradation, which accordingly can accumulate in APC-mutant tumor cells (Korinek et al. 1997). The demonstration that β-catenin is the main effector of the canonical Wnt pathway and that oncogenic β-catenin mutations can be found in colorectal tumors without APC mutations brought this molecule into the limelight of cancer research.

Expression of β-Catenin in Colorectal Carcinomas. Given the two important, contrary roles of β-catenin in either the E-cadherin-dependent determination of the epithelial phenotype (membranous localization) or as transcriptional regulator and main effector of the Wnt pathway (nuclear localization), its function as a oncoprotein in APC-mutant colorectal cancers takes shape. Indeed, nuclear β-catenin is detectable in human FAP-associated and sporadic colon adenomas and adenocarcinomas (Inomata et al. 1996). However, the amount of nuclear β-catenin increases from early adenomas to adeno-

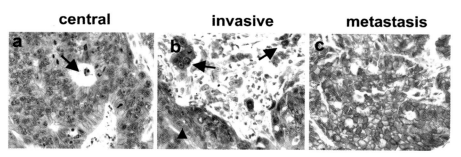

Fig. 4a–c. Expression of β-catenin in colorectal carcinomas. Shown is an immunohistochemical staining of β-catenin (specific staining in *brown*, nuclear counterstaining in *blue*) in a colorectal adenocarcinoma (**a** and **b**) and a corresponding liver metastasis (**c**). In central tumor areas (**a**) an epithelial differentiation, characterized by polarized tumor cells forming tubular structures (*arrow*), is retained. These tumor cells, like normal colon epithelial cells, express membranous β-catenin. In invasive regions, dedifferentiated tumor cells detach from the primary tumor into the surrounding tissue, lose membranous but accumulate nuclear β-catenin (*arrows*). Tumor cells towards the tumor center still lack nuclear β-catenin (*arrowhead*). In growing metastases tumor cells regain their epithelial phenotype, lack nuclear but reexpress membranous β-catenin (**c**)

carcinomas (Brabletz et al. 2000), and its distribution within an individual tumor is very heterogeneous (Fig. 4). In most well-differentiated to moderately differentiated colon adenocarcinomas, nuclear β-catenin is predominantly accumulated in dedifferentiated tumor cells at the invasive front, whereas in central differentiated areas it is located at the membrane, and nuclear accumulation is hardly detectable (Brabletz et al. 1998; Kirchner et al. 2000). Since all tumor cells in an individual tumor harbor APC mutations, a nuclear accumulation of β-catenin can not be due to this alteration alone, but its intracellular distribution within different tumor areas has to be explained by additional events.

Three questions arise from these observations, which may have strong impact on tumor progression:

1. What are the effects of nuclear β-catenin in the dedifferentiating carcinoma cells and is there a direct influence on invasion and metastasis?
2. What regulates the heterogeneous intracellular distribution of β-catenin within the tumors?
3. What is the role of β-catenin in early colorectal carcinogenesis?

Effects of β-Catenin in Carcinoma Cells. Since nuclear β-catenin is a transcriptional activator, the identification of its target genes, characterized by Tcf-binding elements in their promoters, is of particular interest to understand its role in tumor progression (Table 1). The first identified genes regulated by β-catenin/TCF in cancers are the well-defined oncogenes *c-myc* (Batsche

Table 1. β-Catenin target genes relevant for cancer

Target gene	Function
c-myc	Proliferation
Cyclin D1	
c-jun	Oncogenic transcription factors
ets2	
fra-1	
ITF-2	
MMP-7	Protein degradation
MMP-26	
MT1-MMP	
UPA-R	
VEGF	Angiogenesis
BMP-4 ephrinB2/B3	Morphogenesis
Laminin γ-2 chain fibronectin	Migration
CD44	Dissemination
Cdx1	Loss of differentiation
Id2	
Enc-1	
Gastrin PPARdelta	Trophic factors
MDR survivin	Cell survival
Conductin/axin-2 Tcf-1	Negative feedback and Tumor suppression

et al. 1998) and *cyclin D1* (Shtutman et al. 1999; Tetsu et al. 1999), linking dysregulated β-catenin activity to dysregulated proliferation. Also, the *gastrin* gene, which is discussed to be a trophic factor for intestinal tumor growth, is activated by nuclear β-catenin (Koh et al. 2000). Overexpression of peroxisome proliferator-activated receptor (PPAR)δ, a member of the nuclear receptor family, which is bound and activated by fatty acids and therefore a potential mediator of dietary effects on colon carcinogenesis, is due to transcriptional activation by β-catenin/T cell factor (TCF) (He et al. 1999). Other target genes are *MDR1* (Yamada et al. 2000) and *survivin* (Zhang et al. 2001a), which are thought to suppress cell death pathways.

Moreover it was shown that activated β-catenin is involved in dedifferentiation of epithelial cells (Mariadason et al. 2001; Naishiro et al. 2001), which indicates a causal role of nuclear β-catenin in the phenotypical switch towards dedifferentiated tumor cells at the invasive front. This view is supported by an increasing number of target genes, which are known to code for regulators of differentiation and effectors supporting invasion and dissemination (Fig. 3). Recently *Cdx-1*, encoding a homeobox-factor (Lickert et al. 2000), *Id2* (inhibitor of differentiation-2) (Rockman et al. 2001), and *ENC1* (Fujita et al. 2001) were identified as β-catenin/TCF target genes. All three proteins inhibit epithelial differentiation and keep cells in a less differentiated, stem cell-like state. An activation of such genes by nuclear β-catenin could explain the dedifferentiated phenotype of nuclear β-catenin-expressing colon cancer cells at the invasive front of colorectal adenocarcinomas.

Other genes regulated by nuclear β-catenin code for direct effectors of colon cancer progression, like *urokinase-receptor (uPAR)* (Dihlmann et al. 1999), matrix metalloproteinase (MMP)-7/*matrilysin* (Brabletz et al. 1999; Crawford et al. 1999), *MT1-MMP* (Takahashi et al. 2002), *MMP-26* (Marchenko et al. 2002), *c-jun* (Dihlmann et al. 1999), *ets2* (Barker et al. 2001), *VEGF* (Zhang et al. 2001b), *fibronectin* (Gradl et al. 1999), *laminin-5 γ2 chain* (Hlubek et al. 2001), *bone morphogenetic protein 4* (Kim et al. 2002), *ITF-2* (Kolligs et al. 2002), and *CD44* (Wielenga et al. 1999). Both uPAR and matrilysin are overexpressed by the tumor cells and facilitate extracellular matrix proteolysis, which allows detachment and motility enhancement of the tumor cells. The isolated γ2 chain of laminin-5 is overexpressed selectively in dedifferentiated carcinoma cells at the invasive front and is known to be one of the most potent inducers of epithelial cell migration, e.g., in wound healing and embryonic development. The known oncoprotein c-Jun, a component of the transcription factor AP-1, is itself another strong transcriptional activator of invasion factors like uPAR, matrilysin, and laminin-5 γ2. A similar role is described for Ets-transcription factors (Gaire et al. 1994). Another important process in tumor growth and invasion is the generation of surrounding tumor stroma. The stroma and the stromal cells participate directly in tumor growth and invasion by producing various degrading enzymes like MMPs, by storing cytokines, and also by supplying the tumor with blood vessels (Liotta et al. 2001). Both vascular endothelial growth factor (VEGF) and basic fibroblast growth factor (bFGF), activated by nuclear β-catenin, are cytokines involved in the generation of the tumor stroma and tumor angiogenesis (Galzie et al. 1997). Splice variants of CD44 (e.g., v6) are known to directly support dissemination of isolated tumor cells and are associated with the presence of metastases and an unfavorable prognosis of colorectal cancer (Ropponen et al. 1998). Finally, a translocation of membranous β-catenin to the nucleus leads to a loss of E-cadherin function. This further allows detachment of tumor cells from epithelial cell complexes and supports the loss of epithelial features.

It was further shown that β-catenin/Tcf inversely control the expression of the EphB2/EphB3 receptors and their ligand ephrin-B1 in colorectal cancer and along the crypt-villus axis. Disruption of *EphB2* and *EphB3* genes reveals that their gene products restrict cell intermingling and allocate cell populations within the intestinal epithelium. In *EphB2/EphB3*-null mice, the proliferative and differentiated populations intermingle. In adult $EphB3^{(-/-)}$ mice, Paneth cells do not follow their downward migratory path, but scatter along crypt and villus. This means that in the intestinal epithelium β-catenin/Tcf couple proliferation and differentiation to the sorting of cell populations through the EphB/ephrin-B system (Batlle et al. 2002).

Two potential Wnt target genes do not fit to the oncogenic function of all the others: *Tcf1* and *axin-2/conductin*. The most abundant Tcf1 isoforms lack a β-catenin interaction domain. Tcf1 may act as a feedback repressor of β-catenin/Tcf4 target genes and thus may cooperate with APC to suppress malignant transformation of epithelial cells (Roose et al. 1999). The same role as a feedback repressor has axin-2/conductin, which is also activated by β-catenin (Lustig et al. 2002).

There is also indirect evidence that β-catenin participates in the activation and expression of cyclooxygenase 2 (COX-2)(Howe et al. 1999). COX-2 activation is considered a main molecular event for colorectal carcinogenesis, and inhibition of COX-2 in APC delta716 knockout mice led to a suppression of intestinal polyposis (Oshima et al. 1996). These data imply that COX-2 also has a function in supporting the oncogenic role of β-catenin.

Taking all together, the strong oncogenic potency of nuclear β-catenin becomes evident. A cluster of genes, associated with a phenotypical switch from differentiated, epithelial towards dedifferentiated, mesenchyme-like tumor cells, are regulated by β-catenin/Tcf. Similar phenotypical transitions and subsequent migration are induced by nuclear β-catenin in epithelial cells in the blastula during epithelial to mesenchymal transition processes in embryonic gastrulation (Huang et al. 2000), and it is suggested that strong nuclear accumulation of β-catenin leads to dedifferentiation of epithelial cells (Mariadason et al. 2001) and might give the tumor cells a competence similar to embryonic epithelial cells (Kirchner et al. 2000). In addition, nuclear accumulation of β-catenin in tumor cells at the invasive front might be directly involved in the acquisition of the necessary abilities to detach, migrate, and disseminate in the body by increasing the morphogenetic competence of the tumor cells.

Regulators of Intracellular β-Catenin Localization and Their Potential Role in Malignant Tumor Progression. In well to moderately differentiated colorectal adenocarcinomas, a strong nuclear expression of β-catenin is found predominantly in dedifferentiated tumor cells at the invasive front (Brabletz et al. 1998; Kirchner et al. 2000). Another feature of these tumors is their retained epithelial differentiation, characterized by polarized tumor cells building up tubular structures, in central tumor areas. A strikingly similar differentiated phenotype of the primary tumor is found in corresponding lymph node or distant metastases (Brabletz et al. 2001) (Fig. 4), which means that the dedifferentiated phenotype, allowing the tumor cells to disseminate in the body, can only be transient and thus cannot be fixed by alterations in the genome of the tumor cells. Therefore, a main driving force inducing the obvious phenotypical dedifferentiation–redifferentiation switches, and thus potentially invasion and metastasis formation, must be the tumor environment acting on the genetically altered tumor cells. Recent observations demonstrated that the expression and localization of the two

Table 2. External factors modulating the intracellular distribution of β-catenin

Factors	Molecular effects	Consequences
Cytokines (TFFs, IGFs, EGF, HGF)	Tyrosine phosphorylation of β-catenin	Disruption of adherens junctions Nuclear accumulation of β-catenin
Extracellular matrix	Activation of ILK	Nuclear accumulation of β-catenin
MMPs	Activation of snail Cleavage of e.g. E-cadherin	Inhibition of E-cadherin transcription Disruption of adherens junctions

decisive molecules E-cadherin and β-catenin are coupled to these changing phenotypes (Brabletz et al. 2001). Membranous E-cadherin and β-catenin are found in both differentiated areas of primary colorectal carcinomas and their metastases, whereas the invasive areas of primary tumor and metastases show decreasing E-cadherin expression and nuclear β-catenin. This demonstrates a stronger difference between distinct morphogenetic areas within a tumor than between primary tumor and corresponding metastases. Since the enormous potential of nuclear β-catenin becomes evident (see above), the search for the driving force of the phenotypical switches may be focused on environmental factors which induce nuclear translocation of β-catenin.

Cell culture experiments revealed a direct or indirect role of environmental factors, including cytokines and extracellular matrix, on the intracellular β-catenin distribution and function (see Table 2). Intestinal trefoil factor (TFF3) (Liu et al. 1997), insulin-like growth factors (IGF I and IGF II) (Freier et al. 1999), epidermal growth factor (EGF) (Oyama et al. 1994), and hepatocyte growth factor, scatter factor (HGF) (Oyama et al. 1994) bind to receptors on the tumor cells. Activation of receptor tyrosine kinases leads to a tyrosine phosphorylation of β-catenin with subsequent perturbation of E-cadherin binding, loss of intercellular adhesion, and promotion of cell motility. By affecting the function of β-catenin, an overexpression of these cytokines in colon carcinomas is thought to modulate tumor cell adhesion and migration. Overexpression of IGF II enhanced colon tumor growth in a mouse model (Hassan et al. 2000) and induced a nuclear translocation of β-catenin coupled with an epithelial to mesenchymal transition in bladder and mammary carcinoma cell lines (Morali et al. 2001). However, the significance of these cytokines in regulating β-catenin function in human colorectal carcinomas has not been demonstrated yet.

Epithelial–mesenchymal interactions are decisive for intestinal development (Kedinger et al. 1998). Thus, mesenchymal factors, in particular components of the surrounding extracellular matrix (EM), could have a potent regulatory effect on tumor cells, which might have a reactive competence similar to embryonic epithelial cells, due to APC mutations affecting the Wnt pathway (Kirchner et al. 2000). In this context the perhaps most significant regulator of intracellular β-catenin distribution is the integrin linked kinase (ILK). ILK is a serine/threonine kinase which binds to intracellular domains of β1- and β3-integrins. After binding of EM proteins to their integrin receptors, ILK is activated and exerts various intracellular effects. One is the induction of a nuclear translocation of β-catenin and subsequent activation of the β-catenin/TCF transcriptional activator (Novak et al. 1998). Moreover, it was shown that ILK activation leads to an inhibition of E-cadherin transcription by stimulating the transcriptional repressor snail (Marotta et al. 2001). Thus ILK may directly be involved in the acquisition of the dedifferentiated phenotype of nuclear β-catenin expressing tumor cells at the invasive front. However, the relevant external stimulators are currently not known, and it will be of interest what particular EM proteins stimulate ILK activation. Also, other pathways known to be altered in colorectal cancers, like the PTEN/Akt (Persad et al. 2001) or the transforming growth factor (TGF)-β pathway (Nishita et al. 2000), are thought to interfere with the Wnt pathway and possibly the nuclear accumulation of β-catenin. However, the relevance of these interactions for colorectal carcinogenesis is still unclear.

Indirect effects by modulating β-catenin associated proteins also could be relevant for the intratumorous heterogeneity of β-catenin distribution. For instance, colon carcinomas show an ectopic nuclear overexpression of LEF1, which belongs to the Tcf/LEF-family of DNA-binding proteins. Like Tcf-4, LEF1 binds β-catenin, and it is thought that increasing nuclear LEF1 can trap β-catenin in the nucleus (Hovanes et al. 2001). E-cadherin may also be the direct target for environmental factors like MMP-7, which cleaves membranous E-cadherin (Noe et al. 2001), and of ILK activators, leading to a repression of E-cadherin transcription (Marotta et al. 2001). E-cadherin mutations are not common in colorectal cancers (Schuhmacher et al. 1999), but the decreased E-cadherin expression observed at the invasive front may be due to such environmental factors. Indirectly, this could increase the cytoplasmic free pool of β-catenin in addition to APC mutations, which subsequently can be translocated and trapped in the nucleus of the carcinoma cells.

Redifferentiation of Tumor Cells in Metastases. The dedifferentiated phenotype with nuclear β-catenin at the invasive front is not found in the central areas of most metastases, but tumor cells show the same differentiated epithelial phenotype as the primary tumor. This includes the generation of tumor stroma in the metastases, which might promote a reappearance of the epithelial tumor cell phenotype. An apparent drawback of the regain of the epithelial growth pattern and a retranslocation of β-catenin from the nucleus to the cytoplasm and membrane is the existence of a second phenotypic transition step during metastasis formation. Why should dedifferentiated, disseminating tumor cells undergo a redifferentiation? A reduction of proliferative activity in dissociating tumor cells expressing high amounts of nuclear β-ca-

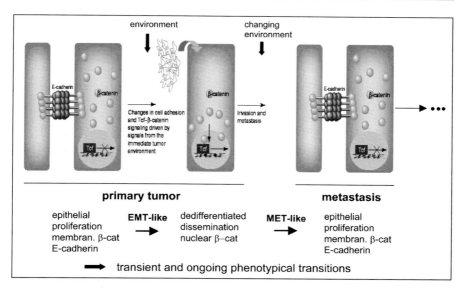

Fig. 5. Model of malignant progression of colorectal carcinomas. Environmental signals lead to a dedifferentiation similar to an epithelial–mesenchymal transition (*EMT-like*) at the invasive front, allowing tumor cells to disseminate in the body. Environmental signals at the invasive site induce an epithelial redifferentiation (*MET-like*), which leads to a metastasis, growing again in the same phenotype as the primary tumor. (Modified from Barker and Clevers 2001)

tenin was found (Brabletz et al. 2001). Although these tumor cells overexpress the β-catenin target gene *cyclin D1*, associated with proliferation, a parallel overexpression of the cell cycle inhibitor p16 was described, which could explain a proliferation arrest (Palmqvist et al. 2000; Jung et al. 2001). Interestingly, overexpression of p16 also seems to be initiated by nuclear β-catenin in tumor cells at the invasive front. Obviously, in well-differentiated tumors, a loss of epithelial capabilities is coupled with a shut-down of proliferation. Accordingly, in order to expand metastatic growth, disseminated dedifferentiated tumor cells of well-differentiated carcinomas must regain their epithelial function. In this view, tumor progression is driven by two forces: an increasing number of genetic alterations and a regulatory role of the changing tumor environment (Fig. 5).

Colorectal Cancers with Normal APC. Like the FAP-type of hereditary colorectal cancer, most sporadic colorectal carcinomas arise from adenomas and have APC mutations as initial genetic alterations. The second major group are the replication error (RER)-positive carcinomas, associated with microsatellite instability (MSI). Hereditary non-polyposis colorectal cancer (HNPCC) syndrome and about 10% of sporadic colorectal carcinomas fall into this group, which is also characterized by a different morphology and clinical prognosis. About half of these carcinomas show normal APC genes (Huang et al. 1996); however, the general importance of a dysregulated Wnt pathway and its main effector β-catenin in colorectal cancer formation is indicated by the fact that mutations in other components of this pathway, leading to enhanced β-catenin activity, are found in a high percentage of such tumors. In particular, dominant mutations at the target serine and threonine residues for GSK3-β in the β-catenin gene itself are found in 27% of microsatellite instable carcinomas, leading to a stabilization of the molecule (Shitoh et al. 2001).

Loss of function mutations in the *conductin/axin-2* gene were demonstrated in 25% of MSI carcinomas, which similar to APC mutations prevent degradation of β-catenin (Liu et al. 2000). Moreover, inactivating frameshift mutations in the Tcf-4 gene were found in 39% of human microsatellite instable colon carcinomas (Duval et al. 1999). The important role of the dysregulated Wnt pathway in colon carcinogenesis is also indicated by the fact that ulcerative colitis-associated colon carcinomas also show APC gene mutations and nuclear accumulation of β-catenin (Aust et al. 2001). Thus, a dysregulation of the Wnt pathway by targeting APC, β-catenin, or other components is found in almost all colorectal carcinomas.

Therapeutical Options. COX-2 function is inhibited by nonsteroidal antiinflammatory drugs (NSAIDs), including aspirin. An impressive body of epidemiological data suggests an inverse relationship between colorectal cancer risk and regular use of NSAIDS. Equally exciting are opportunities for effective chemoprevention with selective COX-2 inhibitors including celecoxib and rofecoxib in a variety of preclinical models of colon cancer (Reddy et al. 2002).

Another therapeutical option is to target the oncogenic functions of β-catenin in cancer. The core regions of β-catenin for protein–protein interaction are its 12 central armadillo repeats, through which it interacts with cadherins, Tcf/LEF family members, and APC in a mutually exclusive way (Gottardi et al. 2001) (Fig. 3). Since an interaction of β-catenin with APC would have a tumor-suppressive effect, an efficient inhibition of its oncogenic function could be achieved by a selective interference of β-catenin binding to Tcf/LEF. However, given the remarkable similarity between the β-catenin–E-cadherin and the β-catenin–TCF binding complexes, it might be anticipated that designing a highly specific inhibitor of the latter complex will be difficult (Daniels et al. 2001). More detailed thermodynamic and structural data on ligand recognition by β-catenin will aid this process. In addition, other components of the Wnt pathway could be alternative therapeutic targets, in particular because a lot of different cancers (see Sect. 15) have defects in any of the Wnt pathway components.

2.3.1.2
Other Cancers Associated with β-Catenin or Wnt Pathway Components

FAP, due to a germ line mutation in the APC gene, is often associated with various extracolonic manifestations: desmoid tumors and other soft tissue tumors (fibromas, lipomas, epidermoid cysts), osteomas of the skull and long bones. Of great concern, patients with FAP can develop other cancers including endometrial, thyroid, duodenal, pancreatic, and hepatoblastoma.

This indicates that disturbances in Wnt pathway components and adherens junction molecules may also be associated with many sporadic tumors (Polakis 2000; Hajra et al. 2002) (Table 3). In many of such extracolonic tumors, β-catenin-activating mutations (Fig. 3) are detectable: endometrial carcinomas (45%), hepatocellular carcinomas (25%), intestinal type gastric carcinomas (25%) (diffuse-type 0%), endometrioid-type ovarian carcinomas (25%), anaplastic thyroid carcinomas (65%), hepatoblastoma (50%), pilomatricoma (75%), medulloblastoma, desmoid tumor (50%).

But also sporadic APC gene mutations can be found in different tumors, such as breast, gastric, pancreatic, hepatocellular, desmoid tumor, medulloblastoma, and hepatoblastoma. Recently, inactivating mutations in the axin-2/conductin gene were also detected in some of hepatocellular carcinomas (5%) and ovarian carcinomas, apparently without mutations in the APC- or β-catenin gene.

Table 3. Adhesive protein gene alterations in human cancers

Gene coding for	Genetic alteration	Tumor type (rare if frequency not noted)
β-Catenin	Activating	Anaplastic thyroid (65%), endometrial (45%), hepatocellular (25%), endometrioid type ovarian (25%), colorectal (5%–10%) hepatoblastoma (50%), desmoid tumor (50%), pilomatricoma (75%), melanoma, medulloblastoma, squamous cell
Other Wnt-pathway members	APC inactivation	Colorectal (70%–80%), desmoid tumor, breast, gastric, ovarian, pancreatic, hepatocellular, medulloblastoma
	Axin/conductin	Hepatocellular (5%), colorectal, ovarian
α-Catenin	Inactivating	Cell lines only: lung, prostate, ovarian, colon
γ-Catenin	Activating	Cell lines only: gastric
p120 Catenin	Inactivating, activating?	Cell lines only: colonic
E-cadherin	Inactivating	Diffuse type gastric (50%), lobular breast (50%), invasive ductal breast, prostate, signet-ring cell type gastric, endometrial, ovarian, colorectal

As has been already described for colorectal cancer, these data show that at least one important component of the Wnt pathway or adherens junctions is mutated in many different tumors.

2.3.2
Other Diseases Associated with β-Catenin

2.3.2.1
Neurodegenerative Diseases

In contrast to cancer development, where genetic alterations in Wnt pathway components lead to an overactivation of β-catenin transcriptional activity, a reduction of its function may be involved in the development in neurodegenerative diseases, in particular Alzheimer's disease (Baki et al. 2001). Alzheimer's disease is a neurodegenerative disease with progressive dementia accompanied by three main structural changes in the brain: diffuse loss of neurons, intracellular protein deposits termed neurofibrillary tangles (NFT), and extracellular protein deposits termed amyloid or senile plaques, surrounded by dystrophic neurites. Two major hypotheses have been proposed in order to explain the molecular hallmarks of the disease: the "amyloid cascade" hypothesis and the "neuronal cytoskeletal degeneration" hypothesis. While the former is supported by genetic studies of the early-onset familial forms of Alzheimer's disease (FAD), the latter revolves around the observation in vivo that cytoskeletal changes—including the abnormal phosphorylation state of the microtubule-associated protein tau—may precede the deposition of senile plaques. Recent studies have suggested that the trafficking process of membrane-associated proteins is modulated by the FAD-linked presenilin (PS) proteins, and that amyloid beta-peptide deposition may be initiated intracellularly, through the secretory pathway. Current hypotheses concerning presenilin function are based upon its cellular localization and its putative interaction with β-catenin and GSK-3β. Developmental studies have shown that PS proteins also function as components in the Notch signal transduction cascade. Both Notch and Wnt pathways are thought to have an important role in brain development, and they have been connected through dishevelled (dsh) protein, a known transducer of the Wnt pathway. It was shown that a sustained loss of function of Wnt signaling components, like β-catenin, triggers a series of misrecognition events and apoptosis, determining the onset and development of Alzheimer's disease (De Ferrari et al. 2000).

2.3.2.2
Therapeutical Options

Thus, in contrast to therapeutical strategies in tumors, which aim at reducing the nuclear activity of β-catenin, the therapeutical aim in neurodegenerative diseases is to enhance its function. Lithium has long been used to treat psychiatric disorders like mood disease. Lithium has also been demonstrated to inhibit GSK-3β, which itself marks β-catenin for degradation. Consistent with the inhibition of GSK-3β, lithium has been demonstrated to exert robust effects against diverse insults both in vitro and in vivo. These findings suggest that lithium may exert some of its long-term beneficial effects in the treatment of mood disorders via underappreciated neuroprotective effects. In the absence of other adequate treatments, the potential efficacy of lithium in the long-term treatment of certain neurodegenerative disorders may be warranted (Manji et al. 1999). Moreover, drugs like SB-216763 and SB-415286 are novel, potent, and selective cell permeable inhibitors of GSK-3β. Therefore, such compounds represent valuable pharmacological tools and may be of future therapeutical use in disease states associated with elevated GSK-3β activity such as Alzheimer's disease (Coghlan et al. 2000).

3
Physiological and Pathological Roles of Other Catenins

Like for β-catenin, targeted deletion of other catenins in mice leads to very early embryonic lethality (α-catenin at day 4 with defective pre-implantation and γ-catenin at day 10.5 with heart malformation), indicating their necessary function already in early development (Savagner 2001). However, in contrast to the enormous knowledge about the involvement of β-catenin in development and disease, fewer data are available for the other catenins.

3.1
α-Catenin

Genetic alterations of the α-catenin gene have not yet been found in human cancers, deletions were only detected in some colon (Giannini et al. 2000), lung (Hirano et al. 1992), and ovarian cancer cell lines (Hirano et al. 1992). The latter report describes the characterization of a human ovarian carcinoma-derived cell line, Ov2008, which expresses a novel mutant form of the α-catenin protein lacking the extreme N-terminus of the wild-type protein. The altered form of α-catenin fails to bind efficiently to β-catenin and is localized in the cytoplasm. Deletion mapping has localized the β-catenin binding site on α-catenin between amino acids 46 and 149, which encompasses the same region of the protein that is deleted in the Ov2008 variant. Restoration of inducible expression of the wild-type α-catenin protein in these cells caused them to assume the morphology typical of an epithelial sheet and retarded their growth in vitro. Additionally, the induction of α-catenin expression in Ov2008 cells injected into nude mice attenuated the ability of these cells to form tumors. These observations support the classification of α-catenin as a growth-regulatory and candidate tumor suppressor gene. Although no mutations were found directly in human cancers, altered expression of α-catenin was detected in many tumors, mainly by immunohistochemistry. Of course, these can be indirect effects due to other mutations in adherens junction or Wnt pathway molecules. Abnormal α-catenin expression in invasive breast cancer correlates with poor patient survival (Nakopoulou et al. 2002). A co-downregulation of cell adhesion proteins α- and β-catenin, p120 catenin, E-cadherin, and CD44 was detected in prostatic adenocarcinomas (Kallakury et al. 2001). Furthermore, a deficient α-catenin expression is associated with melanoma progression (Zhang et al. 1999), and reduced expression with poor prognosis in colorectal carcinoma (Ropponen et al. 1999). This may be explained by an additional function of α-catenin described recently: loss of α-catenin expression in a colon cancer cell line correlates with increased Tcf-dependent transcription. The presence of α-catenin in colon cancer cell nuclei suggests that it inhibits transcription directly, and, in agreement with this, ectopic expression of α-catenin in the nucleus represses Tcf-dependent transcription. Furthermore, recombinant α-catenin disrupts the interaction between the β-catenin/Tcf complex and DNA. In conclusion, α-catenin might inhibit β-catenin signaling in the nucleus by interfering with the formation of a β-catenin/Tcf/DNA complex (Giannini et al. 2000).

Animal models also support the role of α-catenin as a potential tumors suppressor. Targeted deletion in mice skin leads to defects in epidermal and hair follicle differentiation and associated hyperproliferation and precancerous lesions. Thereby, epidermal morphogenesis was dramatically affected, with defects in adherens junction formation, intercellular adhesion, and epithelial polarity. Differentiation occurred, but epidermis displayed hyperproliferation, suprabasal mitoses, and multinucleated cells. In vitro, α-catenin-null keratinocytes were poorly contact inhibited and grew rapidly (Vasioukhin et al. 2001).

αT-catenin is a novel member of the α-catenin family, which shows most abundant expression in cardiomyocytes and in peritubular myoid cells of the testis, pointing to a specific function for αT-catenin in particular muscle tissues. Like other α-catenins, αT-catenin provides an indispensable link between the cadherin-based cell–cell adhesion complex and the cytoskeleton, to mediate cell–cell adhesion. Assessment of the αT-catenin gene on chromosome 10q21 suggests an involvement in dilated cardiomyopathy (Janssens et al. 2003).

3.2
γ-Catenin

γ-Catenin (plakoglobin) is the only component common to both the desmosomal plaque and the cadherin–catenin cell adhesion complex in adherens junctions (Zhurinsky et al. 2000). It is highly homologous to β-catenin and may, like this protein, be also involved in signaling pathways. In adherens junctions, γ-catenin binds to E-cadherin and to α-catenin and anchors adherens junctions to actin. In desmosomes, γ-catenin binds to the desmosomal cadherins desmocollin and desmoglein and to desmoplakin and keratins, providing a link to the intermediate filament skeleton. Interestingly, although γ-catenin can anchor E-cadherin via α-catenin to actin in adherens junctions, it loses this ability when incorporated into desmosomes. γ-Catenin can interact with all partners of β-catenin in the Wnt pathway, including Tcf/LEF factors; however, it is still unclear if γ-catenin can activate β-catenin target genes. For instance, it cannot compensate for the absence of β-catenin in knockout mice. Also, phenotypes of γ-catenin and β-catenin transgenic mice are different and γ-catenin seems not to be involved in human carcinogenesis to the same extend as β-catenin. In contrast to β-catenin, mutations in the γ-catenin gene are almost never found in human cancers.

γ-Catenin-null mutant embryos died from embryonic day 10.5 onward, due to severe heart defects. Some mutant embryos developed further, especially on a C57BL/6 genetic background, and died around birth, presumably due to cardiac dysfunction, and with skin blistering and subcorneal acantholysis. Ultrastructural analysis revealed that here desmosomes were greatly reduced in number and structurally altered. Thus, γ-catenin is an essential structural component for desmosome function. The skin phenotype in γ-catenin-deficient mice is reminiscent of the human blistering disease, epidermolytic hyperkeratosis (Bierkamp et al. 1996). Moreover, γ-catenin is essential for myocardial compliance but dispensable for myofibril insertion into adherens junctions (Isac et al. 1999).

Also, in humans γ-catenin is associated with various skin and heart diseases and apparently not very often with cancer. γ-Catenin plays a central role in the autoimmune disease pemphigus vulgaris (Caldelari et al. 2001). Arrhythmogenic right ventricular cardiomyopathy (ARVC) is an autosomal dominant heart muscle disorder that causes arrhythmia, heart failure, and sudden death. The finding of a deletion in the γ-catenin gene in ARVC suggests that the proteins involved in cell–cell adhesion play an important part in maintaining myocyte integrity, and when junctions are disrupted, cell death, and fibrofatty replacement occur. Therefore, the discovery of a mutation in a protein with functions in maintaining cell junction integrity has important implications for other dominant forms of ARVC, related cardiomyopathies, and other cutaneous diseases (McKoy et al. 2000). In addition, γ-catenin has been shown to be defective in Naxos disease, which results in a cardiomyopathy and growth of abnormal hair (McMillan et al. 2001).

3.3
p120 Catenin

p120 Catenin is the prototypic member of a growing subfamily of Armadillo-domain proteins found at cell–cell junctions and in nuclei. Several splice variants of p120 catenin with unknown differential function exist. In contrast to the functions of the classical catenins, which have been studied extensively, the first clues to p120 catenin's biological functions have only recently emerged, and its role remains controversial.

Unlike β-catenin and γ-catenin, which interact with the C-terminal region of E-cadherin, p120 catenin binds to the so-called juxtamembrane domain of E-cadherin and does not bind to α-catenin. Also, the interaction with E-cadherin is weaker and therefore it was postulated that the main function of this interaction is a regulation of p120 catenin by E-cadherin. In the current model, p120 catenin is an important modulator of cell adhesion and might be a main regulator in the switch between cell–cell adhesion and cell motility, but it is not an essential core component of adherens junctions. In this context, p120 catenin is considered as a regulator of the small GTPases RhoA, cdc42, and rac, which exert different functions in cell–cell-adhesion and cell motility (Anastasiadis et al. 2001) (Fig. 6). Nuclear p120 catenin has also been shown to regulate transcription via its binding partner the transcriptional repressor Kaiso (Myster et al. 2003).

These features make p120 catenin an interesting candidate to regulate the phenotypical switches of tumor cell during invasion and metastasis formation (see Sect. 8), however investigations concerning the role of p120 catenin in cancer and other diseases are only starting, and results are controversial. p120 Catenin expression is frequently altered and/or lost in tumors of the colon, bladder, stomach, breast, prostate, lung, and pancreas. Moreover, in some cases, p120 catenin loss appears to be an early event in tumor progression, possibly preceding loss of E-cadherin (Thoreson et al. 2002). A loss of p120 catenin in human colorectal cancer predicts metastasis and poor survival (Gold et al. 1998). In contrast, abnormal expression of p120 catenin correlates with poor survival in patients with bladder cancer (Syrigos et al. 1998).The p120 catenin gene is mutated in SW48 colorectal cancer cells, and the cadherin adhesion system is impaired as a direct consequence of p120 catenin insufficiency. Restoring normal levels of p120 catenin caused a striking reversion from poorly differentiated to cobblestone-like epithelial morphology, indicating a crucial role for p120 catenin in reactivation of E-cadherin function. Surprisingly, the rescue was associated with substantially increased levels of E-cadherin. E-cadherin mRNA levels were unaffected by p120 catenin expression, but E-cadherin half-life was more than doubled. Direct p120 catenin–E-cadherin interaction was crucial, as p120 catenin deletion analysis revealed a perfect correlation between E-cadherin binding and rescue of epithelial morphology (Ireton et al. 2002).

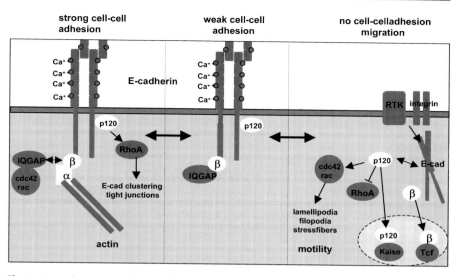

Fig. 6. Hypothetical model for the role of p120 catenin in modulating the balance between adhesive and motile cellular phenotypes. *Weak cell–cell adhesion*: In an intermediate state, p120 catenin (*p120*) and β-catenin (*β*) are bound to E-cadherin in adherens junction. β-catenin is bound by IQGAP, which inhibits binding to α-catenin. *Strong cell–cell adhesion*: E-cadherin bound p120 catenin can activate the small GTPase RhoA, which induces E-cadherin clustering and formation of tight junctions. IQGAP separates from β-catenin, binds and inhibits the small GTPases cdc42 and rac. β-Catenin can now bind α-catenin and connect adherens junctions to the actin cytoskeleton. *No cell–cell adhesion*: Signals, e.g., from integrins or receptor bound tyrosine kinases (RTK), induce a cytoplasmic retention of E-cadherin. p120 Catenin and β-catenin separate from E-cadherin. Cytoplasmic p120 catenin binds and activates cdc42 and rac, but inhibits RhoA. This leads to the formation of lamellipodia, filopodia, and stress fibers and induces cell motility. Nuclear p120 catenin binds to the transcriptional repressor Kaiso with unknown target genes. In this model the availability of membranous E-cadherin determines the intracellular location and function of p120 catenin

All these data support a role of p120 catenin either as a tumor suppressor or metastasis promoter and therefore make p120 catenin a potential target for future cancer therapy.

4
Conclusions

Catenins are not only intracellular molecules of junctional complexes, but are also components of intracellular signaling pathways. Most is known about β-catenin and its function in either cell–cell adhesion or as a component and main effector of the Wnt pathway. The linkage of this pathway to embryonic development and adult tissue homeostasis, and its role in human diseases, in particular cancer, determines β-catenin as a major player in all of these processes and as a potent oncoprotein. In addition, the other catenins are increasingly connected to other functions and various diseases. This knowledge will increase and will make catenins significant and promising targets for future therapeutic approaches.

References

Anastasiadis PZ, Reynolds AB (2001) Regulation of Rho GTPases by p120-catenin. Curr Opin Cell Biol 13:604–610

Aust DE, Terdiman JP, Willenbucher RF, Chew K, Ferrell L, Florendo C, Molinaro-Clark A, Baretton GB, Lohrs U, Waldman FM (2001) Altered distribution of beta-catenin, and its binding proteins E-cadherin and APC, in ulcerative colitis-related colorectal cancers. Mod Pathol 14:29–39

Baki L, Marambaud P, Efthimiopoulos S, Georgakopoulos A, Wen P, Cui W, Shioi J, Koo E, Ozawa M, Friedrich VL Jr, Robakis NK (2001) Presenilin-1 binds cytoplasmic epithelial cadherin, inhibits cadherin/p120 association, and regulates stability and function of the cadherin/catenin adhesion complex. Proc Natl Acad Sci U S A 98:2381–2386

Barker N, Huls G, Korinek V, Clevers H (1999) Restricted high level expression of Tcf-4 protein in intestinal and mammary gland epithelium. Am J Pathol 154:29–35

Barker N, Hurlstone A, Musisi H, Miles A, Bienz M, Clevers H (2001) The chromatin remodelling factor Brg-1 interacts with beta-catenin to promote target gene activation. EMBO J 20:4935–4943

Barth AI, Nathke IS, Nelson WJ (1997) Cadherins, catenins and APC protein: interplay between cytoskeletal complexes and signaling pathways. Curr Opin Cell Biol 9:683–690

Batlle E, Henderson JT, Beghtel H, van den Born MM, Sancho E, Huls G, Meeldijk J, Robertson J, van de Wetering M, Pawson T, Clevers H (2002) Beta-catenin and TCF mediate cell positioning in the intestinal epithelium by controlling the expression of EphB/ephrinB. Cell 111:251–263

Batsche E, Muchardt C, Behrens J, Hurst HC, Cremisi C (1998) RB and c-Myc activate expression of the E-cadherin gene in epithelial cells through interaction with transcription factor AP-2. Mol Cell Biol 18:3647–3658

Behrens J (1999) Cadherins and catenins: role in signal transduction and tumor progression. Cancer Metastasis Rev 18:15–30

Behrens J, von Kries JP, Kuhl M, Bruhn L, Wedlich D, Grosschedl R, Birchmeier W (1996) Functional interaction of beta-catenin with the transcription factor LEF-1. Nature 382:638–642

Bienz M, Clevers H (2000) Linking colorectal cancer to Wnt signaling. Cell 103:311–320

Bienz M, Clevers H (2003) Armadillo/beta-catenin signals in the nucleus—proof beyond a reasonable doubt? Nat Cell Biol 5:179–182

Bierkamp C, McLaughlin KJ, Schwarz H, Huber O, Kemler R (1996) Embryonic heart and skin defects in mice lacking plakoglobin. Dev Biol 180:780–785

Brabletz T, Jung A, Hermann K, Gunther K, Hohenberger W, Kirchner T (1998) Nuclear overexpression of the oncoprotein beta-catenin in colorectal cancer is localized predominantly at the invasion front. Pathol Res Pract 194:701–704

Brabletz T, Jung A, Dag S, Hlubek F, Kirchner T (1999) β-catenin regulates the expression of the matrix metalloproteinase-7 in human colorectal cancer. Amer J Pathol 155:1033–1038

Brabletz T, Herrmann K, Jung A, Faller G, Kirchner T (2000) Expression of nuclear beta-catenin and c-myc is correlated with tumor size but not with proliferative activity of colorectal adenomas. Am J Pathol 156:865–870

Brabletz T, Jung A, Reu S, Porzner M, Hlubek F, Kunz-Schughart L, Knuechel R, Kirchner T (2001) Variable beta-catenin expression in colorectal cancer indicates a tumor progression driven by the tumor environment. Proc Natl Acad Sci U S A 98:10356–10361

Brault V, Moore R, Kutsch S, Ishibashi M, Rowitch DH, McMahon AP, Sommer L, Boussadia O, Kemler R (2001) Inactivation of the beta-catenin gene by Wnt1-Cre-mediated deletion results in dramatic brain malformation and failure of craniofacial development. Development 128:1253–1264

Cadigan KM, Nusse R (1997) Wnt signaling: a common theme in animal development. Genes Dev 11:3286–3305

Caldelari R, de Bruin A, Baumann D, Suter MM, Bierkamp C, Balmer V, Muller E (2001) A central role for the armadillo protein plakoglobin in the autoimmune disease pemphigus vulgaris. J Cell Biol 153:823–834

Cavallo RA, Cox RT, Moline MM, Roose J, Polevoy GA, Clevers H, Peifer M, Bejsovec A (1998) Drosophila Tcf and Groucho interact to repress Wingless signalling activity. Nature 395:604–608

Clevers H, van de Wetering M (1997) TCF/LEF factor earn their wings. Trends Genet 13:485–489

Coghlan MP, Culbert AA, Cross DA, Corcoran SL, Yates JW, Pearce NJ, Rausch OL, Murphy GJ, Carter PS, Roxbee Cox L, Mills D, Brown MJ, Haigh D, Ward RW, Smith DG, Murray KJ, Reith AD, Holder JC (2000) Selective small molecule inhibitors of glycogen synthase kinase-3 modulate glycogen metabolism and gene transcription. Chem Biol 7:793–803

Crawford HC, Fingleton BM, Rudolph-Owen LA, Heppner Goss KJ, Rubinfeld B, Polakis P, Matrisian LM (1999) The metalloproteinase Matrilysin is a target of β-catenin transactivation in intestinal tumors. Oncogene 18:2883–2891

Daniels DL, Eklof Spink K, Weis WI (2001) Beta-catenin: molecular plasticity and drug design. Trends Biochem Sci 26:672–678

De Ferrari GV, Inestrosa NC (2000) Wnt signaling function in Alzheimer's disease. Brain Res Brain Res Rev 33:1–12

Dihlmann S, Gebert J, Siermann A, Herfarth C, von Knebel Doeberitz M (1999) Dominant negative effect of the APC1309 mutation: a possible explanation for genotype-phenotype correlations in familial adenomatous polyposis. Cancer Res 59:1857–1860

Duval A, Gayet J, Zhou XP, Iacopetta B, Thomas G, Hamelin R (1999) Frequent frameshift mutations of the TCF-4 gene in colorectal cancers with microsatellite instability. Cancer Res 59:4213–4215

Eger A, Stockinger A, Schaffhauser B, Beug H, Foisner R (2000) Epithelial mesenchymal transition by c-Fos estrogen receptor activation involves nuclear translocation of beta-catenin and upregulation of beta-catenin/lymphoid enhancer binding factor-1 transcriptional activity. J Cell Biol 148:173–188

Enomoto-Iwamoto M, Kitagaki J, Koyama E, Tamamura Y, Wu C, Kanatani N, Koike T, Okada H, Komori T, Yoneda T, Church V, Francis-West PH, Kurisu K, Nohno T, Pacifici M, Iwamoto M (2002) The Wnt antagonist Frzb-1 regulates chondrocyte maturation and long bone development during limb skeletogenesis. Dev Biol 251:142–156

Fodde R, Edelmann W, Yang K, van Leeuwen C, Carlson C, Renault B, Breukel C, Alt E, Lipkin M, Khan PM, et al (1994) A targeted chain-termination mutation in the mouse Apc gene results in multiple intestinal tumors. Proc Natl Acad Sci U S A 91:8969–8973

Freier S, Weiss O, Eran M, Flyvbjerg A, Dahan R, Nephesh I, Safra T, Shiloni E, Raz I (1999) Expression of the insulin-like growth factors and their receptors in adenocarcinoma of the colon. Gut 44:704–708

Fuchs E, Cleveland DW (1998) A structural scaffolding of intermediate filaments in health and disease. Science 279:514–519

Fujita M, Furukawa Y, Tsunoda T, Tanaka T, Ogawa M, Nakamura Y (2001) Up-regulation of the ectodermal-neural cortex 1 (ENC1) gene, a downstream target of the beta-catenin/T-cell factor complex, in colorectal carcinomas. Cancer Res 61:7722–7726

Gaire M, Magbanua Z, McDonnell S, McNeil L, Lovett DH, Matrisian LM (1994) Structure and expression of the human gene for the matrix metalloproteinase matrilysin. J Biol Chem 269:2032–2040

Galzie Z, Fernig DG, Smith JA, Poston GJ, Kinsella AR (1997) Invasion of human colorectal carcinoma cells is promoted by endogenous basic fibroblast growth factor. Int J Cancer 71:390–395

Gat U, DasGupta R, Degenstein L, Fuchs E (1998) De Novo hair follicle morphogenesis and hair tumors in mice expressing a truncated beta-catenin in skin. Cell 95:605–614

Giannini AL, Vivanco M, Kypta RM (2000) Alpha-catenin inhibits beta-catenin signaling by preventing formation of a beta-catenin T-cell factor*DNA complex. J Biol Chem 275:21883–21888

Gold JS, Reynolds AB, Rimm DL (1998) Loss of p120ctn in human colorectal cancer predicts metastasis and poor survival. Cancer Lett 132:193–201

Gong Y, Slee RB, Fukai N, Rawadi G, Roman-Roman S, Reginato AM, Wang H, Cundy T, Glorieux FH, Lev D, Zacharin M, Oexle K, Marcelino J, Suwairi W, Heeger S, Sabatakos G, Apte S, Adkins WN, Allgrove J, Arslan-Kirchner M, Batch JA, Beighton P, Black GC, Boles RG, Boon LM, Borrone C, Brunner HG, Carle GF, Dallapiccola B, De Paepe A, Floege B, Halfhide ML, Hall B, Hennekam RC, Hirose T, Jans A, Juppner H, Kim CA, Keppler-Noreuil K, Kohlschuetter A, LaCombe D, Lambert M, Lemyre E, Letteboer T, Peltonen L, Ramesar RS, Romanengo M, Somer H, Steichen-Gersdorf E, Steinmann B, Sullivan B, Superti-Furga A, Swoboda W, van den Boogaard MJ, Van Hul W, Vikkula M, Votruba M, Zabel B, Garcia T, Baron R, Olsen BR, Warman ML (2001) LDL receptor-related protein 5 (LRP5) affects bone accrual and eye development. Cell 107:513–523

Gottardi CJ, Gumbiner BM (2001) Adhesion signaling: how beta-catenin interacts with its partners. Curr Biol 11:R792–794

Gradl D, Kuhl M, Wedlich D (1999) The Wnt/Wg signal transducer beta-catenin controls fibronectin expression. Mol Cell Biol 19:5576–5587

Groden J, Thliveris A, Samowitz W, Carlson M, Gelbert L, Albertsen H, Joslyn G, Stevens J, Spirio L, Robertson M, et al (1991) Identification and characterization of the familial adenomatous polyposis coli gene. Cell 66:589–600

Haegel H, Larue L, Ohsugi M, Fedorov L, Herrenknecht K, Kemler R (1995) Lack of beta-catenin affects mouse development at gastrulation. Development 121:3529–3537

Hajra KM, Fearon ER (2002) Cadherin and catenin alterations in human cancer. Genes Chromosomes Cancer 34:255–268

Hassan AB, Howell JA (2000) Insulin-like growth factor II supply modifies growth of intestinal adenoma in Apc(Min/+) mice. Cancer Res 60:1070–1076

He TC, Chan TA, Vogelstein B, Kinzler KW (1999) PPARdelta is an APC-regulated target of nonsteroidal anti-inflammatory drugs. Cell 99:335–345

Hirano S, Kimoto N, Shimoyama Y, Hirohashi S, Takeichi M (1992) Identification of a neural alpha-catenin as a key regulator of cadherin function and multicellular organization. Cell 70:293–301

Hirohashi S (1998) Inactivation of the E-cadherin-mediated cell adhesion system in human cancers. Am J Pathol 153:333–339

Hlubek F, Jung A, Kotzor N, Kirchner T, Brabletz T (2001) Expression of the invasion factor laminin γ2 in colorectal carcinomas is regulated by β-catenin. Cancer Res 61:8089–8093

Hovanes K, Li TW, Munguia JE, Truong T, Milovanovic T, Lawrence Marsh J, Holcombe RF, Waterman ML (2001) Beta-catenin-sensitive isoforms of lymphoid enhancer factor-1 are selectively expressed in colon cancer. Nat Genet 28:53–57

Howe LR, Subbaramaiah K, Chung WJ, Dannenberg AJ, Brown AM (1999) Transcriptional activation of cyclooxygenase-2 in Wnt-1-transformed mouse mammary epithelial cells. Cancer Res 59:1572–1577

Huang J, Papadopoulos N, McKinley AJ, Farrington SM, Curtis LJ, Wyllie AH, Zheng S, Willson JK, Markowitz SD, Morin P, Kinzler KW, Vogelstein B, Dunlop MG (1996) APC mutations in colorectal tumors with mismatch repair deficiency. Proc Natl Acad Sci U S A 93:9049–9054

Huang L, Li X, El-Hodiri HM, Dayal S, Wikramanayake AH, Klein WH (2000) Involvement of Tcf/Lef in establishing cell types along the animal-vegetal axis of sea urchins. Dev Genes Evol 210:73–81

Huelsken J, Behrens J (2002) The Wnt signalling pathway. J Cell Sci 115:3977–3978

Huelsken J, Vogel R, Brinkmann V, Erdmann B, Birchmeier C, Birchmeier W (2000) Requirement for beta-catenin in anterior-posterior axis formation in mice. J Cell Biol 148:567–578

Huelsken J, Vogel R, Erdmann B, Cotsarelis G, Birchmeier W (2001) Beta-catenin controls hair follicle morphogenesis and stem cell differentiation in the skin. Cell 105:533–545

Inomata M, Ochiai A, Akimoto S, Kitano S, Hirohashi S (1996) Alteration of beta-catenin expression in colonic epithelial cells of familial adenomatous polyposis patients. Cancer Res 56:2213–2217

Ireton RC, Davis MA, van Hengel J, Mariner DJ, Barnes K, Thoreson MA, Anastasiadis PZ, Matrisian L, Bundy LM, Sealy L, Gilbert B, van Roy F, Reynolds AB (2002) A novel role for p120 catenin in E-cadherin function. J Cell Biol 159:465–476

Isac CM, Ruiz P, Pfitzmaier B, Haase H, Birchmeier W, Morano I (1999) Plakoglobin is essential for myocardial compliance but dispensable for myofibril insertion into adherens junctions. J Cell Biochem 72:8–15

Ishikawa T, Tamai Y, Zorn AM, Yoshida H, Seldin MF, Nishikawa S, Taketo MM (2001) Mouse Wnt receptor gene Fzd5 is essential for yolk sac and placental angiogenesis. Development 128:25–33

Janssens B, Mohapatra B, Vatta M, Goossens S, Vanpoucke G, Kools P, Montoye T, Van Hengel J, Bowles NE, Van Roy F, Towbin JA (2003) Assessment of the CTNNA3 gene encoding human alphaT-catenin regarding its involvement in dilated cardiomyopathy. Hum Genet 112:227–236

Jung A, Schrauder M, Oswald U, Knoll C, Sellberg P, Palmqvist R, Niedobitek G, Brabletz T, Kirchner T (2001) The invasion front of human colorectal adenocarcinomas shows co-localization of nuclear beta-catenin, cyclin D(1), and p16(INK4A) and is a region of low proliferation. Am J Pathol 159:1613–1617

Kallakury BV, Sheehan CE, Ross JS (2001) Co-downregulation of cell adhesion proteins alpha- and beta-catenins, p120CTN, E-cadherin, and CD44 in prostatic adenocarcinomas. Hum Pathol 32:849–855

Kedinger M, Duluc I, Fritsch C, Lorentz O, Plateroti M, Freund JN (1998) Intestinal epithelial-mesenchymal cell interactions. Ann N Y Acad Sci 859:1–17

Kim JS, Crooks H, Dracheva T, Nishanian TG, Singh B, Jen J, Waldman T (2002) Oncogenic beta-catenin is required for bone morphogenetic protein 4 expression in human cancer cells. Cancer Res 62:2744–2748

Kinzler KW, Vogelstein B (1996) Lessons from hereditary colorectal cancer. Cell 87:159–170

Kinzler KW, Nilbert MC, Su LK, Vogelstein B, Bryan TM, Levy DB, Smith KJ, Preisinger AC, Hedge P, McKechnie D, et al (1991) Identification of FAP locus genes from chromosome 5q21. Science 253:661–665

Kirchner T, Brabletz T (2000) Patterning and nuclear beta-catenin expression in the colonic adenoma- carcinoma sequence: analogies with embryonic gastrulation. Am J Pathol 157:1113–1121

Koh TJ, Bulitta CJ, Fleming JV, Dockray GJ, Varro A, Wang TC (2000) Gastrin is a target of the beta-catenin/TCF-4 growth-signaling pathway in a model of intestinal polyposis. J Clin Invest 106:533–539

Kolligs FT, Nieman MT, Winer I, Hu G, Van Mater D, Feng Y, Smith IM, Wu R, Zhai Y, Cho KR, Fearon ER (2002) ITF-2, a downstream target of the Wnt/TCF pathway, is activated in human cancers with beta-catenin defects and promotes neoplastic transformation. Cancer Cell 1:145–155

Korinek V, Barker N, Morin PJ, van Wichen D, de Weger R, Kinzler KW, Vogelstein B, Clevers H (1997) Constitutive transcriptional activation by a beta-catenin-Tcf complex in APC−/− colon carcinoma [see comments]. Science 275:1784–1787

Korinek V, Barker N, Moerer P, van Donselaar E, Huls G, Peters PJ, Clevers H (1998) Depletion of epithelial stem-cell compartments in the small intestine of mice lacking Tcf-4. Nat Genet 19:379–383

Kucherlapati R, Lin DP, Edelmann W (2001) Mouse models for human familial adenomatous polyposis. Semin Cancer Biol 11:219–225

Lickert H, Domon C, Huls G, Wehrle C, Duluc I, Clevers H, Meyer BI, Freund JN, Kemler R (2000) Wnt/(beta)-catenin signaling regulates the expression of the homeobox gene Cdx1 in embryonic intestine. Development 127:3805–3813

Liotta LA, Kohn EC (2001) The microenvironment of the tumour-host interface. Nature 411:375–379

Liu D, el-Hariry I, Karayiannakis AJ, Wilding J, Chinery R, Kmiot W, McCrea PD, Gullick WJ, Pignatelli M (1997) Phosphorylation of beta-catenin and epidermal growth factor receptor by intestinal trefoil factor. Lab Invest 77:557–563

Liu W, Dong X, Mai M, Seelan RS, Taniguchi K, Krishnadath KK, Halling KC, Cunningham JM, Boardman LA, Qian C, Christensen E, Schmidt SS, Roche PC, Smith DI, Thibodeau SN (2000) Mutations in AXIN2 cause colorectal cancer with defective mismatch repair by activating beta-catenin/TCF signalling. Nat Genet 26:146–147

Logan CY, Miller JR, Ferkowicz MJ, McClay DR (1999) Nuclear beta-catenin is required to specify vegetal cell fates in the sea urchin embryo. Development 126:345–357

Lustig B, Jerchow B, Sachs M, Weiler S, Pietsch T, Karsten U, van de Wetering M, Clevers H, Schlag PM, Birchmeier W, Behrens J (2002) Negative feedback loop of Wnt signaling through upregulation of conductin/axin2 in colorectal and liver tumors. Mol Cell Biol 22:1184–1193

Manji HK, Moore GJ, Chen G (1999) Lithium at 50: have the neuroprotective effects of this unique cation been overlooked? Biol Psychiatry 46:929–940

Marchenko GN, Marchenko ND, Leng J, Strongin AY (2002) Promoter characterization of the novel human matrix metalloproteinase-26 gene: regulation by the T-cell factor-4 implies specific expression of the gene in cancer cells of epithelial origin. Biochem J 363:253–262

Mariadason JM, Bordonaro M, Aslam F, Shi L, Kuraguchi M, Velcich A, Augenlicht LH (2001) Down-regulation of beta-catenin TCF signaling is linked to colonic epithelial cell differentiation. Cancer Res 61:3465–3471

Marotta A, Tan C, Gray V, Malik S, Gallinger S, Sanghera J, Dupuis B, Owen D, Dedhar S, Salh B (2001) Dysregulation of integrin-linked kinase (ILK) signaling in colonic polyposis. Oncogene 20:6250–6257

McKoy G, Protonotarios N, Crosby A, Tsatsopoulou A, Anastasakis A, Coonar A, Norman M, Baboonian C, Jeffery S, McKenna WJ (2000) Identification of a deletion in plakoglobin in arrhythmogenic right ventricular cardiomyopathy with palmoplantar keratoderma and woolly hair (Naxos disease). Lancet 355:2119–2124

McMillan JR, Shimizu H (2001) Desmosomes: structure and function in normal and diseased epidermis. J Dermatol 28:291–298

Molenaar M, van de Wetering M, Oosterwegel M, Peterson-Maduro J, Godsave S, Korinek V, Roose J, Destree O, Clevers H (1996) XTcf-3 transcription factor mediates beta-catenin-induced axis formation in Xenopus embryos. Cell 86:391–399

Morali OG, Delmas V, Moore R, Jeanney C, Thiery JP, Larue L (2001) IGF-II induces rapid beta-catenin relocation to the nucleus during epithelium to mesenchyme transition. Oncogene 20:4942–4950

Myster SH, Cavallo R, Anderson CT, Fox DT, Peifer M (2003) Drosophila p120catenin plays a supporting role in cell adhesion but is not an essential adherens junction component. J Cell Biol 160:433–449

Naishiro Y, Yamada T, Takaoka AS, Hayashi R, Hasegawa F, Imai K, Hirohashi S (2001) Restoration of epithelial cell polarity in a colorectal cancer cell line by suppression of beta-catenin/T-cell factor 4-mediated gene transactivation. Cancer Res 61:2751–2758

Nakopoulou L, Gakiopoulou-Givalou H, Karayiannakis AJ, Giannopoulou I, Keramopoulos A, Davaris P, Pignatelli M (2002) Abnormal alpha-catenin expression in invasive breast cancer correlates with poor patient survival. Histopathology 40:536–546

Nishita M, Hashimoto MK, Ogata S, Laurent MN, Ueno N, Shibuya H, Cho KW (2000) Interaction between Wnt and TGF-beta signalling pathways during formation of Spemann's organizer. Nature 403:781–785

Noe V, Fingleton B, Jacobs K, Crawford HC, Vermeulen S, Steelant W, Bruyneel E, Matrisian LM, Mareel M (2001) Release of an invasion promoter E-cadherin fragment by matrilysin and stromelysin-1. J Cell Sci 114:111–118

Novak A, Hsu SC, Leung-Hagesteijn C, Radeva G, Papkoff J, Montesano R, Roskelley C, Grosschedl R, Dedhar S (1998) Cell adhesion and the integrin-linked kinase regulate the LEF-1 and beta-catenin signaling pathways. Proc Natl Acad Sci U S A 95:4374–4379

Oshima M, Oshima H, Kitagawa K, Kobayashi M, Itakura C, Taketo M (1995) Loss of Apc heterozygosity and abnormal tissue building in nascent intestinal polyps in mice carrying a truncated Apc gene. Proc Natl Acad Sci U S A 92:4482–4486

Oshima M, Dinchuk JE, Kargman SL, Oshima H, Hancock B, Kwong E, Trzaskos JM, Evans JF, Taketo MM (1996) Suppression of intestinal polyposis in Apc delta716 knockout mice by inhibition of cyclooxygenase 2 (COX-2). Cell 87:803–809

Oyama T, Kanai Y, Ochiai A, Akimoto S, Oda T, Yanagihara K, Nagafuchi A, Tsukita S, Shibamoto S, Ito F, et al (1994) A truncated beta-catenin disrupts the interaction between E-cadherin and alpha-catenin: a cause of loss of intercellular adhesiveness in human cancer cell lines. Cancer Res 54:6282–6287

Palmqvist R, Rutegard JN, Bozoky B, Landberg G, Stenling R (2000) Human colorectal cancers with an intact p16/Cyclin D1/pRb pathway have up-regulated p16 expression and decreased proliferation in small invasive tumor clusters. Am J Pathol 157:1947–1953

Perez-Moreno M, Jamora C, Fuchs E (2003) Sticky business: orchestrating cellular signals at adherens junctions. Cell 112:535–548

Perl AK, Wilgenbus P, Dahl U, Semb H, Christofori G (1998) A causal role for E-cadherin in the transition from adenoma to carcinoma. Nature 392:190–193

Persad S, Troussard AA, McPhee TR, Mulholland DJ, Dedhar S (2001) Tumor suppressor PTEN inhibits nuclear accumulation of beta-catenin and T cell/lymphoid enhancer factor 1-mediated transcriptional activation. J Cell Biol 153:1161–1174

Polakis P (2000) Wnt signaling and cancer. Genes Dev 14:1837–1851

Reddy BS, Rao CV (2002) Novel approaches for colon cancer prevention by cyclooxygenase-2 inhibitors. J Environ Pathol Toxicol Oncol 21:155–164

Riese J, Yu X, Munnerlyn A, Eresh S, Hsu SC, Grosschedl R, Bienz M (1997) LEF-1, a nuclear factor coordinating signaling inputs from wingless and decapentaplegic. Cell 88:777–787

Rockman SP, Currie SA, Ciavarella M, Vincan E, Dow C, Thomas RJ, Phillips WA (2001) Id2 is a target of the beta-catenin/TCF pathway in colon carcinoma. J Biol Chem 25:25

Roose J, Clevers H (1999) TCF transcription factors: molecular switches in carcinogenesis. Biochim Biophys Acta 1424:M23–37

Ropponen KM, Eskelinen MJ, Lipponen PK, Alhava E, Kosma VM (1998) Expression of CD44 and variant proteins in human colorectal cancer and its relevance for prognosis. Scand J Gastroenterol 33:301–309

Ropponen KM, Eskelinen MJ, Lipponen PK, Alhava EM, Kosma VM (1999) Reduced expression of alpha catenin is associated with poor prognosis in colorectal carcinoma. J Clin Pathol 52:10–16

Ross SE, Hemati N, Longo KA, Bennett CN, Lucas PC, Erickson RL, MacDougald OA (2000) Inhibition of adipogenesis by Wnt signaling. Science 289:950–953

Rubinfeld B, Souza B, Albert I, Muller O, Chamberlain SH, Masiarz FR, Munemitsu S, Polakis P (1993) Association of the APC gene product with beta-catenin. Science 262:1731–1734

Savagner P (2001) Leaving the neighborhood: molecular mechanisms involved during epithelial-mesenchymal transition. Bioessays 23:912–923

Schuhmacher C, Becker I, Oswald S, Atkinson MJ, Nekarda H, Becker KF, Mueller J, Siewert JR, Hofler H (1999) Loss of immunohistochemical E-cadherin expression in colon cancer is not due to structural gene alterations. Virchows Arch 434:489–495

Shibata H, Toyama K, Shioya H, Ito M, Hirota M, Hasegawa S, Matsumoto H, Takano H, Akiyama T, Toyoshima K, Kanamaru R, Kanegae Y, Saito I, Nakamura Y, Shiba K, Noda T (1997) Rapid colorectal adenoma formation initiated by conditional targeting of the Apc gene. Science 278:120–123

Shitoh K, Furukawa T, Kojima M, Konishi F, Miyaki M, Tsukamoto T, Nagai H (2001) Frequent activation of the beta-catenin-Tcf signaling pathway in nonfamilial colorectal carcinomas with microsatellite instability. Genes Chromosomes Cancer 30:32–37

Shtutman M, Zhurinsky J, Simcha I, Albanese C, D'Amico M, Pestell R, Ben-Ze'ev A (1999) The cyclin D1 gene is a target of the beta-catenin/LEF-1 pathway. Proc Natl Acad Sci U S A 96:5522–5527

Sieber OM, Tomlinson IP, Lamlum H (2000) The adenomatous polyposis coli (APC) tumour suppressor—genetics, function and disease. Mol Med Today 6:462–469

Staal FJ, Meeldijk J, Moerer P, Jay P, van de Weerdt BC, Vainio S, Nolan GP, Clevers H (2001) Wnt signaling is required for thymocyte development and activates Tcf-1 mediated transcription. Eur J Immunol 31:285–293

Su LK, Kinzler KW, Vogelstein B, Preisinger AC, Moser AR, Luongo C, Gould KA, Dove WF (1992) Multiple intestinal neoplasia caused by a mutation in the murine homolog of the APC gene. Science 256:668–670

Syrigos KN, Karayiannakis A, Syrigou EI, Harrington K, Pignatelli M (1998) Abnormal expression of p120 correlates with poor survival in patients with bladder cancer. Eur J Cancer 34:2037–2040

Taipale J, Beachy PA (2001) The Hedgehog and Wnt signalling pathways in cancer. Nature 411:349–354

Takahashi M, Tsunoda T, Seiki M, Nakamura Y, Furukawa Y (2002) Identification of membrane-type matrix metalloproteinase-1 as a target of the beta-catenin/Tcf4 complex in human colorectal cancers. Oncogene 21:5861–5867

Tetsu O, McCormick F (1999) Beta-catenin regulates expression of cyclin D1 in colon carcinoma cells. Nature 398:422–426

Thoreson MA, Reynolds AB (2002) Altered expression of the catenin p120 in human cancer: implications for tumor progression. Differentiation 70:583–589

Van Aken E, De Wever O, Correia da Rocha AS, Mareel M (2001) Defective E-cadherin/catenin complexes in human cancer. Virchows Arch 439:725–751

Vasioukhin V, Bauer C, Degenstein L, Wise B, Fuchs E (2001) Hyperproliferation and defects in epithelial polarity upon conditional ablation of alpha-catenin in skin. Cell 104:605–617

Wielenga VJ, Smits R, Korinek V, Smit L, Kielman M, Fodde R, Clevers H, Pals ST (1999) Expression of CD44 in Apc and Tcf mutant mice implies regulation by the WNT pathway. Am J Pathol 154:515–523

Yamada T, Takaoka AS, Naishiro Y, Hayashi R, Maruyama K, Maesawa C, Ochiai A, Hirohashi S (2000) Transactivation of the multidrug resistance 1 gene by T-cell factor 4/beta-catenin complex in early colorectal carcinogenesis. Cancer Res 60:4761–4766

Zhang T, Otevrel T, Gao Z, Ehrlich SM, Fields JZ, Boman BM (2001a) Evidence that APC regulates survivin expression: a possible mechanism contributing to the stem cell origin of colon cancer. Cancer Res 61:8664–8667

Zhang X, Gaspard JP, Chung DC (2001b) Regulation of vascular endothelial growth factor by the WNT and K-ras pathways in colonic neoplasia. Cancer Res 61:6050–6054

Zhang XD, Hersey P (1999) Expression of catenins and p120cas in melanocytic nevi and cutaneous melanoma: deficient alpha-catenin expression is associated with melanoma progression. Pathology 31:239–246

Zhurinsky J, Shtutman M, Ben-Ze'ev A (2000) Plakoglobin and beta-catenin: protein interactions, regulation and biological roles. J Cell Sci 113:3127–3139

The Molecular Composition and Function of Desmosomes

L. M. Godsel · S. Getsios · A. C. Huen · K. J. Green (✉)

Department of Pathology, Northwestern University Medical School,
303 E. Chicago Avenue, Chicago, IL 60611, USA
kgreen@northwestern.edu

1	Desmosome Ultrastructure, Tissue Distribution, and Composition	138
2	The Desmosomal Cadherins	141
3	Armadillo Proteins of the Desmosome	146
3.1	Plakoglobin	146
3.2	Plakophilins	148
3.3	p0071 (Plakophilin 4)	149
4	Plakins	150
4.1	Desmoplakin	151
4.2	Plectin	153
4.3	Envoplakin and Periplakin	154
5	Other Desmosome Proteins	155
6	Diseases Involving Desmosomes	156
6.1	Autoimmune and Infectious Disease	156
6.1.1	Pemphigus Vulgaris and Pemphigus Foliaceus	159
6.1.2	Staphylococcal Scalded Skin Syndrome	160
6.1.3	Paraneoplastic Pemphigus	160
6.2	Genetic Diseases	161
6.2.1	Mutations in the Desmosomal Cadherins	161
6.2.2	Involvement of Armadillo Proteins in Genetic Disease	162
6.2.3	Mutations in Desmoplakin	162
7	Animal Models Involving Desmosomes	163
7.1	Desmosomal Cadherin Models	166
7.2	Armadillo Protein Models	168
7.3	Desmoplakin Models	169
8	Regulation of Desmosomes	170
8.1	Desmosome Assembly	171
8.2	Dependence of Desmosomes on Adherens Junction Formation	173
8.3	Transcriptional and Post-translational Mechanisms for Modulating Desmosome Assembly State	174
8.4	The Role of Desmosome Components in Signaling and Transcriptional Regulation	175
9	Future Directions and Concluding Remarks	177
	References	178

Abstract Desmosomes are intercellular adhesive junctions that are particularly prominent in tissues experiencing mechanical stress, such as the heart and epidermis. Whereas the related adherens junction links actin to calcium-dependent adhesion molecules known as classical cadherins, desmosomes link intermediate filaments (IF) to the related subfamily of desmosomal cadherins. By tethering these stress-bearing cytoskeletal filaments to the plasma membrane, desmosomes serve as integrators of the IF cytoskeleton throughout a tissue. Recent evidence suggests that IF attachment in turn strengthens desmosomal adhesion. This collaborative arrangement results in formation of a supracellular network, which is critical for imparting mechanical integrity to tissues. Diseases and animal models targeting desmosomal components highlight the importance of desmosomes in development and tissue integrity, while the downregulation of individual protein components in cancer metastasis and wound healing suggests their importance in cell homeostasis. This chapter will provide an update on desmosome composition, function, and regulation, and will also discuss recent work which raises the possibility that desmosome proteins do more than play a structural role in tissues where they reside.

Keywords Desmosome · Cadherins · Armadillo · Plakins · Junction

1
Desmosome Ultrastructure, Tissue Distribution, and Composition

The desmosome is a symmetrical structure comprising two adjacent plasma membranes that sandwich a 30-nm intercellular space bisected by a central dense stratum or midline and are flanked by mirror-image tripartite electron-dense plaques (see North et al. 1999 and references therein and Burdett 1998 for review). Each half plaque comprises a total width of up to 50 nm (Fig. 1). The so-called "core" of the desmosome, containing the plasma membranes and intervening material, contains transmembrane and extracellular domains of the desmosomal cadherins (see Sect. 3). The outer dense plaque, a 15- to 20-nm thick electron-dense region adjacent to the plasma membranes, is separated by less than 10 nm from a 20-nm thick inner dense plaque, the latter of which is intimately associated with a meshwork of intermediate filaments (IFs). The outer and inner plaques contain the cytoplasmic tails of the transmembrane cadherins and associated cytoplasmic proteins.

Desmosomes are present in the epithelia of many tissues, within the intercalated disks of the heart, follicular dendritic cells of the lymph nodes, and in the arachnoid plexus of the meninges (reviewed in Schwarz et al. 1990; Kowalczyk et al. 1999a). The fact that desmosomes are associated with each of the IF family members expressed in these tissues, including keratins, vimentin, and desmin, highlights the versatility of their IF-binding function. Likewise, the molecular composition of desmosome plaque and core components varies among tissues and within complex stratified epithelial tissues. Desmosomes are present throughout the epidermis, but increase in abundance and composition as keratinocytes differentiate (White and Gohari

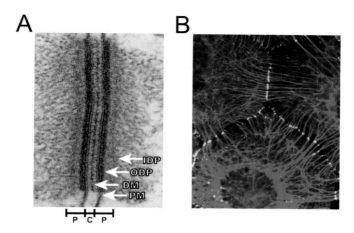

Fig. 1. Ultrastructure of the desmosome. **A** The junction is composed of two opposing plasma membranes which sandwich a 30-nm intercellular space referred to as the core (*C*) that contains the central dense midline (*DM*). On either side of the plasma membranes (*PM*) lie mirror image tripartite plaques (*P*), each composed of an outer dense plaque (*ODP*) and an inner dense plaque (*IDP*), which anchors the IF cytoskeleton. **B** Immunofluorescence of epidermal keratinocytes. Keratin 18 is stained in red and desmoplakin, an IF-anchoring protein spanning the outer and inner dense plaques of the desmosome, is stained in green. IF tonofilaments reach out to the plasma membrane, colocalizing with the desmoplakin at the cell–cell border. (The electron micrograph is reprinted from Kowalczyk et al. 1994 with permission from Elsevier Science)

1984). A related structure, the complexus adhaerentes, has been described in the lymphatic endothelium and in the sinus of lymph nodes. This structure is unique, containing a classical cadherin and the desmosome components plakoglobin and desmoplakin (Schmelz and Franke 1993).

Early procedures developed to enrich for desmosomes from complex stratified epithelia facilitated the identification of a number of protein and glycoprotein components (Skerrow and Matoltsy 1974), which subsequently have been isolated and characterized as belonging to three major groups: desmosomal cadherins, armadillo proteins, and plakin proteins (Fig. 2) (see Kowalczyk et al. 1999a for review). Each of these major families comprises multiple members, many of which are possible constituents of desmosomes. Other minor components, which cannot be categorized into these gene families, add to desmosome complexity. In the following sections (Sects. 3 through 5), each class of desmosomal components will be discussed in detail.

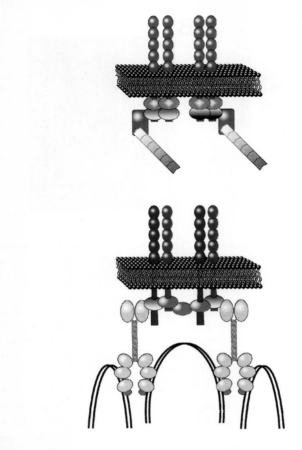

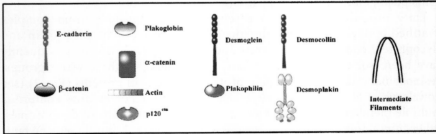

Fig. 2. Structural comparison of desmosomal and adherens junction plaques. In an adherens junction (*top*), the extracellular domains of the transmembrane classic cadherins are involved in homophilic adhesion, whereas the cytoplasmic tails associate with a complex of proteins including p120 catenin and either β-catenin or plakoglobin. These latter armadillo proteins associate with α-catenin, which anchors the actin cytoskeleton. In the desmosome (*bottom*), the extracellular domains of the desmosomal cadherins, desmogleins, and desmocollins mediate adhesion, possibly through heterophilic interactions.

2
The Desmosomal Cadherins

The transmembrane components of desmosomes include members of the cadherin superfamily of calcium-dependent cell adhesion molecules (reviewed in Nollet et al. 2000). The best-studied members of this family are the classical cadherins, which are typically localized to adherens junctions. Comparison of the molecular domain structure and amino acid sequences among classical and desmosomal cadherins suggests that the desmosomal cadherins belong to a distinct subfamily. The desmosomal cadherins can be further subdivided into the desmogleins (Dsg) and desmocollins (Dsc) (Fig. 3). In humans, there are four distinct desmoglein isoforms (Dsg1–4) and three desmocollin isoforms (Dsc1–3) that are the products of closely linked genes on chromosome 18 (Kljuic et al. 2003; Whittock and Bower 2003; reviewed in Nollet et al. 2000). Two novel Dsg1-related genes, termed Dsg1-β and Dsg1-γ have been identified in mice (Kljuic and Christiano 2003; Pulkkinen et al. 2003). Their functional significance and the reason behind the absence of additional Dsg1-related genes in humans remains to be determined. Adding to the complexity of the subfamily, Dsc transcripts are spliced to form two distinct isoforms—an "a" form and a "b" form with a shorter cytoplasmic domain (Collins et al. 1991). Desmocollin "a" isoforms reach further into the outer dense plaque than do the shorter desmocollin "b" isoforms, while the desmogleins span the outer dense plaque (North et al. 1999).

The classical and desmosomal cadherins possess several common structural characteristics. The proteins are translated as precursor proteins further processed by a common family of endoproteases, the subtilisin-like proprotein convertases (Posthaus et al. 1998, 2003). However, the size and composition of the precursor peptides vary among cadherins, with Dsg peptides containing a shorter precursor peptide than classical and Dsc isoforms (reviewed in Nollet et al. 2000). Five ectodomains with conserved calcium-binding motifs are present in the mature extracellular region of all cadherins, including four cadherin repeats (EC1–EC4) and a less-well conserved membrane proximal subdomain, sometimes referred to as the extracellular anchor (EA) or EC5. Calcium stabilizes the structure of classical cadherin ectodomains, possibly contributing to the calcium-dependence of cadherin-

---

Their cytoplasmic tails associate with armadillo family members plakoglobin and plakophilin, which link to desmoplakin. Desmoplakin in turn binds to the IF cytoskeleton, affording strength to this junction. Other proteins found in the inner dense plaque include the plakin proteins plectin, envoplakin, and periplakin. The precise spatial organization of pinin, desmocalmin, keratocalmin, corneodesmosin, and polycystin-1 is uncertain

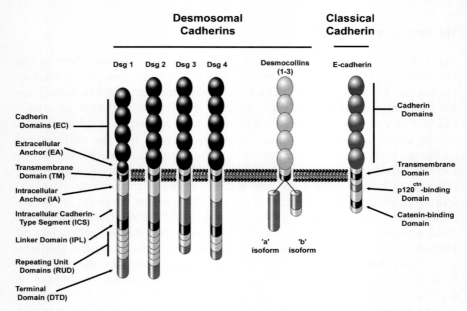

Fig. 3. The desmosomal cadherins. A schematic diagram of the desmosomal cadherins, desmoglein, and desmocollin, compared with the classic cadherin E-cadherin. Cadherins are type I transmembrane proteins, with an amino terminal extracellular domain and an intracellular cytoplasmic domain. Desmocollins can be further broken down into two alternatively spliced forms the "a" and "b" forms. The extracellular domain contains four repeat domains (EC1–4) involved in calcium binding and adhesion between cells. An extracellular and intracellular anchor (EA or EC5 and IA) flank the transmembrane domain (TM). The ICS domains of Dsg and Dsc "a" bind the armadillo family member, plakoglobin. The Dsg tail has several unique domains, the proline-rich linker domain (IPL) a repeated unit domain (RUD) and a terminal domain (DTD). The precise binding sites for the PKPs have not been elucidated but in Dsg may involve sequences in both the ICS and downstream regions. The "b" form of Dsc can bind to PKP3

mediated adhesion (Pertz et al. 1999). A new molecular model derived from electron tomography data suggests a flexible cadherin interface in which the desmosomal cadherins interact via their tips in the region of the EC1 domains generating both *cis* and *trans* interactions (He et al. 2003). These observations are consistent with interactions predicted by the crystal structure of the classical cadherin, C-cadherin.

Functional analysis of the desmosomal cadherin ectodomains has lagged behind that of the classical cadherins. Studies in which desmosomal cadherins were introduced into normally non-adherent L cell fibroblasts and allowed to aggregate in suspension are consistent with the idea that, unlike classical cadherins, a single desmosomal cadherin is insufficient to mediate robust calcium-dependent adhesion. Both desmocollins and desmogleins are necessary (Amagai et al. 1994; Chidgey et al. 1996; Kowalczyk et al. 1996; Marcozzi et al. 1998; Tselepis et al. 1998). In addition, disruption of desmosomal adhesion requires the presence of cell-adhesion recognition (CAR) peptides specific for both Dsg and Dsc isoforms (Tselepis et al. 1998; Runswick et al. 2001). The possibility that Dsg and Dsc interact heterophilically is supported by ectopic transfection studies in which Dsg:Dsc complexes formed between adjacent cells were isolated from a human fibrosarcoma cell line (Chitaev and Troyanovsky 1997). Heterophilic complexes also form in a calcium-dependent manner in in vitro bacterial expression systems using recombinant EC1–2 subdomains of Dsg1/Dsc2 or Dsg2/Dsc2 (Chitaev and Troyanovsky 1997; Syed et al. 2002). Collectively, these studies emphasize important differences between the organization of classical and desmosomal cadherin-based junctions.

Like classic cadherins, the desmosomal cadherin cytoplasmic domains serve as scaffolding for proteins that mediate linkage with the cytoskeleton (Fig. 3). All share an intracellular anchor (IA) subdomain proximal to the plasma membrane and an intracellular cadherin-type sequence (ICS), which lies just downstream and is required for binding to β-catenin or plakoglobin. However, the desmosomal cadherins also exhibit unique features not found in other cadherins. In desmogleins, the ICS is followed by an extended cytoplasmic domain including an intracellular proline-rich linker region (IPL), a variable number of 29 amino acid repeated unit domains (RUD), and a glycine-rich desmoglein terminal domain (DTD) (Koch et al. 1990) (and see Nollet et al. 2000 for review). The desmocollins lack this desmoglein-specific extension. The desmocollin "a" form contains a typical ICS domain, but the "b" form lacks this domain and instead terminates in 11 residues absent in the desmocollin "a" form. Ectopic expression of the Dsg3 or Dsg1 cytoplasmic domains impaired desmosome formation, as did a connexin-Dsg1 chimera and even full length Dsg1 expressed out of its normal cellular context (Troyanovsky et al. 1993; Norvell and Green 1998; Hanakawa et al. 2000; Serpente et al. 2000). Although the mechanism underlying this disruption is not well understood, it may be due to sequestration of the cytoplasmic binding partners plakoglobin and plakophilin (PKP).

The functions of the desmosomal cadherin IA domains are poorly understood. In the classical cadherins, this region has been shown to provide a binding site for p120ctn, which modulates adhesive function by facilitating cadherin stabilization and clustering (see Anastasiadis and Reynolds 2000 for review). Although the IA subdomain of Dsc1 appears to play a role in recruiting desmoplakin to the membrane, the comparable region in Dsg3 is not sufficient to mediate its incorporation into the desmosomes of a human epidermoid cell line (Troyanovsky et al. 1994b; Andl and Stanley 2001).

While classical cadherins interact with β-catenin or plakoglobin via their ICS domains, the desmosomal cadherins normally bind exclusively to plakoglobin. However, under certain circumstances, such as when desmoglein 1 is expressed out of its normal context (Norvell and Green 1998), or in plakoglobin-null keratinocytes (Bierkamp et al. 1999), promiscuous association of β-catenin with desmogleins has been observed (Wahl et al. 2000). Expression of a Dsc3 extracellular domain deletion mutant which was shown to affect adherens junction formation in addition to desmosome assembly, also associated promiscuously with both plakoglobin and β-catenin (Hanakawa et al. 2000). The authors suggested that Dsc3 might be involved in desmosome nucleation downstream of adherens junction assembly. Using a series of β-catenin/plakoglobin chimeras, Wahl et al. (2000) showed that the first 26 resides of β-catenin along with its C-terminal tail interfere with its ability to bind to Dsg2. The shorter Dsc "b" isoforms contain an incomplete ICS and therefore do not bind plakoglobin and failed to mediate desmosome plaque assembly in vitro (Troyanovsky et al. 1993, 1994b). In vivo, however, desmosomes are normal in mutant mice expressing a truncated Dsc1 lacking both the Dsc1a- and Dsc1b-specific domains (Cheng et al. 2004). Together these observations are consistent with the idea that desmosome formation and function are regulated both by the cadherin isoform expressed as well as the cadherin ICS and its armadillo binding partners. The fact that desmosome and adherens junction components can partner with each other in certain cases raises the possibility that these interactions are dynamic and regulated intracellularly during junction assembly and disassembly.

The function(s) of the desmoglein-specific regions of the cytoplasmic tail are for the most part a complete mystery. Homophilic association between RUD domains in rotary shadowed preparations of recombinant proteins has been observed, and it is tempting to speculate that the RUD region may initiate clustering, which triggers cytoplasmic structural or signaling pathways (Rutman et al. 1994). Although data suggest that PKP1's association with Dsg1 may depend in part on sequences in this extended tail, very little is known about other putative interacting proteins (Hatzfeld et al. 2000). Identification of these partner molecules will be critical for our future understanding of structural and possible signaling functions of these domains.

Beginning in the mid-1980s literature began to emerge which suggested that multiple desmosomal cadherin isoforms exist, and, furthermore, that these isoforms exhibit a tissue- and differentiation-specific pattern. The historical accounting of this important literature can be accessed in the reviews by Burdett (1998) and Garrod et al. (2002). Too new for inclusion in these reviews is the identification of desmoglein 4 (Kljuic et al. 2003). Dsg2 and Dsc2 have a widespread distribution in simple epithelia and the basal layer of stratified tissues. In the epidermis the cadherins are distributed as an inverse gradient of expression with Dsg1, Dsc1, and Dsg4, concentrated in the most terminally differentiated layers of the skin, while Dsg3 and Dsc3 are expressed in the basal and suprabasal layers (Fig. 4). The consequent expression pattern results in isoform intermixing within the intermediate epidermal layers; and in at least one instance, immunogold localization demon-

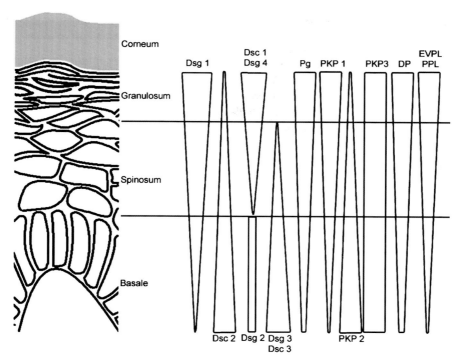

Fig. 4. Desmosome components in stratified epithelium. The distribution and relative level of expression of individual desmosome components is indicated to the *right* of a cartoon showing the various layers of the epidermis

strated that Dsc1 and Dsc3 can be present in a single desmosome (North et al. 1996). Factors that are important for establishing the cell-type and tissue-specific patterns of desmosomal cadherin expression have not been well defined. However, it was shown that protein kinase C (PKC) downregulation suppresses Dsg1 expression in vitro, consistent with its co-regulation with other granular layer markers during keratinocyte differentiation (Denning et al. 1998). Furthermore, recent evidence suggests that activation of the RhoA effector ROCK-II in human keratinocytes inhibits cell cycle progression and increases expression of differentiation-specific genes, including desmosomal proteins (McMullan et al. 2003). The specific mechanisms by which RhoA-mediated signaling, which is known to regulate the actin cytoskeleton and adhesion, controls keratinocyte cell fate decisions have not been defined.

3
Armadillo Proteins of the Desmosome

Armadillo, a protein that regulates segment polarity in *Drosophila*, was later revealed as the homolog of the dual junctional and signaling molecule β-catenin (Peifer et al. 1992). Armadillo proteins are characterized by a series of imperfect 42 amino acid repeats, called arm repeats (Figure 5), which facilitate protein–protein interactions. Members of this family include the adherens junction components β-catenin and p120ctn and the desmosomal armadillo proteins, plakoglobin and PKPs (reviewed in Hatzfeld 1999; Anastasiadis and Reynolds 2000). One of the primary functions attributed to these proteins is to mediate the interaction between desmoplakin and the cadherins, and in the case of PKPs, to facilitate lateral clustering of plaque proteins. However, like other armadillo family members, plakoglobin and PKP are alsonuclear proteins and may play roles in basal or regulated transcriptional machinery.

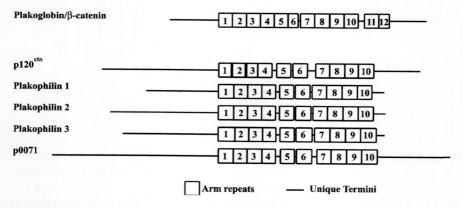

Fig. 5. Armadillo proteins in desmosomes. A schematic diagram of plakoglobin and the p120 family including three plakophilin isoforms (1–3) and p0071. All the proteins contain unique N- and C-terminal domains flanking armadillo repeat domains, which are imperfect 42 amino acid repeats

3.1
Plakoglobin

Plakoglobin (85 kDa) is encoded by a gene on human chromosome 17q21 (Aberle et al. 1995) and is a constituent of both adherens junctions and desmosomes (Cowin et al. 1986). Like β-catenin, plakoglobin is a member of the armadillo family characterized by 12 arm repeat domains flanked by distinct N- and C-termini (Peifer et al. 1992; Huber et al. 1997). Based on the high-resolution structure of β-catenin, plakoglobin is predicted to have arm domains comprising three small α-helices interconnected by loops that coordinate to form a superhelix of helices. This superhelix creates a shallow, positively charged groove thought to facilitate protein–protein interactions (Huber et al. 1997).

The central plakoglobin arm repeats bind directly to the cytoplasmic domains of the classical cadherins (Troyanovsky et al. 1996; Wahl et al. 1996; Witcher et al. 1996; Chitaev and Troyanovsky 1998), facilitating the formation of a complex with a 1:1 stoichiometry (Ozawa and Kemler 1992). In contrast, desmosomal cadherin binding depends heavily on the N-terminal plakoglobin arm repeats, with further cooperation from downstream repeats, particularly in the case of Dscs (Troyanovsky et al. 1996; Wahl et al. 1996; Witcher et al. 1996; Chitaev et al. 1998). Early work suggesting that plakoglobin binds to the Dsg1 isoform in a 6:1 stoichiometry (Kowalczyk et al. 1996; Witcher et al. 1996) has recently been re-visited and the data suggest that whereas plakoglobin binds Dsg2 and Dsg3 with 1:1 a stoichiometry, the plakoglobin:Dsg1 ratio does not appear to exceed 2:1 (Bannon et al. 2001). The plakoglobin armadillo domains also mediate desmoplakin binding, and are thus critical for anchoring the IF cytoskeleton to the plaque (Kowalczyk et al. 1997; Smith and Fuchs 1998; Bornslaeger et al. 2001). The adherens junction component α-catenin associates with plakoglobin; its binding site, which is near the first armadillo repeat, overlaps with sequences important for desmosomal cadherin binding (Sacco et al. 1995). This observation may explain why plakoglobin bound to desmosomal cadherins does not complex with α-catenin, but can still associate with desmoplakin (Plott et al. 1994; Wahl et al. 1996; Witcher et al. 1996; Kowalczyk et al. 1997; Chitaev et al. 1998).

The N- and C-terminal domains of plakoglobin regulate its protein interactions and desmosome assembly. Deletion of the plakoglobin N-terminus had no effect on desmosome morphology, while deletion of the C-terminus caused desmosomes to assemble into long continuous structures, suggesting its importance in limiting the size of the individual plaques (Palka and Green 1997). As the plakoglobin C-terminus has been reported to associate intramolecularly with the central arm repeats, its loss may reveal cryptic binding sites that enhance interactions with components such as desmoplakin, thus contributing to elongated plaques (Troyanovsky et al. 1996). The plakoglobin C-terminus also contains multiple phosphorylation sites, which regulate its interactions with binding partners and may affect its involvement in transcriptional regulation, subjects which will be dealt with in Sect.28) below.

3.2
Plakophilins

PKPs belong to a subfamily of armadillo proteins related to p120ctn (Bonne et al. 1999; Schmidt et al. 1999; and reviewed in Hatzfeld 1999; Anastasiadis and Reynolds 2000). Like other armadillo family members, PKPs exhibit both junctional and nuclear localizations (Mertens et al. 1996; Schmidt et al. 1997; Bonne et al. 1999). Three desmosomal PKPs (1–3) have been described and PKP1 and PKP2 each have two isoforms termed "a" and "b" that result from alternative splicing (Mertens et al. 1996; Schmidt et al. 1997). A fourth protein called p0071, also known as PKP4, falls into a distinct subclass more highly related to p120ctn and δ-catenin (Hatzfeld and Nachtsheim 1996). Each PKP is composed of 10 central arm repeats flanked by N- and C-termini, which are considerably more divergent than the arm repeats (reviewed in Hatzfeld 1999).

PKP1 (~75 kDa) corresponds to Band 6 of desmosome-enriched preparations and was first characterized as a keratin-binding protein (Kapprell et al. 1988). The PKP1 gene, located on human chromosome 1q32 (Schmidt et al. 1999), is widely expressed and present in nuclei even in cells that do not assemble desmosomes (Hatzfeld et al. 1994; Heid et al. 1994; Schmidt et al. 1994). However, junctional PKP1 is localized primarily within suprabasally located desmosomes of stratified epithelia (Moll et al. 1997; Schmidt et al. 1997). The PKP1b form is restricted to the nuclei of stratified epithelia (Schmidt et al. 1997).

The PKP1 head domain, which contains the binding sites for all known binding partners, targets PKP1 to desmosomes and the nucleus (Kowalczyk et al. 1999b; Hatzfeld et al. 2000). PKP1 binds to Dsg1, keratins, Dsc1, and the N-terminus of desmoplakin, the latter interaction resulting in enhanced recruitment of desmoplakin to desmosomes (Smith and Fuchs 1998; Kowalczyk et al. 1999b; Hatzfeld et al. 2000; Hofmann et al. 2000; Bornslaeger et al. 2001). It has been proposed that PKP1 strengthens the desmosome by promoting lateral protein interactions (Kowalczyk et al. 1999b). In support of this idea, cells lacking PKP1 exhibit reduced desmosome stability, size, and number, correlated with a decreased content of desmosome proteins (McMillan et al. 2003; South et al. 2003). A conserved motif within the arm repeat domain appears to be involved in the formation of filopodia and long cellular protrusions, suggesting that, like other p120 family members, PKP1 may regulate the actin cytoskeleton (Hatzfeld et al. 2000).

PKP2a and PKP2b (~92 kDa and ~97 kDa, respectively) are alternatively spliced products of a single gene on human chromosome 12p11 (Mertens et al. 1996; Schmidt et al. 1999). In contrast to PKP1, PKP2 is found in desmosomes of all epithelia as well as the myocardium and lymph nodes, and is concentrated in basal layers of stratified epithelia (Mertens et al. 1996). The PKP2a N-terminal head interacts directly with the plakoglobin arm domain, the desmoplakin N-terminus, Dsg1, Dsg2, Dsc1a, and Dsc2a, and perhaps indirectly with Dsg3 (Chen et al. 2002). Although, like PKP1, PKP2a recruits desmoplakin to cell borders, it is less efficient at doing so, perhaps due to a weaker interaction between the two proteins. A subset of PKP2 in the nucleus is found in complexes with RNA polymerase III holoenzyme, interacting directly with the RPC155 subunit, although its potential role in regulating this complex has not been elucidated (Mertens et al. 2001). In addition, PKP2 associates with non-junctional β-catenin and regulates its signaling activity (Chen et al. 2002), consistent with the possibility that like plakoglobin, the PKP family of armadillo proteins engages in crosstalk with the Wnt signaling pathway.

PKP3 (~87 kDa), the most recently described member of the subfamily, is encoded by a gene on human chromosome 11p15 (Bonne et al. 1999; Schmidt et al. 1999). PKP3 is detected in desmosomes of most simple epithelia and throughout the layers of all stratified epithelial except for hepatocytes (Bonne et al. 1999; Schmidt et al. 1999). Yeast two-hybrid (Y2H) and co-immunoprecipitation analyses suggested that PKP3 binds to plakoglobin, desmoplakin, Dsg1–3, Dsc3a, and Dsc3b, with possible binding to Dsc1a, Dsc1b, Dsc2a, and Dsc2b observed only using the Y2H approach (Bonne et al. 2003). The association with Dsc3b is the first reported for any Dsc "b" isoform. PKP3 also associated with keratin 18, consistent with previous reports that PKP1 and 2 bind IF in vitro (Kapprell et al. 1988; Hatzfeld et al. 2000; Hofmann et al. 2000). The physiological relevance of PKP's keratin-binding properties has been questioned, as its location within the junctional plaque suggests it is inaccessible to IFs (North et al. 1999). However, others have proposed that in the absence of desmoplakin, PKP may play a role in anchoring keratin filaments (Norgett et al. 2000; Vasioukhin et al. 2001; Bonne et al. 2003).

PKP3, like PKP1 and PKP2, is able to recruit desmoplakin to cell–cell borders. Interestingly, two PKP3–desmoplakin binding sites were uncovered, and these depend on different sequences within the head domains of each protein (Bonne et al. 2003). This observation lends support to the hypothesis that by recruiting desmoplakin to desmosomes, PKPs facilitate lateral interactions within the plaque, increasing desmosome strength and keratin recruitment and size in the upper layers of the epithelium (Kowalczyk et al. 1999b; Bornslaeger et al. 2001; Bonne et al. 2003; McMillan et al. 2003).

3.3
p0071 (Plakophilin 4)

The protein p0071, which maps to 2q23-q31 in humans, is predicted to be 135 kDa in size, but migrates at 150 kDa by sodium dodecyl sulfate polyacrylamide gel electrophoresis (SDS-PAGE) (Hatzfeld and Nachtsheim 1996; Bonne et al. 1998). RT-PCR and Southern blot studies confirm the expression of two splice variants of p0071. p0071 is widely expressed in tissues and predominant in epithelia. Although it was shown to colocalize with desmoplakin in desmosomes of HeLa and A431 cells, it colocalizes with E-cadherin in the adherens junctions of MDBK cells, thus setting it apart from PKP1–3. p0071 is also found in the complexus adhaerentes of microvascular endothelial cells, associating with the VE-cadherin juxtamembrane domain via its armadillo domains and with desmoplakin via its N-terminal head domain (Calkins et al. 2003). Thus, p0071 may act as a facilitator of desmoplakin binding to VE-cadherin in a scenario analogous to PKPs facilitating the incorporation of desmoplakin into desmosomes. The p0071 N-terminus also interacts with Dsc3a and plakoglobin, whereas the armadillo repeat domain binds to the classical cadherin, E-cadherin. These domains may mediate localization to either the desmosome or the adherens junction depending on the cell type (Hatzfeld et al. 2003).

p0071 interacts with presenilin1, which is thought to be important in both Alzheimer's disease and in activating transmembrane receptors of the Notch pathway (Levesque et al. 1999; Stahl et al. 1999). Together with the fact that presenilin mutations previously linked to Alzheimer's disease prevent nuclear β-catenin translocation in response to Wnt, this finding raises the possibility that multiple armadillo proteins may be involved in disease pathogenesis (Soriano et al. 2001). p0071 binds to two PDZ proteins, erbin and PAPIN, suggesting it may also be involved in regulating or maintaining cell polarity in neurons and epithelial cells. (Deguchi et al. 2000; Jaulin-Bastard et al. 2002).

4
Plakins

The growing plakin family of cytolinkers currently comprises seven members that are characterized by the presence of combinations of modules, which perform different cytoskeleton linking functions (see Ruhrberg and Watt 1997; Leung et al. 2001 for review). This section will focus on the four plakins that have been localized to desmosomes: desmoplakin, plectin, envoplakin, and periplakin. Each contains an N-terminal plakin domain, a coiled-coil rod domain, a linker subdomain, and all but periplakin contain one or multiple plakin repeat domains (Fig. 6). While all may help to link the IF cytoskeleton to desmosomes, the evidence for desmoplakin's anchoring function is most compelling.

The Molecular Composition and Function of Desmosomes 151

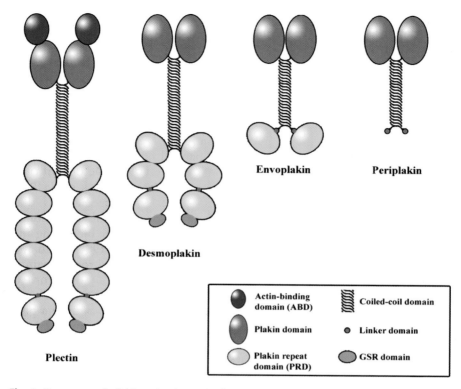

Fig. 6. Desmosomal plakins. A schematic diagram of the plakin family members present in the desmosome. All of the proteins contain an N-terminal plakin domain containing a series of antiparallel alpha helical bundles. The plakin domain is followed by a coiled-coil rod domain, which facilitates protein homodimerization (and heterodimerization in the case of envoplakin and periplakin). The rod is followed by a C-terminal series of plakin repeat domains (PRDs) in all but one family member, periplakin, and a linker domain which is present in all family members. The C-terminal plakin repeat domains and the linker region are involved in IF binding. Plectin also contains an actin binding domain (ABD) at the amino terminus and a glycine-serine-arginine (GSR) domain at the C-terminus, which binds to microtubules

4.1
Desmoplakin

Desmoplakin is an obligate constituent of the inner plaque of the desmosome (North et al. 1999) and is believed to be the most abundant component of the desmosome (Mueller and Franke 1983). The human gene for desmoplakin is located on chromosome 6p24 and encodes the alternatively spliced proteins, desmoplakin I, with a size of ~332 kDa, and desmoplakin II at ~260 kDa, which is missing 599 residues in the rod domain (Arnemann et al. 1991; Virata et al. 1992). Desmoplakin I and II are expressed in a wide variety of tissues; however desmoplakin II is absent from heart and less abundant in simple epithelia (Angst et al. 1990). Desmoplakin associates with keratins in epithelia, desmin in the heart, and vimentin in meningeal cells and follicular dendritic cells of lymph nodes (reviewed in Schwarz et al. 1990; Kowalczyk et al. 1999a). It is also present in specialized non-desmosomal junctions known as syndesmos or complexus adhaerentes in endothelial cells (Schmelz and Franke 1993; Kowalczyk et al. 1998).

The amino terminal plakin domain comprises a number of heptad repeats with a predicted organization of antiparallel bundles. This domain is important for localizing desmoplakin to the plaque via interaction with the armadillo binding proteins plakoglobin and PKPs (Stappenbeck et al. 1993; Kowalczyk et al. 1997; Bornslaeger et al. 2001). Evidence suggests that desmoplakin might also bind directly with at least two of the desmosomal cadherins, Dsg1 and Dsc1, thus providing a mechanism for increasing the complexity of interactions in the plaque (Troyanovsky et al. 1993; Smith and Fuchs 1998; Bornslaeger et al. 2001). The central alpha helical rod domain is thought to mediate homodimerization of the monomers in a parallel orientation (O'Keefe et al. 1989; Green et al. 1990; Virata et al. 1992; North et al. 1999). Together these observations suggest that the plaque is built of a protein meshwork bolstered by linear and lateral protein–protein interactions. The resulting structure is likely important for enhancing tethering to the IF cytoskeleton, which in turn increases adhesive strength (Kowalczyk et al. 1999b; Huen et al. 2002). Consistent with this idea, a desmoplakin mutant which retains the plakoglobin and PKP binding sites, but lacks the C-terminal IF binding domain, was shown to effectively uncouple the IF network from the plaque (Bornslaeger et al. 1996), and resulted in markedly decreased adhesive strength between cells (Huen et al. 2002).

The desmoplakin C-terminus contains three plakin repeat domains (PRDs) termed A, B, and C, each connected by intervening sequences. Each of the PRDs consists of 4.5 copies of a 38 residue "plakin repeat" (PR) (Green et al. 1990; Choi et al. 2002). High-resolution structural analysis of domains B and C revealed that, in contrast to other proteins built from tandem repeat motifs, the plakin repeats participate in unique inter-repeat packing that results in the formation of a globular, rather than linear, structure (Choi et al. 2002). Multiple repeats are likely to take on a "beads-on-a-string" arrangement. Each globular repeat domain contains a shallow groove 10–15 Å wide and 22–26 Å long, lined with conserved, positively charged residues. This groove lies between the two halves of the plakin repeat domain formed by repeats 1–2 and 3–4 and is surrounded by surface that is neutral or acidic. It was suggested that this groove could interact with the face formed by approximately four turns of an alpha helix such as those present in the rod domains of IFs.

That the desmoplakin C-terminus physically associates with vimentin, desmin, and simple epithelial and epidermal keratins has been demonstrated by direct protein binding assays and co-sedimentation analysis, yeast two- and three-hybrid interaction approaches, and co-alignment of ectopically expressed protein in a variety of cell types (see Kowalczyk et al. 1999a for review). Each of the plakin repeat domains were found by in vitro co-sedimentation to exhibit specific, albeit low affinity, interactions with vimentin. B plus C, or a protein containing all three PRDs, exhibited considerably greater binding. In these studies, addition of the "linker" between the B and C subdomains did not increase binding to vimentin, in spite of the reported importance of this region in association with vimentin and K8/18 using a yeast three-hybrid approach (Fontao et al. 2003). In contrast to earlier work demonstrating that the K5 head domain alone associates in vitro with recombinant desmoplakin, more recent yeast three-hybrid analyses suggested that the keratin heterodimer is required for interactions with desmoplakin (Kouklis et al. 1994; Meng et al. 1997; Fontao et al. 2003). Collectively, these studies also suggest that different IF types require overlapping but distinct sets of sequences in the desmoplakin C-terminus for optimal binding. A glycine–serine–arginine (GSR) domain is found at the extreme C-terminus that has been shown to mediate microtubule binding in some plakins. While desmoplakin does contain a C-terminal GSR domain, it does not appear to bind microtubules (Sun et al. 2001). Together, these data underscore the complexity of plakin–IF interactions and suggest that future work will be required to sort out the precise molecular mechanisms involved.

4.2
Plectin

Plectin is widely expressed in epithelial and non-epithelial cells. It is abundantly expressed in many tissues, such as stratified and simple epithelia, muscle, intercalated discs of cardiac muscle and cells forming the blood–brain barrier (reviewed in Wiche 1998). It functions as a versatile cytolinker protein that crosslinks filament networks in the cytoplasm and participates in linking the cytoskeleton to the plasma membrane in desmosomes, hemidesmosomes, and focal contacts (reviewed in Wiche 1998). Plectin, which has been shown to be the same as IFAP300 (Clubb et al. 2000), has a molecular weight of greater than 500 kDa and is encoded by a gene found on human chromosome 8q24 (reviewed in Pulkkinen and Uitto 1999). Plectin has many isoforms generated from multiple transcriptional start sites and differential splicing, primarily involving the N-terminus. Plectin 1a is the most prominently expressed isoform in human keratinocytes (Andra et al. 2003).

Plectin mediates interactions among the cytoskeletal networks by interacting directly with IFs via its C-terminal plakin repeats, actin filaments via the N-terminal actin binding domain (ABD) and microtubules via the C-terminal GSR domain (reviewed in Wiche 1998; Leung et al. 2001). The presence of plectin in desmosomes may be due in part to its interactions with desmoplakin (Eger et al. 1997), while its presence in hemidesmosomes is mediated by interactions with $\beta 4$ integrin (Rezniczek et al. 1998).

Homozygous mutations in the gene encoding plectin are responsible for epidermolysis bullosa simplex, a blistering disease of the basal layer, associated with muscular dystrophy (EBS-MD) (Pulkkinen and Uitto 1999), and disruption of the plectin gene in mice likewise results in blistering and muscle deficiency (Andra et al. 1997). Together, the human and mouse data suggest that plectin may play a more prominent role in hemidesmosomes than desmosomes.

4.3
Envoplakin and Periplakin

Envoplakin and periplakin are found within and between desmosomes (Ma and Sun 1986; Ruhrberg et al. 1996a, 1997; DiColandrea et al. 2000) and have been shown to be building blocks of the cornified envelope in stratified epithelia. Neither component is present in simple epithelia, mesenchymal tissue, or the heart (Ruhrberg et al. 1996a, 1997); thus, these plakins are not obligate constituents of desmosomes. Envoplakin is a 210-kDa molecule encoded by a gene on chromosome 17q25, while periplakin is a protein of 195 kDa, encoded by a gene on chromosome 16p13 (Ma and Sun 1986; Ruhrberg et al. 1996b, 1997; Aho et al. 1998).

Like other plakin family members, envoplakin and periplakin each contain coiled-coil rod domains and amino terminal plakin domains. The C-terminus of envoplakin contains a single C plakin repeat and the linker domain (L) implicated in IF binding, while the C-terminus of periplakin contains only the linker domain (L). Unlike other plakins, periplakin and envoplakin may not only be able to homodimerize, but also heterodimerize, via their rod domains (Ruhrberg et al. 1997). Periplakin also binds to actin via its N-terminus, although the exact binding sites have not been defined and it does not contain the conserved actin-binding domain present in other plakins (DiColandrea et al. 2000). When expressed together, periplakin and envoplakin align with the IF cytoskeleton, a function requiring the linker domains; furthermore, periplakin may be important for the proper localization of envoplakin to cell–cell borders (Ruhrberg et al. 1997; DiColandrea et al. 2000; Karashima and Watt 2002). Gene deletion studies of envoplakin demonstrated a delay in cornified envelope formation and a corresponding delay in acquisition of barrier function (Maatta et al. 2001).

5
Other Desmosome Proteins

A number of additional proteins have been localized to the desmosome, but their functions are poorly understood. Pinin, a 140-kDa protein, is expressed in many tissues including heart, brain, placenta, lung, liver, and pancreas. Based on the facts that pinin is associated preferentially with mature desmosomes, and that its ectopic expression enhanced cell–cell adhesion and increased desmosome number (Ouyang and Sugrue 1996), it has been proposed that pinin plays a role in stabilizing IF at the plaque. However, other investigators have raised the question as to whether pinin is a desmosome constituent, reporting that it is found exclusively in the nucleus and is also present in cells devoid of desmosomes (Brandner et al. 1997).

Corneodesmosin is in desmosomes of the cornified layers of the epidermis and possibly the inner root sheath of hair follicles (Serre et al. 1991; Mils et al. 1992). This glycoprotein, which has been demonstrated to engage in homophilic adhesion, is processed from 52–56 kDa into smaller molecules as it progresses to the upper cornified layers, possibly contributing to desquamation. (Simon et al. 2001; Jonca et al. 2002; Levy-Nissenbaum et al. 2003). Nonsense mutations in the CDSN gene have been linked to hypotrichosis simplex of the scalp, highlighting the importance of corneodesmosin in hair follicle morphogenesis (Levy-Nissenbaum et al. 2003).

Another protein recently associated with desmosomes is a large membrane-spanning glycoprotein called polycystin-1. Autosomal dominant mutations in the polycystic kidney disease 1 gene result in a devastating cystic disease of kidney as well as other organs (The European Polycystic Kidney Disease Consortium 1994). Polycystin-1 is a proposed regulator of G protein signaling (Kim et al. 1999). Based on the observation that polycystin-1 colocalized with desmosomes after desmoplakin incorporation into the plaque (Scheffers et al. 2000), it is speculated that this protein may play an accessory role in regulating adhesive strength and/or cell signaling. More recently, Roitbak et al. discovered that polycystin-1 forms a complex with E-cadherin in normal kidney cells, and that this complex is disrupted in cells isolated from patients with autosomal dominant polycystic kidney disease (Roitbak et al. 2004). The depletion of these proteins from the membrane correlates with a switch to N-cadherin expression, which has been associated with a more de-differentiated phenotype in other systems. Whether polycystin-1 plays a more direct role in adherens junctions than desmosomes is a question which will require further investigation.

Desmocalmin was purified from bovine muzzle epidermis using a calmodulin affinity column and shown to bind keratin filaments in a Mg^{2+}-dependent manner (Tsukita 1985). Keratocalmin, isolated from human epidermis and localized to cell–cell borders in keratinocytes (Fairley et al. 1991) could be the human equivalent of desmocalmin. Although the calmodulin binding activity of desmocalmin suggests a potential role in the regulation of these calcium-dependent junctions, such a function has not yet been identified.

6
Diseases Involving Desmosomes

The desmosome is a target for autoimmune, infectious, and genetic disorders, which often exhibit common features such as skin blistering and hyperkeratosis. Although many are apparent during infancy, some genetic diseases are linked to fatal cardiomyopathies that do not manifest until the late teens and early adulthood. Table 1 lists details about the diseases involving desmosomes.

6.1
Autoimmune and Infectious Disease

Four autoimmune diseases have been linked to the desmosome: pemphigus foliaceus (PF), pemphigus vulgaris (PV), herpetiformis pemphigus (HP), and paraneoplastic pemphigus (PNP). Staphylococcal scalded skin syndrome (SSSS) and a more common condition, bullous impetigo, are caused by the infectious agent *Staphylococcus aureus*, which similarly targets the desmosome. In contrast to diseases of the intracellular IF cytoskeleton which lead to cell fragility and cell lysis, blistering of the skin and/or mucous membranes in these diseases is caused by acantholysis, i.e., a loss of cell–cell adhesion (reviewed in Nousari and Anhalt 1999; Stanley 2001). This realization led investigators to suggest early on that pemphigus antibodies interfere with the intercellular adhesion machinery.

The discovery that patients with pemphigus have circulating antibodies directed against members of the desmoglein family supported the idea that desmosomes are the pathogenic target in pemphigus. Although the accumulated evidence supports the hypothesis that pemphigus lesions are caused by pathogenic IgG antibodies directed against the desmosomal cadherins, alternative targets such as cholinergic receptors have also been proposed (reviewed in Amagai 1999; Nguyen et al. 2000). While efforts are underway toward understanding how pathogenic antibodies arise and how to combat the disease (reviewed in Amagai 1999; Scully and Challacombe 2002), this chapter will focus primarily on the desmogleins as targets, and the possible mechanisms by which their adhesive function is impaired.

Table 1. Diseases involving desmosomes

	Mutated gene/target antigen	Phenotype	Selected references
Genetic Diseases	Plakoglobin	*Naxos disease* Autosomal recessive arrhythmogenic right ventricular cardiomyopathy (ARVC), palmoplantar keratoderma, and woolly hair	McKoy et al. 2000; Protonotarios et al. 1986, 2001b
	Plakophilin 1	Autosomal recessive ectodermal dysplasia, skin fragility syndrome, hair loss, and nail dystrophy	McGrath 1999; McGrath et al. 1997; Whittock et al. 2000
	Desmoglein 1	*Striate palmoplantar keratoderma (SPPK)* Lesions of the palms and soles exacerbated by mechanical trauma	Hunt et al. 2001; Rickman et al. 1999
	Desmoglein 4	*Autosomal recessive hypotrichosis (LAH)* Loss of hair on scalp, chest, arms, legs, and sparse facial hair due to hair follicle abnormalities	Kljuic et al. 2003
	Desmoplakin (haploinsufficiency)	*Striate palmoplantar keratoderma (SPPK)* Lesions of the palms and soles exacerbated by mechanical trauma Acantholysis and keratin retraction	Armstrong et al. 1999; Whittock et al. 1999
	Desmoplakin (N-terminal missense mutation)	Autosomal dominant arrhythmogenic right ventricular cardiomyopathy (ARVD/C)	Rampazzo et al. 2002
	Desmoplakin (C-terminal missense mutation)	Autosomal dominant arrhythmogenic right ventricular cardiomyopathy (ARVD/C)	Alcalai et al. 2003
	Desmoplakin (C-terminal truncation)	Wooly hair and lesions of the palms and soles Dilated left ventricular cardiomyopathy, SPPK, woolly hair	Norgett et al. 2000
	Desmoplakin (compound heterozygosities)	*Palmoplantar keratoderma* One missense mutation in the N-terminus and one nonsense mutation resulting in a C-terminal deletion Alopecia, hyperkeratosis, acantholysis, keratin retraction	Whittock et al. 2002
	SERCA2	*Darier Disease* Mutation in a sarcoplasmic reticulum Ca^{2+} ATPase pump Acantholysis, blistering, mucosal lesions and abnormal keratinization (dyskeratosis), and suprabasal clefting	Sakuntabhai et al. 1999

Table 1. (continued)

	Mutated gene/target antigen	Phenotype	Selected references
	ATP2C1	*Hailey-Hailey Disease* Mutation in a Ca^{2+} ATPase pump Acantholysis, blistering, mucosal lesions and abnormal keratinization (dyskeratosis), and suprabasal clefting	Dobson-Stone et al. 2002
Autoimmune Diseases	Desmoglein 3	*Pemphigus vulgaris* Blistering of the oral cavity caused by Dsg3-specific autoantibodies Presence of antibodies against both Dsg3 and Dsg1 cause blisters in the oral cavity and the deep epidermis	Reviewed in Amagai 1999; Nousari and Anhalt 1999
	Desmoglein 1	*Pemphigus foliaceus* Blistering of the superficial epidermis caused by circulating autoantibodies directed against Dsg1	Reviewed in Amagai 1999; Nousari and Anhalt 1999
	Desmoglein 1 Desmoglein 3 Multiple plakins	*Paraneoplastic Pemphigus* Presence of antibodies against Dsg1, Dsg3, and multiple plakins Blisters of the skin and mucous membranes, more variable and severe in PV Subset of individuals presenting with lesions in the bronchial epithelium The disease is associated with a number of neoplasms	Reviewed in Amagai 1999; Nousari and Anhalt 1999; Hashimoto 2001
Infectious Disease	Desmoglein 1	*Staphylococcal scalded skin syndrome* (SSSS) Also known as Ritter disease. Bullous impetigo is a localized form *Staphylococcus aureus* exfoliative toxin A or B cleaves Dsg1 Superficial blister formation	Amagai et al. 2000, 2002; Hanakawa et al. 2002b, 2003

6.1.1
Pemphigus Vulgaris and Pemphigus Foliaceus

PV is the most common form of pemphigus. It is a chronic progressive disease, resulting in death for 5%–10% of the afflicted individuals within 5 years of onset, in many cases due to sepsis and a loss of bodily fluids (reviewed in Nousari and Anhalt 1999). A mucosal form of PV with limited skin involvement correlates with the presence of anti-Dsg3 antibodies. A more severe, mucocutaneous form is associated with the presence of anti-Dsg3 and anti-Dsg1 autoantibodies and is characterized by both mucosal lesions and deep epidermal acantholysis (reviewed in Amagai 1999). Pemphigus foliaceus is a less devastating disease characterized by the presence of anti-Dsg1 antibodies and superficial cutaneous lesions (Stanley et al. 1986; Amagai 1995; Kowalczyk et al. 1995; Ishii et al. 1997). The characteristic blistering for each form of pemphigus can be explained by the differential expression of Dsg1 in the upper layers of the skin and Dsg3 in the mucosa and the basal skin layers (reviewed in Amagai 1999). Passive transfer studies in neonatal mice demonstrated that both anti-Dsg1 and anti-Dsg3 antibodies are required for efficient blister formation in tissues such as epidermis, which express both cadherins (Mahoney et al. 1999). These observations led to the proposal that Dsg1 and Dsg3 are capable of compensating for one another in adhesion, and raise the interesting question regarding what distinct roles these cadherins play in adhesion and tissue morphogenesis. The recent observation that pemphigus sera also contain antibodies directed against the newly described Dsg4 molecule suggests the possibility that it too is a target in pemphigus, and raises questions regarding the contribution of these antibodies to the clinical phenotypes (Kljuic et al. 2003).

How do pathogenic antibodies against desmogleins cause blistering? It is possible that the antibodies limit the accessibility of desmoglein extracellular domains, thus interfering with their adhesive function. Such an idea is supported by mapping of pathogenic epitopes on Dsg3 to a calcium-dependent conformational epitope comprising residues that form the predicted adhesive interface (Tsunoda et al. 2003). The antibody epitopes without pathogenic activity were mapped to a more C-terminal extracellular region of Dsg3, not predicted to engage in adhesive interactions. These observations are consistent with the idea that direct inhibition of adhesive interaction of Dsg may contribute to blister formation. However, several reports suggest that antibodies may trigger intracellular signals within the cells upon binding the cadherin. Treating cells with patient sera from either PF or PV resulted in a rapid release of intracellular Ca^{2+} from intracellular stores, a process in which PLC was implicated (Esaki et al. 1995; Seishima et al. 1995). Treatment of keratinocytes with PV sera resulted in the translocation of a number of PKC isoenzymes to the membrane (Osada et al. 1997) and the phosphorylation and degradation of Dsg3 with a corresponding dissociation of plakoglobin from the cadherin (Aoyama et al. 1999). More recent work from this group using pulse chase immunoelectron microscopy suggested that PV IgG triggers the endocytic internalization of small, non-keratin associated clusters of Dsg3, presumably impairing their incorporation into maturing desmosomes (Sato et al. 2000). Finally, an in vitro study using plakoglobin-null mouse keratinocytes demonstrated that plakoglobin is required for the loss of adhesion and keratin retraction observed in control keratinocytes treated with PV IgG (Caldelari et al. 2001). Together these data suggest that pemphigus antibodies could both interfere with adhesion and/or trigger pathways that promote desmosomal cadherin turnover and interfere with normal desmosome homeostasis.

6.1.2
Staphylococcal Scalded Skin Syndrome

SSSS (or Ritter disease) and a more localized disease, bullous impetigo, present as superficial blisters caused by exfoliative toxin A, B, or D (ETA, ETB or ETD) produced by *Staphylococcus aureus* (Amagai et al. 2000, 2002). The recognition that blisters in infected patients bear a striking resemblance to those observed in pemphigus foliaceus patients prompted investigators to examine the mechanism of action of these toxins. They found that exfoliative toxins are proteases with exquisite specificity for human Dsg1, but not canine Dsg1 or human Dsg3 (Amagai et al. 2000, 2002; Hanakawa et al. 2003). ETA cleaves Dsg1 extracellularly at a single peptide bond located between EC domain 3 and 4 (Hanakawa et al. 2002b). As is the case in pemphigus, the mechanism by which cleavage leads to blister formation is not fully understood; for instance, the fate of the cleaved protein in cells is unknown. But this elegant series of observations nicely illustrates how two very different agents, which target the same protein, can lead to similar phenotypes.

6.1.3
Paraneoplastic Pemphigus

PNP is characterized by the presence of anti-plakin antibodies directed against almost every reported member of the family; desmoplakin I, desmoplakin II, BPAG1, plectin, periplakin, and envoplakin (reviewed in Hashimoto 2001). PNP blistering is more variable and severe than that observed in PV, and the disease is associated with a number of neoplasms (reviewed in 7Nousari and Anhalt 1999). Antibody depletion and passive transfer studies have shown that PNP is linked to pathogenic antibodies that recognize Dsg 1 and Dsg3, but perhaps not Dsg 2 (Ota et al. 2003 and reviewed in Amagai 1999), but it is not yet clear what role the anti-plakin antibodies play, if any, in the disease pathogenesis.

6.2
Genetic Diseases

The importance of desmosomes in tissue integrity is underscored by the more recent emergence of gene defects in desmosomal cadherins, plakins, and armadillo proteins. Although many of the resulting human disorders share common phenotypic traits, the variations that do exist may reveal important cell type and tissue-specific functions of the target proteins.

6.2.1
Mutations in the Desmosomal Cadherins

Multiple cases of autosomal dominantly inherited Dsg1 mutations, which map to chromosome 18q12.1, have been reported (Hunt et al. 1999; Rickman et al. 1999). Individuals harboring these mutations have striate palmoplantar keratoderma (SPPK), characterized by stripes and plaque-like islands of hyperkeratosis of the palms of the hands and soles of the feet that appear in areas experiencing high mechanical stress. Unlike patients with plakoglobin and desmoplakin mutations (see Sections 6.2.2 and 6.2.3), no heart defects were noted in any of the patients, consistent with the lack of Dsg1 expression in this tissue.

While the presence of mutant mRNA transcript for Dsg1 has been documented, protein expression has not been demonstrated in any case; thus it seems likely that at least certain cases represent haploinsufficiencies. However, in one instance, in which the mutation would result in removal of part of the extracellular domain, the authors suggested that if this protein were expressed, the resulting deletion could impair cadherin dimer formation or act as a dominant-negative protein by sequestering important cytoplasmic binding partners. Although ultrastructural analysis revealed no obvious reductions in desmosome numbers or desmosome abnormalities, it is possible that desmosome adhesive strength is still compromised (Rickman et al. 1999). In any case, these ultrastructural findings are in striking contrast to those from pemphigus patients in which acantholysis leads to a splitting between the two halves of the desmosome. Whether the reason for these differences is due to the incomplete loss of function in the inherited diseases or a fundamental difference in the mechanism of pathogenesis remains to be determined.

Recently, a family with inherited hypotrichosis was shown to have mutations in the newly identified desmosomal cadherin, desmoglein 4, which is abundantly expressed in hair follicles (Kljuic et al. 2003). Although human Dsg4 was also expressed in the upper layers of the interfollicular epidermis, the phenotypes appeared to be restricted to hair follicles. Histologically, a swelling of the precortical region, which resulted in the generation of a bulbous region within the base of the hair shaft, was observed. The authors discovered that a corresponding mouse model, lanceolate, also harbored Dsg4 mutations, as discussed below in Sect. 7.1.

6.2.2
Involvement of Armadillo Proteins in Genetic Disease

Plakoglobin was found to be linked to Naxos disease, named after the Greek Island harboring the first afflicted population described (Protonotarios et al. 1986). This autosomal recessive arrhythmogenic right ventricular cardiomyopathy (ARVC) is frequently fatal and 100% penetrant by adolescence (Protonotarios et al. 2001a). Homozygous individuals also exhibit palmoplantar keratoderma and wooly hair. The mutation, which maps to chromosome 17q21 (Coonar et al. 1998), is a homozygous 2-bp deletion in the plakoglobin gene, resulting in production of a truncated protein ending near the final armadillo repeat (McKoy et al. 2000). It is likely this truncation affects the binding and/or regulatory functions of plakoglobin. The final armadillo repeat has been shown to contribute to cadherin binding (Chitaev et al. 1996; Troyanovsky et al. 1996; Wahl et al. 1996; Witcher et al. 1996). Furthermore, the plakoglobin C-terminus may participate in intramolecular interactions that regulate the availability of binding sites in the central armadillo repeat or be involved in regulating protein interactions through tyrosine phosphorylation (Troyanovsky et al. 1996; Palka and Green 1997; Gaudry et al. 2001; Miravet et al. 2003).

Several autosomal recessive PKP1 mutations lead to premature terminations that result in loss of PKP1 expression in the epidermis (McGrath et al. 1997; Whittock et al. 2000; Hamada et al. 2002 and reviewed in McGrath 1999). Individuals with these mutations were afflicted with trauma-induced skin fragility and ectodermal dysplasia involving the skin, hair, and nails (reviewed in McGrath 1999). Widened intercellular spaces and poorly formed desmosomes were accompanied by partial retraction of keratin filaments (Whittock et al. 2000 and reviewed in McGrath 1999). Heterozygous family members did not exhibit any abnormalities (Whittock et al. 2000); therefore haploinsufficiency for PKP1 does not appear to be problematic, in contrast to other desmosome haploinsufficiencies. Immortalized keratinocytes from patients lacking PKP1 exhibited reduced desmosome stability and increased migration in vitro (South et al. 2003). These data suggest that patients lacking PKP1 may have alterations in wound healing responses in vivo.

6.2.3
Mutations in Desmoplakin

Recent reports describing both dominant and recessive mutations in desmoplakin have revealed its importance in mechanical integrity and morphogenesis of heart, skin, and skin appendages. Two desmoplakin haploinsufficiencies mapping to chromosome 6p21 result from amino terminal deletions. The observed absence of mRNAs encoding the mutations led to the proposal they are subject to nonsense-mediated RNA decay (Armstrong et al. 1999;

Whittock et al. 1999). The affected individuals exhibit SPPK, with no other reported abnormalities, and histological findings of widened intercellular spaces accompanied by decreased keratin IF connections to desmosomes and decreased desmosome number (Armstrong et al. 1999). Mechanical trauma caused by heavy manual labor was an important factor in the delayed manifestation of symptoms, which appeared in young adulthood.

In contrast to the dominant pattern of inheritance exhibited by desmoplakin-null alleles, a nucleotide deletion resulted in a recessively inherited disorder with production of a truncated protein and a more severe phenotype including SPPK, wooly hair, and dilated left ventricular cardiomyopathy (Norgett et al. 2000). The fact that this protein was missing much of the linker region and the C subdomain previously shown to be critical for association with IF suggested that the observed widened intercellular spaces and a collapsed IF network might be due to impaired association with the cytoskeleton. Consistent with this idea, in vitro adhesion assays demonstrated that patient-derived keratinocytes exhibit adhesive defects similar to cultured cells engineered to express a dominant-negative form of desmoplakin lacking the IF binding domain (Huen et al. 2002).

Two families with distinct missense mutations in desmoplakin, one at the N-terminus and the other at the C-terminus, were both found to exhibit arrythmogenic right ventricular cardiomyopathy (Rampazzo et al. 2003; Alcalai et al. 2003). The N-terminal mutation, located in the armadillo protein binding region, perhaps could have an effect on desmoplakin binding to partner proteins such as plakoglobin and PKP2, while the C-terminal mutation, located in the PRD B, could affect desmoplakin binding to IF. Recently, compound heterozygosities with nonsense and missense mutations in the N- or C-terminus of desmoplakin have been described, which result in palmoplantar keratoderma, hyperkeratosis, and alopecia, with no reported cardiac involvement (Whittock et al. 2002). Although the genotype–phenotype correlations are not yet understood, the overlapping spectrum of phenotypic characteristics in these cases may reflect cell-type specific functions of different desmoplakin domains and sequences.

7
Animal Models Involving Desmosomes

Animal models have provided an important complement to human diseases in revealing the importance of desmosome components in development, epidermal morphogenesis, and function. The availability of powerful tissue and differentiation-specific promoters has further afforded investigators the opportunity to mis-express junctional components in a physiologically relevant context. The animal models involving desmosome components are listed in Table 2.

Table 2. Animal models involving desmosome components

Mutated gene/target antigen	Phenotype	Selected references
Desmocollin 1	Gene disruption	Chidgey et al. 2001
Desmocollin 1	Scaly, flaky, fragile skin with ulcerations and acantholysis	
Desmocollin 1	Human protein misexpression in the basal layers of the skin using the keratin 14 promoter	Henkler et al. 2001
Desmocollin 1	Incorporation of Dsc1 in desmosomes, but no phenotypic abnormality was observed	
Desmocollin 1	Mice express a truncated Dsc1 receptor, present in desmosomes, cannot bind to Pg or PKP	Cheng et al. 2004
Desmocollin 1	Dsc2 expression increased in suprabasal layers of newborn skin compared to wildtype	
Desmoglein 2	Protein inactivation	Eshkind et al. 2002
Desmoglein 2	Desmoglein 2-null and heterozygous embryonal stem cells die at implantation	
Desmoglein 2	Blastocysts showed that desmoplakin was perturbed in the trophectoderm	
Desmoglein 3	Protein misexpression in the upper layers of the skin using the involucrin promoter	Elias et al. 2001
Desmoglein 3	Thin stratum corneum, similar to mucosal epithelium, dysadhesion, and severe dehydration	
Desmoglein 3	Protein misexpression in upper layers of the skin using the keratin 1 promoter	Merritt et al. 2002
Desmoglein 3	Flaky skin, pustules, hyperkeratosis, parakeratosis, acantholysis, abnormal hair follicle formation, hyperproliferation	
Desmoglein 3	Amino terminal deletion mutant expressed in the basal layer using the keratin 14 promoter	Allen et al. 1996
Desmoglein 3	Flaky skin, swollen paws, acantholysis, thick skin, parakeratosis, and inflammation	
Desmoglein 3	Targeted Dsg3 mutations or spontaneous mouse mutations mapped to Dsg3	Koch et al. 1997; Pulkkinen et al. 2002
Desmoglein 3	Runting, cycles of hair loss and regrowth, mucosal erosions, acantholysis, and trauma-induced skin lesions	
Desmoglein 3/P-cadherin	Desmoglein 3 and P-cadherin double deletion	Lenox et al. 2000
Desmoglein 3/P-cadherin	Phenotype similar to pemphigus vulgaris with severe oral lesions leading to malnutrition	
Desmoglein 3/Desmoglein 1	Dsg 3-null animals crossed with Dsg1 transgenic animals, transgene expressed from keratin 14 promoter	Hanakawa et al. 2002a
Desmoglein 3/Desmoglein 1	Hair loss and oral lesions were observed	
Desmoglein 4	Autosomal recessive mutation in lanceolate hair (lah and lah^J) animals	Kljuic et al. 2003
Desmoglein 4	Sparse to absent hair and vibrissae resulting from abnormal hair follicles and shafts	
Plakoglobin	Plakoglobin-null animal	Ruiz et al. 1996
Plakoglobin	Embryonic lethality E12–E16 with observation of severe heart defects	

Table 2. (continued)

Mutated gene/target antigen	Phenotype	Selected references
Plakoglobin	Plakoglobin-null animal Embryonic lethality E10.5 Few animals survive to birth, dying shortly thereafter Skin blistering, no stratum corneum, acantholysis, keratin retraction Severely enlarged hearts and abnormal, disorganized heart tissue	Bierkamp et al. 1996
Plakoglobin	Transgenic animals expressing plakoglobin or an N-terminal deletion mutant of plakoglobin transgenic animals, both used by keratin 14 promoter Stunted hair growth	Charpentier et al. 2000
Desmoplakin	Desmoplakin-null animals Embryonic lethality E6.5, and failure of egg cylinder formation	Gallicano et al. 1998
Desmoplakin	Desmoplakin-null animals with tetraploid rescue Embryonic lethality E10–E12.5 Heart, neuroepithelium, skin epidermis, and microvasculature abnormalities	Gallicano et al. 2001
Desmoplakin	Conditional ablation of desmoplakin in the developing mouse epidermis using Cre/loxP system under control of keratin 14 promoter Fragile skin, abnormal terminal differentiation of skin, basal layer acantholysis, and keratin retraction	Vasioukhin et al. 2001

7.1
Desmosomal Cadherin Models

The graded, differentiation-dependent expression of desmosomal cadherins is likely to be a central contributor to the epidermal differentiation program. However, the importance of these expression patterns is unclear. One possibility is that different cadherin pairs provide tissues with varying levels of adhesive strength. For instance, if Dsg1/Dsc1 were to mediate stronger adhesion than Dsg3/Dsc3, expression of this pair could enhance adhesion in the upper epidermal layers, which experience more mechanical stress. On the other hand, Dsg3/Dsc3 might allow for more dynamic regulation of basal cell keratinocyte adhesion, which would facilitate their upward migration during the course of stratification. Unfortunately, such a scenario is purely speculative at this point, as assays to test relative adhesive strengths for different desmosomal cadherin pairings have not been adequately developed. Furthermore, other evidence suggests that tissues in which one desmoglein is inactivated retain their integrity when another isoform is still present or ectopically expressed (reviewed in Ishii and Green 2001). Based on these observations, many researchers favor the hypothesis that graded cadherin expression is not simply important for conferring different adhesive properties on cells, but may play roles in complex epithelial differentiation and/or signaling.

Observations from Dsc1-null mice are consistent with its importance in adhesion, as animals exhibited fragile skin with acantholytic ulcerations leading to decreased barrier function (Chidgey et al. 2001). Although desmosome numbers were normal, these data raise the possibility that desmosomal adhesion was compromised. Based on the observed epidermal thickening and aberrant expression of hyperproliferation markers in these mice, the authors also concluded that loss of Dsc1 compromises differentiation. Interestingly, a recent study reported that a truncated Dsc1 retaining the first 107 residues common to both Dsc1a and Dsc1b but lacking the plakoglobin/PKP binding sites, integrates into desmosomes without causing any epidermal fragility or other detrimental phenotypes (Cheng et al. 2004). Together with observations of null animals, this somewhat surprising finding suggests that the Dsc1 extracellular, but not downstream portions of the intracellular, domain is important for adhesion in the suprabasal epidermis. Mis-expression of Dsc1 in desmosomes of basal keratinocytes also had no effect on desmosomes or adhesive strength, nor did it induce obvious changes in the differentiation status of basal keratinocytes (Henkler et al. 2001).

Whereas information from animal models for the Dsc family is limited, more plentiful information is available from animal models for the Dsg family, including Dsg2, 3, and the newly identified Dsg4. Animals which are null for the most widely expressed desmoglein, Dsg2, are embryonic lethal

(Eshkind et al. 2002). Null and heterozygote animals died at or shortly after implantation. Although the severity of the Dsg2-null phenotype is not surprising based on its early onset of expression, Dsg2-null blastocysts did develop a trophectoderm layer, which is the first tissue expressing desmosomes during development. Dsg2 appeared to be required for propagating normal ES cells in vitro, prompting the authors to conclude this desmoglein plays a role in cell growth control.

Mice with engineered or spontaneous mutations, which are effectively null for desmogleins with more restricted distribution, i.e., Dsg3 and 4, exhibit less severe phenotypes. In the case of the Dsg3 knockout, animals exhibited mucosal erosions, similar to those present in mucocutaneous PV, and displayed cycles of hair growth and subsequent loss in clumps (Koch et al. 1997; Koch et al. 1998). Histologic observation revealed acantholysis in the mucosa and around hair shafts. Koch et al. recognized that the phenotype of the null animals was similar to the phenotype observed in two spontaneous mutant balding mice (*bal*), Dsg3bal-2 J and Dsg3$^{bal-Pas}$, now known to contain mutations in the Dsg3 gene (Koch et al. 1997; Pulkkinen et al. 2002). Interestingly, forced expression of Dsg1 driven by the keratin 14 promoter compensated for the lack of Dsg3 in the hair follicle; however, due to poor transgene expression in the oral mucosa, blistering was still present (Hanakawa et al. 2002a). Crossing the P-cadherin-null allele, from animals exhibiting no phenotype themselves, onto the Dsg3 knockouts resulted in a more severe post-natal lethal phenotype, leading the authors to suggest that compromising both adherens junctions and desmosomes put additional stress on the mucosal tissue during suckling (Radice et al. 1997). Transgenic animals expressing the cytoplasmic domain of Dsg3 in basal cells of the skin using the keratin 14 promoter displayed thickened, flaky skin, desmosome disruption, acantholysis, and increased cell proliferation (Allen et al. 1996). The authors surmise that the Dsg3 cytoplasmic domain could interact with other desmosome components and inhibit their incorporation into desmosomes, thus acting as a dominant-negative protein. Together, these data further support the hypothesis that Dsg3 is important for epidermal integrity.

Two transgenic models have been used to analyze the consequences of forced expression of Dsg3 in the upper layers of the epidermis (Elias et al. 2001; Merritt et al. 2002). For reasons that are poorly understood, the two models differ in the severity and details of the resulting phenotype; however, they both provide potentially important information about the wider range of functions in which desmogleins may participate. In Elias et al. (2001), Dsg3 expression was driven suprabasally using the involucrin promoter, resulting in marked transepidermal water loss and post-natal death due to severe dehydration. The authors made the interesting observation that the Dsg3/Dsg1 ratio resembled that in the oral mucosa, and further observed structural features of the transgenic epidermis that resembled mucosal epithelium, such as a compact stratum corneum. They concluded that the

Dsg3/Dsg1 ratio in a complex epithelium might regulate barrier structure and function. In contrast to these animals, in which alterations in the differentiation pathway were not observed, animals in which Dsg3 was driven suprabasally off the keratin 1 promoter exhibited signs of hyperproliferation and alterations in differentiation marker expression. Taken together, the two reports demonstrate that Dsg3 expression affects both adhesion and hair follicle growth, in support of the mouse models described above. They further raise the possibility that Dsg3's roles extend beyond simple adhesion to regulation of barrier function and differentiation.

Mutations in the newly discovered Dsg4 gene result in a defect in hair follicle morphogenesis, giving rise to lance-shaped, broken hair shafts, thus leading to the mutant name "lanceolate" (Kljuic et al. 2003). The authors suggest that the phenotype results from the combined effects of compromising adhesion and a premature switch from proliferation to differentiation in the hair follicle. Whether the observed defects in the hair follicle differentiation program can be attributed solely to an adhesion defect, or other functions of Dsg4 is not yet clear. The mutant mouse epidermis also exhibited a hyperproliferative phenotype. Interestingly, the ultrastructure of mutant desmosomes did not resemble those from pemphigus patients, but instead, defects appeared to reside in the plaque.

7.2
Armadillo Protein Models

Animal models for the desmosomal armadillo family members hold particular promise in unraveling potential junctional versus non-junctional roles for these proteins. Together these models suggest that plakoglobin plays a role in maintaining the homeostasis of the heart and skin, and may additionally function in regulating keratinocyte proliferation and survival.

Two plakoglobin-null animal models were generated, using different genetic backgrounds (Bierkamp et al. 1996; Ruiz et al. 1996). In one case, embryonic lethality occurred between 12 and 16 days due to severe heart dysfunction (Ruiz et al. 1996). Desmosomes were lacking in the intercalated discs of the heart; however, an extended adherens junction containing desmoplakin was present in epithelial organs, suggesting that plakoglobin might be involved in the segregation of desmoplakin from adherens junctions to desmosomes. In another case, null mice survived until near the time of birth, displaying a thin, fragile epidermis with no stratum corneum (Bierkamp et al. 1996). Keratin tonofilaments were retracted from the desmosomes, which were greatly reduced in number. The hearts were also severely affected with rips and tears in the tissue. Therefore, plakoglobin appears to be necessary for the strength of desmosome adhesion, consistent with in vitro work demonstrating that plakoglobin association with desmosomal cadherins is necessary to anchor the keratin cytoskeleton at the junctional plaque (Troyanovsky et al. 1993, 1994a). Although the underlying basis for differences in severity of phenotypes is not well understood, the authors noted that β-catenin was localized ultrastructurally to desmosomes in animals living until birth, suggesting that it may compensate in part for the loss of plakoglobin (Bierkamp et al. 1999).

It is noteworthy that the phenotypes exhibited by plakoglobin-null animals appeared to be related to defects in junction structure, leading investigators to conclude that, unlike β-catenin, plakoglobin does not participate in signaling functions during development. However, another report raises the possibility that plakoglobin may indeed play a role in regulating cell proliferation and survival in epidermis. Full-length plakoglobin or a mutant missing the first 80 amino acids of the protein, was expressed in the epidermis and hair follicles of mice under the control of the keratin 14 promoter (Charpentier et al. 2000). The resulting animals exhibited stunted hair growth, which, interestingly, is similar to the phenotype of mice expressing Wnt3 and Dvl2, members of the Wnt signaling pathway of which β-catenin is also a member (Millar et al. 1999). This growth defect was attributed to premature termination of the growth phase (anagen) of the hair cycle. Decreased proliferation of epidermal keratinocytes was also observed, as were apoptotic changes that are typically associated with the regressive phase (catagen) of the hair cycle. These findings are in contrast to those reported in mice expressing a constitutively active form of β-catenin, in which there is an increased hair follicle number and the appearance of hair follicle tumors called pilomatrixomas were observed (Gat et al. 1998). Together these observations suggest that plakoglobin and β-catenin may have distinct roles in regulating cell proliferation, survival, and morphogenesis, possibly mediated through different Wnt-related signaling pathways.

Work on plakoglobin's role in animal development has not been restricted to mice. Recent work showed that wound healing is impaired in plakoglobin-depleted Xenopus blastula, due to the loss of the cortical actin cytoskeleton, an effect that was not due to an interaction between plakoglobin and α-catenin, but instead appears to involve the signaling molecule cdc42 (Kofron et al. 2002). Although the mechanism is unknown at this point, these data suggest additional roles for plakoglobin in cytoskeletal function that go beyond desmosome-dependent adhesion.

7.3
Desmoplakin Models

Mouse models underscore the importance of desmoplakin in embryogenesis and development. A desmoplakin-null mouse died during early embryogenesis (Gallicano et al. 1998) and tetraploid rescue increased embryonic survival by several days, allowing the inspection of developing somatic tissues (Gallicano et al. 2001). Marked abnormalities in heart, neuroepithelium, and skin epidermis were observed, as well as abnormality in the complexus adherens junction in the microvasculature, consistent with the widespread importance of desmoplakin in tissue integrity during development. Fragile skin and cell–cell adhesive defects were obvious and were largely restricted to cells that would normally express desmoplakin. Few desmosomes were observed and those present were not linked to keratin filaments.

Using a Cre/loxP system in which Cre expression was driven by the basally active keratin 14 promoter, desmoplakin was specifically ablated in the developing mouse epidermis (Vasioukhin et al. 2001). The resulting animals exhibited extreme skin fragility and pronounced intercellular separations in the basal and spinous layers. Desmosomes were only slightly reduced in number or size, but they lacked an inner plaque and IF attachment. In areas of cell separation the complete desmosome was linked to one of the two cells, suggesting desmosomal cadherin-mediated adhesion was not affected, but rather the entire junction was extracted. Interestingly, adherens junction number was reduced in the basal layer. Maturation of adherens junctions and the associated cortical actin network was also impaired in keratinocytes isolated from null animals, and cells exhibited defects in sealing of intercellular membranes. Unlike keratinocytes in vivo, desmosomes were rare in cultured cells, but re-introduction of the desmoplakin N-terminus at least partially rescued effects on membrane sealing. Thus, adherens junction maturation appears to require certain desmosomal functions, which are dependent on desmoplakin. This finding is particularly intriguing in light of data to be discussed below, suggesting that desmosome assembly also depends on structural and/or chemical signals triggered by adherens junction formation.

8
Regulation of Desmosomes

The rapid dissolution and formation of desmosomes during wound healing and development suggest that although desmosomes are major contributors to the mechanical integrity of tissues, they are also dynamic in nature. Although it is clear that these junctions are regulated by cell contact, calcium, and growth factors, downstream effector signaling pathways have not been well-delineated (Watt et al. 1984; Jones and Goldman 1985; Duden and Franke 1988; Pasdar and Nelson 1988a,b; Savagner et al. 1997). Recent data have begun to reveal the role of transcription factors such as Slug, as well as kinases and phosphatases in these pathways and set the stage for a more mechanistic understanding of desmosome regulation (Fuchs et al. 1996; Savagner et al. 1997; Gaudry et al. 2001; Miravet et al. 2003).

8.1
Desmosome Assembly

Intercellular junction formation is triggered by extracellular calcium concentrations typically above 0.1 mM and involves a temporal sequence in which adherens junctions begin to form within 5–10 min followed by desmosome formation, which begins during the first 15 min and continues over several hours (Jones et al. 1984; Watt et al. 1984; and reviewed in Kitajima 2002). Both adherens junctions and desmosomes undergo a maturation phase that can last for hours or days after junction proteins are first detected at cell–cell borders. Calcium is likely to trigger a number of events that promote junction assembly. The conformation of the extracellular domains as well as their normal adhesive function requires calcium (Pertz et al. 1999). The stability and solubility of cytoplasmic plaque components is dramatically altered in response to extracellular calcium as well. Whereas newly synthesized desmosomal plaques and membrane proteins are unstable in low calcium, they are stabilized after a calcium switch and recruited from a detergent soluble to insoluble pool (Penn et al. 1987; Pasdar and Nelson 1989). This latter process is likely to involve complex formation with other partner proteins during the assembly of desmosomal precursors. Seemingly at odds with these observations is the finding that desmosomal half plaques can form and be observed at the plasma membrane in cells maintained in long-term low-calcium conditions. However, these half plaques are rapidly endocytosed in the absence of stabilizing calcium-dependent intercellular interactions (Demlehner et al. 1995).

Autosomal dominant disorders like Darier's and Hailey–Hailey's disease, in which calcium pump mutations interfere with normal junction homeostasis (Sakuntabhai et al. 1999; Dobson-Stone et al. 2002), underscore the importance of regulating intracellular calcium levels. It has been suggested that these mutations may result in an imbalance in the intracellular calcium concentrations and, as a result, junction proteins may not be properly trafficked from the ER and Golgi to cell borders (Hashimoto et al. 1995; Sakuntabhai et al. 1999). Consistent with this idea is the observation that the SERCA pump inhibitor, thapsigargin, results in abnormal adhesive junction formation and a delay in the transport of desmoplakin to the plasma membrane (Stuart et al. 1996). Alternatively, the decrease in intracellular calcium could adversely affect the glycosylation and subsequent sorting of the cadherin molecules (Dobson-Stone et al. 2002).

The sensitivity of intercellular junctions to levels of extracellular calcium appears to be controlled by a maturation process. During the early stages of assembly and at the edges of wounds or cell sheets, desmosomes are rapidly internalized by depleting calcium in the medium. In contrast, highly stable and mature junctions are calcium-independent, a phenomenon that is reversed in cells at the edges of wounded epithelial cell sheets and further, may involve PKC-α activation (Watt et al. 1984; Wallis et al. 2000). Rapid reversion from the calcium-independent to the calcium-dependent state is a response that may be important for facilitating reepithelialization.

How the interaction of junctional components is coordinated during calcium-induced desmosome assembly is still not well understood. Evidence suggests that plakoglobin associates with the desmosomal cadherins early in their biosynthetic pathway, and that they translocate together to the plasma membrane in a separate compartment from desmoplakin (Pasdar et al. 1991; Pasdar et al. 1995b; Burdett and Sullivan 2002). Desmocollins and desmogleins may also reside in separate membrane compartments (Watt et al. 1984; Burdett and Sullivan 2002). In low calcium conditions, desmoplakin is present in the form of cytoplasmic particles associated with the IF network (Jones and Goldman 1985). Based on a time course immunofluorescence analysis of fixed cells, the cytoplasm is cleared of these particles following the calcium switch, which appear to redistribute over time to the plasma membrane. Cells maintained at intermediate levels of calcium exhibit desmoplakin particles that appear to be "stuck" in intermediate positions on their way to the membrane (Jones and Grelling 1989). Based on these data, it has been hypothesized that these particles are desmosome precursors delivered to the forming cell–cell border to build the organelle (Jones and Goldman 1985; Jones and Grelling 1989). However, others have suggested that the particles are endocytosed desmosome remnants targeted for degradation (Mattey and Garrod 1986; Duden and Franke 1988; Burdett 1993; Burdett and Sullivan 2002). Preliminary data from high-resolution analysis using fluorescently tagged desmosome proteins suggests that at least some desmoplakin particles are indeed desmosome precursors (L.M. Godsel et al., unpublished observations).

Several observations suggest that desmoplakin does not require association with IF to incorporate into desmosomes. C-terminally truncated desmoplakin can incorporate into desmosomes (Stappenbeck et al. 1993; Bornslaeger et al. 1996; Huen et al. 2002) and embryonic stem cells lacking keratin 8 and 18 do not assemble IF networks but still form desmosomes (Baribault and Oshima 1991). On the other hand, live cell imaging suggests that an IF-uncoupled desmoplakin peptide exhibits a more random than linear pathway to the membrane (L.M. Godsel et al., unpublished observations). In addition, mutation of a PKA-dependent phosphorylation site in desmoplakin delays its incorporation into desmosomes by increasing its affinity for IF (L.M. Godsel et al., unpublished observations). One possibility is that, although not absolutely essential for desmosome formation, IF may regulate delivery of desmoplakin to the plaque during physiological processes in which the amount of desmoplakin incorporated into junctions requires tight control. In addition, it is possible that multiple mechanisms exist for delivery of desmoplakin to the plasma membrane, and that in the absence of IF, alternative pathways come into play.

8.2
Dependence of Desmosomes on Adherens Junction Formation

Inhibition of classical cadherin binding or cell surface localization through expression of naturally occurring mutations, dominant-negative constructs, or blocking antibodies not only interferes with adherens junction formation but also inhibits or delays desmosome assembly (Gumbiner et al. 1988; Fujimori and Takeichi 1993; Lewis et al. 1994; Amagai et al. 1995). These studies suggest that there is interdependence between the two junctions. Plakoglobin, which, unlike β-catenin can associate with both classical and desmosomal cadherins, has been proposed as one candidate for mediating such crosstalk. Supporting this idea are reports that plakoglobin appears to drive the expression of desmosomes in the presence of E-cadherin or N-cadherin (Lewis et al. 1997; Parker et al. 1998).

Investigators have observed the association of adherens junction and desmosomal proteins under certain conditions. For instance, in low calcium lateral association of classic and desmosomal cadherins can occur (Troyanovsky et al. 1999), and when junction homeostasis is impaired Dsg1 and Dsc3 can bind to β-catenin (Norvell and Green 1998; Bierkamp et al. 1999; Hanakawa et al. 2000). One possibility is that such complexes represent transitional stages of desmosome assembly which resolve later in the process to yield fully segregated junctions. Consistent with this idea, junction mixing can occur under certain circumstances when junctions are disrupted—for example when plakoglobin is absent (Bierkamp et al. 1999) or upon expression of a dominant-negative desmoplakin (Bornslaeger et al. 1996). These observations suggest that not only junction assembly, but junction segregation requires plakoglobin or its associated proteins. The interdependence of the two junctions is further highlighted by the conditional ablation of desmoplakin in the skin of an animal model generated by Vasioukhin et al. in which desmosomes and adherens junction numbers are both decreased (Vasioukhin et al. 2001).

What is the underlying basis of the interdependence between adherens junctions and desmosomes? It is possible that the role of the adherens junctions is to bring membranes of adjacent cells into close proximity for desmosomal cadherin engagement. The fact that half desmosomes can form in the absence of cell–cell attachment and are subsequently endocytosed supports the scenario in which the requirement for adherens junctions is a structural one (Demlehner et al. 1995). In further support of this scenario, expression of the non-cadherin adhesion molecule, myelin protein P0, led to the formation of desmosomes (Doyle et al. 1995). However, other evidence suggests that desmosome assembly requires more than the close juxtaposition of opposing membranes. Activation of PKC leads to desmoplakin redistribution and desmosome formation without the requirement for either extracellular calcium or assembly of the actin-cytoskeleton linked adherens junction, and calcium is a physiologic activator of PKC (reviewed in Kitajima 2002). Data from another group, however, demonstrated that phorbol esters resulted not in the formation of desmosomes, but instead in the splitting of desmosomes and dislocation of desmosomal components from the cell periphery to the cytoplasm (Amar et al. 1999). Together these observations suggest that desmosome assembly involves a cascade of protein–protein interactions that can be triggered by calcium and may require transitional complexes and/or signals provided by adherens junction components. However, precisely how these events are temporally and spatially coordinated has yet to be determined.

8.3
Transcriptional and Post-translational Mechanisms for Modulating Desmosome Assembly State

The transcription factor Slug has been implicated in the regulation of desmosome dissolution, and may represent an important step in initiating growth factor-induced epithelial to mesenchymal transitions. However, the downstream effectors of Slug, and molecular targets within the desmosome that respond to Slug, are still under investigation (Savagner et al. 1997).

The modification of desmosome proteins by kinases and phosphatases also provides a mechanism for dynamic regulation of desmosomes. Specific targets for modification by protein phosphorylation are now being revealed. As is the case for adherens junctions, the desmosomal armadillo proteins are prime targets for phosphorylation by tyrosine kinases. The plakoglobin C-terminus harbors phosphorylation sites for tyrosine kinases, which have been shown to be important for regulating interactions with its binding partners (Gaudry et al. 2001; Miravet et al. 2003). Plakoglobin is phosphorylated in response to epidermal growth factor (EGF) treatment corresponding with a decrease in cell–cell contact (Gaudry et al. 2001). Phosphorylated plakoglobin was exclusively in the membrane-associated pool and could associate with Dsg2, but not with desmoplakin (Gaudry et al. 2001). Phosphorylation of the plakoglobin C-terminus by Src results in decreased plakoglobin:αcatenin interactions with a concurrent increase in plakoglobin:desmoplakin interactions, while Fer/Fyn phosphorylation of armadillo repeat 7 increases the interaction between plakoglobin and α-catenin (Miravet et al. 2003). Intriguingly, these regulated interactions have an impact that extends beyond differential modulation of intercellular junctions, as increased binding of plakoglobin to adherens junction components increased the transcriptional activity of the β-catenin-Tcf-4 complex (Miravet et al. 2002).

PKP2 has recently been shown to be a substrate for Cdc25C-associated kinase 1 (C-TAK1) and mutation of the 14-3-3 binding motifs on PKP2 phosphorylated by the kinase resulted in an increased nuclear localization of PKP2 (Muller et al. 2003). As PKP1 does not harbor a C-TAK1-dependent

14-3-3 binding site, these results may provide an explanation for previously observed differences in nuclear translocation between PKP2 and PKP1. PKP1 more readily localizes to the nucleus, although its function there is unknown. Phosphorylation of desmoplakin also provides a mechanism for regulating the assembly state of desmosomes. Serine phosphorylation of a PKA consensus sequence in desmoplakin (2849) inhibits its interaction with IF (Stappenbeck et al. 1994) and 12-O-tetradecanoylphorbol 13-acetate (TPA)-dependent serine phosphorylation was associated with desmoplakin's increased solubility (Amar et al. 1999).

It is likely that kinases and phosphatases cooperate to regulate junction assembly and homeostasis. Consistent with this idea, protein kinase B has recently been shown to bind to the plakin family member, periplakin (van den Heuvel et al. 2002). Further support for the idea comes from the finding that the phosphatase inhibitor H-7 interferes with junction disassembly in low-calcium conditions (Citi et al. 1994; Pasdar et al. 1995a), and the protein phosphatase inhibitor okadaic acid interferes with assembly of the desmosomal plaque, without apparent effects on the ability of proteins to traffic to the membrane (Pasdar et al. 1995a). Specific desmosome partners for phosphatases are so far limited to plakoglobin, which, like β-catenin, has been found to complex with human protein tyrosine phosphatase (hPTP)-κ (Fuchs et al. 1996). Plakoglobin could thus serve as scaffolding for signaling modules, which regulate the dynamic state of junction assembly and adhesion. Together these observations suggest that, like adherens junctions, desmosomes are subject to both transcriptional and post-translational controls that govern their assembly state during development, wound healing, and in pathological processes.

8.4
The Role of Desmosome Components in Signaling and Transcriptional Regulation

Like other cell-substrate and cell–cell junction proteins, desmosomal proteins may play roles that extend beyond their junctional functions. The possibility that the differentially expressed desmosomal cadherins play roles in differentiation that go beyond simple adhesion was discussed above, as were potential nuclear roles of the armadillo protein PKP2. The current section will focus primarily on plakoglobin, whose roles in signaling and transcriptional regulation are still poorly understood but have been the focus of much speculation.

Plakoglobin expression has been implicated on the one hand in promoting tumor progression, and on the other hand with decreased proliferation and increased apoptosis. Supporting a positive role in tumor growth, introduction of plakoglobin into plakoglobin-deficient squamous carcinoma cells inhibited apoptosis and promoted growth and foci formation. These changes in cell behavior were correlated with the induction of the anti-apoptotic protein, BCL-2 (Hakimelahi et al. 2000). Both plakoglobin and β-catenin transformed RK3E epithelial cells and this transforming activity involved lymphoid-enhancer binding factor/T cell factor (LEF/Tcf) (Kolligs et al. 2000). However, studies have also reported instances where plakoglobin is lost during cancer progression, suggesting a tumor-suppressor function (Aberle et al. 1995; Simcha et al. 1996; Caca et al. 1999; Amitay et al. 2001). Plakoglobin has been shown to directly interact with c-erbB-2, an oncogenic protein kinase frequently upregulated in malignant cancers, and this interaction may be important for the decreased cell adhesion and invasive growth of these cancers (Ochiai et al. 1994; Kanai et al. 1995). Plakoglobin and β-catenin were shown to upregulate the expression of promyelocytic leukemia (PML), a transcriptional coactivator, also implicated as a tumor suppressor. Interestingly, expression of plakoglobin, β-catenin, or PML could decrease tumorigenicity in $p53^{-/-}$ colon carcinomas (Shtutman et al. 2002). Finally, ectopic expression of plakoglobin in mouse epidermis inhibited hair growth and increased apoptosis, in striking contrast to the appearance of pilomatrixomas and increased hair growth observed in similar transgenics expressing β-catenin (Gat et al. 1998; Charpentier et al. 2000).

The mechanism by which plakoglobin exerts its effects on cell growth and survival is still poorly understood. In the case of β-catenin, strong evidence suggests it plays a positive role in mediating transcription by binding to the Tcf-4/LEF complex (Huber et al. 1996; Hecht et al. 1999). Over-expression of plakoglobin in Xenopus embryos, like β-catenin, results in neural axis duplication coincident with nuclear localization (Karnovsky and Klymkowsky 1995). However, since cytoplasmic forms of plakoglobin induce axis duplication (Merriam et al. 1997), investigators suggested that plakoglobin might function by competing with factors that sequester β-catenin in the adherens junction and/or induce its degradation via the proteasome machinery, allowing it to enter the nucleus to mediate transcription. Consistent with this idea, LEF1 mediated transcription in plakoglobin overexpressing cells was accompanied by an increased translocation of β-catenin into the nucleus (Simcha et al. 1998).

However, some investigators have suggested that Tcf-bound plakoglobin may act as a negative regulator of transcription in the nucleus (Miravet et al. 2002; Hu et al. 2003). Along these lines, it has been shown that plakoglobin and β-catenin can form a ternary complex in vitro suggesting they occupy distinct binding sites on Tcf-4. Furthermore, casein kinase 2-dependent of Tcf-4 phosphorylation reduced the association of plakoglobin but not β-catenin with the transcription factor (Miravet et al. 2002). Other investigators observed that plakoglobin binds to LEF/Tcf-4 relatively inefficiently compared to β-catenin, and this association with the transcription complex is negatively regulated by the N- and C-terminal domains of plakoglobin (Zhurinsky et al. 2000). Thus, the regulation of intramolecular interactions involving the N- and C-terminus might alter the relative binding of plakoglobin and β-catenin in nuclear complexes. We are still left with the question of whether plakoglobin can participate in transcriptional events that are totally independent from β-catenin. The fact that c-myc transcription is increased by plakoglobin expression in R3KE cells, but not by β-catenin expression, suggests that it does have a role in signaling that is distinguished from a β-catenin-dependent pathway (Kolligs et al. 2000).

9
Future Directions and Concluding Remarks

Over the past decade desmosomes have come under increasing scrutiny as the emergence of human and mouse mutations have motivated investigators to determine the molecular basis for the resulting disorders. In addition, new technical advances have pushed forward the in vitro analysis of the structure and function of these junctions. While the importance of desmosomes in mediating strong intercellular adhesion is now clear, a number of exciting challenges remain to be tackled. First, how individual desmosomal components are assembled into precursors and traffic out to the plasma membrane during desmosome assembly, and then how they are stabilized and maintained once at the plasma membrane, are questions which still need to be addressed at a molecular level of detail. Live cell imaging and new biochemical approaches provide a powerful combination to dissect these pathways, and to unravel the spatial and temporal relationships between adherens junction and desmosome components during assembly. Likewise, how whole desmosomes and desmosomal components are turned over in response to environmental signals is still not understood. For instance, are the desmosomal cadherin tails and associated proteins subject to regulation, which triggers internalization of desmosomal cadherins? And do members of the p120 family such as PKPs participate in the regulation of this process, as they do in the case of classic cadherins (Davis et al. 2003; Xiao et al. 2003)?

Another area of future investigation will be to explore how desmosomes integrate adhesive functions with other cellular functions important for cell growth, survival, and differentiation. In the case of adherens junctions, autoregulatory loops which integrate growth factor and β-catenin signaling with E-cadherin expression and control of cell growth have been identified (Conacci-Sorrell et al. 2003). Furthermore, N-, E-, and VE-cadherins have been shown to collaborate with FGF, EGF, and VEGF receptor activity (reviewed in Wheelock and Johnson 2003). The possibility that differentiation-specific desmosomal cadherins participate in such pathways through a distinct set of interactions is an intriguing possibility. Screening approaches designed to identify novel binding partners for these cadherins, particularly their extended tails with no known function, is an exciting direction which may uncover signaling complexes assembled by these domains. The involvement of plakoglobin in coordinating adhesive and signaling functions seems likely, but at this point the data do not allow us to draw a clear picture of how plakoglobin signals. Determining under which circumstances plakoglobin acts to promote or inhibit β-catenin transcription, and under which circumstances it may act independently of β-catenin, will be an important direction for the future.

Acknowledgements The authors would like to thank Andrew Kowalczyk for helpful discussion and critical reading of the manuscript. We regret that the work of many colleagues and authors could not be cited in this review due to space limitations. Work in the authors' lab is supported by NIH grants RO1 AR41836, AR43380, and project #4 of P01 DE12328. L. Godsel was supported by NIH T32 AR07593 and S. Getsios by a postdoctoral fellowship from the CIHR.

References

Aberle H, Bierkamp C, Torchard D, Serova O, Wagner T, Natt E, Wirsching J, Heidkamper C, Montagna M, Lynch HT, et al (1995) The human plakoglobin gene localizes on chromosome 17q21 and is subjected to loss of heterozygosity in breast and ovarian cancers. Proc Natl Acad Sci U S A 92:6384–8

Aho S, McLean WH, Li K, Uitto J (1998) cDNA cloning, mRNA expression, and chromosomal mapping of human and mouse periplakin genes. Genomics 48:242–7

Alcalai R, Metzger S, Rosenheck S, Meiner V, Chajek-Shaul T (2003) A recessive mutation in desmoplakin causes arrhythmogenic right ventricular dysplasia, skin disorder, and woolly hair. J Am Coll Cardiol 42:319–27

Allen E, Yu QC, Fuchs E (1996) Mice expressing a mutant desmosomal cadherin exhibit abnormalities in desmosomes, proliferation, and epidermal differentiation. J Cell Biol 133:1367–82

Amagai M (1995) Adhesion molecules. I: Keratinocyte-keratinocyte interactions; cadherins and pemphigus. J Invest Dermatol 104:146–52

Amagai M (1999) Autoimmunity against desmosomal cadherins in pemphigus. J Dermatol Sci 20:92–102

Amagai M, Karpati S, Klaus-Kovtun V, Udey MC, Stanley JR (1994) Extracellular domain of pemphigus vulgaris antigen (desmoglein 3) mediates weak homophilic adhesion. J Invest Dermatol 103:609–15

Amagai M, Fujimori T, Masunaga T, Shimizu H, Nishikawa T, Shimizu N, Takeichi M, Hashimoto T (1995) Delayed assembly of desmosomes in keratinocytes with disrupted classic-cadherin-mediated cell adhesion by a dominant negative mutant. J Invest Dermatol 104:27–32

Amagai M, Matsuyoshi N, Wang ZH, Andl C, Stanley JR (2000) Toxin in bullous impetigo and staphylococcal scalded-skin syndrome targets desmoglein 1. Nat Med 6:1275–7

Amagai M, Yamaguchi T, Hanakawa Y, Nishifuji K, Sugai M, Stanley JR (2002) Staphylococcal exfoliative toxin B specifically cleaves desmoglein 1. J Invest Dermatol 118:845–50

Amar LS, Shabana AH, Oboeuf M, Martin N, Forest N (1999) Involvement of desmoplakin phosphorylation in the regulation of desmosomes by protein kinase C, in HeLa cells. Cell Adhes Commun 7:125–38

Amitay R, Nass D, Meitar D, Goldberg I, Davidson B, Trakhtenbrot L, Brok-Simoni F, Ben-Ze'ev A, Rechavi G, Kaufmann Y (2001) Reduced expression of plakoglobin correlates with adverse outcome in patients with neuroblastoma. Am J Pathol 159:43–9

Anastasiadis PZ, Reynolds AB (2000) The p120 catenin family: complex roles in adhesion, signaling and cancer. J Cell Sci 113 (Pt 8):1319–34

Andl CD, Stanley JR (2001) Central role of the plakoglobin-binding domain for desmoglein 3 incorporation into desmosomes. J Invest Dermatol 117:1068–74

Andra K, Lassmann H, Bittner R, Shorny S, Fassler R, Propst F, Wiche G (1997) Targeted inactivation of plectin reveals essential function in maintaining the integrity of skin, muscle, and heart cytoarchitecture. Genes Dev 11:3143–56

Andra K, Kornacker I, Jorgl A, Zorer M, Spazierer D, Fuchs P, Fischer I, Wiche G (2003) Plectin-isoform-specific rescue of hemidesmosomal defects in plectin (−/−) keratinocytes. J Invest Dermatol 120:189–97

Angst BD, Nilles LA, Green KJ (1990) Desmoplakin II expression is not restricted to stratified epithelia. J Cell Sci 97 (Pt 2):247–57

Aoyama Y, Owada MK, Kitajima Y (1999) A pathogenic autoantibody, pemphigus vulgaris-IgG, induces phosphorylation of desmoglein 3, and its dissociation from plakoglobin in cultured keratinocytes. Eur J Immunol 29:2233–40

Armstrong DK, McKenna KE, Purkis PE, Green KJ, Eady RA, Leigh IM, Hughes AE (1999) Haploinsufficiency of desmoplakin causes a striate subtype of palmoplantar keratoderma. Hum Mol Genet 8:143–8

Arnemann J, Spurr NK, Wheeler GN, Parker AE, Buxton RS (1991) Chromosomal assignment of the human genes coding for the major proteins of the desmosome junction, desmoglein DGI (DSG), desmocollins DGII/III (DSC), desmoplakins DPI/II (DSP), and plakoglobin DPIII (JUP). Genomics 10:640–5

Bannon LJ, Cabrera BL, Stack MS, Green KJ (2001) Isoform-specific differences in the size of desmosomal cadherin/catenin complexes. J Invest Dermatol 117:1302–6

Baribault H, Oshima RG (1991) Polarized and functional epithelia can form after the targeted inactivation of both mouse keratin 8 alleles. J Cell Biol 115:1675–84

Bierkamp C, McLaughlin KJ, Schwarz H, Huber O, Kemler R (1996) Embryonic heart and skin defects in mice lacking plakoglobin. Dev Biol 180:780–5

Bierkamp C, Schwarz H, Huber O, Kemler R (1999) Desmosomal localization of beta-catenin in the skin of plakoglobin null- mutant mice. Development 126:371–81

Bonne S, van Hengel J, van Roy F (1998) Chromosomal mapping of human armadillo genes belonging to the p120(ctn)/plakophilin subfamily. Genomics 51:452–4

Bonne S, van Hengel J, Nollet F, Kools P, van Roy F (1999) Plakophilin-3, a novel armadillo-like protein present in nuclei and desmosomes of epithelial cells. J Cell Sci 112:2265–76

Bonne S, Gilbert B, Hatzfeld M, Chen X, Green KJ, Van Roy F (2003) Defining desmosomal plakophilin-3 interactions. J Cell Biol 161:403–16

Bornslaeger EA, Corcoran CM, Stappenbeck TS, Green KJ (1996) Breaking the connection: displacement of the desmosomal plaque protein desmoplakin from cell-cell interfaces disrupts anchorage of intermediate filament bundles and alters intercellular junction assembly. J Cell Biol 134:985–1001

Bornslaeger EA, Godsel LM, Corcoran CM, Park JK, Hatzfeld M, Kowalczyk AP, Green KJ (2001) Plakophilin 1 interferes with plakoglobin binding to desmoplakin, yet together with plakoglobin promotes clustering of desmosomal plaque complexes at cell–cell borders. J Cell Sci 114:727–38

Brandner JM, Reidenbach S, Franke WW (1997) Evidence that "pinin", reportedly a differentiation-specific desmosomal protein, is actually a widespread nuclear protein. Differentiation 62:119–27

Burdett ID (1993) Internalisation of desmosomes and their entry into the endocytic pathway via late endosomes in MDCK cells. Possible mechanisms for the modulation of cell adhesion by desmosomes during development. J Cell Sci 106:1115–30

Burdett ID (1998) Aspects of the structure and assembly of desmosomes. Micron 29:309–28

Burdett ID, Sullivan KH (2002) Desmosome assembly in MDCK cells: transport of precursors to the cell surface occurs by two phases of vesicular traffic and involves major changes in centrosome and Golgi location during a Ca(2+) shift. Exp Cell Res 276:296–309

Caca K, Kolligs FT, Ji X, Hayes M, Qian J, Yahanda A, Rimm DL, Costa J, Fearon ER (1999) Beta- and gamma-catenin mutations, but not E-cadherin inactivation, underlie T-cell factor/lymphoid enhancer factor transcriptional deregulation in gastric and pancreatic cancer. Cell Growth Differ 10:369–76

Caldelari R, de Bruin A, Baumann D, Suter MM, Bierkamp C, Balmer V, Muller E (2001) A central role for the armadillo protein plakoglobin in the autoimmune disease pemphigus vulgaris. J Cell Biol 153:823–34

Calkins CC, Hoepner BL, Law CM, Novak MR, Setzer SV, Hatzfeld M, Kowalczyk AP (2003) The Armadillo family protein p0071 is a VE-cadherin- and desmoplakin-binding protein. J Biol Chem 278:1774–83

Charpentier E, Lavker RM, Acquista E, Cowin P (2000) Plakoglobin suppresses epithelial proliferation and hair growth in vivo. J Cell Biol 149:503–20

Chen X, Bonne S, Hatzfeld M, van Roy F, Green KJ (2002) Protein binding and functional characterization of plakophilin 2. Evidence for its diverse roles in desmosomes and beta-catenin signaling. J Biol Chem 277:10512–22

Cheng X, Mihindukulasuriya K, Den Z, Kowalczyk AP, Calkins CC, Ishiko A, Shimizu A, Koch PJ (2004) Assessment of splice variant-specific functions of desmocollin 1 in the skin. Mol Cell Biol 24:154–163

Chidgey M, Brakebusch C, Gustafsson E, Cruchley A, Hail C, Kirk S, Merritt A, North A, Tselepis C, Hewitt J, Byrne C, Fassler R, Garrod D (2001) Mice lacking desmocollin 1 show epidermal fragility accompanied by barrier defects and abnormal differentiation. J Cell Biol 155:821–32

Chidgey MA, Clarke JP, Garrod DR (1996) Expression of full-length desmosomol glycoproteins (desmocollins) is not sufficient to confer strong adhesion on transfected L929 cells. J Invest Dermatol 106:689–95

Chitaev NA, Troyanovsky SM (1997) Direct Ca2+-dependent heterophilic interaction between desmosomal cadherins, desmoglein and desmocollin, contributes to cell–cell adhesion. J Cell Biol 138:193–201

Chitaev NA, Troyanovsky SM (1998) Adhesive but not lateral E-cadherin complexes require calcium and catenins for their formation. J Cell Biol 142:837–46

Chitaev NA, Leube RE, Troyanovsky RB, Eshkind LG, Franke WW, Troyanovsky SM (1996) The binding of plakoglobin to desmosomal cadherins: patterns of binding sites and topogenic potential. J Cell Biol 133:359–69

Chitaev NA, Averbakh AZ, Troyanovsky RB, Troyanovsky SM (1998) Molecular organization of the desmoglein-plakoglobin complex. J Cell Sci 111:1941–9

Choi HJ, Park-Snyder S, Pascoe LT, Green KJ, Weis WI (2002) Structures of two intermediate filament-binding fragments of desmoplakin reveal a unique repeat motif structure. Nat Struct Biol 9:612–20

Citi S, Volberg T, Bershadsky AD, Denisenko N, Geiger B (1994) Cytoskeletal involvement in the modulation of cell–cell junctions by the protein kinase inhibitor H-7. J Cell Sci 107 (Pt 3):683–92

Clubb BH, Chou Y-H, Herrmann H, Svitkina TM, Borisy GG, Goldman RD (2000) The 300-kDa intermediate filament-associated protein (IFAP300) is a hamster plectin ortholog. Biochem Biophys Res Comm 273:183–187

Collins JE, Legan PK, Kenny TP, MacGarvie J, Holton JL, Garrod DR (1991) Cloning and sequence analysis of desmosomal glycoproteins 2 and 3 (desmocollins): cadherin-like desmosomal adhesion molecules with heterogeneous cytoplasmic domains. J Cell Biol 113:381–91

Conacci-Sorrell M, Simcha I, Ben-Yedidia T, Blechman J, Savagner P, Ben-Ze'ev A (2003) Autoregulation of E-cadherin expression by cadherin-cadherin interactions: the roles of {beta}-catenin signaling, Slug, and MAPK. J Cell Biol 163:847–857

Coonar AS, Protonotarios N, Tsatsopoulou A, Needham EW, Houlston RS, Cliff S, Otter MI, Murday VA, Mattu RK, McKenna WJ (1998) Gene for arrhythmogenic right ventricular cardiomyopathy with diffuse nonepidermolytic palmoplantar keratoderma and woolly hair (Naxos disease) maps to 17q21. Circulation 97:2049–58

Cowin P, Kapprell HP, Franke WW, Tamkun J, Hynes RO (1986) Plakoglobin: a protein common to different kinds of intercellular adhering junctions. Cell 46:1063–73

Davis MA, Ireton RC, Reynolds AB (2003) A core function for p120-catenin in cadherin turnover. J Cell Biol 163:525–534

Deguchi M, Iizuka T, Hata Y, Nishimura W, Hirao K, Yao I, Kawabe H, Takai Y (2000) PAPIN. A novel multiple PSD-95/Dlg-A/ZO-1 protein interacting with neural plakophilin-related armadillo repeat protein/delta-catenin and p0071. J Biol Chem 275:29875–80

Demlehner MP, Schafer S, Grund C, Franke WW (1995) Continual assembly of half-desmosomal structures in the absence of cell contacts and their frustrated endocytosis: a coordinated Sisyphus cycle. J Cell Biol 131:745–60

Denning MF, Guy SG, Ellerbroek SM, Norvell SM, Kowalczyk AP, Green KJ (1998) The expression of desmoglein isoforms in cultured human keratinocytes is regulated by calcium, serum, and protein kinase C. Exp Cell Res 239:50–9

DiColandrea T, Karashima T, Maatta A, Watt FM (2000) Subcellular distribution of envoplakin and periplakin: insights into their role as precursors of the epidermal cornified envelope. J Cell Biol 151:573–86

Dobson-Stone C, Fairclough R, Dunne E, Brown J, Dissanayake M, Munro CS, Strachan T, Burge S, Sudbrak R, Monaco AP, Hovnanian A (2002) Hailey-Hailey disease: molecular and clinical characterization of novel mutations in the ATP2C1 gene. J Invest Dermatol 118:338–43

Doyle JP, Stempak JG, Cowin P, Colman DR, D'Urso D (1995) Protein zero, a nervous system adhesion molecule, triggers epithelial reversion in host carcinoma cells. J Cell Biol 131:465–82

Duden R, Franke WW (1988) Organization of desmosomal plaque proteins in cells growing at low calcium concentrations. J Cell Biol 107:1049–63

Eger A, Stockinger A, Wiche G, Foisner R (1997) Polarisation-dependent association of plectin with desmoplakin and the lateral submembrane skeleton in MDCK cells. J Cell Sci 110:1307–16

Elias PM, Matsuyoshi N, Wu H, Lin C, Wang ZH, Brown BE, Stanley JR (2001) Desmoglein isoform distribution affects stratum corneum structure and function. J Cell Biol 153:243–9

Esaki C, Seishima M, Yamada T, Osada K, Kitajima Y (1995) Pharmacologic evidence for involvement of phospholipase C in pemphigus IgG-induced inositol 1,4,5-trisphosphate generation, intracellular calcium increase, and plasminogen activator secretion in DJM-1 cells, a squamous cell carcinoma line. J Invest Dermatol 105:329–33

Eshkind L, Tian Q, Schmidt A, Franke WW, Windoffer R, Leube RE (2002) Loss of desmoglein 2 suggests essential functions for early embryonic development and proliferation of embryonal stem cells. Eur J Cell Biol 81:592–8

Fairley JA, Scott GA, Jensen KD, Goldsmith LA, Diaz LA (1991) Characterization of keratocalmin, a calmodulin-binding protein from human epidermis. J Clin Invest 88:315–22

Fontao L, Favre B, Riou S, Geerts D, Jaunin F, Saurat J-H, Green KJ, Sonnenberg A, Borradori L (2003) Interaction of the Bullous Pemphigoid Antigen 1 (BP230) and Desmoplakin with Intermediate Filaments Is Mediated by Distinct Sequences within Their COOH Terminus. Mol. Biol. Cell 14:1978–1992

Fuchs M, Muller T, Lerch MM, Ullrich A (1996) Association of human protein-tyrosine phosphatase kappa with members of the armadillo family. J Biol Chem 271:16712–16719

Fujimori T, Takeichi M (1993) Disruption of epithelial cell-cell adhesion by exogenous expression of a mutated nonfunctional N-cadherin. Mol Biol Cell 4:37–47

Gallicano GI, Kouklis P, Bauer C, Yin M, Vasioukhin V, Degenstein L, Fuchs E (1998) Desmoplakin is required early in development for assembly of desmosomes and cytoskeletal linkage. J Cell Biol 143:2009–22

Gallicano GI, Bauer C, Fuchs E (2001) Rescuing desmoplakin function in extra-embryonic ectoderm reveals the importance of this protein in embryonic heart, neuroepithelium, skin and vasculature. Development 128:929–41

Garrod DR, Merritt AJ, Nie Z (2002) Desmosomal cadherins. Curr Opin Cell Biol 14:537–45

Gat U, DasGupta R, Degenstein L, Fuchs E (1998) De Novo hair follicle morphogenesis and hair tumors in mice expressing a truncated beta-catenin in skin. Cell 95:605–14

Gaudry CA, Palka HL, Dusek RL, Huen AC, Khandekar MJ, Hudson LG, Green KJ (2001) Tyrosine-phosphorylated plakoglobin is associated with desmogleins but not desmoplakin after epidermal growth factor receptor activation. J Biol Chem 276:24871–80

Green KJ, Parry DA, Steinert PM, Virata ML, Wagner RM, Angst BD, Nilles LA (1990) Structure of the human desmoplakins. Implications for function in the desmosomal plaque. J Biol Chem 265:11406–7

Gumbiner B, Stevenson B, Grimaldi A (1988) The role of the cell adhesion molecule uvomorulin in the formation and maintenance of the epithelial junctional complex. J Cell Biol 107:1575–87

Hakimelahi S, Parker HR, Gilchrist AJ, Barry M, Li Z, Bleackley RC, Pasdar M (2000) Plakoglobin regulates the expression of the anti-apoptotic protein BCL-2. J Biol Chem 275:10905–11

Hamada T, South AP, Mitsuhashi Y, Kinebuchi T, Bleck O, Ashton GH, Hozumi Y, Suzuki T, Hashimoto T, Eady RA, McGrath JA (2002) Genotype-phenotype correlation in skin fragility-ectodermal dysplasia syndrome resulting from mutations in plakophilin 1. Exp Dermatol 11:107–14

Hanakawa Y, Amagai M, Shirakata Y, Sayama K, Hashimoto K (2000) Different effects of dominant negative mutants of desmocollin and desmoglein on the cell–cell adhesion of keratinocytes. J Cell Sci 113:1803–11

Hanakawa Y, Matsuyoshi N, Stanley JR (2002a) Expression of desmoglein 1 compensates for genetic loss of desmoglein 3 in keratinocyte adhesion. J Invest Dermatol 119:27–31

Hanakawa Y, Schechter NM, Lin C, Garza L, Li H, Yamaguchi T, Fudaba Y, Nishifuji K, Sugai M, Amagai M, Stanley JR (2002b) Molecular mechanisms of blister formation in bullous impetigo and staphylococcal scalded skin syndrome. J Clin Invest 110:53–60

Hanakawa Y, Schechter NM, Lin C, Nishifuji K, Amagai M, Stanley JR (2003) Enzymatic and molecular characteristics of the efficiency and specificity of exfoliative toxin cleavage of desmoglein 1. J Biol Chem

Hashimoto K, Fujiwara K, Tada J, Harada M, Setoyama M, Eto H (1995) Desmosomal dissolution in Grover's disease, Hailey-Hailey's disease and Darier's disease. J Cutan Pathol 22:488–501

Hashimoto T (2001) Immunopathology of paraneoplastic pemphigus. Clin Dermatol 19:675–82

Hatzfeld M (1999) The armadillo family of structural proteins. Int Rev Cytol 186:179–224

Hatzfeld M, Nachtsheim C (1996) Cloning and characterization of a new armadillo family member, p0071, associated with the junctional plaque: evidence for a subfamily of closely related proteins. J Cell Sci 109 (Pt 11):2767–78

Hatzfeld M, Kristjansson GI, Plessmann U, Weber K (1994) Band 6 protein, a major constituent of desmosomes from stratified epithelia, is a novel member of the armadillo multigene family. J Cell Sci 107:2259–70

Hatzfeld M, Haffner C, Schulze K, Vinzens U (2000) The function of plakophilin 1 in desmosome assembly and actin filament organization. J Cell Biol 149:209–22

Hatzfeld M, Green KJ, Sauter H (2003) Targeting of p0071 to desmosomes and adherens junctions is mediated by different protein domains. J Cell Sci 116:1219–1233

He W, Cowin P, Stokes DL (2003) Untangling desmosomal knots with electron tomography. Science 302:109–13

Hecht A, Litterst CM, Huber O, Kemler R (1999) Functional characterization of multiple transactivating elements in beta-catenin, some of which interact with the TATA-binding protein in vitro. J Biol Chem 274:18017–25

Heid HW, Schmidt A, Zimbelmann R, Schafer S, Winter-Simanowski S, Stumpp S, Keith M, Figge U, Schnolzer M, Franke WW (1994) Cell type-specific desmosomal plaque proteins of the plakoglobin family: plakophilin 1 (band 6 protein). Differentiation 58:113–31

Henkler F, Strom M, Mathers K, Cordingley H, Sullivan K, King I (2001) Transgenic misexpression of the differentiation-specific desmocollin isoform 1 in basal keratinocytes. J Invest Dermatol 116:144–9

Hofmann I, Mertens C, Brettel M, Nimmrich V, Schnolzer M, Herrmann H (2000) Interaction of plakophilins with desmoplakin and intermediate filament proteins: an in vitro analysis. J Cell Sci 113 (Pt 13):2471–83

Hu P, Berkowitz P, O'Keefe EJ, Rubenstein DS (2003) Keratinocyte adherens junctions initiate nuclear signaling by translocation of plakoglobin from the membrane to the nucleus. J Invest Dermatol 121:242–251

Huber AH, Nelson WJ, Weis WI (1997) Three-dimensional structure of the armadillo repeat region of beta-catenin. Cell 90:871–82

Huber O, Bierkamp C, Kemler R (1996) Cadherins and catenins in development. Curr Opin Cell Biol 8:685–91

Huen AC, Park JK, Godsel LM, Chen X, Bannon LJ, Amargo EV, Hudson TY, Mongiu AK, Leigh IM, Kelsell DP, Gumbiner BM, Green KJ (2002) Intermediate filament-membrane attachments function synergistically with actin-dependent contacts to regulate intercellular adhesive strength. J Cell Biol 159:1005–17

Hunt DM, Sahota VK, Taylor K, Simrak D, Hornigold N, Arnemann J, Wolfe J, Buxton RS (1999) Clustered cadherin genes: a sequence-ready contig for the desmosomal cadherin locus on human chromosome 18. Genomics 62:445–55

Hunt DM, Rickman L, Whittock NV, Eady RA, Simrak D, Dopping-Hepenstal PJ, Stevens HP, Armstrong DK, Hennies HC, Kuster W, Hughes AE, Arnemann J, Leigh IM, McGrath JA, Kelsell DP, Buxton RS (2001) Spectrum of dominant mutations in the desmosomal cadherin desmoglein 1, causing the skin disease striate palmoplantar keratoderma. Eur J Hum Genet 9:197–203

Ishii K, Green KJ (2001) Cadherin function: breaking the barrier. Curr Biol 11:R569–72

Ishii K, Amagai M, Hall RP, Hashimoto T, Takayanagi A, Gamou S, Shimizu N, Nishikawa T (1997) Characterization of autoantibodies in pemphigus using antigen-specific enzyme-linked immunosorbent assays with baculovirus-expressed recombinant desmogleins. J Immunol 159:2010–7

Jaulin-Bastard F, Arsanto JP, Le Bivic A, Navarro C, Vely F, Saito H, Marchetto S, Hatzfeld M, Santoni MJ, Birnbaum D, Borg JP (2002) Interaction between Erbin and a catenin-related protein in epithelial cells. J Biol Chem 277:2869–75

Jonca N, Guerrin M, Hadjiolova K, Caubet C, Gallinaro H, Simon M, Serre G (2002) Corneodesmosin, a component of epidermal corneocyte desmosomes, displays homophilic adhesive properties. J Biol Chem 277:5024–9

Jones JC, Goldman RD (1985) Intermediate filaments and the initiation of desmosome assembly. J Cell Biol 101:506–17

Jones JC, Grelling KA (1989) Distribution of desmoplakin in normal cultured human keratinocytes and in basal cell carcinoma cells. Cell Motil Cytoskeleton 13:181–94

Jones JC, Arnn J, Staehelin LA, Goldman RD (1984) Human autoantibodies against desmosomes: possible causative factors in pemphigus. Proc Natl Acad Sci U S A 81:2781–5

Kanai Y, Ochiai A, Shibata T, Oyama T, Ushijima S, Akimoto S, Hirohashi S (1995) c-erbB-2 gene product directly associates with beta-catenin and plakoglobin. Biochem Biophys Res Commun 208:1067–72

Kapprell HP, Owaribe K, Franke WW (1988) Identification of a basic protein of Mr 75,000 as an accessory desmosomal plaque protein in stratified and complex epithelia. J Cell Biol 106:1679–91

Karashima T, Watt FM (2002) Interaction of periplakin and envoplakin with intermediate filaments. J Cell Sci 115:5027–37

Karnovsky A, Klymkowsky MW (1995) Anterior axis duplication in Xenopus induced by the over-expression of the cadherin-binding protein plakoglobin. Proc Natl Acad Sci U S A 92:4522–6

Kim E, Arnould T, Sellin L, Benzing T, Comella N, Kocher O, Tsiokas L, Sukhatme VP, Walz G (1999) Interaction between RGS7 and polycystin. Proc Natl Acad Sci U S A 96:6371–6

Kitajima Y (2002) Mechanisms of desmosome assembly and disassembly. Clin Exp Dermatol 27:684–90

Kljuic A, Christiano AM (2003) A novel mouse desmosomal cadherin family member, desmoglein 1gamma. Exp Dermatol 12:20–9

Kljuic A, Bazzi H, Sundberg JP, Martinez-Mir A, O'Shaughnessy R, Mahoney MG, Levy M, Montagutelli X, Ahmad W, Aita VM, Gordon D, Uitto J, Whiting D, Ott J, Fischer S, Gilliam TC, Jahoda CA, Morris RJ, Panteleyev AA, Nguyen VT, Christiano AM (2003) Desmoglein 4 in hair follicle differentiation and epidermal adhesion. Evidence from inherited hypotrichosis and acquired pemphigus vulgaris. Cell 113:249–60

Koch PJ, Walsh MJ, Schmelz M, Goldschmidt MD, Zimbelmann R, Franke WW (1990) Identification of desmoglein, a constitutive desmosomal glycoprotein, as a member of the cadherin family of cell adhesion molecules. Eur J Cell Biol 53:1–12

Koch PJ, Mahoney MG, Ishikawa H, Pulkkinen L, Uitto J, Shultz L, Murphy GF, Whitaker-Menezes D, Stanley JR (1997) Targeted disruption of the pemphigus vulgaris antigen (desmoglein 3) gene in mice causes loss of keratinocyte cell adhesion with a phenotype similar to pemphigus vulgaris. J Cell Biol 137:1091–1102

Koch PJ, Mahoney MG, Cotsarelis G, Rothenberger K, Lavker RM, Stanley JR (1998) Desmoglein 3 anchors telogen hair in the follicle. J Cell Sci 111:2529–37

Kofron M, Heasman J, Lang SA, Wylie CC (2002) Plakoglobin is required for maintenance of the cortical actin skeleton in early Xenopus embryos and for cdc42-mediated wound healing. J Cell Biol 158:695–708

Kolligs FT, Kolligs B, Hajra KM, Hu G, Tani M, Cho KR, Fearon ER (2000) Gamma-catenin is regulated by the APC tumor suppressor and its oncogenic activity is distinct from that of beta-catenin. Genes Dev 14:1319–31

Kouklis PD, Hutton E, Fuchs E (1994) Making a connection: direct binding between keratin intermediate filaments and desmosomal proteins. J Cell Biol 127:1049–60

Kowalczyk AP, Stappenbeck TS, Parry DA, Palka HL, Virata ML, Bornslaeger EA, Nilles LA, Green KJ (1994) Structure and function of desmosomal transmembrane core and plaque molecules. Biophys Chem 50:97–112

Kowalczyk AP, Anderson JE, Borgwardt JE, Hashimoto T, Stanley JR, Green KJ (1995) Pemphigus sera recognize conformationally sensitive epitopes in the amino-terminal region of desmoglein-1. J Invest Dermatol 105:147–52

Kowalczyk AP, Borgwardt JE, Green KJ (1996) Analysis of desmosomal cadherin-adhesive function and stoichiometry of desmosomal cadherin-plakoglobin complexes. J Invest Dermatol 107:293–300

Kowalczyk AP, Bornslaeger EA, Borgwardt JE, Palka HL, Dhaliwal AS, Corcoran CM, Denning MF, Green KJ (1997) The amino-terminal domain of desmoplakin binds to plakoglobin and clusters desmosomal cadherin-plakoglobin complexes. J Cell Biol 139:773–84

Kowalczyk AP, Navarro P, Dejana E, Bornslaeger EA, Green KJ, Kopp DS, Borgwardt JE (1998) VE-cadherin and desmoplakin are assembled into dermal microvascular endothelial intercellular junctions: a pivotal role for plakoglobin in the recruitment of desmoplakin to intercellular junctions. J Cell Sci 111:3045–57

Kowalczyk AP, Bornslaeger EA, Norvell SM, Palka HL, Green KJ (1999a) Desmosomes: intercellular adhesive junctions specialized for attachment of intermediate filaments. Int Rev Cytol 185:237–302

Kowalczyk AP, Hatzfeld M, Bornslaeger EA, Kopp DS, Borgwardt JE, Corcoran CM, Settler A, Green KJ (1999b) The head domain of plakophilin-1 binds to desmoplakin and enhances its recruitment to desmosomes. Implications for cutaneous disease. J Biol Chem 274:18145–8

Lenox JM, Koch PJ, Mahoney MG, Lieberman M, Stanley JR, Radice GL (2000) Postnatal lethality of P-cadherin/desmoglein 3 double knockout mice: demonstration of a cooperative effect of these cell adhesion molecules in tissue homeostasis of stratified squamous epithelia. J Invest Dermatol 114:948–52

Leung CL, Liem RK, Parry DA, Green KJ (2001) The plakin family. J Cell Sci 114:3409–10

Levesque G, Yu G, Nishimura M, Zhang DM, Levesque L, Yu H, Xu D, Liang Y, Rogaeva E, Ikeda M, Duthie M, Murgolo N, Wang L, VanderVere P, Bayne ML, Strader CD, Rommens JM, Fraser PE, St George-Hyslop P (1999) Presenilins interact with armadillo proteins including neural-specific plakophilin-related protein and beta-catenin. J Neurochem 72:999–1008

Levy-Nissenbaum E, Betz RC, Frydman M, Simon M, Lahat H, Bakhan T, Goldman B, Bygum A, Pierick M, Hillmer AM, Jonca N, Toribio J, Kruse R, Dewald G, Cichon S, Kubisch C, Guerrin M, Serre G, Nothen MM, Pras E (2003) Hypotrichosis simplex of the scalp is associated with nonsense mutations in CDSN encoding corneodesmosin. Nat Genet 34:151–3

Lewis JE, Jensen PJ, Wheelock MJ (1994) Cadherin function is required for human keratinocytes to assemble desmosomes and stratify in response to calcium. J Invest Dermatol 102:870–7

Lewis JE, Wahl JK, 3rd, Sass KM, Jensen PJ, Johnson KR, Wheelock MJ (1997) Cross-talk between adherens junctions and desmosomes depends on plakoglobin. J Cell Biol 136:919–34

Ma AS, Sun TT (1986) Differentiation-dependent changes in the solubility of a 195-kD protein in human epidermal keratinocytes. J Cell Biol 103:41–8

Maatta A, DiColandrea T, Groot K, Watt FM (2001) Gene targeting of envoplakin, a cytoskeletal linker protein and precursor of the epidermal cornified envelope. Mol Cell Biol 21:7047–53

Mahoney MG, Wang ZH, Stanley JR (1999) Pemphigus vulgaris and pemphigus foliaceus antibodies are pathogenic in plasminogen activator knockout mice. J Invest Dermatol 113:22–5

Marcozzi C, Burdett ID, Buxton RS, Magee AI (1998) Coexpression of both types of desmosomal cadherin and plakoglobin confers strong intercellular adhesion. J Cell Sci 111:495–509

Mattey DL, Garrod DR (1986) Calcium-induced desmosome formation in cultured kidney epithelial cells. J Cell Sci 85:95–111

McGrath JA (1999) Hereditary diseases of desmosomes. J Dermatol Sci 20:85–91

McGrath JA, McMillan JR, Shemanko CS, Runswick SK, Leigh IM, Lane EB, Garrod DR, Eady RA (1997) Mutations in the plakophilin 1 gene result in ectodermal dysplasia/skin fragility syndrome. Nat Genet 17:240–4

McKoy G, Protonotarios N, Crosby A, Tsatsopoulou A, Anastasakis A, Coonar A, Norman M, Baboonian C, Jeffery S, McKenna WJ (2000) Identification of a deletion in plakoglobin in arrhythmogenic right ventricular cardiomyopathy with palmoplantar keratoderma and woolly hair (Naxos disease). Lancet 355:2119–24

McMillan JR, Haftek M, Akiyama M, South AP, Perrot H, McGrath JA, Eady RA, Shimizu H (2003) Alterations in desmosome size and number coincide with the loss of keratinocyte cohesion in skin with homozygous and heterozygous defects in the desmosomal protein plakophilin 1. J Invest Dermatol 121:96–103

McMullan R, Lax S, Robertson VH, Radford DJ, Broad S, Watt FM, Rowles A, Croft DR, Olson MF, Hotchin NA (2003) Keratinocyte differentiation is regulated by the Rho and ROCK signaling pathway. Curr Biol 13:2185–9

Meng JJ, Bornslaeger EA, Green KJ, Steinert PM, Ip W (1997) Two-hybrid analysis reveals fundamental differences in direct interactions between desmoplakin and cell type-specific intermediate filaments. J Biol Chem 272:21495–503

Merriam JM, Rubenstein AB, Klymkowsky MW (1997) Cytoplasmically anchored plakoglobin induces a WNT-like phenotype in Xenopus. Dev Biol 185:67–81

Merritt AJ, Berika MY, Zhai W, Kirk SE, Ji B, Hardman MJ, Garrod DR (2002) Suprabasal desmoglein 3 expression in the epidermis of transgenic mice results in hyperproliferation and abnormal differentiation. Mol Cell Biol 22:5846–58

Mertens C, Kuhn C, Franke WW (1996) Plakophilins 2a and 2b: constitutive proteins of dual location in the karyoplasm and the desmosomal plaque. J Cell Biol 135:1009–25

Mertens C, Hofmann I, Wang Z, Teichmann M, Sepehri Chong S, Schnolzer M, Franke WW (2001) Nuclear particles containing RNA polymerase III complexes associated with the junctional plaque protein plakophilin 2. Proc Natl Acad Sci U S A 98:7795–800

Millar SE, Willert K, Salinas PC, Roelink H, Nusse R, Sussman DJ, Barsh GS (1999) WNT signaling in the control of hair growth and structure. Dev Biol 207:133–49

Mils V, Vincent C, Croute F, Serre G (1992) The expression of desmosomal and corneodesmosomal antigens shows specific variations during the terminal differentiation of epidermis and hair follicle epithelia. J Histochem Cytochem 40:1329–37

Miravet S, Piedra J, Miro F, Itarte E, Garcia de Herreros A, Dunach M (2002) The transcriptional factor Tcf-4 contains different binding sites for beta-catenin and plakoglobin. J Biol Chem 277:1884–91

Miravet S, Piedra J, Castano J, Raurell I, Franci C, Dunach M, Garcia de Herreros A (2003) Tyrosine phosphorylation of plakoglobin causes contrary effects on its association with desmosomes and adherens junction components and modulates beta-catenin-mediated transcription. Mol Cell Biol 23:7391–402

Moll I, Kurzen H, Langbein L, Franke WW (1997) The distribution of the desmosomal protein, plakophilin 1, in human skin and skin tumors. J Invest Dermatol 108:139–46

Mueller H, Franke WW (1983) Biochemical and immunological characterization of desmoplakins I and II, the major polypeptides of the desmosomal plaque. J Mol Biol 163:647–71

Muller J, Ritt DA, Copeland TD, Morrison DK (2003) Functional analysis of C-TAK1 substrate binding and identification of PKP2 as a new C-TAK1 substrate. Embo J 22:4431–42

Nguyen VT, Ndoye A, Shultz LD, Pittelkow MR, Grando SA (2000) Antibodies against keratinocyte antigens other than desmogleins 1 and 3 can induce pemphigus vulgaris-like lesions. J Clin Invest 106:1467–79

Nollet F, Kools P, van Roy F (2000) Phylogenetic analysis of the cadherin superfamily allows identification of six major subfamilies besides several solitary members. J Mol Biol 299:551–72

Norgett EE, Hatsell SJ, Carvajal-Huerta L, Cabezas JC, Common J, Purkis PE, Whittock N, Leigh IM, Stevens HP, Kelsell DP (2000) Recessive mutation in desmoplakin disrupts desmoplakin-intermediate filament interactions and causes dilated cardiomyopathy, woolly hair and keratoderma. Hum Mol Genet 9:2761–6

North AJ, Chidgey MA, Clarke JP, Bardsley WG, Garrod DR (1996) Distinct desmocollin isoforms occur in the same desmosomes and show reciprocally graded distributions in bovine nasal epidermis. Proc Natl Acad Sci U S A 93:7701–5

North AJ, Bardsley WG, Hyam J, Bornslaeger EA, Cordingley HC, Trinnaman B, Hatzfeld M, Green KJ, Magee AI, Garrod DR (1999) Molecular map of the desmosomal plaque. J Cell Sci 112:4325–36

Norvell SM, Green KJ (1998) Contributions of extracellular and intracellular domains of full length and chimeric cadherin molecules to junction assembly in epithelial cells. J Cell Sci 111:1305–18

Nousari HC, Anhalt GJ (1999) Pemphigus and bullous pemphigoid. Lancet 354:667–72

O'Keefe EJ, Erickson HP, Bennett V (1989) Desmoplakin I and desmoplakin II. Purification and characterization. J Biol Chem 264:8310–8

Ochiai A, Akimoto S, Kanai Y, Shibata T, Oyama T, Hirohashi S (1994) c-erbB-2 gene product associates with catenins in human cancer cells. Biochem Biophys Res Commun 205:73–8

Osada K, Seishima M, Kitajima Y (1997) Pemphigus IgG activates and translocates protein kinase C from the cytosol to the particulate/cytoskeleton fractions in human keratinocytes. J Invest Dermatol 108:482–7

Ota T, Amagai M, Watanabe M, Nishikawa T (2003) No involvement of IgG autoantibodies against extracellular domains of desmoglein 2 in paraneoplastic pemphigus or inflammatory bowel diseases. J Dermatol Sci 32:137–41

Ouyang P, Sugrue SP (1996) Characterization of pinin, a novel protein associated with the desmosome-intermediate filament complex. J Cell Biol 135:1027–42

Ozawa M, Kemler R (1992) Molecular organization of the uvomorulin-catenin complex. J Cell Biol 116:989–96

Palka HL, Green KJ (1997) Roles of plakoglobin end domains in desmosome assembly. J Cell Sci 110:2359–71

Parker HR, Li Z, Sheinin H, Lauzon G, Pasdar M (1998) Plakoglobin induces desmosome formation and epidermoid phenotype in N- cadherin-expressing squamous carcinoma cells deficient in plakoglobin and E-cadherin. Cell Motil Cytoskeleton 40:87–100

Pasdar M, Nelson WJ (1988a) Kinetics of desmosome assembly in Madin-Darby canine kidney epithelial cells: temporal and spatial regulation of desmoplakin organization and stabilization upon cell–cell contact. I. Biochemical analysis. J Cell Biol 106:677–85

Pasdar M, Nelson WJ (1988b) Kinetics of desmosome assembly in Madin-Darby canine kidney epithelial cells: temporal and spatial regulation of desmoplakin organization and stabilization upon cell–cell contact. II. Morphological analysis. J Cell Biol 106:687–95

Pasdar M, Nelson WJ (1989) Regulation of desmosome assembly in epithelial cells: kinetics of synthesis, transport, and stabilization of desmoglein I, a major protein of the membrane core domain. J Cell Biol 109:163–77

Pasdar M, Krzeminski KA, Nelson WJ (1991) Regulation of desmosome assembly in MDCK epithelial cells: coordination of membrane core and cytoplasmic plaque domain assembly at the plasma membrane. J Cell Biol 113:645–55

Pasdar M, Li Z, Chan H (1995a) Desmosome assembly and disassembly are regulated by reversible protein phosphorylation in cultured epithelial cells. Cell Motil Cytoskeleton 30:108–21

Pasdar M, Li Z, Chlumecky V (1995b) Plakoglobin: kinetics of synthesis, phosphorylation, stability, and interactions with desmoglein and E-cadherin. Cell Motil Cytoskeleton 32:258–72

Peifer M, McCrea PD, Green KJ, Wieschaus E, Gumbiner BM (1992) The vertebrate adhesive junction proteins beta-catenin and plakoglobin and the Drosophila segment polarity gene armadillo form a multigene family with similar properties. J Cell Biol 118:681–91

Penn EJ, Burdett ID, Hobson C, Magee AI, Rees DA (1987) Structure and assembly of desmosome junctions: biosynthesis and turnover of the major desmosome components of Madin-Darby canine kidney cells in low calcium medium. J Cell Biol 105:2327–34

Pertz O, Bozic D, Koch AW, Fauser C, Brancaccio A, Engel J (1999) A new crystal structure, Ca2+ dependence and mutational analysis reveal molecular details of E-cadherin homoassociation. Embo J 18:1738–47

Plott RT, Amagai M, Udey MC, Stanley JR (1994) Pemphigus vulgaris antigen lacks biochemical properties characteristic of classical cadherins. J Invest Dermatol 103:168–72

Posthaus H, Dubois CM, Laprise MH, Grondin F, Suter MM, Muller E (1998) Proprotein cleavage of E-cadherin by furin in baculovirus over- expression system: potential role of other convertases in mammalian cells. FEBS Lett 438:306–10

Posthaus H, Dubois CM, Muller E (2003) Novel insights into cadherin processing by subtilisin-like convertases. FEBS Lett 536:203–8

Protonotarios N, Tsatsopoulou A, Patsourakos P, Alexopoulos D, Gezerlis P, Simitsis S, Scampardonis G (1986) Cardiac abnormalities in familial palmoplantar keratosis. Br Heart J 56:321–6

Protonotarios N, Tsatsopoulou A, Anastasakis A, Sevdalis E, McKoy G, Stratos K, Gatzoulis K, Tentolouris K, Spiliopoulou C, Panagiotakos D, McKenna W, Toutouzas P (2001a) Genotype-phenotype assessment in autosomal recessive arrhythmogenic right ventricular cardiomyopathy (Naxos disease) caused by a deletion in plakoglobin. J Am Coll Cardiol 38:1477–84

Protonotarios N, Tsatsopoulou A, Fontaine G (2001b) Naxos disease: keratoderma, scalp modifications, and cardiomyopathy. J Am Acad Dermatol 44:309–11

Pulkkinen L, Uitto J (1999) Mutation analysis and molecular genetics of epidermolysis bullosa. Matrix Biol 18:29–42

Pulkkinen L, Choi YW, Simpson A, Montagutelli X, Sundberg J, Uitto J, Mahoney MG (2002) Loss of cell adhesion in Dsg3bal-Pas mice with homozygous deletion mutation (2079del14) in the desmoglein 3 gene. J Invest Dermatol 119:1237–43

Pulkkinen L, Choi YW, Kljuic A, Uitto J, Mahoney MG (2003) Novel member of the mouse desmoglein gene family: Dsg1-beta. Exp Dermatol 12:11–9

Radice GL, Ferreira-Cornwell MC, Robinson SD, Rayburn H, Chodosh LA, Takeichi M, Hynes RO (1997) Precocious mammary gland development in P-cadherin-deficient mice. J Cell Biol 139:1025–32

Rampazzo A, Nava A, Malacrida S, Beffagna G, Bauce B, Rossi V, Zimbello R, Simionati B, Basso C, Thiene G, Towbin JA, Danieli GA (2002) Mutation in human desmoplakin domain binding to plakoglobin causes a dominant form of arrhythmogenic right ventricular cardiomyopathy. Am J Hum Genet 71:1200–6

Rampazzo A, Beffagna G, Nava A, Occhi G, Bauce B, Noiato M, Basso C, Frigo G, Thiene G, Towbin J, Danieli GA (2003) Arrhythmogenic right ventricular cardiomyopathy type 1 (ARVD1): confirmation of locus assignment and mutation screening of four candidate genes. Eur J Hum Genet 11:69–76

Rezniczek GA, de Pereda JM, Reipert S, Wiche G (1998) Linking integrin alpha6beta4-based cell adhesion to the intermediate filament cytoskeleton: direct interaction between the beta4 subunit and plectin at multiple molecular sites. J Cell Biol 141:209–25

Rickman L, Simrak D, Stevens HP, Hunt DM, King IA, Bryant SP, Eady RA, Leigh IM, Arnemann J, Magee AI, Kelsell DP, Buxton RS (1999) N-terminal deletion in a desmosomal cadherin causes the autosomal dominant skin disease striate palmoplantar keratoderma. Hum Mol Genet 8:971–6

Roitbak T, Ward CJ, Harris PC, Bacallao R, Ness SA, Wandinger-Ness A (2004) A polycystin-1 multiprotein complex is disrupted in polycystic kidney disease cells. Mol Biol Cell 15:1334–1346

Ruhrberg C, Watt FM (1997) The plakin family: versatile organizers of cytoskeletal architecture. Curr Opin Genet Dev 7:392–7

Ruhrberg C, Hajibagheri MA, Simon M, Dooley TP, Watt FM (1996a) Envoplakin, a novel precursor of the cornified envelope that has homology to desmoplakin. J Cell Biol 134:715-29

Ruhrberg C, Williamson JA, Sheer D, Watt FM (1996b) Chromosomal localisation of the human envoplakin gene (EVPL) to the region of the tylosis oesophageal cancer gene (TOCG) on 17q25. Genomics 37:381-5

Ruhrberg C, Hajibagheri MA, Parry DA, Watt FM (1997) Periplakin, a novel component of cornified envelopes and desmosomes that belongs to the plakin family and forms complexes with envoplakin. J Cell Biol 139:1835-49

Ruiz P, Brinkmann V, Ledermann B, Behrend M, Grund C, Thalhammer C, Vogel F, Birchmeier C, Gunthert U, Franke WW, Birchmeier W (1996) Targeted mutation of plakoglobin in mice reveals essential functions of desmosomes in the embryonic heart. J Cell Biol 135:215-25

Runswick SK, O'Hare MJ, Jones L, Streuli CH, Garrod DR (2001) Desmosomal adhesion regulates epithelial morphogenesis and cell positioning. Nat Cell Biol 3:823-30

Rutman AJ, Buxton RS, Burdett ID (1994) Visualisation by electron microscopy of the unique part of the cytoplasmic domain of a desmoglein, a cadherin-like protein of the desmosome type of cell junction. FEBS Lett 353:194-6

Sacco PA, McGranahan TM, Wheelock MJ, Johnson KR (1995) Identification of plakoglobin domains required for association with N-cadherin and alpha-catenin. J Biol Chem 270:20201-6

Sakuntabhai A, Ruiz-Perez V, Carter S, Jacobsen N, Burge S, Monk S, Smith M, Munro CS, O'Donovan M, Craddock N, Kucherlapati R, Rees JL, Owen M, Lathrop GM, Monaco AP, Strachan T, Hovnanian A (1999) Mutations in ATP2A2, encoding a Ca2+ pump, cause Darier disease. Nat Genet 21:271-7

Sato M, Aoyama Y, Kitajima Y (2000) Assembly pathway of desmoglein 3 to desmosomes and its perturbation by pemphigus vulgaris-IgG in cultured keratinocytes, as revealed by time-lapsed labeling immunoelectron microscopy. Lab Invest 80:1583-92

Savagner P, Yamada KM, Thiery JP (1997) The zinc-finger protein slug causes desmosome dissociation, an initial and necessary step for growth factor-induced epithelial-mesenchymal transition. J Cell Biol 137:1403-19

Scheffers MS, van der Bent P, Prins F, Spruit L, Breuning MH, Litvinov SV, de Heer E, Peters DJ (2000) Polycystin-1, the product of the polycystic kidney disease 1 gene, co-localizes with desmosomes in MDCK cells. Hum Mol Genet 9:2743-50

Schmelz M, Franke WW (1993) Complexus adhaerentes, a new group of desmoplakin-containing junctions in endothelial cells: the syndesmos connecting retothelial cells of lymph nodes. Eur J Cell Biol 61:274-89

Schmidt A, Heid HW, Schafer S, Nuber UA, Zimbelmann R, Franke WW (1994) Desmosomes and cytoskeletal architecture in epithelial differentiation: cell type-specific plaque components and intermediate filament anchorage. Eur J Cell Biol 65:229-45

Schmidt A, Langbein L, Rode M, Pratzel S, Zimbelmann R, Franke WW (1997) Plakophilins 1a and 1b: widespread nuclear proteins recruited in specific epithelial cells as desmosomal plaque components. Cell Tissue Res 290:481-99

Schmidt A, Langbein L, Pratzel S, Rode M, Rackwitz HR, Franke WW (1999) Plakophilin 3—a novel cell-type-specific desmosomal plaque protein. Differentiation 64:291-306

Schwarz MA, Owaribe K, Kartenbeck J, Franke WW (1990) Desmosomes and hemidesmosomes: constitutive molecular components. Annu Rev Cell Biol 6:461-91

Scully C, Challacombe SJ (2002) Pemphigus vulgaris: update on etiopathogenesis, oral manifestations, and management. Crit Rev Oral Biol Med 13:397-408

Seishima M, Esaki C, Osada K, Mori S, Hashimoto T, Kitajima Y (1995) Pemphigus IgG, but not bullous pemphigoid IgG, causes a transient increase in intracellular calcium and inositol 1,4,5-triphosphate in DJM-1 cells, a squamous cell carcinoma line. J Invest Dermatol 104:33–7

Serpente N, Marcozzi C, Roberts GA, Bao Q, Angst BD, Hirst EM, Burdett ID, Buxton RS, Magee AI (2000) Extracellularly truncated desmoglein 1 compromises desmosomes in MDCK cells. Mol Membr Biol 17:175–83

Serre G, Mils V, Haftek M, Vincent C, Croute F, Reano A, Ouhayoun JP, Bettinger S, Soleilhavoup JP (1991) Identification of late differentiation antigens of human cornified epithelia, expressed in re-organized desmosomes and bound to cross-linked envelope. J Invest Dermatol 97:1061–72

Shtutman M, Zhurinsky J, Oren M, Levina E, Ben-Ze'ev A (2002) PML is a target gene of beta-catenin and plakoglobin, and coactivates beta-catenin-mediated transcription. Cancer Res 62:5947–54

Simcha I, Geiger B, Yehuda-Levenberg S, Salomon D, Ben-Ze'ev A (1996) Suppression of tumorigenicity by plakoglobin: an augmenting effect of N-cadherin. J Cell Biol 133:199–209

Simcha I, Shtutman M, Salomon D, Zhurinsky J, Sadot E, Geiger B, Ben-Ze'ev A (1998) Differential nuclear translocation and transactivation potential of beta-catenin and plakoglobin. J Cell Biol 141:1433–48

Simon M, Jonca N, Guerrin M, Haftek M, Bernard D, Caubet C, Egelrud T, Schmidt R, Serre G (2001) Refined characterization of corneodesmosin proteolysis during terminal differentiation of human epidermis and its relationship to desquamation. J Biol Chem 276:20292–9

Skerrow CJ, Matoltsy AG (1974) Chemical characterization of isolated epidermal desmosomes. J Cell Biol 63:524–30

Smith EA, Fuchs E (1998) Defining the interactions between intermediate filaments and desmosomes. J Cell Biol 141:1229–41

Soriano S, Kang DE, Fu M, Pestell R, Chevallier N, Zheng H, Koo EH (2001) Presenilin 1 negatively regulates beta-catenin/T cell factor/lymphoid enhancer factor-1 signaling independently of beta-amyloid precursor protein and notch processing. J Cell Biol 152:785–794

South AP, Wan H, Stone MG, Dopping-Hepenstal PJ, Purkis PE, Marshall JF, Leigh IM, Eady RA, Hart IR, McGrath JA (2003) Lack of plakophilin 1 increases keratinocyte migration and reduces desmosome stability. J Cell Sci 116:3303–14

Stahl B, Diehlmann A, Sudhof TC (1999) Direct interaction of Alzheimer's disease-related presenilin 1 with armadillo protein p0071. J Biol Chem 274:9141–8

Stanley JR (2001) Pathophysiology and therapy of pemphigus in the 21st century. J Dermatol 28:645–6

Stanley JR, Koulu L, Klaus-Kovtun V, Steinberg MS (1986) A monoclonal antibody to the desmosomal glycoprotein desmoglein I binds the same polypeptide as human autoantibodies in pemphigus foliaceus. J Immunol 136:1227–30

Stappenbeck TS, Bornslaeger EA, Corcoran CM, Luu HH, Virata ML, Green KJ (1993) Functional analysis of desmoplakin domains: specification of the interaction with keratin versus vimentin intermediate filament networks. J Cell Biol 123:691–705

Stappenbeck TS, Lamb JA, Corcoran CM, Green KJ (1994) Phosphorylation of the desmoplakin COOH terminus negatively regulates its interaction with keratin intermediate filament networks. J Biol Chem 269:29351–4

Stuart RO, Sun A, Bush KT, Nigam SK (1996) Dependence of epithelial intercellular junction biogenesis on thapsigargin-sensitive intracellular calcium stores. J Biol Chem 271:13636–41

Sun D, Leung CL, Liem RK (2001) Characterization of the microtubule binding domain of microtubule actin crosslinking factor (MACF): identification of a novel group of microtubule associated proteins. J Cell Sci 114:161–172

Syed SE, Trinnaman B, Martin S, Major S, Hutchinson J, Magee AI (2002) Molecular interactions between desmosomal cadherins. Biochem J 362:317–27

Troyanovsky RB, Chitaev NA, Troyanovsky SM (1996) Cadherin binding sites of plakoglobin: localization, specificity and role in targeting to adhering junctions. J Cell Sci 109:3069–78

Troyanovsky RB, Klingelhofer J, Troyanovsky S (1999) Removal of calcium ions triggers a novel type of intercadherin interaction. J Cell Sci 112:4379–87

Troyanovsky SM, Eshkind LG, Troyanovsky RB, Leube RE, Franke WW (1993) Contributions of cytoplasmic domains of desmosomal cadherins to desmosome assembly and intermediate filament anchorage. Cell 72:561–74

Troyanovsky SM, Troyanovsky RB, Eshkind LG, Krutovskikh VA, Leube RE, Franke WW (1994a) Identification of the plakoglobin-binding domain in desmoglein and its role in plaque assembly and intermediate filament anchorage. J Cell Biol 127:151–60

Troyanovsky SM, Troyanovsky RB, Eshkind LG, Leube RE, Franke WW (1994b) Identification of amino acid sequence motifs in desmocollin, a desmosomal glycoprotein, that are required for plakoglobin binding and plaque formation. Proc Natl Acad Sci U S A 91:10790–4

Tselepis C, Chidgey M, North A, Garrod D (1998) Desmosomal adhesion inhibits invasive behavior. Proc Natl Acad Sci U S A 95:8064–9

Tsukita S (1985) Desmocalmin: a calmodulin-binding high molecular weight protein isolated from desmosomes. J Cell Biol 101:2070–80

Tsunoda K, Ota T, Aoki M, Yamada T, Nagai T, Nakagawa T, Koyasu S, Nishikawa T, Amagai M (2003) Induction of pemphigus phenotype by a mouse monoclonal antibody against the amino-terminal adhesive interface of desmoglein 3. J Immunol 170:2170–2178

van den Heuvel AP, de Vries-Smits AM, van Weeren PC, Dijkers PF, de Bruyn KM, Riedl JA, Burgering BM (2002) Binding of protein kinase B to the plakin family member periplakin. J Cell Sci 115:3957–3966

Vasioukhin V, Bowers E, Bauer C, Degenstein L, Fuchs E (2001) Desmoplakin is essential in epidermal sheet formation. Nat Cell Biol 3:1076–1085

Virata ML, Wagner RM, Parry DA, Green KJ (1992) Molecular structure of the human desmoplakin I and II amino terminus. Proc Natl Acad Sci U S A 89:544–8

Wahl JK, Sacco PA, McGranahan-Sadler TM, Sauppe LM, Wheelock MJ, Johnson KR (1996) Plakoglobin domains that define its association with the desmosomal cadherins and the classical cadherins: identification of unique and shared domains. J Cell Sci 109 (Pt 5):1143–54

Wahl JKr, Nieset JE, Sacco-Bubulya PA, Sadler TM, Johnson KR, Wheelock MJ (2000) The amino- and carboxyl-terminal tails of (beta)-catenin reduce its affinity for desmoglein 2. J Cell Sci 113 (Pt 10):1737–45

Wallis S, Lloyd S, Wise I, Ireland G, Fleming TP, Garrod D (2000) The alpha isoform of protein kinase C is involved in signaling the response of desmosomes to wounding in cultured epithelial cells. Mol Biol Cell 11:1077–92

Watt FM, Mattey DL, Garrod DR (1984) Calcium-induced reorganization of desmosomal components in cultured human keratinocytes. J Cell Biol 99:2211–5

Wheelock MJ, Johnson KR (2003) Cadherin-mediated cellular signaling. Curr Opin Cell Biol 15:509–514

White FH, Gohari K (1984) Desmosomes in hamster cheek pouch epithelium: their quantitative characterization during epithelial differentiation. J Cell Sci 66:411–29

Whittock NV, Bower C (2003) Genetic evidence for a novel human desmosomal cadherin, desmoglein 4. J Invest Dermatol 120:523–30

Whittock NV, Ashton GH, Dopping-Hepenstal PJ, Gratian MJ, Keane FM, Eady RA, McGrath JA (1999) Striate palmoplantar keratoderma resulting from desmoplakin haploinsufficiency. J Invest Dermatol 113:940–6

Whittock NV, Haftek M, Angoulvant N, Wolf F, Perrot H, Eady RA, McGrath JA (2000) Genomic amplification of the human plakophilin 1 gene and detection of a new mutation in ectodermal dysplasia/skin fragility syndrome. J Invest Dermatol 115:368–74

Whittock NV, Wan H, Morley SM, Garzon MC, Kristal L, Hyde P, McLean WH, Pulkkinen L, Uitto J, Christiano AM, Eady RA, McGrath JA (2002) Compound heterozygosity for non-sense and mis-sense mutations in desmoplakin underlies skin fragility/woolly hair syndrome. J Invest Dermatol 118:232–8

Wiche G (1998) Role of plectin in cytoskeleton organization and dynamics. J Cell Sci 111 (Pt 17):2477–86

Witcher LL, Collins R, Puttagunta S, Mechanic SE, Munson M, Gumbiner B, Cowin P (1996) Desmosomal cadherin binding domains of plakoglobin. J Biol Chem 271:10904–9

Xiao K, Allison DF, Buckley KM, Kottke MD, Vincent PA, Faundez V, Kowalczyk AP (2003) Cellular levels of p120 catenin function as a set point for cadherin expression levels in microvascular endothelial cells. J Cell Biol 163:535–45

Zhurinsky J, Shtutman M, Ben-Ze'ev A (2000) Differential mechanisms of LEF/TCF family-dependent transcriptional activation by beta-catenin and plakoglobin. Mol Cell Biol 20:4238–52

Part II
Integrins and Extracellular Matrix

Regulation of Signal Transduction by Integrins

Y. Miyamoto · P. Reddig · R. L. Juliano (✉)

Department of Pharmacology, University of North Carolina, CB 7365, Chapel Hill, NC 27599-7365, USA
arjay@med.unc.edu

1	Overview	197
2	Direct Signaling by Integrins	198
2.1	Focal Adhesion Kinase and Other Focal Contact Proteins	198
2.2	Direct Activation of MAP Kinase Cascades by Integrins	201
2.3	Integrins and Rho GTPases	202
3	Integrin Regulation of RTK Signaling Cascades	205
4	Integrin Regulation of Signaling by GPCRs	209
5	Summary	210
References		210

Abstract The integrin family of cell membrane receptors plays an important role in signal transduction cascades. Ligation of integrins by extracellular matrix proteins can lead to direct activation of Rho-family GTPases and MAP kinase pathways. However, perhaps the most significant signaling function of integrins is to modulate signal transduction events initiated by receptor tyrosine kinases and G protein-coupled receptors. This probably plays a role in coordinating information about cell shape and position with information about the availability of soluble growth factors.

Keywords Integrins · Signaling · MAP kinase · FAK · Rho GTPases

1
Overview

It is now widely appreciated that transmembrane cell adhesion receptors such as integrins play an important role in signal transduction processes. Integrins and associated cytoskeletal proteins can interact directly with key molecules in signaling cascades to activate or regulate those cascades. In addition, integrins and the cytoskeleton modulate signaling pathways that are triggered by growth factors, cytokines, and many other ligands. Thus, understanding the interplay between signaling processes and cell adhesion proteins has emerged as a vital aspect of current cell biology. In this chapter we will review several facets of integrin-regulated signaling with the emphasis on new findings in this area that we deem particularly intriguing. Since this field of investigation has grown enormously in recent years we cannot undertake a comprehensive review of the literature and thus, with regret, may pass over many interesting and worthy articles.

2
Direct Signaling by Integrins

Integrin subunits (except for the $\beta 4$ subunit) have rather short cytoplasmic domains and thus it is somewhat surprising that they play such a key role in signaling events. However, it is clear that binding of a ligand to an integrin, accompanied by integrin clustering, can recruit various structural and signaling proteins, leading to activation of important signaling cascades, most notably pathways leading to activation of mitogen-activated protein (MAP) kinases, but other pathways as well. There is an abundance of literature on direct integrin signaling with several good reviews on this topic (Aplin et al. 1998; Giancotti and Ruoslahti 1999; Juliano 2002). Here we will briefly outline the basic ideas and discuss some intriguing recent findings.

2.1
Focal Adhesion Kinase and Other Focal Contact Proteins

Integrin-mediated adhesion or clustering (or both) leads to enhanced tyrosine phosphorylation of several intracellular proteins (Kornberg et al. 1991). This is due to activation of a non-receptor tyrosine kinase now known as FAK (focal adhesion kinase) (Hanks et al. 1992; Schaller et al. 1992). This protein consists of a central kinase domain flanked by large amino-terminal and carboxy terminal extensions. There is a region of the C-terminus known as the 'FAT' (focal adhesion targeting) sequence that is responsible for recruiting FAK to the integrin-rich adhesion structures known as focal adhesions or focal contacts. FAK can bind to a number of other intracellular proteins including c-Src, PI-3-Kinase, GRAF (a Rho-GAP), paxillin, talin, and p130 Crk-associated substrate (p130Cas). While activation and tyrosine phosphorylation of FAK accompanies integrin-mediated adhesion, de-phosphorylation occurs almost immediately when cells are detached (Schaller 1996; Aplin et al. 1998; Parsons et al. 2000). The mechanisms regulating these events remain largely undefined. The so-called FERM (protein 4.1, ezrin, radixin, moesin) domain in the amino terminus region of FAK has been proposed to interact with integrins (Schaller et al. 1995; Sieg et al. 2000), but it has been difficult to demonstrate direct interactions in vivo. In some cell types, FAK phosphorylation seems to be a relatively late event that is downstream of actin filament assembly (Gao et al. 1997). Nonetheless, certain mutations in the integrin β subunit cytoplasmic region can strongly affect FAK activation, suggesting specific, if indirect, linkages (Wennerberg et al. 2000). FAK plays important roles in the regulation of cell motility, and in control of apoptosis (Gabarra-Niecko et al. 2003). As discussed below, FAK has also been implicated in integrin-mediated activation of MAP kinase pathways. Recent studies also indicate a complex role for FAK in the regulation of cell migration and tumor cell invasion (Hsia et al. 2003).

Recently, it was demonstrated that transmembrane clustering and cytoplasmic dimerization of FAK mediates tyrosine phosphorylation of FAK (Katz et al. 2002). The phosphorylation and activation of ERK2 was demonstrated by expressing chimeric proteins fused to either $\beta1$ integrin cytoplasmic domains or FAK domains. The clustering was facilitated with the use of antibodies to the extracellular domains of the chimeric proteins. The results demonstrated that FAK clustering at the cell membrane is required for FAK phosphorylation and subsequent downstream signaling, and dimerization between FAK proteins alone is sufficient for auto-phosphorylation of FAK. Since integrin clustering and lateral associations (Li et al. 2003) occur upon ligand binding and integrin activation, the resulting inside-out signaling and downstream signaling events can facilitate FAK dimerization and phosphorylation.

In addition to the role of FAK as a kinase and scaffold protein, FAK also functions as a protease targeting protein. This targeting function is regulated by p42ERK (p42 extracellular signal regulated kinase) and c-Src (Carragher et al. 2003). FAK associates with calpain in chick embryo fibroblasts and this is enhanced in v-src-transformed cells; this interaction is not dependent on FAK kinase activity. However, the proline sequence of FAK is necessary, and mutations in the NH_2-terminal region of FAK prevents calpain 2 binding and inhibits the interaction between FAK, calpain, and the ERK complex. Both FAK and calpain localize to focal adhesions, and the turnover of FA involves calpain-mediated proteolysis of FAK (Cooray et al. 1996). The ERK/MAPK-mediated activation of calpain (Glading et al. 2000) and the complex formed by FAK, calpain 2, and p42ERK in FA demonstrates an important role for FAK in promoting FA turnover.

Paxillin is another prominent integrin-associated focal adhesion protein involved in cell migration and signaling. Paxillin is an adaptor protein with a multitude of binding partners and has been shown to associate with both $\alpha4$ and the closely related $\alpha9$ integrin subunit via their cytoplasmic domains (Liu et al. 1999; Turner 2000; Young et al. 2001). The association between paxillin and $\alpha4$ or $\alpha9$ enhanced cell migration but reduced cell spreading and formation of focal adhesions and stress fibers. The $\alpha4$ paxillin interacting site has been identified to be a conserved consensus site for protein kinase A (PKA) (Goldfinger et al. 2003; Han et al. 2003). Upon phosphorylation of S988 on $\alpha4$ integrin in Jurkat T cells, phosphorylated $\alpha4$ becomes localized along the leading edge of the cells; however, only the unphosphorylated $\alpha4$ colocalizes with paxillin (Goldfinger et al. 2003). Upon PKA inhibition with specific pharmacological inhibitors of PKA, phosphorylation and subsequent extension of lamellipodia is reduced. When the phosphorylation site is mutated to aspartic acid (S988D), this mutation abolishes the delay in cell spreading and migration across transwell filters, while inhibiting interaction with paxillin, and decreasing Pyk2 phosphorylation. In contrast, when the phosphorylation site was mutated to alanine, the interaction with paxillin increases. The importance of the interaction of $\alpha4$ integrin and paxillin was further confirmed by constitutively fusing paxillin to the $\alpha4$ COOH terminus. The constitutive association between paxillin and $\alpha4$ reduces cell migration, and the ability to maintain ruffles and extend protrusions is also compromised (Han et al. 2003). The role of the $\alpha4$ integrin phosphorylation site was further investigated. This study emphasized that optimal cell migration requires both de-phosphorylation and phosphorylation of the S988, which allows the requisite association and dissociation with paxillin. A role for paxillin as a substrate for c-Jun N-terminal kinase (JNK) has also been recently identified (Huang et al. 2003). Through the use of phospho-amino acid analysis and mass spectrometry, Ser 178 of paxillin was found to be the target site for JNK. Mutation of this site inhibits migration and alters the intracellular distribution of paxillin.

Talin is a large cytoskeletal protein that directly interacts with signaling proteins such as FAK, vinculin, actin, and the cytoplasmic domain of integrin β subunits (Liu et al. 2000). With the applications of current structural techniques, novel aspects of talin function have been identified. First, the interaction between integrin and talin has been characterized with the use of crystallography and nuclear magnetic resonance (NMR). Second, the head domain of talin, containing the conserved FERM and phosphotyrosine-binding (PTB) motifs, was found important for integrin activation and for interaction between other signaling proteins. The head domain of talin binds to several β integrin tails and is predicted to contain a FERM domain, which is composed of three sub-domains F1, F2, and F3. The F2 and F3 domains were shown to bind to the $\beta3$ integrin tail, with the talin F3 subdomain important for the activation of the integrin. The F3 module of talin is similar to PTB domains that recognize peptide ligands containing β-turns (Forman-Kay 1999; Pearson et al. 2000; Ulmer 2003). Recent work demonstrated the NPXY motifs conserved in integrin β tails interact with talin F2 and F3 subdomains, and this interaction is important for the activation of integrins (Calderwood et al. 2002; Garcia-Alvarez 2003). The talin PTB-domain interaction with integrins was used as a model to identify other proteins that interact with integrin cytoplasmic domains (Calderwood 2003). Numb (a negative regulator of Notch signaling) and Dok-1 (signaling adaptor involved in cell migration), Dab (a downstream target of c-Abl), EPS8 (regulator of Rac signaling), and tensin are among the proteins containing PTB domains identified to associate with integrin cytoplasmic domains in binding assays and studies involving integrin-related functions.

Structural studies in combination with mutational analysis have revealed that conformational changes resulting from integrin ligand binding can activate and signal to the cytoplasmic domains in a bi-directional manner (Vinogradova et al. 2002; Shimaoka et al. 2003). Thus, talin head domain binding to a $\beta 3$ peptide disrupts the interaction between the C-terminal transmembrane (TM) domain with the α subunit domain. The interaction between talin and the cytoplasmic domain of $\beta 2$ integrins is similarly important in the activation of LFA-1. Studies utilizing fluorescence resonance energy transfer (FRET) investigated the cytoplasmic conformational changes occurring between αL and $\beta 2$ integrins (Kim et al. 2003). The findings suggest that: (1) activation of $\alpha L \beta 2$ integrins causes a spatial separation between the α and β cytoplasmic domains and (2) the interaction between talin and the $\beta 2$ cytoplasmic domain activates $\alpha L \beta 2$ binding to intercellular adhesion molecule (ICAM)-1. Thus, the role of talin with $\beta 2$ corroborates the results from the studies done with talin and $\beta 3$ (Calderwood et al. 1999; Liu et al. 2000; Vinogradova et al. 2002).

In summary, the interplay between integrins and focal contact proteins, including FAK, paxillin, and talin, both regulates downstream signaling events and affects the state of integrin activation, with both processes having implications for regulation of signal transduction and cell motility.

2.2
Direct Activation of MAP Kinase Cascades by Integrins

There is substantial evidence that integrin-mediated adhesion itself can directly activate the Erk and Jnk MAP kinase pathways. Several distinct mechanisms have been proposed to account for this phenomenon. One mechanism postulates that FAK acts similarly to a receptor tyrosine kinase in activating the canonical tyrosine kinase–Ras–Erk cascade. Thus, when integrins engage extracellular matrix (ECM) proteins, FAK is recruited to focal contacts and is autophosphorylated at Y397, a potential binding site for the Src SH2 domain. This recruits Src, which then phosphorylates FAK at additional sites. Phosphorylation of Y925 provides a consensus binding site for the SH2 domain of the Grb-2 adaptor protein, which is constitutively associated with SOS, an exchange factor for the Ras guanosine triphosphatase (GTPase). This allows activation of Ras followed by activation of the downstream kinases Raf-1, mitogen-activated extracellular signal regulated kinase activating kinase (MEK), and Erk (Aplin et al. 1998; Schlaepfer et al. 1999). Support of this mechanism comes from studies where various activated or dominant-negative versions of FAK or Src were over-expressed. However, there are also studies suggesting that Erk activation can take place independently of FAK activation (Wary et al. 1996; Lin et al. 1997b).

A second model suggests that integrin-mediated activation of the Erk cascade involves the TM protein caveolin-1, the Src-family kinase Fyn, and the adaptor protein Shc. A sub-set of integrin α subunits seems to activate Fyn, thus causing tyrosine phosphorylation of Shc and subsequent recruitment of the Grb-2/Sos complex. This is followed by Ras activation and triggering of the downstream kinase cascade leading to Erk activation (Wary et al. 1998; Giancotti and Ruoslahti 1999). In support of this model, depletion of caveolin using antisense has been reported to disrupt integrin-mediated adhesion and signaling (Wei et al. 1999). There is also evidence for Ras-independent mechanisms of integrin signaling. For example, integrin-mediated adhesion can activate a mutant form of Raf-1 that is defective in its ability to bind Ras (Howe and Juliano 1998).

The existence of several models for integrin-mediated Erk activation may reflect the fact that signaling processes are often highly cell-context dependent. One notion is that the relative contribution of FAK-dependent and Shc-dependent pathways may be related to the levels of Raf-1 and B-Raf in different cell types. In addition to the FAK/Grb-2/Sos association, FAK binds and phosphorylates p130Cas, which then provides binding sites for the CrkII-C3G adaptor protein-exchange factor complex (Vuori 1998). C3G can activate the Rap1 GTPase, which can activate B-Raf, which in turn can activate MEK and Erk. Thus, depending on the levels of Raf isoforms, integrin signaling to the Erk pathway might be predominantly through Shc (where Raf-1 is high) or through FAK (where B-Raf is high) (Barberis et al. 2000). Integrin engagement can also directly activate other branches of the MAP Kinase pathway, leading to JNK or to the p38 kinase. Integrin-mediated activation of JNK depends on FAK and may play a role in cell cycle traverse (Oktay et al. 1999) and cell survival (Almeida et al. 2000). Interestingly, stimulation via the $\alpha2\beta1$ integrin has been reported to selectively activate p38 (Ivaska et al. 1999).

The physiological significance of direct integrin activation of MAP Kinase cascades remains unclear. Erk activation mediated solely through integrins is not sufficient to move cells through the cell cycle; additional signals provided by growth factors are still required. An important potential role for integrin activation of the Erk pathway is that it provides a mechanism for local control of contractility and cell movement (Cheresh et al. 1999).

2.3
Integrins and Rho GTPases

Integrins are capable of bi-directional signaling with the small GTPase superfamily. It is clear that integrin engagement can regulate the activation status of Rho, Rac, and CDC42, while several Ras family members influence the activation of integrins, as recently reviewed elsewhere (Parise et al. 2000). The Rho-family of small GTPases modulate many key cellular processes, but are particularly vital in regulation of the actin cytoskeleton (Kjoller and Hall 1999; Schoenwaelder and Burridge 1999). Thus Rac and

CDC 42 regulate lamellipodia and filopodia respectively, while Rho A promotes stress fibers. The mechanisms linking Rho–GTPases to control of actin filament assembly are becoming better known. For example, both the Rho-stimulated kinase ROCK and the Rac-stimulated kinase PAK can activate LIM-kinase, which can then phosphorylate and inactivate cofilin, an actin de-polymerizing protein (Edwards et al. 1999; Maekawa et al. 1999; Sumi et al. 1999). However, there are also other linkages between Rac and/or CDC42 and actin organization, including those mediated through the Arp 2,3 complex (Higgs and Pollard 1999; Miki et al. 2000), whereas Rho also functions through mDiα1 in forming stress fibers (Watanabe et al. 1999).

Integrin engagement controls several Rho family GTPases in a complex fashion (Bourdoulous et al. 1998; Clark et al. 1998; Price et al. 1998; Ren et al. 1999; O'Connor et al. 2000), while the GTPases then link to downstream effectors. For example, it was found that constitutively active Rac is not sufficient to activate PAK in cells held in suspension, while in adherent cells, Rac translocates to the membrane where it interacts with and activates PAK (del Pozo et al. 2000). The mechanism(s) of integrin-mediated modulation of Rho family GTPases have not yet been fully delineated; most likely exchange factors and/or GTPase activating proteins are important. The most detailed information is available for RhoA, which undergoes a complex response to integrin-mediated cell adhesion, first dipping in activity and then showing increased activity as integrins are further engaged. The early dip in RhoA activity may be linked to either FAK activation (Ren et al. 2000) or Src-mediated tyrosine phosphorylation and activation of p190 RhoGAP (Arthur et al. 2000). Although many details have yet to emerge, it is clear that integrin modulation of Rho GTPases likely plays a significant role in cytoskeletal organization and cell motility.

Recently, several new aspects of the integrin-Rho GTPase connection have emerged. As discussed above, adhesion regulates the localization of Rac to membrane domains where it can interact with effectors like PAK. This is a critical step in activation of the Rac signaling cascade, since an active Rac is unable to signal to PAK in non-adherent cells and inhibition of Rac membrane localization in adherent cells dominantly inhibits Rac signal transduction (del Pozo et al. 2000). Using FRET with the Rac binding domain of PAK to mark the location of active Rac, it was demonstrated that the interaction of Rac with its effectors is localized in growing lamellipodia and is dependent on Rac membrane targeting. The localized activation of integrins using fibronectin or anti-integrin β1-coated beads also stimulated localized Rac effector coupling. This effect appears to be mediated in part by Rho-guanine nucleotide dissociation inhibitor (GDI). A mutant active Rac that does not interact with Rho-GDI binds effectors in the cytoplasm, indicating that Rho-GDI binds to active Rac and inhibits coupling to effectors. The GTP loading of Rac is not necessarily localized, but its coupling to effectors after dissociation of Rho-GDI appears to be. Thus, integrin-mediated adhesion stimulates the dissociation from Rho-GDI and membrane translocation of Rac which facilitates its coupling to effector proteins (del Pozo et al. 2002).

The interaction with Rho-GDI is also critical for Cdc42-mediated transformation. Mutation of Arg 66 or 68 in the switch 2 region of Cdc42 disrupts its interaction with Rho-GDI. The disruption of Rho-GDI binding did not alter effector interactions, but allowed movement of all active Cdc42 to the membrane. This GDI-binding mutation in an active CDC42 (F28L) led to flattening of the cells, multinucleation, and disruption of the actin cytoskeleton without disruption of the focal adhesions. Rho-GDI may be a shuttle protein that transfers Cdc42 to its different membrane sites in the cell from the Golgi. Vesicle transport may be important for Cdc42 transformation because interaction with the Coatmer (COP) complex is essential for CDC42 transformation (Qiong et al. 2003).

The tyrosine phosphorylation state of the adapter protein Crk also contributes to the regulation of Rac membrane localization. Adhesion to fibronectin leads to the phosphorylation of Y-221 on CrkII. This may lead to intramolecular association of the CrkII SH2 domain with the Y-P residue and inhibition of CrkII coupling to other effectors. A non-phosphorylatable mutant of CrkII, Y221F, displayed enhanced binding to paxillin as might be expected, but inhibited adhesion-dependent JNK and PAK activation and haptotactic migration to fibronectin. CrkII-Y221F did not affect adhesion-dependent Rac-GTP loading, but reduced its membrane localization upon adhesion to fibronectin. Again, GTP loading was not affected by CrkII alterations (Abassi and Vuori 2002). Thus, the tyrosine phosphorylation of CrkII appears to be important for the adhesion-dependent translocation of Rac to the membrane.

An example of the importance of compartmentalized control of Rac activity is in developing chemotactic pseudopods where active Rac localizes to the pseudopod independently of integrin ligation and Cas/Crk coupling. The active Rac induces polymerization of the actin cytoskeleton and membrane protrusion. Once the developing pseudopod forms, integrin-stimulated coupling of Cas and Crk sustains Rac activity. Then Rac positively feeds back to maintain this coupling and the pseudopod (Cho and Klemke 2002).

Another interesting example of the importance of the localized, integrin-mediated activation of Rho GTPases is in endothelial cells under shear stress. Shear stress can increase integrin occupancy on fibronectin and vitronectin. This is accompanied by an increased tyrosine phosphorylation of Shc, association of Shc with the integrin subunits $\beta 1$ or $\beta 5$, and activation of the MAPK and JNK signaling pathways (Chen et al. 1999). New integrin-ligand connections formed during shear stress are necessary for the activation of these signaling pathways (Jalali et al. 2001).

The integrin $\alpha v \beta 3$ is activated by shear stress in endothelial cells, and this allows for the formation of new integrin-ligand contacts. Rho activity is altered in a biphasic manner, with an initial repression and then stimulation after the formation of new adhesive contacts. This biphasic modulation of Rho activity positively correlates with the activation state of $\alpha v\beta 3$ after shear stress. The initial depression of Rho activity appears necessary to allow cells to align along the direction of flow, because a constitutive V14Rho inhibits the realignment (Tzima et al. 2001).

Rac and Cdc42 are also activated by shear stress in endothelial cells. Their activation is dependent on the formation of new integrin contacts. The active Rac/Cdc 42 are located downstream of the direction of flow. Localized, not global, activation of Rac provides the spatial information for the orientation of cells in the direction of flow (Tzima et al. 2002). Localized activation of Cdc42 appears to be necessary for the reorientation of the microtubule-organizing center to the direction of flow. This Cdc42 MTOC reorientation is dependent on the Par6/PKCz polarity-signaling complex (Tzima et al. 2003).

Thus, integrins interact in complex ways with Rho GTPase. These interactions play an important role in regulating both cytoskeletal organization and cellular signaling.

3
Integrin Regulation of RTK Signaling Cascades

There has been a substantial amount of work concerning the interconnections between integrins, cytoskeletal components, and the receptor tyrosine kinase/Ras/MAP kinase cascade. Integrin-mediated cell anchorage, and the resultant formation of cytoskeletal complexes, can regulate the receptor tyrosine kinase (RTK)/Ras/MAPK pathway at three distinct loci. The first is at the level of activation of the RTK. The second concerns the coupling between upstream and downstream events in the pathway. Finally, a third locus is at the transmission of the signal between the cytoplasm and nucleus (Fig. 1).

Early work demonstrated that ligand binding and integrin aggregation were required for triggering activation of the EGF, PDGF, and FGF receptors (Miyamoto et al. 1996). Enhanced RTK activation was then paralleled by increased Erk/MAPK activity. In some cases integrin activation of RTKs can take place in the absence of growth factors (Sundberg and Rubin 1996; Moro et al. 1998). A more common situation, however, is that both integrin engagement and the presence of growth factor are essential for efficient RTK activation. Direct association of the $\alpha v\beta 3$ integrin and platelet-derived growth factor (PDGF)βR, insulin receptor, and vascular endothelial growth factor (VEGF)R2 have been observed (Schneller et al. 1997; Soldi et al. 1999). It seems obvious that formation of direct or indirect complexes between the RTKs and the integrins could result in enhanced RTK dimerization and cross-phosphorylation. Evidence has emerged indicating that inte-

Integrin Regulation of the RTK/Ras/MAPK Pathway

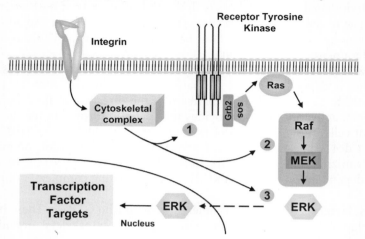

Fig. 1. Integrin/cytoskeletal control of signaling into RTK/Ras/MAPK pathways. *1–3*: Known sites of regulation: (*1*) RTK activation; (*2*) transmission of signal through the cytoplasmic kinase cascade; (*3*) entry of ERK into the nucleus

grin associated cytoskeletal components play a key role in these putative complexes. Thus, an association between actin and the *Neu* RTK has been observed (Li et al. 1999), while another report has proposed a coupling between integrins and RTKs via FAK (Sieg et al. 2000). Thus, there is substantial evidence for a role for integrins in regulating RTK activity.

A second locus of integrin regulation of the RTK/Ras/Erk cascade is at the level of coupling between the upstream and downstream components of the pathway. Thus, in fibroblasts the loss of integrin-mediated cell anchorage blocks signal propagation from Ras to Raf-1. In this system, growth factor stimulation results in normal autophosphorylation of RTKs and GTP loading of Ras in cells in suspension. However, the activation of Raf-1 and of the downstream Mek and Erk kinases are significantly impaired (Lin et al. 1997a; Le Gall et al. 1998). Somewhat similar results were observed in another study (Renshaw et al. 1997), with the exception that the locus of anchorage regulation was placed between Raf and Mek. Thus, there seems to be an anchorage-dependent step between Ras and its downstream kinases in the signaling cascade triggered by peptide mitogens.

Integrin regulation of Erk activation involves the actin cytoskeleton (Chen et al. 1994). In particular, it seems that it is cortical actin filaments rather than focal contacts and stress fibers that are important (Aplin and Juliano 1999). Consistent with this, expression of active CDC42, which promotes cortical actin assembly, partially rescued Erk activation in suspended cells (Aplin and Juliano 1999). Thus, integrin-mediated cell anchorage, and subsequent actin cytoskeletal organization, can influence upstream to downstream coupling of the Ras/Raf/Mek/Erk intracytoplasmic signaling cascade in various cells. However, the precise mechanism of regulation has not been consistently defined.

A third locus of integrin regulation concerns the transmission of the signal from cytoplasm to nucleus. Evidence for the existence of this aspect of regulation first came from studies showing that forced activation of Erk is insufficient to drive cells into the cell cycle (Le Gall et al. 1998). This led to study of the role of integrin-mediated anchorage in the trafficking of Erk between the cytoplasm and nucleus (Aplin et al. 2001). The subcellular localization of Erk plays an important role in growth control (Brunet et al. 1999). The concept is that inactive Erk is held in the cytoplasm through its association with Mek (Fukuda et al. 1997; Adachi et al. 2000). When activated, Erk is phosphorylated, dissociates from Mek, and enters the nucleus, probably as a dimer (Khokhlatchev et al. 1998). After dephosphorylation by nuclear phosphatases, inactive nuclear Erk associates with Mek and is exported from the nucleus (Adachi et al. 2000). To this picture one must now add a level of regulation by integrins and the actin cytoskeleton. Thus, in non-anchored or cytochalasin D-treated cells, the normal trafficking of Erk is disrupted. Despite being activated, Erk fails to enter the nucleus and thus cannot phosphorylate its key immediate targets, such as the Elk1 transcription factor (Danilkovitch et al. 2000; Aplin et al. 2001). The mechanism underlying this integrin/actin modulation of Erk trafficking is undefined at this point.

The potential importance of differential sites of integrin regulation of the MAPK pathway was illustrated recently in examination of the signaling downstream of growth factors important for angiogenesis. On chick chorioallantoic membranes, basic fibroblast growth factor (bFGF) stimulates Ras, Raf, and Erk activity. The activation of Raf and Erk by bFGF, but not Ras, were dependent on $\alpha v \beta 3$ but not $\alpha v \beta 5$. Prolonged VEGF stimulation of Ras, Raf, and Erk activity was dependent on $\alpha v \beta 5$ but not $\alpha v \beta 3$. Inhibition of $\alpha v \beta 3$ with blocking antibodies could prevent active Ras, but not active Raf, from inducing angiogenesis. The integrin $\alpha v \beta 3$ regulation of this pathway is at or above the level of Raf, but below Ras. However, inhibition of $\alpha v \beta 5$ with antibodies did not inhibit angiogenesis by active Ras or Raf, so its control is above Ras (Hood et al. 2003). Thus, the point of integrin regulation of the MAPK pathway may depend on the growth factor receptor and integrin involved.

As discussed above (Aplin and Juliano 1999), it has been demonstrated that the translocation of activated Erk to the nucleus is dependent on cell adhesion and the integrity of the actin cytoskeleton. This adhesion regulation of nuclear transport was not a generalized effect, since the nuclear transport of the JNK and p38 MAPKs and cyclin D1 were not affected (Aplin et al. 2001, 2002). The importance of this finding was recently confirmed in embryonic fibroblasts carrying homozygous mutations in the variable domain of the β1 integrin. The homozygous mice were embryonic lethal and cultured embryonic fibroblasts cells exhibited defective G1/S transition and reduced survival. Interestingly, the phosphorylation of Erk is not deficient in these cells; however, the translocation of Erk to the nucleus is impaired in the β1 mutant cells. Rac activation was impaired in the mutant cells; this appears to be important for regulation of nuclear trafficking of MAPK, since activated Rac rescues the nuclear translocation of Erk (Hirsch et al. 2002).

The activation of B-Raf also contributes to the activation and nuclear translocation of ERK in cells that express this isoform of Raf, such as melanocytes. Normal melanocytes are adhesion-dependent in FGF signaling to Erk1/2. The presence of an activated B-Raf V599E in melanoma cells subverts the anchorage regulation of this pathway. B-Raf V599E allows anchorage-independent activation of Erk1/2 and translocation of Erk1/2 to the nucleus. The B-Raf isoform is frequently mutated in melanoma, and this mutation could be critical in the transformation of melanocytes (Conner et al. 2003). This report also reinforces the importance of B-Raf, which had been demonstrated to be downstream of the FAK/Src and Rap-1, in adhesion-dependent signaling to the MAPK pathway (Barberis et al. 2000).

Several recent studies have shed further light on how adhesion regulates the MAPK pathway at the level of MEK. The activation of Erk by Mek requires their direct association that is facilitated by an ERK binding domain in the N-terminus of MEK (Fukuda et al. 1997; Bardwell et al. 2001). Examination of this interaction in suspended versus adherent cells demonstrated that adhesion regulates the interaction of Mek1 and Erk2 in COS-1 cells. However, the association of Mek2 and Erk2 remains stable regardless of the state of adhesion (Eblen et al. 2002). Additional studies demonstrated that the phosphorylation of MEK1 at serine 298 is critical for adhesion-dependent activation of MEK1, and this phosphorylation is downstream of FAK and Src. Downstream of these kinases, PAK1 directly mediates the adhesion-dependent phosphorylation of serine 298 on MEK1, but not MEK2 [which is a poor PAK substrate (Frost et al. 1996)] and this phosphorylation is necessary for adhesion-dependent activation of the MAPK pathway (Slack-Davis et al. 2003). Interestingly, homozygous deletion of MEK1 leads to embryonic lethality due to angiogenesis defects in placental ontogeny. Additionally, the MEK1-deficient mouse embryonic fibroblasts are defective in fibronectin-stimulated haptotaxis, but Erk2 phosphorylation is retained after adhesion (Giroux 1999). However, MEK2 is dispensable for mouse growth and development, suggesting the relative importance of MEK1 (Belanger et al. 2003).

The maintenance of Erk in an active form is necessary for the induction of cyclin D1 mRNA and protein in mid G1. Activation of the Rho–Rho kinase pathway is needed for this sustained Erk activation. However, inhibition of the Rho–Rho-kinase pathway allows cyclin D1 production independent of Erk activation in murine fibroblasts. This activation appears to be dependent on Rac and Cdc42, which can stimulate cyclin D1 production in early G1 phase or in non-adherent cells independent of Erk activity. Active Rac/Cdc42 can thus stimulate cyclin D1 without integrin ligation or the formation of the focal adhesion complex. Rho suppresses this early induction of cyclin D1 downstream of Rac activation and this is not dependent on the extent of cell spreading or the presence of $p21^{cip}$ (Welsh et al. 2001).

Downstream of Rho, the inhibition of Rho kinase, LIM kinase, and myosin light chain (MLC) kinase blocks stress fiber formation and sustained Erk activation. The inhibition of Rho Kinase or LIM kinase allows early cyclin D1 expression and accelerated entry into S phase independent of Erk activation. However, the inhibition of MLC kinase blocks entry into S phase and cyclin D1 expression. The overexpression of cyclin D1 can overcome stress fiber dependence in MLC kinase inhibited cells and allow entry into S phase. Thus, the regulation of stress fibers formation by Rho kinase, LIM kinase, and MLC kinase is important for sustained Erk activation and the timing and duration of cyclin D1 expression (Roovers and Assoian 2003).

Thus newer aspects of integrin signaling have implicated the timing of Erk activation, as controlled by Rho GTPases, as an important aspect of cell cycle control.

4
Integrin Regulation of Signaling by GPCRs

Interesting connections have emerged between integrins, the cytoskeleton, and G protein-coupled receptors (GPCRs). Thus, several agonists for GPCRs, including bombesin, gastrin, endothelin, and various muscarinic agents, can trigger activation of FAK (Rozengurt 1998; Slack 1998). FAK activation in response to bombesin depends on a functional actin cytoskeleton and requires the Rho GTPase (Rozengurt 1998). Peptides that block integrin binding to extracellular matrix proteins also block activation of FAK by muscarinic acetylcholine receptors. It has been suggested that muscarinic signaling activates integrins, which then activate FAK (Slack 1998).

Integrin-mediated cell anchorage also impacts on the ability of GPCRs to signal to the Erk/MAPK pathway. Thus, cells deprived of integrin-mediated anchorage, or having a disrupted actin cytoskeleton, are less able to activate this pathway (Renshaw et al. 1997; Short et al. 1998; Della Rocca et al. 1999). The role of integrins in the pathway leading from P2Y receptors to PLCβ and then to Erk activation has recently been explored. Thus, all upstream aspects of this signaling cascade were anchorage independent up to the level of inositol 1,4,5-trisphosphate (IP3) generation and calcium release. However, downstream events including activation of MEK and Erk were highly anchorage dependent (Short et al. 2000). Thus, a number of studies have shown that integrin-mediated anchorage modulates signaling between various GPCRs and the Erk/MAPK module, but the precise molecular mechanism underlying these events remains to be defined.

An interesting series of studies has linked an integrin-associated protein (IAP, CD47) with heterotrimeric G proteins. IAP is a member of the Ig-superfamily that has several TM helices. IAP associates strongly with $\beta 3$ integrins as well as with with Gi (Frazier et al. 1999). Recent evidence suggests that the IAP, Gi, $\beta 3$ integrin complex is associated in a cholesterol-rich membrane domain or "raft" that may serve as a multiprotein signaling complex (Green et al. 1999).

These various observations linking integrins and the actin cytoskeleton with GPCR signaling are very consistent with the concept that sub-cellular localization of signaling components and the formation of multimeric complexes are critical for GPCR signaling (Edwards et al. 2000). Given emerging information, one could easily visualize interesting possibilities for the association of integrins, cytoskeletal adaptor proteins, and G protein-based signaling proteins into multi-protein signaling complexes localized at specific sites on the cell membrane.

5
Summary

As delineated above, there are multiple connections between integrins, the cytoskeleton, and signal transduction pathways. Integrin-mediated cell anchorage and accompanying actin filament re-organization dramatically modulates the efficiency of various signaling processes, with important implications for cell cycle control, survival, and differentiation.

References

Abassi YA, Vuori K (2002) Tyrosine 221 in Crk regulates adhesion-dependent membrane localization of Crk and Rac and activation of Rac signaling. EMBO J 21:4571–4582

Adachi M, Fukuda M, Nishida E (2000) Nuclear export of MAP kinase (ERK) involves a MAP kinase kinase (MEK)-dependent active transport mechanism. J Cell Biol 148:849–856

Almeida EA, Ilic D, Han Q, Hauck CR, Jin F, Kawakatsu H, Schlaepfer DD, Damsky CH (2000) Matrix survival signaling: from fibronectin via focal adhesion kinase to c-Jun NH(2)-terminal kinase. J Cell Biol 149:741–754

Aplin AE, Juliano RL (1999) Integrin and cytoskeletal regulation of growth factor signaling to the MAP kinase pathway. J Cell Sci 112:695–706

Aplin A, Howe A, Alahari S, Juliano RL (1998) Signal transduction and signal modulation by cell adhesion receptors: the role of integrins, cadherins, immunoglobulin-cell adhesion molecules, and selectins. Pharmacol Rev 50:197–263

Aplin AE, Stewart SA, Assoian RK, Juliano RL (2001) Integrin-mediated adhesion regulates ERK nuclear translocation and phosphorylation of Elk-1. J Cell Biol 153:273–282

Aplin AE, Hogan BP, Tomeu J, Juliano RL (2002) Cell adhesion differentially regulates the nucleocytoplasmic distribution of active MAP kinases. J Cell Sci 115:2781–2790

Arthur WT, Petch LA, Burridge K (2000) Integrin engagement suppresses RhoA activity via a c-Src-dependent mechanism. Curr Biol 10:719–722

Barberis L, Wary KK, Fiucci G, Liu F, Hirsch E, Brancaccio M, Altruda F, Tarone G, Giancotti FG (2000) Distinct roles of the adaptor protein shc and focal adhesion kinase in integrin signaling to ERK. J Biol Chem 275:36532–36540

Bardwell AJ, Flatauer LJ, Matsukuma K, Thorner J, Bardwell L (2001) A conserved docking site in MEKs mediates high-affinity binding to MAP kinases and cooperates with a scaffold protein to enhance signal transmission. J Biol Chem 276:10374–10386

Belanger L-F, Roy S, Tremblay M, Brott B, Steff A-M, Mourad W, Hugo P, Erikson R, Charron J (2003) Mek2 is dispensable for mouse growth and development. Mol Cell Biol 23:4778–4787

Bourdoulous S, Orend G, MacKenna DA, Pasqualini R, Ruoslahti E (1998) Fibronectin matrix regulates activation of RHO and CDC42 GTPases and cell cycle progression. J Cell Biol 143:267–276

Brunet A, Roux D, Lenormand P, Dowd S, Keyse S, Pouyssegur J (1999) Nuclear translocation of p42/p44 mitogen-activated protein kinase is required for growth factor-induced gene expression and cell cycle entry. EMBO J 18:664–674

Calderwood DA, Zent R, Grant R, Rees DJ, Hynes RO, Ginsberg MH (1999) The Talin head domain binds to integrin beta subunit cytoplasmic tails and regulates integrin activation. J Biol Chem 274:28071–28074

Calderwood DA, Yan B, de Pereda JM, Alvarez BG, Fujioka Y, Liddington RC, Ginsberg MH (2002) The Phosphotyrosine Binding-like Domain of Talin Activates Integrins. J Biol Chem 277:21749–21758

Calderwood DA, Fujioka Y, de Pereda JM, Garcia-Alvarez B, Nakamoto T, Margolis B, McGlade CJ, Liddington RC, Ginsberg MH (2003) Integrin beta cytoplasmic domain interactions with phosphotyrosine-binding domains: a structural prototype for diversity in integrin signaling. Proc Natl Acad Sci U S A 100:2272–2277

Carragher NO, Westhoff MA, Fincham VJ, Schaller MD, Frame MC (2003) A novel role for FAK as a protease-targeting adaptor protein: regulation by p42 ERK and Src. Curr Biol 13:1442–1450

Chen KD, Li YS, Kim M, Li S, Yuan S, Chien S, Shyy JY (1999) Mechanotransduction in response to shear stress. Roles of receptor tyrosine kinases, integrins, and Shc. J Biol Chem 274:18393–18400

Chen Q, Kinch MS, Lin TH, Burridge K, Juliano RL (1994) Integrin-mediated cell adhesion activates mitogen-activated protein kinases. J Biol Chem 269:26602–26605

Cheresh DA, Leng J, Klemke RL (1999) Regulation of cell contraction and membrane ruffling by distinct signals in migratory cells. J Cell Biol 146:1107–1116

Cho SY, Klemke RL (2002) Purification of pseudopodia from polarized cells reveals redistribution and activation of Rac through assembly of a CAS/Crk scaffold. J Cell Biol 156:725–736

Clark EA, King WG, Brugge JS, Symons M, Hynes RO (1998) Integrin-mediated signals regulated by members of the Rho family of GTPases. J Cell Biol 142:573–586

Conner SR, Scott G, Aplin AE (2003) Adhesion-dependent activation of the ERK1/2 cascade is by-passed in melanoma cells. J Biol Chem 278:34548–34554

Cooray P, Yuan Y, Schoenwaelder SM, Mitchell CA, Salem HH, Jackson SP (1996) Focal adhesion kinase (pp125FAK) cleavage and regulation by calpain. Biochem J 318:41–47

Danilkovitch A, Donley S, Skeel A, Leonard EJ (2000) Two independent signaling pathways mediate the antiapoptotic action of macrophage-stimulating protein on epithelial cells. Mol Cell Biol 20:2218–2227

del Pozo MA, Price LS, Alderson NB, Ren XD, Schwartz MA (2000) Adhesion to the extracellular matrix regulates the coupling of the small GTPase Rac to its effector PAK. EMBO J 19:2008–2014

del Pozo MA, Kiosses WB, Alderson NB, Meller N, Hahn KM, Schwartz MA (2002) Integrins regulate GTP-Rac localized effector interactions through dissociation of Rho-GDI. Nat Cell Biol 4:232–239

Della Rocca GJ, Maudsley S, Daaka Y, Lefkowitz RJ, Luttrell LM (1999) Pleiotropic coupling of G protein-coupled receptors to the mitogen-activated protein kinase cascade. J Biol Chem 274:13978–13984

Eblen ST, Slack JK, Weber MJ, Catling AD (2002) Rac-PAK Signaling Stimulates Extracellular Signal-Regulated Kinase (ERK) Activation by Regulating Formation of MEK1-ERK Complexes. Mol Cell Biol 22:6023–6033

Edwards DC, Sanders LC, Bokoch GM, Gill GN (1999) Activation of LIM-kinase by Pak1 couples Rac/Cdc42 GTPase signalling to actin cytoskeletal dynamics. Nat Cell Biol 1:253–259

Edwards SW, Tan CM, Limbird LE (2000) Localization of G-protein-coupled receptors in health and disease. Trends Pharmacol Sci 21:304–308

Forman-Kay JD PT (1999) Diversity in protein recognition by PTB domains. Curr Opin Struct Biol 9:690–695

Frazier WA, Gao AG, Dimitry J, Chung J, Brown EJ, Lindberg FP, Linder ME (1999) The thrombospondin receptor integrin-associated protein (CD47) functionally couples to heterotrimeric Gi. J Biol Chem 274:8554–8560

Frost JA, Xu S, Hutchison MR, Marcus S, Cobb MH (1996) Actions of Rho family small G proteins and p21-activated protein kinases on mitogen-activated protein kinase family members. Mol Cell Biol 16:3707–3713

Fukuda M, Gotoh Y, Nishida E (1997) Interaction of MAP kinase with MAP kinase kinase: its possible role in the control of nucleocytoplasmic transport of MAP kinase. EMBO J 16:1901–1908

Gabarra-Niecko V, Schaller MD, Dunty JM (2003) FAK regulates biological processes important for the pathogenesis of cancer. Cancer Metastasis Rev 22:359–374

Gao J, Zoller KE, Ginsburg MH, Brugge JS, Shattil SJ (1997) Regulation of the $pp72^{syk}$ protein tyrosine kinase by platelet integrin $\alpha IIb\beta 3$. EMBO J 16:6414–6425

Garcia-Alvarez B dPJ, Calderwood DA, Ulmer TS, Critchley D, Campbell ID, Ginsberg MH, Liddington RC (2003) Structural determinants of integrin recognition by talin. Mol Cell 11:49–58

Giancotti FG, Ruoslahti E (1999) Integrin signaling. Science 285:1028–1032

Giroux S TM, Bernard D, Cardin-Girard JF, Aubry S, Larouche L, Rousseau S, Huot J, Landry J, Jeannotte L, Charron J (1999) Embryonic death of Mek1-deficient mice reveals a role for this kinase in angiogenesis in the labyrinthine region of the placenta. Curr Biol 9:369–372

Glading A, Chang P, Lauffenburger DA, Wells A (2000) Epidermal growth factor receptor activation of calpain is required for fibroblast motility and occurs via an ERK/MAP kinase signaling pathway. J Biol Chem 275:2390–2398

Goldfinger LE, Han J, Kiosses WB, Howe AK, Ginsberg MH (2003) Spatial restriction of alpha4 integrin phosphorylation regulates lamellipodial stability and alpha4beta1-dependent cell migration. J Cell Biol 162:731–741

Green JM, Zhelesnyak A, Chung J, Lindberg FP, Sarfati M, Frazier WA, Brown EJ (1999) Role of cholesterol in formation and function of a signaling complex involving alphavbeta3, integrin-associated protein (CD47), and heterotrimeric G proteins. J Cell Biol 146:673–682

Han J, Rose DM, Woodside DG, Goldfinger LE, Ginsberg MH (2003) Integrin alpha 4 beta 1-dependent T cell migration requires both phosphorylation and dephosphorylation of the alpha 4 cytoplasmic domain to regulate the reversible binding of paxillin. J Biol Chem 278:34845–34853

Hanks SK, Calalb MB, Harper MC, Patel SK (1992) Focal adhesion protein-tyrosine kinase phosphorylated in response to cell spreading on fibronectin. Proc Natl Acad Sci USA 89:8487–8491

Higgs HN, Pollard TD (1999) Regulation of actin polymerization by Arp2/3 complex and WASp/Scar proteins. J Biol Chem 274:32531–32534

Hirsch E, Barberis L, Brancaccio M, Azzolino O, Xu D, Kyriakis JM, Silengo L, Giancotti FG, Tarone G, Fassler R, Altruda F (2002) Defective Rac-mediated proliferation and survival after targeted mutation of the beta1 integrin cytodomain. J Cell Biol 157:481–492

Hood JD, Frausto R, Kiosses WB, Schwartz MA, Cheresh DA (2003) Differential alphav integrin-mediated Ras-ERK signaling during two pathways of angiogenesis. J Cell Biol 162:933–943

Howe AK, Juliano RL (1998) Distinct mechanisms mediate the initial and sustained phases of integrin-mediated activation of the Raf/MEK/mitogen-activated protein kinase cascade. J Biol Chem 273:27268–27274

Hsia DA, Mitra SK, Hauck CR, Streblow DN, Nelson JA, Ilic D, Huang S, Li E, Nemerow GR, Leng J, Spencer KSR, Cheresh DA, Schlaepfer DD (2003) Differential regulation of cell motility and invasion by FAK. J Cell Biol 160:753–767

Huang C, Rajfur Z, Borchers C, Schaller MD, Jacobson K (2003) JNK phosphorylates paxillin and regulates cell migration. Nature 424:219–223

Ivaska J, Reunanen H, Westermarck J, Koivisto L, Kahari VM, Heino J (1999) Integrin alpha2beta1 mediates isoform-specific activation of p38 and upregulation of collagen gene transcription by a mechanism involving the alpha2 cytoplasmic tail. J Cell Biol 147:401–416

Jalali S, del Pozo MA, Chen K, Miao H, Li Y, Schwartz MA, Shyy JY, Chien S (2001) Integrin-mediated mechanotransduction requires its dynamic interaction with specific extracellular matrix (ECM) ligands. Proc Natl Acad Sci U S A 98:1042–1046

Juliano RL (2002) Signal transduction by cell adhesion receptors and the cytoskeleton: functions of integrins, cadherins, selectins, and immunoglobulin-superfamily members. Annu Rev Pharmacol Toxicol 42:283–323

Katz BZ, Miyamoto S, Teramoto H, Zohar M, Krylov D, Vinson C, Gutkind JS, Yamada KM (2002) Direct transmembrane clustering and cytoplasmic dimerization of focal adhesion kinase initiates its tyrosine phosphorylation. Biochim Biophys Acta 1592:141–152

Khokhlatchev AV, Canagarajah B, Wilsbacher J, Robinson M, Atkinson M, Goldsmith E, Cobb MH (1998) Phosphorylation of the MAP kinase ERK2 promotes its homodimerization and nuclear translocation. Cell 93:605–615

Kim M, Carman CV, Springer TA (2003) Bidirectional transmembrane signaling by cytoplasmic domain separation in integrins. Science 301:1720–1725

Kjoller L, Hall A (1999) Signaling to Rho GTPases. Exp Cell Res 253:166–179

Kornberg LJ, Earp HS, Turner CE, Prockop C, Juliano RL (1991) Signal transduction by integrins: increased protein tyrosine phosphorylation caused by clustering of beta 1 integrins. Proc Natl Acad Sci USA 88:8392–8396

Le Gall M, Grall D, Chambard JC, Pouyssegur J, Van Obberghen-Schilling E (1998) An anchorage-dependent signal distinct from p42/44 MAP kinase activation is required for cell cycle progression. Oncogene 17:1271–1277

Li R, Mitra N, Gratkowski H, Vilaire G, Litvinov R, Nagasami C, Weisel JW, Lear JD, DeGrado WF, Bennett JS (2003) Activation of integrin alphaIIbbeta3 by modulation of transmembrane helix associations. Science 300:795–798

Li Y, Hua F, Carraway KL, Carraway CA (1999) The p185(neu)-containing glycoprotein complex of a microfilament-associated signal transduction particle. Purification, reconstitution, and molecular associations with p58(gag) and actin. J Biol Chem 274:25651–25658

Lin T, Chen Q, Howe A, Juliano R (1997a) Cell anchorage permits efficient signal transduction between Ras and its downstream kinases. J Biol Chem 272:8849–8852

Lin TH, Aplin AE, Shen Y, Chen Q, Schaller M, Romer L, Aukhil I, Juliano RL (1997b) Integrin-mediated activation of MAP kinase is independent of FAK: evidence for dual integrin signaling pathways in fibroblasts. J Cell Biol 136:1385–1395

Liu S, Thomas SM, Woodside DG, Rose DM, Kiosses WB, Pfaff M, Ginsberg MH (1999) Binding of paxillin to alpha4 integrins modifies integrin-dependent biological responses. Nature 402:676–681

Liu S, Calderwood DA, Ginsberg MH (2000) Integrin cytoplasmic domain-binding proteins. J Cell Sci 113:3563–3571

Maekawa M, Ishizaki T, Boku S, Watanabe N, Fujita A, Iwamatsu A, Obinata T, Ohashi K, Mizuno K, Narumiya S (1999) Signaling from Rho to the actin cytoskeleton through protein kinases ROCK and LIM-kinase. Science 285:895–898

Miki H, Yamaguchi H, Suetsugu S, Takenawa T (2000) IRSp53 is an essential intermediate between Rac and WAVE in the regulation of membrane ruffling. Nature 408:732–735

Miyamoto S, Teramoto H, Gutkind J, Yamada K (1996) Integrins can collaborate with growth factors for phosphorylation of receptor tyrosine kinases and MAP kinase activation: roles of integrin aggregation and occupancy of receptors. J Cell Biol 135:1633–1642

Moro L, Venturino M, Bozzo C, Silengo L, Altruda F, Beguinot L, Tarone G, Defilippi P (1998) Integrins induce activation of EGF receptor: role in MAP kinase induction and adhesion-dependent cell survival. EMBO J 17:6622–6632

O'Connor KL, Nguyen BK, Mercurio AM (2000) RhoA function in lamellae formation and migration is regulated by the alpha6beta4 integrin and cAMP metabolism. J Cell Biol 148:253–258

Oktay M, Wary KK, Dans M, Birge RB, Giancotti FG (1999) Integrin-mediated activation of focal adhesion kinase is required for signaling to Jun NH2-terminal kinase and progression through the G1 phase of the cell cycle. J Cell Biol 145:1461–1469

Parise LV, Lee JW, Juliano RL (2000) New aspects of integrin signaling in cancer. Semin Cancer Biol 10:407–414

Parsons JT, Martin KH, Slack JK, Taylor JM, Weed SA (2000) Focal adhesion kinase: a regulator of focal adhesion dynamics and cell movement. Oncogene 19:5606–5613

Pearson MA, Reczek D, Bretscher A, Karplus PA (2000) Structure of the ERM protein moesin reveals the FERM domain fold masked by an extended actin binding tail domain. Cell 101:259–270

Price LS, Leng J, Schwartz MA, Bokoch GM (1998) Activation of Rac and Cdc42 by integrins mediates cell spreading. Mol. Biol Cell 9:1863–1871

Qiong L, Reina NF, Wannian Y, Cerione CA (2003) RhoGDI Is Required for Cdc42-Mediated Cellular Transformation. Curr Biol 13:1469–1479

Ren X, Kiosses WB, Sieg DJ, Otey CA, Schlaepfer DD, Schwartz MA (2000) Focal adhesion kinase suppresses Rho activity to promote focal adhesion turnover. J Cell Sci 113:3673–3678

Ren XD, Kiosses WB, Schwartz MA (1999) Regulation of the small GTP-binding protein Rho by cell adhesion and the cytoskeleton. EMBO J 18:578–585

Renshaw M, Ren X-D, Schwartz M (1997) Growth factor activation of MAP kinase requires cell adhesion. EMBO J 16:5592–5599

Roovers K, Assoian RK (2003) Effects of rho kinase and actin stress fibers on sustained extracellular signal-regulated kinase activity and activation of G(1) phase cyclin-dependent kinases. Mol Cell Biol 23:4283–4294

Rozengurt E (1998) Signal transduction pathways in the mitogenic response to G protein-coupled neuropeptide receptor agonists. J Cell Physiol 177:507–517

Schaller MD (1996) The focal adhesion kinase. J Endocrinol 150:1-7

Schaller MD, Borgman CA, Cobb BS, Vines RR, Reynolds AB, Parsons JT (1992) pp125^{FAK}, a structurally distinctive protein-tyrosine kinase associated with focal adhesions. Proc Natl Acad Sci USA 89:5192–5196

Schaller MD, Otey CA, Hildebrand JD, Parsons JT (1995) Focal adhesion kinase and paxillin bind to peptides mimicking beta integrin cytoplasmic domains. J Cell Biol 130:1181–1187

Schlaepfer DD, Hauck CR, Sieg DJ (1999) Signaling through focal adhesion kinase. Prog Biophys Mol Biol 71:435–478

Schneller M, Vuori K, Ruoslahti E (1997) $\alpha v \beta 3$ integrin associates with activated insulin and PDGFβ receptors and potentiates the biological activity of PDGF. EMBO J 16:5600–5607

Schoenwaelder SM, Burridge K (1999) Bidirectional signaling between the cytoskeleton and integrins. Curr Opin Cell Biol 11:274–286

Shimaoka M, Xiao T, Liu JH, Yang Y, Dong Y, Jun CD, McCormack A, Zhang R, Joachimiak A, Takagi J, Wang JH, Springer TA (2003) Structures of the alpha L I domain and its complex with ICAM-1 reveal a shape-shifting pathway for integrin regulation. Cell 112:99–111

Short S, Talbot G, Juliano RL (1998) Integrin mediated signaling events in human endothelial cells. Mol Biol Cell 9:169–180

Short SM, Boyer JL, Juliano RL (2000) Integrins regulate the linkage between upstream and downstream events in G protein-coupled receptor signaling to mitogen-activated protein kinase. J Biol Chem 275:12970–12977

Sieg DJ, Hauck CR, Ilic D, Klingbeil CK, Schaefer E, Damsky CH, Schlaepfer DD (2000) FAK integrates growth-factor and integrin signals to promote cell migration. Nat Cell Biol 2:249–256

Slack BE (1998) Tyrosine phosphorylation of paxillin and focal adhesion kinase by activation of muscarinic m3 receptors is dependent on integrin engagement by the extracellular matrix. Proc Natl Acad Sci USA 95:7281–7286

Slack-Davis JK, Eblen ST, Zecevic M, Boerner SA, Tarcsafalvi A, Diaz HB, Marshall MS, Weber MJ, Parsons JT, Catling AD (2003) PAK1 phosphorylation of MEK1 regulates fibronectin-stimulated MAPK activation. J Cell Biol 162:281–291

Soldi R, Mitola S, Strasly M, Defilippi P, Tarone G, Bussolino F (1999) Role of alphavbeta3 integrin in the activation of vascular endothelial growth factor receptor-2. EMBO J 18:882–892

Sumi T, Matsumoto K, Takai Y, Nakamura T (1999) Cofilin phosphorylation and actin cytoskeletal dynamics regulated by rho- and Cdc42-activated LIM-kinase 2. J Cell Biol 147:1519–1532

Sundberg C, Rubin K (1996) Stimulation of β1 integrins on fibroblasts induces PDGF independent tyrosine phosphorylation of PDGF β-receptors. J Cell Biol 132:741–752

Turner CE (2000) Paxillin interactions. J Cell Sci 113:4139–4140

Tzima E, del_Pozo MA, Shattil SJ, Chien S, Schwartz MA (2001) Activation of integrins in endothelial cells by fluid shear stress mediates Rho-dependent cytoskeletal alignment. EMBO J 20:4639–4647

Tzima E, Del Pozo MA, Kiosses WB, Mohamed SA, Li S, Chien S, Schwartz MA (2002) Activation of Rac1 by shear stress in endothelial cells mediates both cytoskeletal reorganization and effects on gene expression. EMBO J 21:6791–6800

Tzima E, Kiosses WB, Del Pozo MA, Schwartz MA (2003) Localized cdc42 activation, detected using a novel assay, mediates microtubule organizing center positioning in endothelial cells in response to fluid shear stress. J Biol Chem 278:31020–31023

Ulmer TS CD, Ginsberg MH, Campbell ID (2003) Domain-specific interactions of talin with the membrane-proximal region of the integrin beta3 subunit. Biochemistry 42:8307–8312

Vinogradova O, Velyvis A, Velyviene A, Hu B, Haas T, Plow E, Qin J (2002) A structural mechanism of integrin alpha(IIb)beta(3) "inside-out" activation as regulated by its cytoplasmic face. Cell 110:587–597

Vuori K (1998) Integrin signaling: tyrosine phosphorylation events in focal adhesions. J Membr Biol 165:191–199

Wary K, Mainiero F, Isakoff S, Marcantonio E, Giancotti F (1996) The adaptor protein Shc couples a class of integrins to the control of cell cycle progression. Cell 87:733–743

Wary KK, Mariotti A, Zurzolo C, Giancotti FG (1998) A requirement for caveolin-1 and associated kinase Fyn in integrin signaling and anchorage-dependent cell growth. Cell 94:625–634

Watanabe N, Kato T, Fujita A, Ishizaki T, Narumiya S (1999) Cooperation between mDia1 and ROCK in Rho-induced actin reorganization. Nat Cell Biol 1:136–143

Wei Y, Yang X, Liu Q, Wilkins JA, Chapman HA (1999) A role for caveolin and the urokinase receptor in integrin-mediated adhesion and signaling. J Cell Biol 144:1285–1294

Welsh CF, Roovers K, Villanueva J, Liu Y, Schwartz MA, Assoian RK (2001) Timing of cyclin D1 expression within G1 phase is controlled by Rho. Nat Cell Biol 3:950–957

Wennerberg K, Armulik A, Sakai T, Karlsson M, Fassler R, Schaefer EM, Mosher DF, Johansson S (2000) The cytoplasmic tyrosines of integrin subunit beta1 are involved in focal adhesion kinase activation. Mol Cell Biol 20:5758–5765

Young BA, Taooka Y, Liu S, Askins KJ, Yokosaki Y, Thomas SM, Sheppard D (2001) The cytoplasmic domain of the integrin alpha9 subunit requires the adaptor protein paxillin to inhibit cell spreading but promotes cell migration in a paxillin-independent manner. Mol Biol Cell 12:3214–3225

Integrins and Extracellular Matrix in Animal Models

U. Müller

Department of Cell Biology, Institute for Childhood and Neglected Disease,
The Scripps Research Institute, 10550 N. Torrey Pines Road, La Jolla, CA 92037, USA
umueller@scripps.edu

1	Introduction	218
2	Integrin–ECM Interactions and Early Embryonic Development	222
3	Integrin–ECM Interactions During Mesoderm Formation and in the Heart	223
4	Integrin–ECM Interactions During Epithelial Tubulogenesis	224
5	Integrin–ECM Interactions and Angiogenesis	225
6	Integrin–ECM Interactions During Skin Development	226
7	Integrin–ECM Interactions in Skeletal Development	227
8	Integrins, ECM Molecules, and Inner Ear Development	227
9	Integrin–ECM Interactions in Skeletal Muscle Development	228
10	Integrin–ECM Interactions in Peripheral Nerve and at the Neuromuscular Junction	230
11	Integrin–ECM Interactions in the CNS	232
12	Summary	234
	References	234

Abstract Integrins are a family of transmembrane receptors that mediate interactions of cells with extracellular matrix (ECM) constituents and cell surface counter receptors. Each integrin mediates interactions with specific sets of ligands and regulates distinct aspects of cellular function including attachment to and organization of ECM assemblies, cell migration, proliferation and survival, and mechanical force transmission. Integrins exert their versatile functions by establishing a transmembrane link between the cell exterior and the cytoskeleton, and by activating intracellular second messenger systems. In addition, cellular signals can modulate integrin activity and ligand interactions, enabling transduction of information from the inside of the cell to the outside. Many of the basic functions of integrins and their ECM ligands have been uncovered by studying them biochemically or with cells in culture. Integrin and ECM functions have also been determined genetically, defining their essential roles in the organism. The ongoing challenge is to integrate cell biological, biochemical, and genetical evidence into a coherent picture. I will discuss here genetic findings, focusing on the murine system, that have shed light on the developmental functions of integrins and their ECM ligands. Where suitable information is available, I will relate the genetical finding to results obtained with cell biological and biochemical approaches.

Keywords Integrin · Extracellular matrix · Laminin · Development · Organogenesis · Central nervous system · Synapse · Muscle

1
Introduction

The development of mechanisms that allow cells to engage in specific and dynamically regulated adhesive interactions was an important evolutionary step essential for the emergence of metazoans. Adhesive interactions play a pivotal role throughout metazoan life. At conception, species-specific recognition mechanisms mediate adhesion between sperm and egg. During development, modulation of adhesive interactions allows a cell to switch between a stationary adhesive and migratory state. In the adult, adhesive interactions control tissue integrity and remodeling, and immune responses. Modulation of adhesive strength is important to regulate the formation, function, and plasticity of synaptic connections, therefore affecting learning, memory, and behavior.

One important class of cell adhesion molecules are the integrins, a family of receptors that are widely expressed in the organism (Fig. 1). Each integrin consists of an α and a β subunit. Both subunits are type I transmembrane glycoproteins consisting of large extracellular domains, a transmembrane domain, and a short cytoplasmic domain. The integrin $\beta 4$ subunit is an ex-

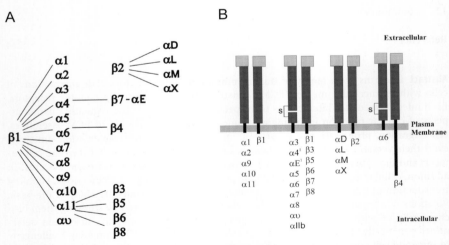

Fig. 1A, B. Vertebrate integrins. Integrins are cell surface receptors consisting of α and β subunits. Each α and β subunit is encoded by a separate gene. **A** The different heterodimer combinations that have been described are indicated. **B** The cytoplasmic domains of all α and β subunits except $\beta 4$ are short. Some of the α subunits are posttranslationally cleaved into two fragments that are connected by disulfide bonds (indicated by a *bracket*)

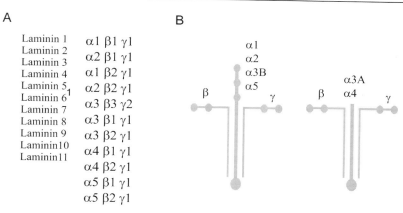

Fig. 2A, B. Laminin molecules. Laminins are heterotrimeric molecules consisting of α, β, and γ subunits. A List of known laminin molecules and their subunit composition.[1] Note that laminin 5 can contain the α3A or α3B subunit. B Diagram of laminin heterotrimers. Note that the α3B and α4 subunits are shorter than other α subunits

ception. It contains a cytoplasmic domain of approx. 1,000 amino acids (reviewed in Hynes 1992; Hynes and Zhao 2000; Bokel and Brown 2002). Integrins mediate interactions with many ligands, including ECM components, cell surface counter receptors, proteases, and pathogens (Table 1). Many ECM molecules that serve as integrin ligands are not freely diffusible but are assembled into highly organized three-dimensional networks such as basement membranes (BMs). BMs contain laminins (Fig. 2), collagen IV, nidogen/entactin, and perlecan. BM formation requires self-assembly of laminin and collagen IV into two independent networks that are connected by nidogen/entactin (reviewed in Yurchenco and O'Rear 1994; Timpl and Brown 1996; Li et al. 2003).

Integrins not only mediate adhesion, they regulate the activity of second messenger systems, modulating many aspects of cellular behavior including actin dynamics, proliferation, survival/apoptosis, polarity, motility, and gene expression (reviewed in Hynes 1992; Giancotti and Ruoslahti 1999; Hynes 2002b; Schwartz and Ginsberg 2002; Brakebusch and Fassler 2003). Based on biochemical and structural studies, integrins have been described as "bidirectional allosteric signaling machines" (Hynes 2002b). This term reflects the ability of integrins to transmit signals into the cell that are coupled to conformational changes in the extracellular and intracellular receptor domains. Likewise, signals from within the cell trigger conformational changes across the cell membrane, thereby affecting integrin ligand binding activity (reviewed in Hynes 2002b; Liddington and Ginsberg 2002; Giancotti 2003).

Many of the basic functions of integrins and their ECM ligands have been uncovered by studying them biochemically or with cells in culture. Integrin and ECM functions have also been determined genetically. The ongoing

Table 1. Integrin ligands. Integrins interact with secreted molecules, counter receptors, and pathogens. Many of the counter receptors are members of the Ig-superfamily. ADAM (a disintegrin and metalloproteinase) show homology to disintegrins and metalloproteinase

Integrin	Alternative names	ECM/secreted ligands	Receptor molecules	Viruses/pathogens
$\alpha 1\beta 1$	CD49a/CD29, VLA-1	Co I, IV, Ln		
$\alpha 2\beta 1$	CD49b/CD29, VLA-2, GPIa-GPIIa	Co I, IV, Ln, Tn-C, dis		Echovirus
$\alpha 3\beta 1$	CD49c/CD29, VLA-3, ECMRI, Gapb3	Co I, Ln, Tsp-1		Invasin
$\alpha 4\beta 1$	CD49d/CD29, VLA-4	Fn, Opn, dis	VCAM-1, LW-glycoprot	Invasin
$\alpha 4\beta 7$	CD49d/CD-, LPAM-1	Fn, dis	VCAM-1, MAdCAM-1	
$\alpha E\beta 7$	CD103/CD-, M290 IEL, HML-1		E-Cadherin	
$\alpha 5\beta 1$	CD49e/CD29, VLA-5	Fn, Fb, Opn, dis	mL1, ADAM15	Invasin, *Shigella flexneri*, *Borrelia burgdorferi*, *Bordetella pertussis*
$\alpha 6\beta 1$	CD49f/CD29, VLA-6	Ln	ADAM-2,-9	Invasin
$\alpha 6\beta 4$	CD49f/CD104	Ln		
$\alpha 7\beta 1$	CD-/CD29	Ln		
$\alpha 8\beta 1$	CD-/CD29	Fn, Opn, Tn-C, Vn, Nn		
$\alpha 9\beta 1$	CD-/CD29	Opn, Tn-C, dis	VCAM-1, ADAM-12,-15	
$\alpha 10\beta 1$	CD-/CD29	Co II		
$\alpha 11\beta 1$	CD-/CD29	Co I		
$\alpha v\beta 1$	CD51/CD29	Fn, Fb, Opn, VN, agrin		Invasin, HIV-Tat
$\alpha v\beta 3$	CD51/CD61, VNR	Vn, Fn, Fb, Bsp-1, Ln, Opn, pthr, Tn-C, Tsp-1, vWF, MMP2, dis	PECAM-1, L1, ADAM-15,-23	Adenovirus (Ad) penton base, HIV-TAT, *Borrelia b.*
$\alpha v\beta 5$	CD51/CD	Fn, Vn, Bsp-1		Ad penton base, HIV-TAT
$\alpha v\beta 6$	CD51/CD	Fn, Tn-C		
$\alpha v\beta 8$	CD51/CD	Fn		
$\alpha D\beta 2$	CD-/CD18		VCAM-1	
$\alpha L\beta 2$	CD11a/CD18, LFA-1		ICAM-1,-2,-3,-4,-5	
$\alpha M\beta 2$	CD11b/CD18, Mac-1, Mo1	Fb, C3bi, factor X, NIF	ICAM-1,-4	*Candida albicans*, *Borrelia b.*, *Bordetella p.*
$\alpha X\beta 2$	CD11c/CD18, p150,95	Fb, C3bi		*Borrelia b.*
$\alpha IIb\beta 3$	CD41/CD61, GPIIb-IIIa	Fn, Vn, Fb, vWF, Tsp-1, pthr, plas, dis		*Borrelia b.*

Bsp-1, bone sialoprotein-1; C3bi, inactivated form of C3b component of complement; Co, collagen; factor x, coagulation factor X; Fb, fibrinogen; Fn, fibronectin; Ln, laminin; MMP2, matrix metalloproteinase 2; NIF, neutrophil inhibitory factor; Nn, nephronectin; Opn, osteopontin; pls, plasminogen; proThr, prothrombin; Tn-C, tenascin-C; Tsp-1, thrombospondine-1; Vn, vitronectin; vWF, von Willebrand factor; dis, disintegrins (integrin antagonists from snake venoms)

Table 2. Integrin knock-out mice. The phenotype of mice with null mutations in integrin subunit genes is indicated. Further details can be found in the text and in the cited references. Phenotypes of double-null mice or of mice with tissue specific knock-outs (CRE/LOX) are omitted, but described in the text

Integrin	Phenotypes of null alleles	Reference(s)
α1	Defects in collagen synthesis, tumor vascularization	Gardner et al. 1996; Pozzi et al. 1998, 2000, 2002
α2	Defects in branching of mammary gland, platelet aggregation	Holtekotter et al. 2002; Chen et al. 2002
α3	Defects in kidney, lung, cortical lamination, skin blisters	Kreidberg et al. 1996; DiPersio et al. 1997; Anton et al. 1999
α4	Defects in placenta, heart, and hematopoiesis	Yang et al. 1995; Arroyo et al. 1996, 1999, 2000
α5	Defects in mesoderm, vasculature, defects in neural crest survival, muscle fiber integrity	Yang et al. 1993; Goh et al. 1997; Taverna et al. 1998
α6	Skin blisters, defects in many epithelia defects in lamination of retina and cortex	Georges-Labouesse et al. 1996, 1998
α7	Defects in myotendinous junctions	Mayer et al. 1997
α8	Defects in kidney development and hair cells	Müller et al. 1997; Littlewood-Evans and Müller 2000
α9	Defect in the lymphatic duct, chylothorax	Huang et al. 2000
αv	Defects in placenta, cerebral vasculature, cleft palate	Bader et al. 1998; McCarty et al. 2002
αL	Defects in leukocyte recruitment, tumor rejection	Schmits et al. 1996
αM	Defects in neutrophil function, mast cells, adipose tissue	Coxon et al. 1996; Dong et al. 1997; Tang et al. 1997
αE	Defects in lymphocyte trafficking, skin inflammation	Schon et al. 1999
αIIb	Defects in platelet aggregation	Tronik-Le Roux et al. 2000
β1	Inner cell mass failure	Fassler and Meyer 1995; Stephens et al. 1995
β2	Defects in leukocyte function; skin infections	Scharffetter-Kochanek et al. 1998
β3	Defects in platelet aggregation, osteosclerosis	Hodivala-Dilke et al. 1999; McHugh et al. 2000; Reynolds et al. 2002
β4	Skin blisters, defects in many epithelia	van der Neut et al. 1996; Dowling et al. 1996
β5	No obvious defect	Huang et al. 2000
β6	Skin and airway inflammation; impaired lung fibrosis	Huang et al. 1996; Munger et al. 1999; Morris et al. 2003
β7	Abnormal Peyer's patches, defects in lymphocyte recruitment	Wagner et al. 1996
β8	Defects in placenta, cerebral vasculature	Zhu et al. 2002

challenge is to integrate cell biological, biochemical, and genetical evidence into a coherent picture. I will discuss here genetic findings, focusing on the murine system, that have shed light on the developmental functions of integrins and their ECM ligands (Table 2). A discussion of integrin function in blood cells and during immune responses has been omitted as several excellent recent reviews cover this area (Alon and Feigelson 2002; Lindbom and Werr 2002; Sims and Dustin 2002; Miyamoto et al. 2003; Nieswandt and Watson 2003).

2
Integrin–ECM Interactions and Early Embryonic Development

Cell–ECM interactions have been proposed to regulate early steps during embryonic development such as implantation and movement of cells within the pregastrulation embryo. To address the function of integrins at this early stage genetically, two laboratories have disrupted the integrin $\beta 1$ subunit gene, thereby inactivating all $\beta 1$ integrins. These studies have shown that $\beta 1$ integrins are dispensable for blastocyst development and implantation, but the embryos die subsequently at embryonic day (E)5.5 due to inner cell mass (ICM) failure (Fassler and Meyer 1995; Stephens et al. 1995). The mechanisms that cause the defect are not entirely clear, but may include defects in interactions of ICM-derived cells with BMs. The first BM in the embryo forms between derivatives of the ICM, the primitive endoderm, and the epiblast. Embryos with a targeted mutation in the laminin $\gamma 1$ gene do not form these BMs and die at a similar stage as $\beta 1$-null embryos (Smyth et al. 1999), suggesting that ICM failure in $\beta 1$-null embryos is caused by defect in interactions of these cells with laminin, and possibly by defects in BM assembly.

To investigate the function of $\beta 1$ integrins in BM assembly, embryoid bodies (EBs) have been derived from genetically modified embryonic stem (ES) cells. In EBs derived from wild-type ES cells, the outer cells of the EBs flatten within 2 days and resemble the primitive endoderm of the blastocyst. A BM is subsequently assembled between the endoderm and remaining cells in the interior of the EB. This BM is likely equivalent to the first BM that forms within the embryo in utero (Smyth et al. 1999; Li et al. 2002). Apoptosis within the EB leads to the formation of a cavity, a process that may be regulated by laminin (Murray and Edgar 2000). The cells that form the rim around the cavity develop a polar morphology and form an epiblast-like cell layer. No BM forms in EBs derived from laminin $\gamma 1$-null ES cells, the ectoderm never polarizes, and apoptosis and cavity formation do not progress properly (Smyth et al. 1999). A similar phenotype is observed in $\beta 1$-null EBs, suggesting that $\beta 1$ integrins regulate BM assembly. However, the laminin $\alpha 1$ chain gene is not expressed in $\beta 1$-null EBs, thereby preventing secretion of functional laminin heterotrimers (Aumailley et al. 2000; Li et al.

2002). Exogenous laminin-1 bypasses the defect in laminin γ1- and integrin β1-null embryoid body, restoring BMs along with epiblast differentiation and cavitation (Li et al. 2002). This suggests that assembly of this early embryonic BM can progress without β1 integrins, and that the defects in the synthesis of the laminin α1 chain is an important factor of the phenotype of β1-null embryos. However, defects in cell adhesion to BM in β1-null EBs is accompanied by apoptosis and delay or loss of endodermal differentiation (Aumailley et al. 2000; Li et al. 2002). Thus, β1 integrins in the early embryo appear to be critical for epithelial/BM adhesion and survival.

3
Integrin–ECM Interactions During Mesoderm Formation and in the Heart

The genetic study of fibronectin (FN) and its integrin receptors has provided several important clues as to the function of these molecules during mesoderm formation. FN-null embryos die between E8 and E8.5 with mesodermal defects (George et al. 1993; Georges-Labouesse et al. 1996b). Several integrins bind FN (Table 1), but only α5-null mice display mesodermal defects similar to those observed in FN-deficient mice (Yang et al. 1993). Studies of the α5-null mice suggest that defects in mesodermal tissue are not caused by defects in lineage commitment, apoptosis, or proliferation of mesodermal cells. Interactions between α5β1 and FN are rather crucial for the maintenance of mesodermal derivatives. The data also show that the integrin α5β1 regulates survival of some neural crest cells (Goh et al. 1997).

α5-Null embryos die about 2 days later than FN-null embryos, suggesting that other receptors may mediate some FN functions in mesodermal tissue. Mutants double null in α3 or α4, and α5 integrin subunit genes do not show synergies in their phenotypic defects, but αv/α5 double mutant embryos died shortly after E7.5, with defects similar to FN-null embryos. This suggests that αv and α5 integrins cooperate to mediate FN functions during mesodermal development. However, αv/α5 double mutant embryos have a more severe phenotype than FN-null mice, suggesting that defects in interactions of these integrins with other ligands contribute to the phenotype (Yang et al. 1999).

The study of mice with mutations in specific FN-binding integrins have revealed that they act in some tissues through other ligands. This is particularly apparent for the integrin α4β1 that binds to FN and vascular cell adhesion molecule (VCAM)1. α4-Null embryos fail to fuse the allantois with the chorion during placentation, show defects in the heart and vasculature, and display abnormalities in cranial and facial structure (Yang et al. 1995). Similar defects are observed in VCAM1-null mice (Kwee et al. 1995). Detailed studies of the mutant mice have shown that interactions between VCAM-1 and α4β1 regulate not the formation but maintenance of the epicardium.

Subsequent studies have shown that $\alpha 4\beta 1$ also regulates migration of epicardial progenitors (Sengbusch et al. 2002), but it is at present unclear whether this earlier function of $\alpha 4\beta 1$ is mediated by VCAM1 or FN.

4
Integrin–ECM Interactions During Epithelial Tubulogenesis

The development of many organs is dependent on the formation of epithelial tubules within mesenchymal tissue, and genetic evidence suggests that integrins and ECM proteins are important for tubulogenesis. Particularly informative results have been obtained by studying kidney development (reviewed in Müller and Brandli 1999). In the kidney, the ureteric bud, an epithelial tubule, invades the metanephric mesenchyme and induces the transformation of mesenchymal cells into epithelial tubules that fuse with the ureter. The metanephric mesenchyme in turn induces the ureteric bud to branch and grow (reviewed in Saxen 1987; Vainio and Müller 1997). Expression of the integrin $\alpha 8$ subunit is induced in mesenchymal cells adjacent to the growing ureter. Ureter growth and branching is disrupted in $\alpha 8$-null mice (Müller et al. 1997). Using expression cloning, a ligand for the integrin $\alpha 8\beta 1$ named nephronectin was identified that is expressed in kidney tubules (Brandenberger et al. 2001). Integrin $\alpha 8\beta 1$ in the mesenchyme may bind to nephronectin on the ureter epithelium to modulate its adhesion, growth, and branching. Other $\alpha 8\beta 1$ ligands, such as FN and osteopontin are also expressed in the kidney (Müller et al. 1997). Osteopontin is not essential for kidney organogenesis (Liaw et al. 1998), but FN has an important function in branching morphogenesis (Sakai et al. 2003). Therefore, several $\alpha 8\beta 1$ ligands may cooperate to mediate $\alpha 8\beta 1$ functions in the kidney.

The integrin $\alpha 3\beta 1$ has also been implicated in regulating growth and/or branching of the ureter epithelium. The integrin is expressed on epithelial tubules in the kidney, including the ureteric bud and its derivatives. $\alpha 3$-deficient mice die shortly after birth. The mice have fewer collecting ducts in the kidney consistent with a defect in the growth of epithelial tubules. The BM that surrounds the epithelial tubules are reduced in thickness, suggesting that interactions between epithelial cells with BM components are important (Kreidberg et al. 1996). Genetic evidence supports the view that laminin 10 or 11 (or both) mediates $\alpha 3\beta 1$ function in the kidney. Accordingly, laminin $\alpha 5$-null mice that lack expression of the laminin 10 and 11 heterotrimers have kidney defects similar to integrin $\alpha 3$-null mice (Miner and Li 2000). Mice with targeted mutations in the laminin $\gamma 1$ gene that disrupt a binding site for nidogen affect growth and branching of the ureter, providing further evidence that BM assembly is important for the development of epithelial kidney tubules (Willem et al. 2002).

The integrin $\alpha3$ subunit and the laminin $\alpha5$ subunit are also expressed in the glomerulus, where excretory nephrons establish contact with blood vessels. Consistent with the expression pattern, integrin $\alpha3$-null and laminin $\alpha5$-null mice have additional prominent defects in the glomerulus, with defects in foot processes of podocytes and an abnormal organization of the glomerular BM (Miner and Li 2000). Mice lacking the laminin $\beta2$ subunit or the collagen $\alpha3$(IV) subunit also develop kidney defects (Noakes et al. 1995b; Cosgrove et al. 1996; Miner and Sanes 1996). Kidney formation progresses normally in these mice, but the composition and function of the glomerular BM is altered (Müller and Brandli 1999).

Taken together, these findings provide evidence that interactions between specific $\beta1$ integrins regulate growth and branching of epithelial structures in the kidney, with additional functions in the assembly/maintenance of the glomerular BM.

5
Integrin–ECM Interactions and Angiogenesis

There is clear evidence demonstrating a function for $\beta1$ integrins in regulating angiogenesis. Integrin $\alpha5$- and FN-null mice not only have mesodermal defects (see above), but also vascular abnormalities (George et al. 1993; Yang et al. 1993). Likewise, antibodies that block the function of the collagen receptors $\alpha1\beta1$ and $\alpha2\beta1$ inhibit angiogenesis in tumors (Senger et al. 2002), and $\alpha1$-null mice support reduced tumor growth and angiogenesis (Pozzi et al. 2000). Integrins containing the αv subunit have also been implicated in angiogenesis. Reagents that interfere with interactions between $\alpha v\beta3$ and $\alpha v\beta5$ and their ligands block angiogenesis in response to growth factors in tumors and in retinal angiogenesis (Brooks et al. 1994; Brooks et al. 1995; Friedlander et al. 1995; Friedlander et al. 1996; Hammes et al. 1996). However, mice and humans with mutations in the integrin $\beta3$ subunit gene, or mice with mutation in the $\beta5$ and $\beta6$ subunit gene, or pairwise combinations of these mutations are viable (Huang et al. 1996; Hodivala-Dilke et al. 1999; Huang et al. 2000; Reynolds et al. 2002). Mice that lack both $\alpha v\beta3$ and $\alpha v\beta5$ integrins show enhanced tumor angiogenesis, providing evidence that these integrins are also not essential for and in fact inhibit pathogenic angiogenesis (Reynolds et al. 2002). Furthermore, αv-null mice show extensive angiogenesis and have selective vascular defects in brain and placenta only (Bader et al. 1998). In the brain, this is likely a consequence of perturbed interaction of brain parenchyma with the cerebral vasculature (McCarty et al. 2002). Integrin $\beta8$-null mice show similar defects, indicating that the phenotype arises from loss of $\alpha v\beta8$ (Zhu et al. 2002).

It is not clear how the contrasting findings obtained by genetic means or using function blocking tools against $\alpha v\beta3/\alpha v\beta5$ can be reconciled; $\alpha v\beta3/\beta5$

could be negative regulators of angiogenesis in some instances, and blocking agents may not block all integrin functions, but rather act as antagonists of a negative regulatory function (Hynes 2002a). Alternatively, unligated αv integrins may promote apoptosis by recruiting caspase-8 (Stupack et al. 2001). As an additional possibility, studies by Diaz-Gonzales et al. (1996) show that blocking $\beta 3$ integrin function leads to transdominant inhibition of $\beta 1$ integrins, including integrins such as $\alpha 5\beta 1$ that have a proangiogenic role.

6
Integrin–ECM Interactions During Skin Development

The epidermis of the skin is multilayered and consists of keratinocytes at different stages of differentiation. Basal keratinocytes are in contact with the BM and form specialized adhesion junctions by interacting with laminin 5 in the BM. The integrin $\alpha 3\beta 1$ establishes a transmembrane link to the actin cytoskeleton, and the $\alpha 6\beta 4$ integrin is an integral part of hemidesmosomes establishing a connection to intermediate filaments (reviewed in Nievers et al. 1999; Fuchs and Raghavan 2002; Watt 2002). Mutations that affect laminin 5, or the integrin $\alpha 6$ and $\beta 4$ lead to skin blisters in mice and humans (reviewed in Watt 2002; Müller 2003). Hemidesmosomes are absent in $\alpha 6$- and $\beta 4$-null mice, and epidermal cell layers detach, but a BM forms (Dowling et al. 1996; Georges-Labouesse et al. 1996a; van der Neut et al. 1996). Similar phenotypes are observed when the $\beta 4$ cytoplasmic domain is deleted (Murgia et al. 1998). Integrin $\alpha 3$-null mice and $\alpha 3/\beta 4$ double-null mice also have skin blisters, but remarkably, in the absence of all known laminin 5-binding integrins keratinocytes proliferate, and skin morphogenesis appears normal prior to blister formation (DiPersio et al. 1997 2000).

Two laboratories have inactivated the integrin $\beta 1$ subunit gene, and therefore all $\beta 1$ integrins, by Cre/Lox-mediated gene ablation in the skin (Brakebusch et al. 2000; Raghavan et al. 2000). The observed defects are more severe than in $\alpha 3\beta 1$-deficient mice, suggesting that additional $\beta 1$ integrins have important functions in the skin. The defects include perturbed cell proliferation, but no defects in cell survival or differentiation. Hair follicle morphogenesis is also affected. Posttranslational processing of laminin 5 and BM deposition is perturbed, and hemidesmosomes are disrupted, probably as a consequence of defects in the BM. The defects in cell proliferation are intriguing. Proliferating keratinocytes in the skin express higher $\beta 1$ integrin levels than differentiating keratinocytes, and skin stem cells can be enriched by selecting cells with the highest $\beta 1$ levels. In transgenic mice, overexpression of integrins in the skin leads to hyperproliferation, suggesting direct roles for $\beta 1$ integrins in regulating proliferation (reviewed in Watt 2002). No proliferative defects have been reported in $\alpha 6\beta 4$-deficient mice (Georges-Labouesse et al. 1996a; van der Neut et al. 1996; DiPersio et al. 2000).

These data show that β1 integrins and α6β4 are required for the integrity of the dermo-epidermal junction, but not for epidermal morphogenesis. β1 integrins and α6β4 cooperate to mediate adhesion and to regulate the assembly/maintenance of BM structures. The data also suggest that signals transmitted via β1 integrins, but not via the integrin α6β4, regulate cell proliferation in the skin.

7
Integrin–ECM Interactions in Skeletal Development

The vertebrate skeleton develops by two mechanisms called endochondral ossification and intramembranous ossification. In the former process, cartilaginous precursors are replaced; in the latter process, bone develops directly within mesenchymal tissue (reviewed in Kronenberg 2003). Two defects have been described in skeletal development in mice with targeted mutations in integrin genes. First, β3-null mice show an increase in bone mass, due to decreased bone resorption by β3-null osteoclasts. β3-Null osteoclasts fail to spread in vitro and to produce membrane ruffles in vivo, suggesting defects in actin remodeling (McHugh et al. 2000). Second, integrin α3/α6 double-null mice, or mice lacking the laminin α5 chain, display skeletal abnormalities associated with shortened and abnormally shaped limbs (Miner et al. 1998; De Arcangelis et al. 1999). These findings show that several integrins are important for skeletal development, but we have only a limited understanding as to the mechanism by which different integrins exert their function in this tissue.

8
Integrins, ECM Molecules, and Inner Ear Development

The vertebrate inner ear contains mechanosensory hair cells for the perception of sound waves and acceleration. At their apical surface, hair cells develop stereocilia that are in close contact with ECM assemblies, including the tectorial membrane in the cochlea. Hair cells are also surrounded by support cells, and the support cells are situated on a BM. Mutations in integrins and ECM components lead to deafness in mice and humans, consistent with a function of these molecules in regulating mechanosensor development and function (reviewed in Littlewood Evans and Müller 2000). Mutations in genes encoding ECM components such as α- and β-tectorin and otogelin affect the tectorial membrane (Legan et al. 2000; Simmler et al. 2000a,b). Mutations in genes for other ECM components such as Usherin, Coch, and several collagens also cause deafness, but the mechanism by which these molecules act is unclear (reviewed in Siemens et al. 2001). Finally, genetic studies have shown that integrins have important functions in the inner ear. Several integrin subunits are expressed in hair cells, and integrin $\alpha 8$-null mice show defects in hair cell stereocilia (Littlewood Evans and Müller 2000). Mutations in the gene encoding the integrin downstream effector and rho target dia-1 lead to deafness in humans (Lynch et al. 1997; Watanabe et al. 1999), raising the possibility that it may act in an integrin $\alpha 8\beta 1$-dependent pathway. Taken together, the studies show that specific integrins and ECM components have important functions in the inner ear, but their mechanism of action remains to be established.

9
Integrin–ECM Interactions in Skeletal Muscle Development

During skeletal muscle development, myoblasts fuse to form skeletal muscle fibers. Myoblasts and muscle fibers express integrin receptors and develop in close contact with ECM components, and ECM components and integrins have been implicated in regulating muscle fiber development (reviewed in Mayer 2003; Sanes 2003).

The function of integrins in myoblast fusion has been addressed with cells in culture, and genetically in mice. Function blocking antibodies to the integrin $\beta 1$ and $\alpha 4$ subunits, and antisense mRNA to the $\alpha 6$ subunit block myoblast fusion in vitro (Menko and Boettiger 1987; Rosen et al. 1992; Sastry et al. 1996). However, $\alpha 4$-, $\alpha 5$-, and $\beta 1$-null myoblasts and EBs form myotubes in vitro, and are incorporated into chimeric muscle fibers in vivo (Fassler and Meyer 1995; Fassler et al. 1996; Yang et al. 1996; Hirsch et al. 1998; Rohwedel et al. 1998; Taverna et al. 1998). To resolve these contrasting findings, the $\beta 1$ integrin subunit has been inactivated in myoblasts by Cre/Lox-mediated gene ablation (Schwander et al. 2003). The analysis of the mutant mice has demonstrated an essential function for $\beta 1$ integrins in myoblast fusion in vivo. Because myotubes can form in $\beta 1$-null EBs in vitro, albeit only under certain culture conditions and with a delay (Hirsch et al. 1998), alternative mechanisms may exists that can substitute for $\beta 1$ integrin function, at least in vitro.

The mechanisms by which $\beta 1$ integrins regulate myoblast fusion has been investigated further. In vitro experiments have shown that fusion defects are rescued when $\beta 1$-null and wild-type myoblasts are mixed, suggesting that heterophilic interactions between $\beta 1$ integrins and a yet to be defined cell surface receptor may be important. Cell surface expression of the integrin-associated tetraspanin CD9 is abolished in $\beta 1$-deficient myoblasts, suggesting that CD9 and $\beta 1$ integrins act in a common pathway (Schwander et al. 2003). In vivo, CD9 is essential for sperm–egg but not myoblast fusion (Kaji et al. 2000). However, another member or members of the tetraspanin family may compensate for a loss of CD9. In fact, antibodies to CD9 and a second tetraspanin, CD81, have additive effects on myotube formation in vitro (Tachibana and Hemler 1999). During fertilization, CD9 is required for fusion at a step subsequent to adhesion (Kaji et al. 2000). $\beta1$ integrins are also required to regulate myoblast fusion at a step subsequent to myoblast adhesion (Schwander et al. 2003). This suggests that both proteins act at the same step in the fusion process.

In muscle, the $\beta1A$ variant is expressed during embryonic stages but is replaced perinatally by the $\beta1D$ variant (van der Flier et al. 1997). Genetically modified mice that only express the $\beta1D$ variant have reduced muscle mass, suggesting that $\beta1A$ has specific functions in myoblasts and/or muscle fibers in the embryo that cannot be carried out by $\beta1D$ (Cachaco et al. 2003).

Genetic studies have also assessed the function of integrin α subunits in muscle. None of the α knock-outs phenocopy the defect of $\beta1$-null myoblasts, suggesting that several $\beta1$ integrins cooperate in regulating myoblast fusion. However, muscles chimeric for cells that lack or express the integrin $\alpha5\beta1$ show dystrophic symptoms, suggesting that $\alpha5\beta1$ is required for muscle fiber integrity (Taverna et al. 1998). In integrin $\alpha7$-null mice, skeletal muscle develops but myotendinous junctions (MTJs) are not maintained (Mayer et al. 1997). Only the maintenance but not the initial formation of MTJs is dependent on $\beta1$ integrins (Mayer et al. 1997). αv-Containing integrins are still localized to the MTJs of $\beta1$-null muscle (Schwander et al. 2003). This suggests that αv integrins may be required for the initial assembly of MTJs, or that $\beta1$ and αv-containing integrins have redundant functions in this process.

Substantial evidence suggests that interactions between laminin in the muscle fiber BM with ECM receptors in muscle fibers are important for maintaining the integrity of skeletal muscle fibers. The BM around muscle fibers contains the laminin2/4 heterotrimers that share a common laminin $\alpha2$ subunit (Fig. 2). Mutations in the laminin $\alpha2$ subunit gene causes muscular dystrophy in humans and mice (Vachon et al. 1996; Miyagoe et al. 1997). Since muscle fibers form in the absence of the laminin $\alpha2$ subunit, the data suggest that laminins have a crucial function for the mechanical stability of muscle fibers. It is at present unclear to what extent integrins are required to mediate laminin 2/4 functions. The predominant laminin-binding integrin in muscle is $\alpha7\beta1$, but $\alpha7$-null muscle fibers have less severe defects than laminin $\alpha2$-null muscle fibers. It is possible that several integrins have redundant functions to mediate interactions with laminin 2/4. In addition, muscle fibers express the dystrophin–glycoprotein complex (DGC) that also mediates interactions with ECM components, including laminins. Mutations in genes that encode components of the DGC or affect its interaction with ECM components cause muscular dystrophy in mice and humans. It has therefore been suggested that the DGC mediates interactions with laminin

2/4 that are important to maintain skeletal muscle fibers (reviewed in Henry and Campbell 1999; Michele and Campbell 2003).

Taken together, the genetic findings show that $\beta1$ integrins have several important function in skeletal muscle, such as during myoblast fusion and in the maintenance of MTJs. Further studies are necessary to clarify the role of integrins and the DGC in regulating interactions of mature muscle fibers with the muscle fiber BM.

10
Integrin–ECM Interactions in Peripheral Nerve and at the Neuromuscular Junction

Integrins and their ECM ligands not only regulate formation of skeletal muscle fibers, they have also been implicated in controlling aspects of peripheral nerve development, as well as the development of synaptic contacts between motor neurons and muscle fibers at the neuromuscular junction (NMJ). While there is at present no genetic evidence demonstrating a function for an integrin in regulating neural crest migration or axon outgrowth during the development of the peripheral nervous system, it is clear that integrins regulate myelination of peripheral nerve by Schwann cells. Schwann cells are in contact with BM components throughout their development, and mutations in the laminin $\alpha2$ subunit gene are associated not only with muscle fiber defects (see above) but also with myelination defects (reviewed in Scherer 2002). Premyelinating Schwann cells express the laminin receptor integrin $\alpha6\beta1$, while myelinating Schwann cells express the laminin receptor integrin $\alpha6\beta4$ (Previtali et al. 2001). The integrin $\alpha6\beta4$ does not appear to be essential for myelination (Frei et al. 1999), but genetic ablation of the integrin $\beta1$ subunit in Schwann cells by Cre-mediated recombination leads to myelination defects. $\beta1$-Null Schwann cells fail to subdivide axon bundles, do not progress past the premyelinating stage, and frequently fail to establish stable contact with the BM. Some axons are not ensheathed, but myelinated axons that appear, although with a delay, seem normal. These findings suggest that $\beta1$ integrins are essential for the first steps during axonal ensheathment (Feltri et al. 2002). Interestingly, the integrin $\alpha5\beta1$ has recently been shown to regulate Schwann cell proliferation, while the integrin $\alpha4\beta1$ regulates Schwann cell survival, providing evidence that $\beta1$ integrins regulate other aspects of Schwann cell development as well (Haack and Hynes 2001).

The presynaptic nerve terminal at the NMJ is capped by the terminal Schwann cell, and pre- and postsynaptic membranes are separated by the synaptic cleft that contains a highly specialized BM (reviewed in Burden 1998; Sanes and Lichtman 2001). Laminin 2 is expressed predominantly extrasynaptically, while laminin chains that form laminin 4, 9, and 11 are expressed synaptically (Patton et al. 1997). During formation of the NMJ, synaptic proteins such as acetylcholine receptors (AChRs) become clustered in the postsynaptic muscle membrane. Clustering of proteins is dependent on the ECM molecule agrin, which is released from motor nerve terminals and activates the Musk receptor tyrosine kinase in muscle (DeChiara et al. 1996; Gautam et al. 1996). Agrin can bind to integrins and associates with laminin in the synaptic basal lamina (Denzer et al. 1997; Martin and Sanes 1997; Burgess et al. 2002). Laminin 1 and 2/4 can induce AChR clustering in myotubes in vitro, and agrin and laminin act synergistically in AChR clustering (Sugiyama et al. 1997; Montanaro et al. 1998; Burkin et al. 2000). The laminin receptor integrin $\alpha7\beta1$ is localized in the postsynaptic membrane at the NMJ, and antibodies against the $\beta1$ or $\alpha7$ integrin subunits modulate laminin- and agrin-induced AChR clustering in myotubes (Martin and Sanes 1997; Burkin et al. 1998). The integrin $\alpha7\beta1$ also colocalizes with laminin-induced AChR clusters, but less with agrin induced clusters (Burkin et al. 1998). It is tempting to speculate that agrin, laminin, and the $\alpha7\beta1$ integrin cooperate to achieve maximal AChR clustering. Genetic evidence confirming a role for $\alpha7\beta1$ integrins and laminins in postsynaptic differentiation and agrin signaling is still missing (Noakes et al. 1995a; Mayer et al. 1997).

However, there is now substantial evidence that laminins regulate presynaptic differentiation. In vitro, synaptic laminins can act as stop signals for growing motor neurons. While $\beta1$-containing laminin trimers promote neurite outgrowth, $\beta2$ fragments or laminin 11 cause motoneurons to stop growing and start differentiating into nerve terminals (Porter et al. 1995). Consistent with these findings, presynaptic differentiation is aberrant in laminin $\beta2$-null mice (Noakes et al. 1995a). Also, apposition of active zones in the presynaptic nerve terminals with postsynaptic junctional folds is perturbed in mice lacking laminin $\alpha4$ that forms with the $\beta2$ and $\gamma1$ subunit synaptic laminin 9 (Patton et al. 2001). The receptors that mediate these effects of laminin on presynaptic nerve terminals are not known, but integrins are good candidates. Intriguingly, agrin may also affect presynaptic differentiation such as growth and branching of nerve terminals, but its precise role in presynaptic differentiation still needs to be defined (reviewed in Burden 1998; Sanes and Lichtman 2001).

Taken together, these results show that integrins and ECM receptors regulate Schwann cell development. The data also demonstrate that ECM components that serve as integrin ligands such as laminins and agrin regulate aspects of pre- and postsynaptic differentiation at the NMJ. However, genetic evidence is missing that demonstrates a function for $\beta1$ integrins at the NMJ.

11
Integrin–ECM Interactions in the CNS

During development of the cerebral cortex, cohorts of neurons migrate along radial glial fibers from the ventricular neuroepithelium towards the pial surface, where they stop to migrate and are assembled into layers. Earlier migrating neurons come to sit in deeper layers of the cortex while later migrating neurons bypass these cells and come to sit in more superficial layers. Cajal Retzius cells that form a layer below the pial surface express reelin that has been proposed to serve as a stop signal for migrating neurons (reviewed in Hatten 1999; Nadarajah and Parnavelas 2002). Recent studies provide strong genetic evidence that integrins affect the formation of cortical layers by regulating the assembly and maintenance of the pial BM that surrounds the brain.

Integrin $\alpha 6$-null mice or $\alpha 3/\alpha 6$ double-null mice show defects in the formation of cortical layers and defects in the pial BM (Georges-Labouesse et al. 1998; De Arcangelis et al. 1999). Similar defects are observed when the integrin $\beta 1$ subunit is inactivated in the precursors of neurons and glia in the CNS by Cre-Lox mediated gene ablation (Graus-Porta et al. 2001). The defects include disruption in the pial BM and defects in the anchorage of glia endfeet at the pial BM. Targeted mutations in genes for BM components such as perlecan or entactin/nidogen, and a mutation that disrupts the entactin-binding site in laminin, also lead to cortical layering defects (Arikawa-Hirasawa et al. 1999; Costell et al. 1999; Dong et al. 2002; Halfter et al. 2002). These findings support the model that interactions between $\beta 1$ integrins in radial glial fibers with the pial BM are important for the formation of the glial scaffold and BM maintenance/remodeling. Formation of the Cajal Retzius cell layer has also been analyzed in some of these studies, and disruptions were consistently observed (Graus-Porta et al. 2001; Halfter et al. 2002). Further studies will be necessary to assess whether Cajal Retzius cells express $\beta 1$ integrins that may anchor these cells at the pial BM.

Previous studies had provided evidence that integrins may regulate neuronal migration along radial glial fibers. The migration of tectal neurons is perturbed in chickens upon infection with retroviruses expressing antisense mRNAs of the integrin $\beta 1$ or $\alpha 6$ subunits (Galileo et al. 1992; Zhang and Galileo 1998). Antibodies that inhibit interactions of $\alpha 3\beta 1$ or αv integrins with their ligands perturb migration of neurons along glial fibers in vitro, and defects in the formation of cortical layers have been reported in mice with targeted mutations in the integrin $\alpha 3$ subunit gene (Anton et al. 1999). Further studies suggest that the integrin $\alpha 3\beta 1$ binds to reelin and transmits a reelin stop signal into migrating neurons (Dulabon et al. 2000). However, recent findings provide strong evidence that reelin does not act as a stop signal for migrating neurons (e.g., Magdaleno et al. 2002). Furthermore, neuron–glia interaction and neuronal migration are not affected in mice that lack β1 integrins in the CNS. In vitro studies also demonstrate that β1-null neurons interact with and migrate along β1-null radial glial fibers (Graus-Porta et al. 2001). These findings show that β1 integrins are not essential for glial-guided migration and that defects in cortical layers in mice with mutations in integrin genes are likely caused by defects in the cortical marginal zone.

Recent studies demonstrate that β1 integrins also regulate the development of oligodendrocytes. Oligodendrocyte precursors arise in restricted areas of the CNS, disperse by migration, and myelinate axonal tracts. Oligodendrocyte proliferation and differentiation is regulated by contact with axonal targets and soluble factors, such as platelet-derived growth factor (PDGF) and neuregulins (NRGs) (reviewed in Barres and Raff 1994; Buonanno and Fischbach 2001). At earlier stages of development, these soluble factors act as mitogens and inhibit differentiation, while later they become crucial in promoting cell survival (Colognato et al. 2002; and references therein). Ffrench-Constant and colleagues have now demonstrated that integrins can determine the signaling readout of NRG. Binding of α6β1 to laminin switches the response of cultured oligodendrocytes to NRG from proliferation to enhanced survival and differentiation. Consistent with this finding, brain stem axonal tracts of α6-null mice contain reduced numbers of oligodendrocytes with a parallel increase in apoptosis (Colognato et al. 2002).

Finally, the function of integrins has been studied during formation and function of synapses in the CNS. Expression of many integrin subunits has been detected in the CNS by in situ hybridization, and several, including the α3, α8, and β8 subunits, have been localized to synaptic sites by immunoelectron microscopy. Studies with antibodies and small molecular weight inhibitors of integrins have provided evidence that β1 integrins and αv integrins are required for the stabilization of long-term potentiation (LTP) in the hippocampus, but not for basic synaptic transmission (reviewed in Clegg 2000). Further in vitro studies have implicated integrins, and in particular the β3 integrin subunit and possibly integrin αvβ3, in the maturation of hippocampal synaptic connections (Chavis and Westbrook 2001). The function of several integrins at CNS synapses has been addressed genetically in vivo only recently. Mice with single, double, and triple mutations in the integrin α3, α5, and α8 subunit genes have been analyzed (Chan et al. 2003). These studies provide evidence that $α3^{+/-}$ mice fail to maintain LTP that is generated in hippocampal CA1 neurons. No such defects have been observed in $α5^{+/-}$ or $α8^{+/-}$ mice. However, LTP is nearly completely abolished in triple heterozygous mice. No structural defects have been observed in the hippocampus. Taken together, these data suggest important roles for these three integrins in the generation and maintenance of LTP. Intriguingly, mice with a mutation in the gene for the integrin-associated protein (IAP) show significant reduction in the magnitude in LTP (Chang et al. 1999; Chang et al. 2001). IAP binds to integrins containing the β3 subunit, but also to α2β1

(Brown and Frazier 2001), raising the possibility that these integrins also regulate synaptic functions in the CNS.

12 Summary

Genetic studies provide conclusive evidence that integrin–ECM interactions are important components of the regulatory network that controls the development of many tissues and organs. One important conclusion that can be drawn from the genetic studies is that integrins and their ECM ligand not only mediate adhesive interactions but also have instructive roles coordinating the behavior of cells and groups of cells. However, we still have only a limited understanding as to the mechanisms by which integrins regulate cellular behavior in tissues and organs. The future challenge will be to define the signaling mechanisms by which integrins exert their function in different cell types, in order to link genetic and cell biological findings into a coherent framework.

Acknowledgements I apologize to those colleagues whose work could not be cited due to space constraints. I would like to thank Anastasia Kralli and Martin Schwander for critical reading of the manuscript, and the NIH for financial support (R01 DC05965-01).

References

Alon R, Feigelson S (2002) From rolling to arrest on blood vessels: leukocyte tap dancing on endothelial integrin ligands and chemokines at sub-second contacts. Semin Immunol 14:93–104

Anton ES, Kreidberg JA, Rakic P (1999) Distinct functions of alpha3 and alpha(v) integrin receptors in neuronal migration and laminar organization of the cerebral cortex. Neuron 22:277–289

Arikawa-Hirasawa E, Watanabe H, Takami H, Hassell JR, Yamada Y (1999) Perlecan is essential for cartilage and cephalic development. Nat Genet 23:354–358

Aumailley M, Pesch M, Tunggal L, Gaill F, Fassler R (2000) Altered synthesis of laminin 1 and absence of basement membrane component deposition in (beta)1 integrin-deficient embryoid bodies. J Cell Sci 113 Pt 2:259–268

Bader BL, Rayburn H, Crowley D, Hynes RO (1998) Extensive vasculogenesis, angiogenesis, and organogenesis precede lethality in mice lacking all alpha v integrins. Cell 95:507–519

Barres BA, Raff MC (1994) Control of oligodendrocyte number in the developing rat optic nerve. Neuron 12:935–942

Bokel C, Brown NH (2002) Integrins in development: moving on, responding to, and sticking to the extracellular matrix. Dev Cell 3:311–321

Brakebusch C, Fassler R (2003) The integrin-actin connection, an eternal love affair. EMBO J 22:2324–2333

Brakebusch C, Grose R, Quondamatteo F, Ramirez A, Jorcano JL, Pirro A, Svensson M, Herken R, Sasaki T, Timpl R, Werner S, Fassler R (2000) Skin and hair follicle integrity is crucially dependent on beta 1 integrin expression on keratinocytes. Embo J 19:3990–4003

Brandenberger R, Schmidt A, Linton J, Wang D, Backus C, Denda S, Müller U, Reichardt LF (2001) Identification and characterization of a novel extracellular matrix protein nephronectin that is associated with integrin alpha8beta1 in the embryonic kidney. J Cell Biol 154:447–458

Brooks PC, Montgomery AM, Rosenfeld M, Reisfeld RA, Hu T, Klier G, Cheresh DA (1994) Integrin alpha v beta 3 antagonists promote tumor regression by inducing apoptosis of angiogenic blood vessels. Cell 79:1157–1164

Brooks PC, Stromblad S, Klemke R, Visscher D, Sarkar FH, Cheresh DA (1995) Antiintegrin alpha v beta 3 blocks human breast cancer growth and angiogenesis in human skin. J Clin Invest 96:1815–1822

Brown EJ, Frazier WA (2001) Integrin-associated protein (CD47) and its ligands. Trends Cell Biol 11:130–135

Buonanno A, Fischbach GD (2001) Neuregulin and ErbB receptor signaling pathways in the nervous system. Curr Opin Neurobiol 11:287–296

Burden SJ (1998) The formation of neuromuscular synapses. Genes Dev 12:133–148

Burgess RW, Dickman DK, Nunez L, Glass DJ, Sanes JR (2002) Mapping sites responsible for interactions of agrin with neurons. J Neurochem 83:271–284

Burkin DJ, Kim JE, Gu M, Kaufman SJ (2000) Laminin and alpha7beta1 integrin regulate agrin-induced clustering of acetylcholine receptors. J Cell Sci 113 (Pt 16):2877–2886

Cachaco AS, Chuva de Sousa Lopes SM, Kuikman I, Bajanca F, Abe K, Baudoin C, Sonnenberg A, Mummery CL, Thorsteinsdottir S (2003) Knock-in of integrin beta 1D affects primary but not secondary myogenesis in mice. Development 130:1659–1671

Chan CS, Weeber EJ, Kurup S, Sweatt JD, Davis RL (2003) Integrin requirement for hippocampal synaptic plasticity and spatial memory. J Neurosci 23:7107–7116

Chang HP, Lindberg FP, Wang HL, Huang AM, Lee EH (1999) Impaired memory retention and decreased long-term potentiation in integrin-associated protein-deficient mice. Learn Mem 6:448–457

Chang HP, Ma YL, Wan FJ, Tsai LY, Lindberg FP, Lee EH (2001) Functional blocking of integrin-associated protein impairs memory retention and decreases glutamate release from the hippocampus. Neuroscience 102:289–296

Chavis P, Westbrook G (2001) Integrins mediate functional pre- and postsynaptic maturation at a hippocampal synapse. Nature 411:317–321

Clegg DO (2000) Novel roles for integrins in the nervous system. Mol Cell Biol Res Commun 3:1–7

Colognato H, Baron W, Avellana-Adalid V, Relvas JB, Baron-Van Evercooren A, Georges-Labouesse E, ffrench-Constant C (2002) CNS integrins switch growth factor signalling to promote target-dependent survival. Nat Cell Biol 4:833–841

Cosgrove D, Meehan DT, Grunkemeyer JA, Kornak JM, Sayers R, Hunter WJ, Samuelson GC (1996) Collagen COL4A3 knockout: a mouse model for autosomal Alport syndrome. Genes Dev 10:2981–2992

Costell M, Gustafsson E, Aszodi A, Morgelin M, Bloch W, Hunziker E, Addicks K, Timpl R, Fassler R (1999) Perlecan maintains the integrity of cartilage and some basement membranes. J Cell Biol 147:1109–1122

De Arcangelis A, Mark M, Kreidberg J, Sorokin L, Georges-Labouesse E (1999) Synergistic activities of alpha3 and alpha6 integrins are required during apical ectodermal ridge formation and organogenesis in the mouse. Development 126:3957–3968

DeChiara TM, Bowen DC, Valenzuela DM, Simmons MV, Poueymirou WT, Thomas S, Kinetz E, Compton DL, Rojas E, Park JS, Smith C, DiStefano PS, Glass DJ, Burden SJ, Yancopoulos GD (1996) The receptor tyrosine kinase MuSK is required for neuromuscular junction formation in vivo. Cell 85:501–512

Denzer AJ, Brandenberger R, Gesemann M, Chiquet M, Ruegg MA (1997) Agrin binds to the nerve-muscle basal lamina via laminin. J Cell Biol 137:671–683

DiPersio CM, Hodivala-Dilke KM, Jaenisch R, Kreidberg JA, Hynes RO (1997) Alpha3beta1 integrin is required for normal development of the epidermal basement membrane. J Cell Biol 137:729–742

DiPersio CM, van der Neut R, Georges-Labouesse E, Kreidberg JA, Sonnenberg A, Hynes RO (2000) Alpha3beta1 and alpha6beta4 integrin receptors for laminin-5 are not essential for epidermal morphogenesis and homeostasis during skin development. J Cell Sci 113:3051–3062

Dong L, Chen Y, Lewis M, Hsieh JC, Reing J, Chaillet JR, Howell CY, Melhem M, Inoue S, Kuszak JR, DeGeest K, Chung AE (2002) Neurologic defects and selective disruption of basement membranes in mice lacking entactin-1/nidogen-1. Lab Invest 82:1617–1630

Dowling J, Yu QC, Fuchs E (1996) Beta4 integrin is required for hemidesmosome formation, cell adhesion and cell survival. J Cell Biol 134:559–572

Dulabon L, Olson EC, Taglienti MG, Eisenhuth S, McGrath B, Walsh CA, Kreidberg JA, Anton ES (2000) Reelin binds alpha3beta1 integrin and inhibits neuronal migration. Neuron 27:33–44

Fassler R, Meyer M (1995) Consequences of lack of beta 1 integrin gene expression in mice. Genes Dev 9:1896–1908

Fassler R, Rohwedel J, Maltsev V, Bloch W, Lentini S, Guan K, Gullberg D, Hescheler J, Addicks K, Wobus AM (1996) Differentiation and integrity of cardiac muscle cells are impaired in the absence of beta 1 integrin. J Cell Sci 109 (Pt 13):2989–2999

Feltri ML, Graus Porta D, Previtali SC, Nodari A, Migliavacca B, Cassetti A, Littlewood-Evans A, Reichardt LF, Messing A, Quattrini A, Müller U, Wrabetz L (2002) Conditional disruption of beta 1 integrin in Schwann cells impedes interactions with axons. J Cell Biol 156:199–209

Frei R, Dowling J, Carenini S, Fuchs E, Martini R (1999) Myelin formation by Schwann cells in the absence of beta4 integrin. Glia 27:269–274

Friedlander M, Brooks PC, Shaffer RW, Kincaid CM, Varner JA, Cheresh DA (1995) Definition of two angiogenic pathways by distinct alpha v integrins. Science 270:1500–1502

Friedlander M, Theesfeld CL, Sugita M, Fruttiger M, Thomas MA, Chang S, Cheresh DA (1996) Involvement of integrins alpha v beta 3 and alpha v beta 5 in ocular neovascular diseases. Proc Natl Acad Sci U S A 93:9764–9769

Fuchs E, Raghavan S (2002) Getting under the skin of epidermal morphogenesis. Nat Rev Genet 3:199–209

Galileo DS, Majors J, Horwitz AF, Sanes JR (1992) Retrovirally introduced antisense integrin RNA inhibits neuroblast migration in vivo. Neuron 9:1117–1131

Gautam M, Noakes PG, Moscoso L, Rupp F, Scheller RH, Merlie JP, Sanes JR (1996) Defective neuromuscular synaptogenesis in agrin-deficient mutant mice. Cell 85:525–535

George EL, Georges-Labouesse EN, Patel-King RS, Rayburn H, Hynes RO (1993) Defects in mesoderm, neural tube and vascular development in mouse embryos lacking fibronectin. Development 119:1079–1091

Georges-Labouesse E, Messaddeq N, Yehia G, Cadalbert L, Dierich A, Le Meur M (1996a) Absence of integrin alpha 6 leads to epidermolysis bullosa and neonatal death in mice. Nat Genet 13:370–373

Georges-Labouesse E, Mark M, Messaddeq N, Gansmuller A (1998) Essential role of alpha 6 integrins in cortical and retinal lamination. Curr Biol 8:983–986

Georges-Labouesse EN, George EL, Rayburn H, Hynes RO (1996b) Mesodermal development in mouse embryos mutant for fibronectin. Dev Dyn 207:145–156

Giancotti FG (2003) A structural view of integrin activation and signaling. Dev Cell 4:149–151

Giancotti FG, Ruoslahti E (1999) Integrin signaling. Science 285:1028–1032

Goh KL, Yang JT, Hynes RO (1997) Mesodermal defects and cranial neural crest apoptosis in alpha5 integrin-null embryos. Development 124:4309–4319

Graus-Porta D, Blaess S, Senften M, Littlewood-Evans A, Damsky C, Huang Z, Orban P, Klein R, Schittny JC, Muller U (2001) Beta1-class integrins regulate the development of laminae and folia in the cerebral and cerebellar cortex. Neuron 31:367–379

Haack H, Hynes RO (2001) Integrin receptors are required for cell survival and proliferation during development of the peripheral glial lineage. Dev Biol 233:38–55

Halfter W, Dong S, Yip YP, Willem M, Mayer U (2002) A critical function of the pial basement membrane in cortical histogenesis. J Neurosci 22:6029–6040

Hammes HP, Brownlee M, Jonczyk A, Sutter A, Preissner KT (1996) Subcutaneous injection of a cyclic peptide antagonist of vitronectin receptor-type integrins inhibits retinal neovascularization. Nat Med 2:529–533

Hatten ME (1999) Central nervous system neuronal migration. Annu Rev Neurosci 22:511–539

Henry MD, Campbell KP (1999) Dystroglycan inside and out. Curr Opin Cell Biol 11:602–607

Hirsch E, Lohikangas L, Gullberg D, Johansson S, Fassler R (1998) Mouse myoblasts can fuse and form a normal sarcomere in the absence of beta1 integrin expression. J Cell Sci 111 (Pt 16):2397–2409

Hodivala-Dilke KM, McHugh KP, Tsakiris DA, Rayburn H, Crowley D, Ullman-Cullere M, Ross FP, Coller BS, Teitelbaum S, Hynes RO (1999) Beta3-integrin-deficient mice are a model for Glanzmann thrombasthenia showing placental defects and reduced survival. J Clin Invest 103:229–238

Huang X, Griffiths M, Wu J, Farese RV Jr, Sheppard D (2000) Normal development, wound healing, and adenovirus susceptibility in beta5-deficient mice. Mol Cell Biol 20:755–759

Huang XZ, Wu JF, Cass D, Erle DJ, Corry D, Young SG, Farese RV Jr, Sheppard D (1996) Inactivation of the integrin beta 6 subunit gene reveals a role of epithelial integrins in regulating inflammation in the lung and skin. J Cell Biol 133:921–928

Hynes RO (1992) Integrins: versatility, modulation, and signaling in cell adhesion. Cell 69:11–25

Hynes RO (2002a) A reevaluation of integrins as regulators of angiogenesis. Nat Med 8:918–921

Hynes RO (2002b) Integrins: bidirectional, allosteric signaling machines. Cell 110:673–687

Hynes RO, Zhao Q (2000) The evolution of cell adhesion. J Cell Biol 150:F89–96

Kaji K, Oda S, Shikano T, Ohnuki T, Uematsu Y, Sakagami J, Tada N, Miyazaki S, Kudo A (2000) The gamete fusion process is defective in eggs of Cd9-deficient mice. Nat Genet 24:279–282

Kreidberg JA, Donovan MJ, Goldstein SL, Rennke H, Shepherd K, Jones RC, Jaenisch R (1996) Alpha 3 beta 1 integrin has a crucial role in kidney and lung organogenesis. Development 122:3537–3547

Kronenberg HM (2003) Developmental regulation of the growth plate. Nature 423:332–336

Kwee L, Baldwin HS, Shen HM, Stewart CL, Buck C, Buck CA, Labow MA (1995) Defective development of the embryonic and extraembryonic circulatory systems in vascular cell adhesion molecule (VCAM-1) deficient mice. Development 121:489–503

Legan PK, Lukashkina VA, Goodyear RJ, Kossi M, Russell IJ, Richardson GP (2000) A targeted deletion in alpha-tectorin reveals that the tectorial membrane is required for the gain and timing of cochlear feedback. Neuron 28:273–285

Li S, Harrison D, Carbonetto S, Fassler R, Smyth N, Edgar D, Yurchenco PD (2002) Matrix assembly, regulation, and survival functions of laminin and its receptors in embryonic stem cell differentiation. J Cell Biol 157:1279–1290

Li S, Edgar D, Fassler R, Wadsworth W, Yurchenco PD (2003) The role of laminin in embryonic cell polarization and tissue organization. Dev Cell 4:613–624

Liaw L, Birk DE, Ballas CB, Whitsitt JS, Davidson JM, Hogan BL (1998) Altered wound healing in mice lacking a functional osteopontin gene (spp1). J Clin Invest 101:1468–1478

Liddington RC, Ginsberg MH (2002) Integrin activation takes shape. J Cell Biol 158:833–839

Lindbom L, Werr J (2002) Integrin-dependent neutrophil migration in extravascular tissue. Semin Immunol 14:115–121

Littlewood Evans A, Müller U (2000) Stereocilia defects in the sensory hair cells of the inner ear in mice deficient in integrin alpha8beta1. Nat Genet 24:424–428

Lynch ED, Lee MK, Morrow JE, Welcsh PL, Leon PE, King MC (1997) Nonsyndromic deafness DFNA1 associated with mutation of a human homolog of the Drosophila gene diaphanous. Science 278:1315–1318

Magdaleno S, Keshvara L, Curran T (2002) Rescue of ataxia and preplate splitting by ectopic expression of Reelin in reeler mice. Neuron 33:573–586

Martin PT, Sanes JR (1997) Integrins mediate adhesion to agrin and modulate agrin signaling. Development 124:3909–3917

Mayer U (2003) Integrins: redundant or important players in skeletal muscle? J Biol Chem 278:14587–14590

Mayer U, Saher G, Fassler R, Bornemann A, Echtermeyer F, von der Mark H, Miosge N, Poschl E, von der Mark K (1997) Absence of integrin alpha 7 causes a novel form of muscular dystrophy. Nat Genet 17:318–323

McCarty JH, Monahan-Earley RA, Brown LF, Keller M, Gerhardt H, Rubin K, Shani M, Dvorak HF, Wolburg H, Bader BL, Dvorak AM, Hynes RO (2002) Defective associations between blood vessels and brain parenchyma lead to cerebral hemorrhage in mice lacking alphav integrins. Mol Cell Biol 22:7667–7677

McHugh KP, Hodivala-Dilke K, Zheng MH, Namba N, Lam J, Novack D, Feng X, Ross FP, Hynes RO, Teitelbaum SL (2000) Mice lacking beta3 integrins are osteosclerotic because of dysfunctional osteoclasts. J Clin Invest 105:433–440

Menko AS, Boettiger D (1987) Occupation of the extracellular matrix receptor, integrin, is a control point for myogenic differentiation. Cell 51:51–57

Michele DE, Campbell KP (2003) Dystrophin-glycoprotein complex: post-translational processing and dystroglycan function. J Biol Chem 278:15457–15460

Miner JH, Li C (2000) Defective glomerulogenesis in the absence of laminin alpha5 demonstrates a developmental role for the kidney glomerular basement membrane. Dev Biol 217:278–289

Miner JH, Sanes JR (1996) Molecular and functional defects in kidneys of mice lacking collagen alpha 3(IV): implications for Alport syndrome. J Cell Biol 135:1403–1413

Miner JH, Cunningham J, Sanes JR (1998) Roles for laminin in embryogenesis: exencephaly, syndactyly, and placentopathy in mice lacking the laminin alpha5 chain. J Cell Biol 143:1713–1723

Miyagoe Y, Hanaoka K, Nonaka I, Hayasaka M, Nabeshima Y, Arahata K, Takeda S (1997) Laminin alpha2 chain-null mutant mice by targeted disruption of the Lama2 gene: a new model of merosin (laminin 2)-deficient congenital muscular dystrophy. FEBS Lett 415:33–39

Miyamoto YJ, Andruss BF, Mitchell JS, Billard MJ, McIntyre BW (2003) Diverse roles of integrins in human T lymphocyte biology. Immunol Res 27:71–84

Montanaro F, Gee SH, Jacobson C, Lindenbaum MH, Froehner SC, Carbonetto S (1998) Laminin and alpha-dystroglycan mediate acetylcholine receptor aggregation via a MuSK-independent pathway. J Neurosci 18:1250–1260

Müller U (2003) Cell adhesion molecules and human disorders. In: Encyclopedia of the Human Genome: Nature Publishing Group

Müller U, Brandli AW (1999) Cell adhesion molecules and extracellular-matrix constituents in kidney development and disease. J Cell Sci 112 (Pt 22):3855–3867

Müller U, Wang D, Denda S, Meneses JJ, Pedersen RA, Reichardt LF (1997) Integrin alpha8beta1 is critically important for epithelial-mesenchymal interactions during kidney morphogenesis. Cell 88:603–613

Murgia C, Blaikie P, Kim N, Dans M, Petrie HT, Giancotti FG (1998) Cell cycle and adhesion defects in mice carrying a targeted deletion of the integrin beta4 cytoplasmic domain. Embo J 17:3940–3951

Murray P, Edgar D (2000) Regulation of programmed cell death by basement membranes in embryonic development. J Cell Biol 150:1215–1221

Nadarajah B, Parnavelas JG (2002) Modes of neuronal migration in the developing cerebral cortex. Nat Rev Neurosci 3:423–432

Nieswandt B, Watson SP (2003) Platelet-collagen interaction: is GPVI the central receptor? Blood 102:449–461

Nievers MG, Schaapveld RQ, Sonnenberg A (1999) Biology and function of hemidesmosomes. Matrix Biol 18:5–17

Noakes PG, Gautam M, Mudd J, Sanes JR, Merlie JP (1995a) Aberrant differentiation of neuromuscular junctions in mice lacking s-laminin/laminin beta 2. Nature 374:258–262

Noakes PG, Miner JH, Gautam M, Cunningham JM, Sanes JR, Merlie JP (1995b) The renal glomerulus of mice lacking s-laminin/laminin beta 2: nephrosis despite molecular compensation by laminin beta 1. Nat Genet 10:400–406

Patton BL, Miner JH, Chiu AY, Sanes JR (1997) Distribution and function of laminins in the neuromuscular system of developing, adult, and mutant mice. J Cell Biol 139:1507–1521

Patton BL, Cunningham JM, Thyboll J, Kortesmaa J, Westerblad H, Edstrom L, Tryggvason K, Sanes JR (2001) Properly formed but improperly localized synaptic specializations in the absence of laminin alpha4. Nat Neurosci 4:597–604

Porter BE, Weis J, Sanes JR (1995) A motoneuron-selective stop signal in the synaptic protein S-laminin. Neuron 14:549–559

Pozzi A, Moberg PE, Miles LA, Wagner S, Soloway P, Gardner HA (2000) Elevated matrix metalloprotease and angiostatin levels in integrin alpha 1 knockout mice cause reduced tumor vascularization. Proc Natl Acad Sci U S A 97:2202–2207

Previtali SC, Feltri ML, Archelos JJ, Quattrini A, Wrabetz L, Hartung H (2001) Role of integrins in the peripheral nervous system. Prog Neurobiol 64:35–49

Raghavan S, Bauer C, Mundschau G, Li Q, Fuchs E (2000) Conditional ablation of beta1 integrin in skin. Severe defects in epidermal proliferation, basement membrane formation, and hair follicle invagination. J Cell Biol 150:1149–1160

Reynolds LE, Wyder L, Lively JC, Taverna D, Robinson SD, Huang X, Sheppard D, Hynes RO, Hodivala-Dilke KM (2002) Enhanced pathological angiogenesis in mice lacking beta3 integrin or beta3 and beta5 integrins. Nat Med 8:27–34

Rohwedel J, Guan K, Zuschratter W, Jin S, Ahnert-Hilger G, Furst D, Fassler R, Wobus AM (1998) Loss of beta1 integrin function results in a retardation of myogenic, but an acceleration of neuronal, differentiation of embryonic stem cells in vitro. Dev Biol 201:167–184

Rosen GD, Sanes JR, LaChance R, Cunningham JM, Roman J, Dean DC (1992) Roles for the integrin VLA-4 and its counter receptor VCAM-1 in myogenesis. Cell 69:1107–1119

Sakai T, Larsen M, Yamada KM (2003) Fibronectin requirement in branching morphogenesis. Nature 423:876–881

Sanes JR (2003) The basement membrane/basal lamina of skeletal muscle. J Biol Chem 278:12601–12604

Sanes JR, Lichtman JW (2001) Induction, assembly, maturation and maintenance of a postsynaptic apparatus. Nat Rev Neurosci 2:791–805

Sastry SK, Lakonishok M, Thomas DA, Muschler J, Horwitz AF (1996) Integrin alpha subunit ratios, cytoplasmic domains, and growth factor synergy regulate muscle proliferation and differentiation. J Cell Biol 133:169–184

Saxen L (1987) Organogenesis of the kidney. Cambridge University Press, Cambridge

Scherer SS (2002) Myelination: some receptors required. J Cell Biol 156:13–15

Schwander M, Leu M, Stumm M, Dorchies OM, Ruegg UT, Schittny J, Muller U (2003) Beta1 integrins regulate myoblast fusion and sarcomere assembly. Dev Cell 4:673–685

Schwartz MA, Ginsberg MH (2002) Networks and crosstalk: integrin signalling spreads. Nat Cell Biol 4:E65–68

Sengbusch JK, He W, Pinco KA, Yang JT (2002) Dual functions of [alpha]4[beta]1 integrin in epicardial development: initial migration and long-term attachment. J Cell Biol 157:873–882

Siemens J, Littlewood Evans A, Senften M, Müller U (2001) Genes, deafness, and balance disorders. Gene Funct Dis 2:76–82

Simmler MC, Cohen-Salmon M, El-Amraoui A, Guillaud L, Benichou JC, Petit C, Panthier JJ (2000a) Targeted disruption of otog results in deafness and severe imbalance. Nat Genet 24:139–143

Simmler MC, Zwaenepoel II, Verpy E, Guillaud L, Elbaz C, Petit C, Panthier JJ (2000b) Twister mutant mice are defective for otogelin, a component specific to inner ear acellular membranes. Mamm Genome 11:961–966

Sims TN, Dustin ML (2002) The immunological synapse: integrins take the stage. Immunol Rev 186:100–117

Smyth N, Vatansever HS, Murray P, Meyer M, Frie C, Paulsson M, Edgar D (1999) Absence of basement membranes after targeting the LAMC1 gene results in embryonic lethality due to failure of endoderm differentiation. J Cell Biol 144:151–160

Stephens LE, Sutherland AE, Klimanskaya IV, Andrieux A, Meneses J, Pedersen RA, Damsky CH (1995) Deletion of beta 1 integrins in mice results in inner cell mass failure and peri-implantation lethality. Genes Dev 9:1883–1895

Stupack DG, Puente XS, Boutsaboualoy S, Storgard CM, Cheresh DA (2001) Apoptosis of adherent cells by recruitment of caspase-8 to unligated integrins. J Cell Biol 155:459–470

Sugiyama JE, Glass DJ, Yancopoulos GD, Hall ZW (1997) Laminin-induced acetylcholine receptor clustering: an alternative pathway. J Cell Biol 139:181–191

Tachibana I, Hemler ME (1999) Role of transmembrane 4 superfamily (TM4SF) proteins CD9 and CD81 in muscle cell fusion and myotube maintenance. J Cell Biol 146:893–904

Taverna D, Disatnik MH, Rayburn H, Bronson RT, Yang J, Rando TA, Hynes RO (1998) Dystrophic muscle in mice chimeric for expression of alpha5 integrin. J Cell Biol 143:849–859

Timpl R, Brown JC (1996) Supramolecular assembly of basement membranes. Bioessays 18:123–132

Vachon PH, Loechel F, Xu H, Wewer UM, Engvall E (1996) Merosin and laminin in myogenesis; specific requirement for merosin in myotube stability and survival. J Cell Biol 134:1483–1497

Vainio S, Müller U (1997) Inductive tissue interactions, cell signaling, and the control of kidney organogenesis. Cell 90:975–978

van der Flier A, Gaspar AC, Thorsteinsdottir S, Baudoin C, Groeneveld E, Mummery CL, Sonnenberg A (1997) Spatial and temporal expression of the beta1D integrin during mouse development. Dev Dyn 210:472–486

van der Neut R, Krimpenfort P, Calafat J, Niessen CM, Sonnenberg A (1996) Epithelial detachment due to absence of hemidesmosomes in integrin beta 4 null mice. Nat Genet 13:366–369

Watanabe N, Kato T, Fujita A, Ishizaki T, Narumiya S (1999) Cooperation between mDia1 and ROCK in Rho-induced actin reorganization. Nat Cell Biol 1:136–143

Watt FM (2002) Role of integrins in regulating epidermal adhesion, growth and differentiation. Embo J 21:3919–3926

Willem M, Miosge N, Halfter W, Smyth N, Jannetti I, Burghart E, Timpl R, Mayer U (2002) Specific ablation of the nidogen-binding site in the laminin gamma1 chain interferes with kidney and lung development. Development 129:2711–2722

Yang JT, Rayburn H, Hynes RO (1993) Embryonic mesodermal defects in alpha 5 integrin-deficient mice. Development 119:1093–1105

Yang JT, Rayburn H, Hynes RO (1995) Cell adhesion events mediated by alpha 4 integrins are essential in placental and cardiac development. Development 121:549–560

Yang JT, Rando TA, Mohler WA, Rayburn H, Blau HM, Hynes RO (1996) Genetic analysis of alpha 4 integrin functions in the development of mouse skeletal muscle. J Cell Biol 135:829–835

Yang JT, Bader BL, Kreidberg JA, Ullman-Cullere M, Trevithick JE, Hynes RO (1999) Overlapping and independent functions of fibronectin receptor integrins in early mesodermal development. Dev Biol 215:264–277

Yurchenco PD, O'Rear JJ (1994) Basement membrane assembly. Methods Enzymol 245:489–518

Zhang Z, Galileo DS (1998) Retroviral transfer of antisense integrin alpha6 or alpha8 sequences results in laminar redistribution or clonal cell death in developing brain. J Neurosci 18:6928–6938

Zhu J, Motejlek K, Wang D, Zang K, Schmidt A, Reichardt LF (2002) Beta8 integrins are required for vascular morphogenesis in mouse embryos. Development 129:2891–2903

Hemidesmosomes: Molecular Organization and Their Importance for Cell Adhesion and Disease

J. Koster[1] · L. Borradori[2] · A. Sonnenberg[1] (✉)

[1] Division of Cell Biology, The Netherlands Cancer Institute, Plesmanlaan 121, 1066 CX, Amsterdam, The Netherlands
a.sonnenberg@nki.nl

[2] Department of Dermatology, University Medical Hospital, Rue Micheli du Crest 22, 1211 Geneva, Switzerland

1	Hemidesmosomes	245
1.1	General Introduction	245
1.2	The $\alpha 6\beta 4$ Integrin	246
1.3	BP180	250
1.4	CD151	251
1.5	Keratins	252
1.6	Laminins	252
2	Plakins	253
2.1	General Introduction	253
2.2	Plectin	255
2.3	BP230	258
2.4	Desmoplakin	259
2.5	Microtubule–Actin Cross-Linking Factor	260
2.6	Envoplakin	260
2.7	Periplakin	261
2.8	Epiplakin	261
3	Molecular Interactions Involved in Hemidesmosome Assembly	262
3.1	The Role of $\beta 4$ in HD Assembly	262
3.2	The Role of BP180 in HD Assembly	263
3.3	The Linkage of Keratins to Hemidesmosomes	264
4	A Hierarchical Interaction Model for the Assembly of HDs	265
5	General Conclusions and Future Perspectives	266
References		267

Abstract In the skin, basal epithelial cells constantly divide to renew the epidermis. The newly formed epithelial cells then differentiate in a process called keratinization, ultimately leading to the death of these cells and a pile-up of cell material containing vast amounts of keratins. The basal keratinocytes in skin are attached to their underlying basement membrane via specialized adhesion complexes termed hemidesmosomes (HDs). These complexes ascertain stable adhesion of the epidermis to the dermis, and mutations in components of these complexes often result in tissue fragility and blistering of the skin. In this review, we will describe the various hemidesmosomal proteins in detail as well as, briefly, the protein families to which they belong. Specifically, we will report the protein–protein interactions involved in the assembly of hemidesmosomes and their molecular organization. Some signaling pathways involving primarily the $\alpha 6\beta 4$ integrin will be discussed, since they appear to profoundly modulate the assembly and function of hemidesmosomes. Furthermore, the importance of these hemidesmosomal components for the maintenance of tissue homeostasis and their involvement in various clinical disorders will be emphasized. Finally, we will present a model for the assembly of HDs, based on our present knowledge.

Keywords Hemidesmosome · Integrin · Laminin-5 · Plakin · Bullous pemphigoid · Intermediate filament

Abbreviations

ABD	Actin-binding domain
ACF7	Actin cross-linking factor 7
ADAM	A disintegrin and metalloprotease
BP180	Bullous pemphigoid 180
BP230	Bullous pemphigoid 230
BPAG	Bullous pemphigoid antigen
CC-ROD	Coiled coil rod domain
CH	Calponin homology
DP	Desmoplakin
DPI	Desmoplakin 1
DPII	Desmoplakin 2
DRG	Dorsal root ganglion
EBS	Epidermolysis bullosa simplex
EBS-MD	Epidermolysis bullosa simplex associated with muscular dystrophy
EGF	Epidermal growth factor
ERBIN	ERBB2 interacting protein
ERB-B2	Erythroblastic leukemia viral oncogene homolog 2
F-actin	Filamentous actin
FNIII	Fibronectin type III
GABEB	Generalized atrophic benign epidermolysis bullosa
GAS2	Growth arrest specific protein 2
HD	Hemidesmosome
IF	Intermediate filament
IFAP300	Intermediate filament-associated protein of 300 kDa
IFBD	Intermediate filament-binding domain
JEB	Junctional epidermolysis bullosa
kDa	Kilodalton
LAD	Linear IgA bullous dermatosis
LAP	Leucine-rich repeats and PDZ domain
MACF	Microtubule–actin cross-linking factor
MARCO	Macrophage receptor with collagenous structure
mRNA	Messenger RNA

MSR1R	Macrophage scavenger receptor 1R
MTBD	Microtubule-binding domain
nm	Nanometer
PA-JEB	Pyloric atresia with junctional epidermolysis bullosa
PI3K	Phosphoinositide-3-OH kinase
PKB	Protein kinase B
SH3	Src homology domain 3
SOS	Son of sevenless
SR	Spectrin repeat
SR-ROD	Spectrin repeat rod domain
TACE	Tumor necrosis factor-α-converting enzyme

1
Hemidesmosomes

1.1
General Introduction

Hemidesmosomes (HDs) are multiprotein complexes that mediate firm adhesion of epithelial cells to the underlying basement membrane via linkage of the intracellular cytoskeleton with extracellular matrix proteins. These multiprotein complexes determine cell–stromal coherence, providing cells with cues critical for their tissue architecture, spatial organization, and polarization (Green and Jones 1996; Gumbiner 1996; Borradori and Sonnenberg 1999). HDs are found in basal cells of stratified- and pseudostratified epithelia, complex epithelia, as well as in myoepithelial cells of glandular epithelia.

Ultrastructurally, HDs appear as small electron-dense structures associated with the plasma membrane at the basal side of keratinocytes (Green and Jones 1996; Borradori and Sonnenberg 1999). They have a tripartite structure consisting of an inner plaque, which serves as the anchorage site for the keratin filaments, and an outer plaque associated with the plasma membrane, separated by an electro-lucent zone. Furthermore, a sub-basal dense plate is present immediately beneath the outer plaque in the basement membrane (Eady 1994). While the inner plaque contains BP230 (BPAG1-e) and plectin, the outer plaque consists of $\alpha 6\beta 4$, BP180, and CD151 (Nievers et al. 1999; Sterk et al. 2000) (Fig. 1). Together with the extracellular matrix protein laminin-5 and the cytoskeleton-associated keratins, these proteins constitute a functional unit, referred to as the hemidesmosomal adhesion complex (Burgeson and Christiano 1997).

In this review we will describe the various hemidesmosomal proteins in detail, as well as, briefly, the protein families to which these components belong. Specifically, we will report the protein–protein interactions involved in

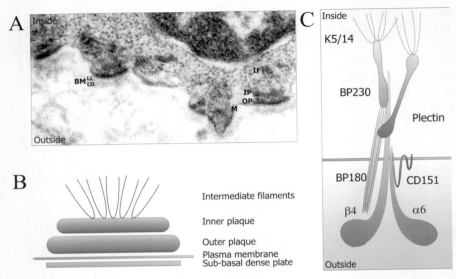

Fig. 1A–C. Three aspects of hemidesmosomes. **A** An electron micrograph of a skin sample, containing hemidesmosomes. *M*, membrane; *OP*, outer plaque; *IP*, inner plaque; *IF*, intermediate filaments; *BM*, basement membrane; *LL*, lamina lucida; *LD*, lamina densa. **B** A schematic representation of an HD as seen with an electron microscope. **C** Schematic representation of the components that reconstitute the hemidesmosome and a general overview of their connections to each other

the assembly of HDs and their molecular organization. Some signaling pathways involving primarily the $\alpha 6\beta 4$ integrin will be discussed, since they appear to profoundly modulate the assembly and function of HDs. Furthermore, the importance of these hemidesmosomal components for the maintenance of tissue homeostasis and their involvement in various clinical disorders will be emphasized. Finally, we will present a model for the assembly of HDs, based on our latest knowledge.

1.2
The $\alpha 6\beta 4$ Integrin

The $\alpha 6\beta 4$ integrin is not only present in tissues that contain HDs (Kajiji et al. 1989; Stepp et al. 1990; Sonnenberg et al. 1991), but also in various cells that lack discernible HDs. These include Schwann cells and perineural fibroblasts of peripheral nerves, endothelial cells of different types of blood vessels, and immature thymocytes (Sonnenberg et al. 1990; Kennel et al. 1992; Wadsworth et al. 1992; Niessen et al. 1994a). The $\alpha 6\beta 4$ integrin can bind several laminins, including laminin-5 (Niessen et al. 1994b), which is a major component of the epidermal basement membranes (Carter et al. 1991; Rousselle et al. 1991). It is primarily involved in maintaining adhesion of

Table 1. Involvement of hemidesmosomal components in human diseases

Morphological structure	Protein	Disease		Animal model
		Autoimmune	Inherited	
Intermediate filaments	Keratin 5 and 14	–	EBS	Yes
Hemidesmosomes	BP230	Bullous pemphigoid	–	Yes
	Plectin	Pemphigoid like disease	EBS-MD	Yes
	α6 Integrin	–	PA-JEB	Yes
	β4 Integrin	Cicatricial pemphigoid	PA-JEB	Yes
	BP180	Bullous pemphigoid Cicatricial pemphigoid Gestational pemphigoid	GABEB	No
Anchoring filaments	Laminin-5	Cicatricial pemphigoid	JEB	Yes

EBS, epidermolysis bullosa simplex; GABEB, generalized atrophic benign epidermolysis bullosa; JEB, junctional epidermolysis bullosa; MD, muscular dystrophy; PA, pyloric atresia.
Autoimmune blistering diseases are associated with autoantibodies directed against constituents of the hemidesmosomal adhesion complex, while mutations result in inherited bullous disorders with often a similar phenotype. In autoimmune blistering diseases, autoantibodies might be directed against additional target antigens.

cells to substrates in epithelia, especially of keratinocytes of the epidermal basal layer, as attested by the following observations. In humans, pathogenic mutations in the genes encoding α6 or β4 (Vidal et al. 1995; Brown et al. 1996; Niessen et al. 1996; Pulkkinen et al. 1997a; Ruzzi et al. 1997; Takizawa et al. 1997) cause junctional epidermolysis bullosa (JEB), an inherited skin disorder, characterized by mechanical fragility and blistering of the skin associated with dermo-epidermal separation and the formation of rudimentary HDs (see also Table 1) (McMillan et al. 1998). Furthermore, in null-mutant mice for the α6 or β4 integrin subunit (Dowling et al. 1996; Georges-Labouesse et al. 1996; van der Neut et al. 1996) widespread epidermal detachments are observed resulting in death of the mice shortly after birth. In the α6 and β4 null-mutant mice, the skin is typically detached at the level of the basement membrane, although separations through the cell, at the level of HDs, have also been demonstrated (van der Neut et al. 1996). These data emphasize the important role of α6β4 as a transmembrane linker that connects the extracellular matrix with the cytoskeleton (Borradori and Sonnenberg 1999; Green and Jones 1996).

The α6 subunit contains a large extracellular domain and a short cytoplasmic domain (Hogervorst et al. 1991; Tamura et al. 1991). The N-terminus of the extracellular domain contains 7 homologous repeat domains, three of which contain putative divalent cation binding sites that are important for integrin function. The extracellular domain of the integrin α6 subunit is cleaved near its C-terminal end, resulting in a disulfide-linked heavy and light chain. The cytoplasmic domain of α6, like all α subunits, contains the conserved GFFKR sequence proximal to the transmembrane region. Two cytoplasmic variants have been described of α6, α6A and α6B (Cooper et al. 1991; Hogervorst et al. 1991, 1993). These variants differ in a sequence after the GFFKR sequence and their expression is tissue-specific. In skin, only the α6A variant is expressed. However, there is no apparent difference between the roles of α6A and α6B in the formation of HDs, since in α6A knockout mice, which express α6B at sites where normally α6A is expressed, HDs appear to be normal (Gimond et al. 1998).

The extracellular domain of the β4 integrin subunit contains four cysteine-rich repeats characteristic of integrin β subunits. These cysteine repeats form intramolecular disulfide bridges, thereby shaping the three-dimensional structure of the extracellular domain. It is likely that the overall folding of the extracellular part of α6β4 resembles that of αvβ3. Of this integrin, the crystal structure of the extracellular part, bound or not bound to ligand (RGD-peptide), has been determined (Xiong et al. 2002). Interestingly, the N-terminal regions of the α and β subunit strikingly resembled Gβ and Gα subunits, respectively, of G proteins, and even contact each other in similar ways (Xiong et al. 2002). Upon binding to its ligand, αvβ3 undergoes multiple conformational changes, leading to an active integrin (Xiong et al. 2002).

Whereas the cytoplasmic domain of other β subunits contains approximately 50 amino acids, that of the β4 subunit consists of over 1,000 residues (Hogervorst et al. 1990; Suzuki and Naitoh 1990). This large cytoplasmic domain is essential for the role of β4 in the assembly of HDs. Proximal to the transmembrane region of β4 there are six cysteine residues that could be palmitoylated. A *Calx-β*, Ca^{2+}-binding motif is present at a distance of 266 amino acids from the transmembrane region (Schwarz et al 1997). Further towards the C-terminus of the cytoplasmic domain, β4 contains two pairs of fibronectin type III repeats (FNIII) that are separated by a connecting segment. These FNIII repeats are involved in multiple protein–protein interactions (see Sect. 20). The three-dimensional folding of the first pair of FNIII repeats of the β4 integrin subunit has been determined (de Pereda et al. 1999), and has contributed to our understanding of the clinical phenotypes observed in JEB patients carrying pathogenic mutations in the β4 gene (Koster et al. 2001).

In addition to the most canonical β4A subunit, at least 5 cytoplasmic variants generated by alternative mRNA splicing have been identified. The β4B variant contains an insertion of 53 amino acids towards the end of the connecting segment. The mRNA encoding this latter variant has been detected in placenta, peripheral nerves, and spleen (Hogervorst et al. 1990; Kennel et al. 1993; Niessen et al. 1994a). β4C contains an insertion of 70 amino acids in the cytoplasmic tail, and the mRNA for this variant is present in a number of carcinomas (Tamura et al. 1990). In β4D, identified in a colon carcinoma cell line, there is a deletion of seven amino acids in the fourth FNIII repeat (Clarke et al. 1994). Finally, the RNA encoding β4E contains an insertion of 37 bases in the cytoplasmic domain, resulting in a frameshift followed by a premature stop codon: it is found in a large number of tissues including the epidermis (van Leusden et al. 1997). It is likely that these different variants have distinct binding activities and specificity for proteins they may interact with, but their significance remains to be elucidated.

Although the involvement of $\alpha 6\beta 4$ in the organization and function of HDs has been well established, recent work has revealed novel functions of $\alpha 6\beta 4$ in the control of cell growth, migration and survival of epithelial- and epithelium-derived tumor cells.

Activation of the $\alpha 6\beta 4$ integrin in response to clustering or cell adhesion results in the phosphorylation of $\beta 4$ and the subsequent recruitment of the adapter proteins Shc and Grb2. In turn, Grb2 associates with the exchange factor mSOS, leading to the activation of the Ras–Erk pathways and subsequent growth (Mainiero et al. 1995). In addition, both tyrosine and serine phosphorylation of $\beta 4$ occurs as a result of the treatment of epithelial cells with epidermal growth factor (EGF). This has been associated with the disassembly of HDs and a reduction of adhesion of the cell, but is not associated with the recruitment of Shc (Mainiero et al. 1996; Rabinovitz et al. 1999).

Furthermore, through its ability to associate with growth factor receptors, $\alpha 6\beta 4$ can modulate their activity and have an impact on cell migration and tumorigenesis. For example, it has been shown that the expression of $\beta 4$ promotes invasiveness of NIH3T3 cells transformed by the Erb-B2 oncogene (Gambaletta et al. 2000). This effect is dependent on the activation of phosphoinositide-3-OH kinase (PI3 K) and on a specific region of the $\beta 4$ cytoplasmic domain (amino acids 854–1183). It is noteworthy that both $\beta 4$ and Erb-B2 can bind to ERBIN, a novel protein belonging to the LAP (for leucine-rich repeats and PDZ domain) protein family (Favre et al. 2001). ERBIN, which is also able to bind to BP230, may contribute to both HD assembly and Erb-B2 receptor signaling. Furthermore, an association between $\alpha 6\beta 4$ and the hepatocyte growth factor (HGF) receptor, Met, has been demonstrated (Trusolino et al. 2001). Specifically, it was shown that in carcinoma cells, signaling by the Met receptor is dependent on $\alpha 6\beta 4$. Only when $\alpha 6\beta 4$ was present could signaling via the Met receptor occur and could tumor cells acquire an invasive phenotype. Similarly, oncogenic transformation of keratinocytes by Ras and IκBα [an inhibitor of nuclear factor (NF)-κB] was shown to be dependent on $\alpha 6\beta 4$ (Dajee et al. 2003).

The effect of $\alpha 6\beta 4$ on cell survival is mediated through the activation of the protein kinase B (PKB)/Akt kinase pathway (Bachelder et al. 1999a). Interestingly, the introduction of $\alpha 6\beta 4$ in carcinomas can also result in the induction of apoptosis, but only when wild-type p53 is present (Bachelder et al. 1999b). Recently, it was shown that ligation of $\alpha 6\beta 4$ and tissue polarity can support cell survival of normal and malignant mammary epithelia through signaling via NF-κB (Weaver et al. 2002).

1.3
BP180

BP180, also known as type XVII collagen, was initially identified as one of the two major target antigens of an autoimmune subepidermal blistering disorder of the skin and mucosae: bullous pemphigoid (Giudice et al. 1992). BP180 is a type II transmembrane protein with its C-terminus located extracellularly. Its N-terminal end resides in the cytoplasm, where it contributes to the formation of the cytoplasmic plaque of HDs (Hopkinson et al. 1992). BP180 is found in stratified, pseudo-stratified, and transitional epithelia.

BP180, collagen XIII, and collagen XXV represent a separate cluster in the collagen family. All three are non-fibrillar collagens located at the cell surface, and they possess a transmembrane domain (Pihlajaniemi and Rehn 1995; Hagg et al. 1998; Snellman et al. 2000; Hashimoto et al. 2002). Although not included in the collagen family, the domain structure of ectodysplasin A, MSR1R (macrophage scavenger receptor 1R), and macrophage receptor with collagenous structure (MARCO) protein is similar, including a collagenous extracellular domain (Krieger 1992; Kere et al. 1996; Ferguson et al. 1997; Elomaa et al. 1998). Interestingly, whereas BP180 is associated with HDs, collagen XIII is found in focal adhesions at the end of actin stress fibers (Hagg et al. 2001).

The extracellular domain of BP180 consists of a series of 15 collagen domains, separated by non-collagenous regions that together constitute a 120-kDa collagenous fragment (Giudice et al. 1992; Li et al. 1993). Results of chemical cross-linking and sedimentation experiments indicate that this collagenous fragment forms a homotrimeric complex (Balding et al. 1997). By rotary shadowing of purified BP180, three distinct regions were recognized: a cytoplasmic globular head domain, a central rod corresponding to the largest collagen repeat (COL15), and a flexible tail consisting of the remaining 14 collagen repeats interrupted by non-collagenous sequences (Hirako et al. 1996). The 500-amino acid intracellular domain of BP180 has no similarity to other proteins. It contains four tandemly arranged 24-amino acid repeats and, close to the membrane, four cysteine residues which may be subject to palmitoylation (Giudice et al. 1992; Hopkinson et al. 1992). The extracellular domain may undergo proteolytic processing resulting in the formation of a 120-kDa fragment, which is incorporated into the basement membrane and may have cell adhesion properties (Tasanen et al. 2000). Recently the enzymes involved in the proteolytic cleavage of the ectodomain of BP180, ADAM-9, ADAM-10, and ADAM-17 (TACE), have been identified (Franzke et al. 2002). Patients suffering from the autoimmune blistering disease linear IgA bullous dermatosis (LAD) characteristically produce IgA autoantibodies directed against the 120-kDa cleaved extracellular domain of BP180 (Roh et al. 2000). The large collagenous extracellular domain serves as a cell surface receptor for extracellular matrix proteins. Although little is known about the ligand of BP180, there are indications that it might be laminin-5 (Reddy et al. 1998).

The role of BP180 in the formation of HDs and in promoting dermo-epidermal cohesion is best attested by clinical observations. Defective expression of BP180 due to mutations in the BP180 gene (COLXVIIA gene) causes a distinct form of non-lethal JEB, previously known as generalized atrophic benign epidermolysis bullosa (GABEB) (see also Table 1). The number of HDs in the keratinocytes from these JEB patients is reduced, and they appear to be hypoplastic (Jonkman et al. 1995; McGrath et al. 1995). Furthermore, in bullous pemphigoid, the presence of autoantibodies directed against extracellular epitopes of BP180 causes tissue damage, leading to the formation of subepidermal blisters (Giudice et al. 1993; Liu et al. 1993).

1.4
CD151

CD151 belongs to the tetraspanin superfamily of proteins. Like the more than 20 highly conserved members of this family, CD151 spans the plasma membrane four times (Wright and Tomlinson 1994; Maecker et al. 1997; Hemler 1998). Tetraspanin proteins all contain one small and one large extracellular loop with short cytoplasmic N- and C-terminal domains. Tetraspanins are characterized by the presence of a conserved CCG motif in the large extracellular loop and a single cysteine residue at a defined distance from the transmembrane regions (Maecker et al. 1997).

For CD81, the crystal structure of the large extracellular loop has been determined. The loop of CD81 contains 5 α-helices that fold into a mushroom-like structure (Kitadokoro et al. 2001). Sequence analysis of tetraspanin proteins has indicated that the key structural features and the fold of the extracellular domain are conserved. Tetraspanins can self-aggregate and form multimeric clusters with other transmembrane components. Their large extracellular loop is involved in binding to a wide variety of proteins, among which is the α subunit of integrins (Imai and Yoshie 1993; Mannion et al. 1996; Lagaudriere-Gesbert et al. 1997; Yauch et al. 1998). CD151 is expressed by the basal keratinocytes of the skin and other epithelia (Sincock et al. 1997). Recently, it has been demonstrated that CD151 is a constituent of HDs and that its recruitment into these structures requires the $\alpha 6$ subunit to be associated with $\beta 4$ (Sterk et al. 2000). However, binding of CD151 to $\alpha 6\beta 4$ is not required for the formation of hemidesmosomal clusters containing plectin, BP180, and BP230, since the expression of an IL2R/$\beta 4$ chimeric construct in $\beta 4$-deficient keratinocytes induced the formation of such clusters, even though it is unable to bind to CD151 (Sterk et al. 2000). The function of CD151 in HDs remains to be elucidated, but it may play a role in the stabilization of HDs by regulating the spatial organization and lateral interactions of HD proteins.

1.5
Keratins

The keratin intermediate filament (IF) cytoskeleton forms an extended network spanning from a perinuclear ring through the cytoplasm to distinct cytoplasmic membrane sites, such as desmosomes, cell–cell adhesion complexes, and HDs (Fuchs and Weber 1994). Intermediate filaments are 10 nm thick, and are found ubiquitously in multicellular eukaryotes (Fuchs and Weber 1994; Steinert et al. 1994). Keratins are the main group of IF proteins (Fuchs and Weber 1994; Parry and Steinert 1999). The epithelial type I keratins form heterodimeric proteins with their type II partners, and together assemble into IFs. Cytokeratins are expressed differentially in most epithelial cells, depending on the stage of development and differentiation. In the epidermis, two major pairs of keratins are expressed: the keratins 5/14 are characteristically present in the basal layers of dividing cells (Fuchs and Weber 1994), whereas the keratins 1/10 are mainly detected in suprabasal differentiating cells (Fuchs and Weber 1994).

Structurally, keratins have a central α-helical rod domain, which is flanked by non-helical head and tail domains. The rod by itself can be further divided into four subdomains, which are separated by three short non-helical linkers. Both the N- and C-terminal regions of the rod domain are highly conserved and essential for filament assembly (Hatzfeld and Weber 1991; Letai et al. 1992). The head and tail regions of keratins can be phosphorylated, a means by which filament assembly and disassembly can be regulated (Heins and Aebi 1994).

1.6
Laminins

Laminins, which consist of 3 subunits (α, β and γ) are important constituents of basement membranes (Timpl and Brown 1994; Delwel et al. 1995). Different isoforms of these three subunits exist, the combination of which determines the type of the various laminin isoforms. Most laminin isoforms have the ability to self polymerize into a network and to strongly interact with nidogen (Yurchenco and O'Rear 1994). Laminin-5, previously known as nicein or kalinin (Verrando et al. 1988; Rousselle et al. 1991), is the laminin isoform associated with the hemidesmosomal adhesion complex. It consists of $\alpha 3$, $\beta 3$, and $\gamma 2$ chains, with molecular weights of 200, 140, and 155 kDa, respectively (Rousselle et al. 1991), that undergo complex proteolytic processing. Pathogenic mutations in one of the genes for the $\alpha 3$, $\beta 3$, and γ2 chains, leading to loss or aberrant expression of laminin-5 (Aberdam et al. 1994; Pulkkinen et al. 1994; Pulkkinen et al. 1997b), underlie distinct forms of JEB, the most severe of which leads to death soon after birth (Verrando et al. 1991; Christiano and Uitto 1996; McGowan and Marinkovich 2000). The formation of HDs in null-mutant mice for the α3 or γ2 chain of laminin-5 is abnormal with defective dermo-epidermal cohesion and post-natal lethality because of widespread blistering (Ryan et al. 1999; Meng et al 2003). Keratinocytes isolated from the α3 null mice also show defects in survival, which could be rescued by antibodies against α6β4 or by exogenous application of laminin-5 (Ryan et al. 1999).

Interestingly, laminin-5 lacks the domains that are responsible for self-polymerization and for the interaction with nidogen. Laminin-5 can bind to laminin-6, which facilitates its incorporation into basement membranes (Marinkovich et al. 1992; Champliaud et al. 1996). The cell-binding site of laminin-5 is located in the G-domain of the α3 chain. In fact, antibodies directed against the C-terminal end of the α3 chain block binding of cells to laminin-5 (Delwel et al. 1994; Niessen et al. 1994b). Furthermore, the G-domain harbors peptide sequences that mediate adhesion via the integrins α3β1 and α6β4 (Ryan et al. 1994; Shang et al. 2001).

2
Plakins

2.1
General Introduction

Plakins are cytolinker proteins that associate with and cross-link components of the various cytoskeletal systems, such as IF, microtubules, and microfilaments. Furthermore, they mediate attachment of IFs to the plasma membrane at specialized sites, i.e. desmosomes and/or HDs. Initially, the N-terminus of plakins was subdivided into homologous subregions based on the presence of α-helices, designated NN, Z, Y, X, W, and V (Green et al. 1992). These subregions together have now been denominated as the plakin domain (Leung et al. 2002). At their C-terminus, most plakins contain one or more so-called PLEC repeats (~300 amino acids long repeat domains), which, based on sequence dissimilarities, can be subdivided into three types (A, B, and C).

The plakin family of proteins is defined by the presence of either a plakin domain and/or PLEC repeats. Plakins also harbor other structural domains that are present in some but not all members of this family: an actin-binding domain (ABD), a coiled-coil rod, a spectrin repeat containing rod-domain, and a growth arrest specific protein 2 (GAS2)-microtubule-binding domain (MTBD). Recently, the name spectra-plakins has been proposed for the pla-

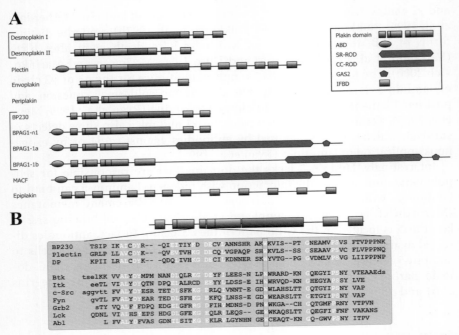

Fig. 2A, B. Structural overview of the plakin family. **A** Scale models of all plakin family members known to date. The different modules are represented on scale in different shapes and shades. The plakin domain is represented with its subregions (NN,Z,Y,X,V,W); *CC-Rod*, coiled-coil rod domain; *SR-ROD*, spectrin repeat rod domain; *GAS2*, growth arrest specific protein 2 (microtubule-binding domain); *IFBD*, intermediate filament-binding domain. **B** Sequence alignment of the region between the Y and X subregion of the plakin domain of BP230, plectin and DP, and of some known SH3 domains of SH3 domain-containing proteins

kins, based on the potential presence of spectrin-like repeats within the plakin-domain in all members of the plakin family (Roper et al. 2002).

To date, 7 members of the plakin family of proteins have been identified: desmoplakin (DP), plectin, BP230 (BPAG1-e), microtubule–actin cross-linking factor (MACF), envoplakin, periplakin, and epiplakin (Fig. 2). Most of the plakins are expressed in tissues that are exposed to mechanical stress, such as muscles and epithelia. In these tissues, the plakins play a critical role in maintaining tissue integrity and imparting mechanical strength. The proper expression of plakin family members is further critical for cell survival, growth, and development of certain cells, such as neurons (Brown et al. 1995a). In humans, acquired or congenital abnormalities of plakin family members cause dramatic clinical disorders, including cardiomyopathies, muscular dystrophies, and skin diseases, such as epidermolysis bullosa or palmo-plantar keratodermas (Pulkkinen et al. 1996; Ruhrberg et al. 1996;

Ruhrberg et al. 1997; Armstrong et al. 1999; DiColandrea et al. 2000; Norgett et al. 2000).

Comparison of the plakin domain of the various plakins with products of other genes revealed the presence of a region between the Z and Y subdomains in BP230, plectin, and desmoplakin that has weak, but significant similarity to the SH3 domain of known SH3-containing proteins (Fig. 2). Future investigations will have to determine whether a genuine SH3 domain is present in plakins.

2.2
Plectin

Plectin (Herrmann and Wiche 1987), also known as intermediate filament-associated protein of 300 kDa (IFAP300) (Clubb et al. 2000) and hemidesmosomal protein 1 (HD1) (Okumura et al. 1999), was initially identified as a major component of IF extracted from cultured cells (Pytela and Wiche 1980). Two years later, it was isolated from rat C6 glioma cells as a vimentin-associated protein (Wiche and Baker 1982).

Plectin is expressed in nearly all tissues and cell types including muscles, epithelia, and neural tissues (Wiche 1998). In human stratified epithelia, plectin is concentrated at the basal surface of basal epithelial cells (Wiche et al. 1984). Plectin is a component of both desmosomes and HDs (Hieda et al. 1992; Gache et al. 1996; Eger et al. 1997). The function of plectin in desmosomes remains to be determined, whereas in HDs this protein directly links IFs to the cytoplasmic domain of the $\beta 4$ integrin subunit and of BP180 (Niessen et al. 1997a; Rezniczek et al. 1998; Geerts et al. 1999; Koster et al. 2003). Plectin is also present in focal contacts in cultured fibroblasts (Seifert et al. 1992; Sanchez-Aparicio et al. 1997), in close association with the Z-lines of the sarcomeric unit of striated muscle cells, and in intercalated discs of cardiac muscle cells (Wiche et al. 1983).

The 500-kDa polypeptide deduced from plectin cDNA exhibits a multi-domain structure, including an N-terminal ABD and a C-terminal intermediate filament-binding domain (IFBD) (Wiche et al. 1991; McLean et al. 1996; Nikolic et al. 1996). When visualized by rotary shadowing electron microscopy, plectin appears as a ~200-nm long central rod domain flanked by globular end domains of ~9 nm (Wiche 1998). These structures represent either dimers formed by two polypeptide chains arranged in parallel (Wiche 1998), or tetramers, in which two parallel dimers are arranged in an antiparallel fashion (Foisner and Wiche 1987).

The structure of the C-terminal globular domain is dominated by 6 PLEC repeats (5B and 1C) (Wiche et al. 1991). These repeat domains are connected to each other by linker sequences of variable length. The repeat itself is composed of 5 tandem repeats of a 38-residue motif (2×19 residues) (Wiche et al. 1991).

The N-terminal ABD consists of a pair of calponin-like (CH) subdomains arranged in tandem (Castresana and Saraste 1995). This type of ABD is found in a large family of actin cross-linking proteins. Plectin, which has been found along actin stress fibers in some cells (Seifert et al. 1992), shows the same binding affinity for actin (Andra et al. 1998; Fontao et al. 2001) as that of other actin-binding proteins, such as dystonin (Yang et al. 1996), tensin (Lo et al. 1994), and dystrophin (Jarrett and Foster 1995). The ABD of plectin binds not only to actin, but also to the integrin $\beta 4$ subunit (see below) (Geerts et al. 1999).

Recently, the crystal structure of the ABD from plectin has been determined (Garcia-Alvarez et al. 2003). The CH domains that make up this ABD domain are folded in a similar manner as those of utrophin and dystrophin (Keep et al. 1999; Norwood et al. 2000). However, there are also some differences. First, the linker region that connects these two CH domains in plectin lacks a defined structure, whereas in utrophin and dystrophin, this region is α-helical. Furthermore, the first α-helix of the N-terminal CH domain is unusually long. The ABD of plectin was crystallized as a monomer, whereas those of utrophin and dystrophin were crystallized as antiparallel dimers, a feature that is attributed to domain swapping within the ABD. Interestingly, it appears that the ABD can undergo a conformational change from a closed to an open state upon interaction with F-actin, a feature that is not essential for the binding of F-actin to the ABD. Binding of $\beta 4$ to the ABD of plectin does not result in such a change (Garcia-Alvarez et al. 2003).

Plectin is a versatile cytolinker that can associate with microtubules (Svitkina et al. 1996), cytokeratins (Nikolic et al. 1996; Steinbock et al. 2000), as well as with filamentous actin (Andra et al. 1998). As such, plectin plays an important role in maintaining the structural integrity of various cells and tissues. This role is best illustrated by the following two observations. First, gene mutations leading to a defective expression of plectin cause a distinct form of epidermolysis bullosa simplex associated with muscular dystrophy (EBS-MD) (see also Table 1) (Chavanas et al. 1996; Gache et al. 1996; McLean et al. 1996; Pulkkinen et al. 1996; Smith et al. 1996; Mellerio et al. 1997). The patients show severe fragility and blistering of the skin as a result of an impaired attachment of keratin filaments to HDs, causing rupture of keratinocytes in the basal cell layers of the epidermis upon application of mechanical stress. In addition, necrotic muscle fibers with disorganized myofibrils and sarcomeres are observed. Second, ablation of plectin in mice resulted in a similar clinical phenotype (Andra et al. 1997). Plectin (−/−) mice, which die 2–3 days after birth, show an increased number of necrotic muscle fibers, focal loss of myofilaments, streaming of Z lines, focal ruptures of the sarcolemmal membrane, and subsarcolemmal accumulation of mitochondria (Andra et al. 1997).

A significant number of the mutations described in EBS-MD patients are located within the rod domain of plectin, and lead to premature stop codons and/or instability of mRNA (Uitto et al. 1996). In these patients, plectin as detected with the monoclonal antibody (MAb) 121, directed against an epitope within the plectin rod domain (Okumura et al. 1999), often appears to be completely absent. However, many cell types, including keratinocytes, express a naturally occurring splice variant of plectin, lacking the rod domain (Elliott et al. 1997). The presence of this variant in EBS-MD patients, therefore, would be unnoticed if the MAb 121 had been used as the only diagnostic test for the absence of plectin (Gache et al. 1996). The occurrence of this rod-less plectin in EBS-MD patients, which can likely fulfill important functions, could explain some differences in the phenotypes observed in plectin-null mice and patients (Chavanas et al. 1996; Gache et al. 1996; McLean et al. 1996; Pulkkinen et al. 1996; Smith et al. 1996; Andra et al. 1997; Mellerio et al. 1997).

In 1996, the sequence, the exon-intron organization, and the chromosomal localization of human plectin was published by two groups (Liu et al. 1996; McLean et al. 1996). The 32 exons coding for plectin are spread over about 32 kb of q24 on chromosome 8 in the human genome. Nearly all exons reside in a region encoding the N-terminal domain of the molecule, while the central rod and the C-terminal domains are encoded by single exons of the unusually large sizes of more than 3 and 6 kb, respectively (Liu et al. 1996; McLean et al. 1996). However, the size and sequence of the first coding exon reported by the two papers was different, indicating the existence of variants. In subsequent studies more than 16 plectin transcripts with alternatively spliced exons in the 5' region of plectin have been identified in man, mouse, and rat (Elliott et al. 1997; Wiche 1998; Fuchs et al. 1999). Eleven of these variants (1–1j) are the result of alternative splicing of different first exons into a common exon 2, whereas three arise by alternative splicing of non-coding exons preceding exon 1c (Fuchs et al. 1999). In addition, exons 2α and 3α are spliced between exons 2 and 3 or between exons 3 and 4, respectively (Fuchs et al. 1999). Of 2α, it has been shown that its introduction in the ABD (exons 2–8) enhances the actin-binding capacity of plectin (Fuchs et al. 1999).

The tissue expression of the various plectin variants is different, suggesting that they have distinct functions. However, as to date the biological function of these variants of the first coding exon remains elusive, it has been postulated that the various first exons could be part of distinct promoters, which may be independently controlled by specific transcriptional regulators (Wiche 1998). Alternatively, the different N-termini may be responsible for the targeting of plectin to distinct subcellular locations, such as HDs, where it may exert specific functions by interacting with distinct protein partners (Wiche 1998; Andra et al. 2003).

2.3
BP230

Epithelial BPAG1 (BP230) is a 2,649 amino acid protein and, like BP180, was first identified as a target antigen of circulating autoantibodies from patients with bullous pemphigoid (see also Table 1) (Stanley 1993). BP230 resides in the inner cytoplasmic plaque of HDs, where it contributes to the tethering of the keratin intermediate filament system via its C-terminal domain (Yang et al. 1996).

The epithelial and most prominent form of BPAG1, BP230, contains a plakin domain, a coiled-coil rod domain and 2 PLEC repeats (B and C). Cell transfection and yeast two-hybrid experiments have shown that the last 768-amino acid tail domain of the BP230 protein contains sequences important for its interaction with IFs (Yang et al. 1996; Fontao et al. 2003). In line with these results, BPAG1 null-mutant mice reveal discrete signs of skin blistering, most likely as a result of an impaired attachment of keratin filaments to HDs, leading to mechanical fragility of basal keratinocytes (Guo et al. 1995). Unexpectedly, these mice also developed a neurological disease characterized by severe neuro-degeneration with dystonia and ataxia (Brown et al. 1995a; Guo et al. 1995). This phenotype appeared to be the result of the concomitant inactivation of neuronal isoforms of BPAG1, that differ with BP230 in both their N-terminal extremities and/or their C-terminal region (Brown et al. 1995a; Guo et al. 1995; Leung et al. 2001). These neuronal isoforms are essential for the maintenance of the cytoarchitecture of neurons (Yang et al. 1996, 1999). They contain an ABD and/or a MTBD (Yang et al. 1996, 1999) and are able to link the three cytoskeletal networks to each other. Specifically, BPAG1-1a differs from BP230 by the presence of an ABD at the extreme N-terminus of the protein. Instead of having a coiled-coil rod domain, this variant contains a spectrin repeat rod domain followed by a distinct C-terminus containing a MTBD and not an IFBD. BPAG1-1a is abundantly expressed in pituitary primordium and dorsal root ganglia (DRG) (Leung et al. 2001). The BPAG1-1b variant resembles the BPAG1-1a variant but has an insertion between the plakin domain and the spectrin repeat rod domain containing a putative IFBD. BPAG1-1b is more restricted to the heart, skeletal muscles, and bone cartilage of the vertebrae in developing mouse embryos (Leung et al. 2001). Additional variants of BPAG1 have been described (Brown et al. 1995b; Guo et al. 1995; Yang et al. 1999), the existence of which remains to be confirmed. In fact, the use of RNA probes and/or antibodies potentially cross-reactive with any other isoform precluded any firm conclusions (Leung et al. 2001). The N-terminus of BP230 is important for its recruitment into HDs. This region can associate with the cytoplasmic regions of two transmembrane components of HDs: BP180 and the $\beta4$ integrin subunit (Hopkinson and Jones 2000; Koster et al. 2003).

2.4
Desmoplakin

The desmoplakins (DPs) are components of intercellular cell–cell adhesion complexes that form adhesive junctions in tissues that are subject to mechanical stress (Franke et al. 1981). The desmoplakin gene encodes two variants that arise as a result of alternative splicing: DPI (322 kDa) and DPII (259 kDa) (Green et al. 1990; Virata et al. 1992). A 599-residue region within the central rod region of DPI is missing in DPII, the functional consequences of which are unknown (Virata et al. 1992). Desmoplakin is a dumb-bell-shaped molecule (O'Keefe et al. 1989). The α-helical rod domain of DPI is predicted to form a 130-nm coiled-coil homodimer, which is flanked by a globular head and a tail domain (O'Keefe et al. 1989; Green et al. 1990; Virata et al. 1992).

The N-terminal plakin domain is important for the targeting of DP to desmosomes by means of interactions with plakoglobin and plakophilin 1 (Kowalczyk et al. 1997; Smith and Fuchs 1998; Hatzfeld et al. 2000; Bornslaeger et al. 2001). Furthermore, DP can bind to itself, to desmocollin 1a and to plectin (Eger et al. 1997; Kowalczyk et al. 1997; Smith and Fuchs 1998). Support for the idea that DP is critical for the proper anchorage of intermediate filaments to the desmosomal plaque was derived from the observation that a dominant-negative mutant of DP was able to uncouple keratin filaments from the plasma membrane, by competing for the desmosomal proteins at the desmosomal plaque (Bornslaeger et al. 1996).

Furthermore, when transiently expressed in COS-7 cells and mouse fibroblasts (with few or no desmosomes, respectively), N-terminally truncated DPI constructs are co-localized with and disrupt the endogenous IF networks (Stappenbeck and Green 1992; Stappenbeck et al. 1993; Bornslaeger et al. 1996). The C-terminal region of DP, consisting of three PLEC repeats (A, B, and C), decorates IFs in cultured keratinocytes (Kouklis et al. 1994). In vitro and yeast two-hybrid assays with a recombinant C-tail of DP have shown that it can directly interact with the head domains of type II epidermal keratins (Kouklis et al. 1994; Meng et al. 1997). Nevertheless, recent studies indicate that the association of DP, with both epidermal and simple keratins, is critically affected by the tertiary structure induced by heterodimerization and involves recognition sites primarily located in the rod domain of these keratins (Fontao et al. 2003).

Consistent with the idea that DP is critical for the maintenance of tissue architecture integrity, DP null-mutant mice die at day 6.5 of embryonic development as a result of the disorganization and fragility of the embryonic ectoderm (Gallicano et al. 1998). Interestingly, loss of DP leads to much earlier defects than the loss of the keratin 8/18, proteins that are normally connected to desmosomes at this stage of development (Baribault et al. 1993; Magin et al. 1998).

The B- and C-domain of the C-terminus of DP has been crystallized and their structure has been determined (Choi et al. 2002). These domains of 4.5×38-amino acid repeats were shown to form discrete globular subdomains that bind IFs (Choi et al. 2002).

2.5
Microtubule–Actin Cross-Linking Factor

The MACF, was identified by a partial cDNA clone (named ACF7) found in a screen for proteins possessing a dystrophin-like ABD (Byers et al. 1995). The complete cDNA encodes a large 806-kDa protein, which is a paralog of BPAG1-1a and is currently regarded as an additional member of the plakin family (Leung et al. 1999; Leung et al. 2001). Although trabeculin-α and macrophin are two cytoskeletal proteins whose cDNAs were cloned undependently, their primary sequences show that they are identical to MACf (San et al. 1999; Okada et al. 1999). Similar to BPAG1-1a, MACF is composed of an ABD, a plakin domain, a 23 spectrin repeat-containing rod domain, and a GAS2-MTBD. The ABD of MACF, like those of BPAG1 and plectin, has two CH domains, but alternative splicing of the 5' end of the MACF gene generates variants with a varying number of CH domains (Bernier et al. 1996). MACF is ubiquitously expressed in the mouse embryo as assessed by in situ hybridization (Leung et al. 1999). Immunohistochemistry has demonstrated that MACF is expressed at high levels in the epidermis of newborn and embryonic mice (Karakesisoglou et al. 2000). Since MACF contains both an ABD and a MTBD, the protein likely functions as a cytoskeletal linker protein. When ectopically expressed in COS-7 cells, MACF is localized with actin filaments and microtubules, while in cultured epidermal keratinocytes it is localized at intersections of the microfilaments and the microtubules (Leung et al. 1999; Karakesisoglou et al. 2000). Upon stimulation of the formation of intercellular adhesions, MACF becomes distributed with microtubules and relocated to sites of cell–cell contact, implying that it modulates microtubules at cellular junctions (Karakesisoglou et al. 2000).

2.6
Envoplakin

Envoplakin, with a molecular mass of 210 kDa, was originally identified as one of the cornified envelope precursor proteins (Ruhrberg et al. 1996). Based on sequence homology, envoplakin belongs to the plakin family. It is associated with the desmosomal plaque and with keratin filaments in the suprabasal differentiated layers of the epidermis (Ruhrberg et al. 1996). There are only subtle changes in the cornified envelope of envoplakin null-mutant mice as compared to that of wild-type mice, suggesting that envoplakin is, at least to a certain extent, dispensable for the assembly of the cornified envelope (Maatta et al. 2001).

2.7
Periplakin

Periplakin, a protein of 195 kDa, was first identified as a protein that becomes incorporated into the cornified envelope of cultured epidermal keratinocytes (Ruhrberg et al. 1997). This plakin is, like envoplakin, associated with the desmosomal plaque and with keratin filaments in the differentiated layers of the epidermis. Periplakin and envoplakin are closely related plakins. The rod domains of both are highly homologous, which suggests that the formation of homodimers as well as heterodimers is energetically favorable. Indeed, the two proteins can be co-immunoprecipitated (Ruhrberg et al. 1997). Confocal immunofluorescent microscopy studies of cultured epidermal keratinocytes revealed that envoplakin and periplakin form a network radiating from desmosomes (Ruhrberg et al. 1997). Even though periplakin does not contain any PLEC repeats, it can be co-aligned with IFs (DiColandrea et al. 2000). The first half of the N-terminus of periplakin (up to the X subregion in the plakin domain) appears to be important for the localization of this protein into desmosomes (DiColandrea et al. 2000). Periplakin can interact with PKB and may thus play a role as localization signal in PKB-mediated signaling (van den Heuvel et al. 2002).

2.8
Epiplakin

Epiplakin was originally identified as a 450-kDa epidermal autoantigen in a patient with a skin blistering disease closely resembling bullous pemphigoid. Unlike all other plakins, epiplakin does not contain a rod domain or any other known motif for dimerization and also lacks an N-terminal plakin domain (Fujiwara et al. 2001). The protein is composed of 13 B domains, the last 5 of which are well conserved. Epiplakin is widely expressed in human tissues and is particularly abundant in the liver, the digestive tract, and the salivary glands, as shown by Northern dot-blot analysis. Immunofluorescence analysis of the skin revealed the presence of epiplakin in the entire epidermis. Future investigations will undoubtedly provide further knowledge about this unusual plakin member.

3
Molecular Interactions Involved in Hemidesmosome Assembly

Recent studies have revealed that the recruitment of the different hemidesmosomal components into HDs is regulated by a hierarchy of interactions. Various hemidesmosomal components can interact with multiple other components. Furthermore, interactions with more than one component are required for the efficient targeting of proteins into HDs. In the following section, we will discuss in detail the different steps involved in the formation of HDs in cultured cells.

3.1
The Role of $\beta 4$ in HD Assembly

Studies on the importance of $\beta 4$ in the formation of HDs have greatly benefited from the use of a cell line derived from a patient suffering from pyloric atresia associated with junctional epidermolysis bullosa (PA-JEB). In this cell line which lacks $\beta 4$, no HDs are assembled. However, upon the introduction of $\beta 4$ in these $\beta 4$-deficient PA-JEB cells, $\alpha 6\beta 4$ is expressed, and clusters of this integrin together with plectin, BP180, and BP230—the clusters resemble HDs—are formed at the basal side. Thus, expression of $\beta 4$ in these cells is capable of restoring their ability to form HDs (Gagnoux-Palacios et al. 1997; Schaapveld et al. 1998).

Transfection studies in PA-JEB, and other cell types using deletion mutants of $\beta 4$, revealed that the first pair of FNIII repeats and a small part of the connecting segment (1328–1355) are essential for the incorporation of plectin into the HD-like structures (Niessen et al. 1997b; Schaapveld et al. 1998). However, efficient recruitment of BP180 and BP230 also required the C-terminal part of $\beta 4$ (Schaapveld et al. 1998). It is noteworthy that in patients with non-lethal forms of JEB, mutations within the second FNIII repeat of $\beta 4$ have been described (Pulkkinen et al. 1998; Nakano et al. 2001). These mutations critically affect the ability of $\beta 4$ to recruit plectin, resulting in impaired dermo-epidermal adhesion and blistering of the skin (Koster et al. 2001).

The first pair of FNIII repeats together with a small part of the connecting segment of $\beta 4$ have also been implicated in the localization of $\beta 4$ into HDs in 804G rat bladder carcinoma cells (Spinardi et al. 1993; Mainiero et al. 1995; Spinardi et al. 1995; Niessen et al. 1997b) and the redistribution of plectin to the basal side of COS-7 (Niessen et al. 1997b) and GD25 cells (Sanchez-Aparicio et al. 1997). The recruitment of plectin to $\alpha 6\beta 4$ requires the presence of the ABD of plectin (Geerts et al. 1999).

Additional binding sites that might stabilize the interaction with plectin are present in the connecting segment and the C-tail of $\beta 4$. The corresponding sites on plectin are located in the PLEC repeats 3–6, located in the C-terminal end of the protein and in the N-terminal region of plectin (Rezniczek et al. 1998). Interestingly, in the absence of ligand binding, the localization of α6β4 in HDs is entirely driven by its interaction with plectin (Nievers et al. 1998, 2000).

Based on the results of both yeast two-hybrid and blot overlay assays, it has been suggested that intramolecular folding occurs within the cytoplasmic domain of β4 (Rezniczek et al. 1998; Schaapveld et al. 1998). The domains involved in this intramolecular interaction reside near those mediating the β4-plectin interaction. Thus, binding of plectin to β4 may change the conformation of β4 in such a way that the binding surfaces for BP180 and BP230 on β4 become accessible.

3.2
The Role of BP180 in HD Assembly

The role of BP180 in HD assembly has become evident by the studies of patients with a non-lethal form of JEB, in whom BP180 is not expressed due to pathogenic mutations in the BP180 (COLXVIIA) gene (Jonkman et al. 1995; McGrath et al. 1995). Although HDs are formed in these patients, they are often rudimentary and lack a sub-basal dense plate. BP180-deficient keratinocytes isolated from these patients are able to form type II HD-like structures (as present in the intestine) that only contain α6β4 and plectin (Uematsu et al. 1994; Orian-Rousseau et al. 1996; Borradori et al. 1998).

The C-terminal part of the cytoplasmic domain of β4 harbors the binding site for BP180, and plays an important role in the recruitment of this molecule into HDs (Borradori et al. 1997; Schaapveld et al. 1998; Aho and Uitto 1998). However, incorporation of BP180 into HDs may not only be dependent on β4 (Nievers et al. 1998; Schaapveld et al. 1998), but also on a direct interaction between BP180 and plectin that is associated with α6β4 (Koster et al. 2003). Indeed, it has been shown that a β4 subunit that is unable to bind plectin cannot efficiently recruit BP180 into HDs (Koster et al. 2003). Moreover, in plectin-deficient keratinocytes, BP180 localization in HDs is diminished (Koster et al. 2003) or even completely lost (Gache et al. 1996). The BP180-binding site on plectin has been located in its plakin domain (Koster et al. 2003), although its C-terminal region has also been implicated (Aho and Uitto 1997). Furthermore, there is data suggesting that an interaction between the NC16a domain of BP180 and α6 contributes to an efficient localization of BP180 into HDs (Hopkinson et al. 1995).

The localization of BP230 in HDs is dependent on the presence of BP180. Upon expression of BP180 in BP180-deficient keratinocytes, both BP180 and BP230 become co-distributed with α6β4 and plectin (Borradori et al. 1998; Koster et al. 2003). Yeast two-hybrid studies have shown that BP230 can directly bind to BP180 (Hopkinson and Jones 2000; Koster et al. 2003). BP230 can also directly interact with β4. However, this interaction is not sufficiently strong for recruiting BP230 into HDs, since BP230 is not co-localized with α6β4 in BP180-deficient keratinocytes (Koster et al. 2003). Taken together, the data favor a model in which the efficient localization of BP180 into HDs is dependent on its interaction with both β4 and plectin, whereas that of BP230 depends on binding to BP180 and β4.

3.3
The Linkage of Keratins to Hemidesmosomes

Both plectin and BP230 are able to bind to keratins as has been demonstrated in in vitro studies and in yeast interaction assays (Foisner et al. 1988; Yang et al. 1996; Fontao et al. 1997, Fontao et al. 2003). Interaction of keratin IFs with plectin and BP230 is not essential for the formation of HDs, because HDs appeared to be unaffected in patients lacking keratin 14 due to mutations in the KRT14 gene (Bonifas et al. 1991; Coulombe et al. 1991). Similarly, no effect of the absence of keratin 5 or keratin 14 on HDs was observed in the respective null-mutant mice (Lloyd et al. 1995; Peters et al. 2001). Absence of keratin 5 or keratin 14 gives rise to epidermolysis bullosa simplex, characterized by fragility of keratinocytes and rupturing of cells upon application of mechanical force. The deletion of keratin 5 renders the basal epidermis deficient of keratin IFs, whereas in the keratin 14-null mice and patients a loose keratin IF network is observed. This is due to the association of keratin 5 with keratin 15, which is another keratin protein expressed in the basal epidermis. Nevertheless, keratin 15 does not compensate for the loss of keratin 14.

As is evident from studies with null-mutant mice, both plectin and BP230 contribute to the linkage of IFs to the hemidesmosomal plaque. However, while BP230 appeared to be essential for IF attachment, plectin is not. There were no IFs associated with HDs in the BPAG1 null-mutant mice, and many HDs lacked an inner plaque. In contrast, in the plectin-deficient mice, IFs still associated with HDs, although they appeared to be more loosely distributed, particularly at the sites of their insertion into the inner plaque.

In plectin, the site of interaction with IFs has been mapped to a stretch of 50 amino acids in the fifth PLEC repeat located within the C-terminal region (Nikolic et al. 1996). Different IF proteins can bind to plectin, including various types of keratins, vimentin, and desmin (Nikolic et al. 1996; Geerts et al. 1999; Hijikata et al. 2003). On the other hand, BP230 can only bind to the keratin 5/14 fragment, not to keratin 8/18 or vimentin, as assessed in cell transfection and yeast three-hybrid experiments (Fontao et al. 2003). A fragment, encompassing the B and C domain plus the C-extremity, has been identified as the region on BP230 that interacts directly with the keratins 5/14. This interaction occurs in the absence of the head or tail domain of the keratins, implicating an important role for the rod domain in this interaction (Fontao et al. 2003).

4
A Hierarchical Interaction Model for the Assembly of HDs

By the use of different approaches, various interactions among proteins of HDs have been dissected (Fig. 3). This knowledge has allowed us to define different steps in the assembly of these complexes in cultured keratinocytes. From the body of data available, a model has emerged according to which the integrin α6β4 first interacts with plectin. Subsequently, these two proteins together recruit BP180. Finally, BP230 becomes incorporated into the complex via interactions with BP180 as well as β4. The interaction of plectin with α6β4 is essential for initiating the formation of HDs.

In addition to providing an essential binding site for the recruitment of BP180, plectin may in another way be essential for the formation of HDs in culture, because it may help to unfold the cytoplasmic domain of β4. Recent studies have indicated that one of the binding sites for plectin in the connecting segment of β4 overlaps with the site involved in the intramolecular

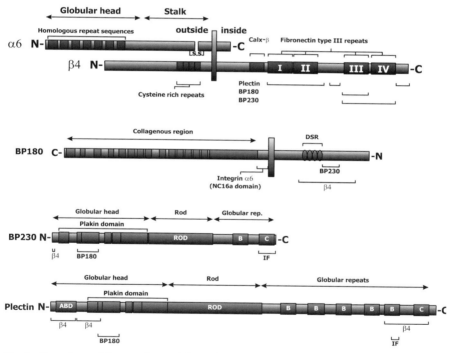

Fig. 3. Overview of interactions of hemidesmosomal components. Interactions of the components of hemidesmosomes that have been published are indicated below the schematic representations. See text for further details. *Calx-β*, high-affinity calcium binding motif; *DSR*, direct sequence repeats; *IF*, intermediate filaments; *N*, N-terminus; *C*, C terminus

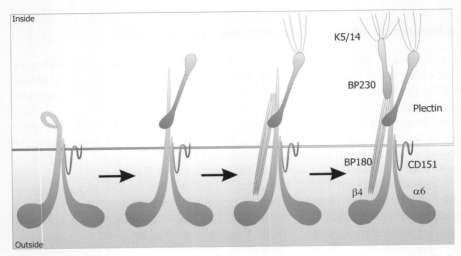

Fig. 4. Model for the assembly of hemidesmosomes. First plectin interacts with $\beta 4$, which may result in the unfolding of the cytoplasmic domain of $\beta 4$, rendering the binding sites for BP180 and BP230 available. BP180 is recruited into the complex by binding to both plectin and $\beta 4$. Finally, BP230 is incorporated via binding to BP180 as well as $\beta 4$, making the hemidesmosome complete

folding of the $\beta 4$ cytoplasmic domain. Hence, binding of plectin to $\beta 4$ may either unfold this part of $\beta 4$ or ensure that $\beta 4$ is maintained in an open conformation, thus making the interaction sites for BP180 and BP230 in the C-terminal half of the $\beta 4$ cytoplasmic domain accessible for binding or ensure that they remain accessible (Fig. 4). Future investigations will have to prove or disprove the validity of this model.

5
General Conclusions and Future Perspectives

Great quantities of data have become available about the interactions between proteins within the HD and the hierarchy of its assembly. However, it is still unclear which signals regulate this assembly. Does $\alpha 6\beta 4$ indeed need to be activated before it can bind to plectin, or are HDs assembled simply by the aggregation and clustering of hemidesmosomal components? Furthermore, what is the role of the $\alpha 3\beta 1$ integrin (the other laminin-5 receptor present on keratinocytes that is connected to the actin cytoskeleton) in this process, and how does the binding of plectin to $\beta 4$ influence the stability of the actin cytoskeleton? Equally important is it to learn how HDs are disassembled during cell migration and which signals control this process. Signaling may be regulated by the interaction of $\alpha 6\beta 4$ with laminin-5, with plectin, or both. Cleavage of laminin-5 may weaken the linkage of $\alpha6\beta4$ with laminin-5, allowing cells to become detached, and thus cell migration is promoted. Likewise, when the interaction of $\beta4$ with plectin is broken, the interaction of $\alpha6\beta4$ with laminin-5 may become unstable with a similar effect. Another important question is whether the stability of HDs depends on the amount of laminin-5 that is deposited by the cells into the extracellular matrix. For example, an increased amount of laminin-5 has been found at the leading edge of tumors, and has been associated with their increased invasiveness (Pyke et al. 1994; Chao et al. 1996; Lohi 2001), possibly because it leads to the disassembly of HDs. It should be clear that much has yet to be learned about the dynamics of HDs, and several interesting questions still wait to be answered. Hence, the research on HDs is far from finished; in fact, it is entering a new dimension.

Acknowledgements The work presented in this review by J.K. and A.S. is supported by grants from DEBRA (UK). The work by L.B. is supported by grants from the Swiss National Foundation of Scientific Research and the Office fédéral de l'éducation et de la science.

References

Aberdam D, Galliano MF, Vailly J, Pulkkinen L, Bonifas J, Christiano AM, Tryggvason K, Uitto J, Epstein EH, Ortonne JP (1994) Herlitz's junctional epidermolysis bullosa is linked to mutations in the gene (LAMC2) for the $\gamma2$ subunit of nicein/kalinin (LAMININ-5). Nat Genet 6:299–304

Aho S, Uitto J (1997) Basement membrane zone protein–protein interactions disclosed by yeast two-hybrid system. J Invest Dermatol 108:546a

Aho S, Uitto J (1998) Direct interaction between the intracellular domains of bullous pemphigoid antigen 2 (BP180) and $\beta4$ integrin, hemidesmosomal components of basal keratinocytes. Biochem Biophys Res Commun 243:694–699

Andra K, Lassmann H, Bittner R, Shorny S, Fassler R, Propst F, Wiche G (1997) Targeted inactivation of plectin reveals essential function in maintaining the integrity of skin, muscle, and heart cytoarchitecture. Genes Dev 11:3143–3156

Andra K, Nikolic B, Stocher M, Drenckhahn D, Wiche G (1998) Not just scaffolding: plectin regulates actin dynamics in cultured cells. Genes Dev 12:3442–3451

Andra K, Kornacker I, Jorgl A, Zorer M, Spazierer D, Fuchs P, Fischer I, Wiche G (2003) Plectin-isoform-specific rescue of hemidesmosomal defects in plectin (−/−) keratinocytes. J Invest Dermatol 120:189–197

Armstrong DK, McKenna KE, Purkis PE, Green KJ, Eady RA, Leigh IM, Hughes AE (1999) Haploinsufficiency of desmoplakin causes a striate subtype of palmoplantar keratoderma. Hum Mol Genet 8:143–148

Bachelder RE, Ribick MJ, Marchetti A, Falcioni R, Soddu S, Davis KR, Mercurio AM (1999a) p53 inhibits $\alpha6\beta4$ integrin survival signaling by promoting the caspase 3-dependent cleavage of AKT/PKB. J Cell Biol 147:1063–1072

Bachelder RE, Marchetti A, Falcioni R, Soddu S, Mercurio AM (1999b) Activation of p53 function in carcinoma cells by the $\alpha6\beta4$ integrin. J Biol Chem 274:20733–20737

Balding SD, Diaz LA, Giudice GJ (1997) A recombinant form of the human BP180 ectodomain forms a collagen-like homotrimeric complex. Biochemistry 36:8821–8830

Baribault H, Price J, Miyai K, Oshima RG (1993) Mid-gestational lethality in mice lacking keratin 8. Genes Dev 7:1191–1202

Bernier G, Mathieu M, De Repentigny Y, Vidal SM, Kothary R (1996) Cloning and characterization of mouse ACF7, a novel member of the dystonin subfamily of actin binding proteins. Genomics 38:19–29

Bonifas JM, Rothman AL, Epstein EH Jr (1991) Epidermolysis bullosa simplex: evidence in two families for keratin gene abnormalities. Science 254:1202–5

Bornslaeger EA, Corcoran CM, Stappenbeck TS, Green KJ (1996) Breaking the connection: displacement of the desmosomal plaque protein desmoplakin from cell–cell interfaces disrupts anchorage of intermediate filament bundles and alters intercellular junction assembly. J Cell Biol 134:985–1001

Bornslaeger EA, Godsel LM, Corcoran CM, Park JK, Hatzfeld M, Kowalczyk AP, Green KJ (2001) Plakophilin 1 interferes with plakoglobin binding to desmoplakin, yet together with plakoglobin promotes clustering of desmosomal plaque complexes at cell–cell borders. J Cell Sci 114:727–738

Borradori L, Sonnenberg A (1999) Structure and function of hemidesmosomes: more than simple adhesion complexes. J Invest Dermatol 112:411–418

Borradori L, Koch PJ, Niessen CM, Erkeland S, van Leusden MR, Sonnenberg A (1997) The localization of bullous pemphigoid antigen 180 (BP180) in hemidesmosomes is mediated by its cytoplasmic domain and seems to be regulated by the $\beta 4$ integrin subunit. J Cell Biol 136:1333–1347

Borradori L, Chavanas S, Schaapveld RQ, Gagnoux-Palacios L, Calafat J, Meneguzzi G, Sonnenberg A (1998) Role of the bullous pemphigoid antigen 180 (BP180) in the assembly of hemidesmosomes and cell adhesion–reexpression of BP180 in generalized atrophic benign epidermolysis bullosa keratinocytes. Exp Cell Res 239:463–476

Brown A, Dalpe G, Mathieu M, Kothary R (1995a) Cloning and characterization of the neural isoforms of human dystonin. Genomics 29:777–780

Brown A, Bernier G, Mathieu M, Rossant J, Kothary R (1995b) The mouse dystonia musculorum gene is a neural isoform of bullous pemphigoid antigen 1. Nat Genet 10:301–306

Brown TA, Gil SG, Sybert VP, Lestringant GG, Tadini G, Caputo R, Carter WG (1996) Defective integrin $\alpha 6\beta 4$ expression in the skin of patients with junctional epidermolysis bullosa and pyloric atresia. J Invest Dermatol 107:384–391

Burgeson RE, Christiano AM (1997) The dermal-epidermal junction. Curr Opin Cell Biol 9:651–658

Byers TJ, Beggs AH, McNally EM, Kunkel LM (1995) Novel actin crosslinker superfamily member identified by a two step degenerate PCR procedure. FEBS Lett 368:500–504

Carter WG, Ryan MC, Gahr PJ (1991) Epiligrin, a new cell adhesion ligand for integrin $\alpha 3\beta 1$ in epithelial basement membranes. Cell 65:599–610

Castresana J, Saraste M (1995) Does Vav bind to F-actin through a CH domain? FEBS Lett 374:149–151

Champliaud MF, Lunstrum GP, Rousselle P, Nishiyama T, Keene DR, Burgeson RE (1996) Human amnion contains a novel laminin variant, laminin 7, which like laminin 6, covalently associates with laminin 5 to promote stable epithelial-stromal attachment. J Cell Biol 132:1189–1198

Chao C, Lotz MM, Clarke AC, Mercurio AM (1996) A function for the integrin $\alpha 6\beta 4$ in the invasive properties of colorectal carcinoma cells. Cancer Res 56:4811–4819

Chavanas S, Pulkkinen L, Gache Y, Smith FJ, McLean WH, Uitto J, Ortonne JP, Meneguzzi G (1996) A homozygous nonsense mutation in the PLEC1 gene in patients with epidermolysis bullosa simplex with muscular dystrophy. J Clin Invest 98:2196–2200

Choi HJ, Park-Snyder S, Pascoe LT, Green KJ, Weis WI (2002) Structures of two intermediate filament-binding fragments of desmoplakin reveal a unique repeat motif structure. Nat Struct Biol 9:612–620

Christiano AM, Uitto J (1996) Molecular complexity of the cutaneous basement membrane zone. Revelations from the paradigms of epidermolysis bullosa. Exp Dermatol 5:1–11

Clarke AS, Lotz MM, Mercurio AM (1994) A novel structural variant of the human $\beta 4$ integrin cDNA. Cell Adhes Commun 2:1–6

Clubb BH, Chou YH, Herrmann H, Svitkina TM, Borisy GG, Goldman RD (2000) The 300-kDa intermediate filament-associated protein (IFAP300) is a hamster plectin ortholog. Biochem Biophys Res Commun 273:183–187

Cooper HM, Tamura RN, Quaranta V (1991) The major laminin receptor of mouse embryonic stem cells is a novel isoform of the $\alpha 6 \beta 1$ integrin. J Cell Biol 115:843–850

Coulombe PA, Hutton ME, Letai A, Hebert A, Paller AS, Fuchs E (1991) Point mutations in human keratin 14 genes of epidermolysis bullosa simplex patients: genetic and functional analyses. Cell 66:1301–11

Dajee M, Lazarov M, Zhang JY, Cai T, Green CL, Russell AJ, Marinkovich MP, Tao S, Lin Q, Kubo Y, Khavari PA (2003) NF-κB blockade and oncogenic Ras trigger invasive human epidermal neoplasia. Nature 421:639–643

de Pereda JM, Wiche G, Liddington RC (1999) Crystal structure of a tandem pair of fibronectin type III domains from the cytoplasmic tail of integrin $\alpha 6 \beta 4$. EMBO J 18:4087–4095

Delwel GO, de Melker AA, Hogervorst F, Jaspars LH, Fles DL, Kuikman I, Lindblom A, Paulsson M, Timpl R, Sonnenberg A (1994) Distinct and overlapping ligand specificities of the $\alpha 3A\beta 1$ and $\alpha 6A\beta 1$ integrins: recognition of laminin isoforms. Mol Biol Cell 5:203–215

Delwel GO, Kuikman I, Sonnenberg A (1995) An alternatively spliced exon in the extracellular domain of the human $\alpha 6$ integrin subunit—functional analysis of the $\alpha 6$ integrin variants. Cell Adhes Commun 3:143–161

DiColandrea T, Karashima T, Maatta A, Watt FM (2000) Subcellular distribution of envoplakin and periplakin: insights into their role as precursors of the epidermal cornified envelope. J Cell Biol 151:573–586

Dowling J, Yu QC, Fuchs E (1996) $\beta 4$ integrin is required for hemidesmosome formation, cell adhesion and cell survival. J Cell Biol 134:559–572

Eady RA (1994) The hemidesmosome: a target in auto-immune bullous disease. Dermatology 189 Suppl 1:38–41

Eger A, Stockinger A, Wiche G, Foisner R (1997) Polarisation-dependent association of plectin with desmoplakin and the lateral submembrane skeleton in MDCK cells. J Cell Sci 110:1307–1316

Elliott CE, Becker B, Oehler S, Castanon MJ, Hauptmann R, Wiche G (1997) Plectin transcript diversity: identification and tissue distribution of variants with distinct first coding exons and rodless isoforms. Genomics 42:115–125

Elomaa O, Sankala M, Pikkarainen T, Bergmann U, Tuuttila A, Raatikainen-Ahokas A, Sariola H, Tryggvason K (1998) Structure of the human macrophage MARCO receptor and characterization of its bacteria-binding region. J Biol Chem 273:4530–4538

Favre B, Fontao L, Koster J, Shafaatian R, Jaunin F, Saurat JH, Sonnenberg A, Borradori L (2001) The hemidesmosomal protein bullous pemphigoid antigen 1 and the integrin β4 subunit bind to ERBIN. Molecular cloning of multiple alternative splice variants of ERBIN and analysis of their tissue expression. J Biol Chem 276:32427–32436

Ferguson BM, Brockdorff N, Formstone E, Ngyuen T, Kronmiller JE, Zonana J (1997) Cloning of Tabby, the murine homolog of the human EDA gene: evidence for a membrane-associated protein with a short collagenous domain. Hum Mol Genet 6:1589–1594

Foisner R, Wiche G (1987) Structure and hydrodynamic properties of plectin molecules. J Mol Biol 198:515–531

Foisner R, Leichtfried FE, Herrmann H, Small JV, Lawson D, Wiche G (1988) Cytoskeleton-associated plectin: in situ localization, in vitro reconstitution, and binding to immobilized intermediate filament proteins. J Cell Biol 106:723–733

Fontao L, Dirrig S, Owaribe K, Kedinger M, Launay JF (1997) Polarized expression of HD1: relationship with the cytoskeleton in cultured human colonic carcinoma cells. Exp Cell Res 231:319–327

Fontao L, Geerts D, Kuikman I, Koster J, Kramer D, Sonnenberg A (2001) The interaction of plectin with actin: evidence for cross-linking of actin filaments by dimerization of the actin-binding domain of plectin. J Cell Sci 114:2065–2076

Fontao L, Favre B, Riou S, Geerts D, Jaunin F, Saurat JH, Green KJ, Sonnenberg A, Borradori L (2003) Interaction of the bullous pemphigoid antigen 1 (BP230) and desmoplakin with intermediate filaments is mediated by distinct sequences within their COOH terminus. Mol Biol Cell 14:1978–1992

Franke WW, Schmid E, Grund C, Muller H, Engelbrecht I, Moll R, Stadler J, Jarasch ED (1981) Antibodies to high molecular weight polypeptides of desmosomes: specific localization of a class of junctional proteins in cells and tissue. Differentiation 20:217–241

Franzke CW, Tasanen K, Schacke H, Zhou Z, Tryggvason K, Mauch C, Zigrino P, Sunnarborg S, Lee DC, Fahrenholz F, Bruckner-Tuderman L (2002) Transmembrane collagen XVII, an epithelial adhesion protein, is shed from the cell surface by ADAMs. EMBO J 21:5026–5035

Fuchs E, Weber K (1994) Intermediate filaments: structure, dynamics, function, and disease. Annu Rev Biochem 63:345–382

Fuchs P, Zorer M, Rezniczek GA, Spazierer D, Oehler S, Castanon MJ, Hauptmann R, Wiche G (1999) Unusual 5′ transcript complexity of plectin isoforms: novel tissue-specific exons modulate actin binding activity. Hum Mol Genet 8:2461–2472

Fujiwara S, Takeo N, Otani Y, Parry DA, Kunimatsu M, Lu R, Sasaki M, Matsuo N, Khaleduzzaman M, Yoshioka H (2001) Epiplakin, a novel member of the plakin family originally identified as a 450-kDa human epidermal autoantigen. Structure and tissue localization. J Biol Chem 276:13340–13347

Gache Y, Chavanas S, Lacour JP, Wiche G, Owaribe K, Meneguzzi G, Ortonne JP (1996) Defective expression of plectin/HD1 in epidermolysis bullosa simplex with muscular dystrophy. J Clin Invest 97:2289–2298

Gagnoux-Palacios L, Gache Y, Ortonne JP, Meneguzzi G (1997) Hemidesmosome assembly assessed by expression of a wild-type integrin β4 cDNA in junctional epidermolysis bullosa keratinocytes. Lab Invest 77:459–468

Gallicano GI, Kouklis P, Bauer C, Yin M, Vasioukhin V, Degenstein L, Fuchs E (1998) Desmoplakin is required early in development for assembly of desmosomes and cytoskeletal linkage. J Cell Biol 143:2009–2022

Gambaletta D, Marchetti A, Benedetti L, Mercurio AM, Sacchi A, Falcioni R (2000) Cooperative signaling between α6β4 integrin and ErbB-2 receptor is required to promote phosphatidylinositol 3-kinase-dependent invasion. J Biol Chem 275:10604–10610

Garcia-Alvarez B, Bobkov A, Sonnenberg A, de Pereda JM (2003) Structural and functional analysis of the actin binding domain of plectin suggests alternative mechanisms for binding to F-actin and integrin $\beta 4$. Structure 11:615–625

Geerts D, Fontao L, Nievers MG, Schaapveld RQ, Purkis PE, Wheeler GN, Lane EB, Leigh IM, Sonnenberg A (1999) Binding of integrin $\alpha 6\beta 4$ to plectin prevents plectin association with F-actin but does not interfere with intermediate filament binding. J Cell Biol 147:417–434

Georges-Labouesse E, Messaddeq N, Yehia G, Cadalbert L, Dierich A, Le Meur M (1996) Absence of integrin $\alpha 6$ leads to epidermolysis bullosa and neonatal death in mice. Nat Genet 13:370–373

Gimond C, Baudoin C, van der Neut R, Kramer D, Calafat J, Sonnenberg A (1998) Cre-loxP-mediated inactivation of the $\alpha 6A$ integrin splice variant in vivo: evidence for a specific functional role of $\alpha 6A$ in lymphocyte migration but not in heart development. J Cell Biol 143:253–266

Giudice GJ, Emery DJ, Diaz LA (1992) Cloning and primary structural analysis of the bullous pemphigoid autoantigen BP180. J Invest Dermatol 99:243–250

Giudice GJ, Emery DJ, Zelickson BD, Anhalt GJ, Liu Z, Diaz LA (1993) Bullous pemphigoid and herpes gestationis autoantibodies recognize a common non-collagenous site on the BP180 ectodomain. J Immunol 151:5742–5750

Green KJ, Jones JC (1996) Desmosomes and hemidesmosomes: structure and function of molecular components. FASEB J 10:871–881

Green KJ, Parry DA, Steinert PM, Virata ML, Wagner RM, Angst BD, Nilles LA (1990) Structure of the human desmoplakins. Implications for function in the desmosomal plaque. J Biol Chem 265:2603–2612

Green KJ, Virata ML, Elgart GW, Stanley JR, Parry DA (1992) Comparative structural analysis of desmoplakin, bullous pemphigoid antigen and plectin: members of a new gene family involved in organization of intermediate filaments. Int J Biol Macromol 14:145–153

Gumbiner BM (1996) Cell adhesion: the molecular basis of tissue architecture and morphogenesis. Cell 84:345–357

Guo L, Degenstein L, Dowling J, Yu QC, Wollmann R, Perman B, Fuchs E (1995) Gene targeting of BPAG1: abnormalities in mechanical strength and cell migration in stratified epithelia and neurologic degeneration. Cell 81:233–243

Hagg P, Rehn M, Huhtala P, Vaisanen T, Tamminen M, Pihlajaniemi T (1998) Type XIII collagen is identified as a plasma membrane protein. J Biol Chem 273:15590–15597

Hagg P, Vaisanen T, Tuomisto A, Rehn M, Tu H, Huhtala P, Eskelinen S, Pihlajaniemi T (2001) Type XIII collagen: a novel cell adhesion component present in a range of cell-matrix adhesions and in the intercalated discs between cardiac muscle cells. Matrix Biol 19:727–742

Hashimoto T, Wakabayashi T, Watanabe A, Kowa H, Hosoda R, Nakamura A, Kanazawa I, Arai T, Takio K, Mann DM, Iwatsubo T (2002) CLAC: a novel Alzheimer amyloid plaque component derived from a transmembrane precursor, CLAC-P/collagen type XXV. EMBO J 21:1524–1534

Hatzfeld M, Weber K (1991) Modulation of keratin intermediate filament assembly by single amino acid exchanges in the consensus sequence at the C-terminal end of the rod domain. J Cell Sci 99:351–362

Hatzfeld M, Haffner C, Schulze K, Vinzens U (2000) The function of plakophilin 1 in desmosome assembly and actin filament organization. J Cell Biol 149:209–222

Heins S, Aebi U (1994) Making heads and tails of intermediate filament assembly, dynamics and networks. Curr Opin Cell Biol 6:25–33

Hemler ME (1998) Integrin associated proteins. Curr Opin Cell Biol 10:578–585
Herrmann H, Wiche G (1987) Plectin and IFAP-300 K are homologous proteins binding to microtubule-associated proteins 1 and 2 and to the 240-kilodalton subunit of spectrin. J Biol Chem 262:1320–1325
Hieda Y, Nishizawa Y, Uematsu J, Owaribe K (1992) Identification of a new hemidesmosomal protein, HD1: a major, high molecular mass component of isolated hemidesmosomes. J Cell Biol 116:1497–1506
Hijikata T, Murakami T, Ishikawa H, Yorifuji H (2003) Plectin tethers desmin intermediate filaments onto subsarcolemmal dense plaques containing dystrophin and vinculin. Histochem Cell Biol 119:109–123
Hirako Y, Usukura J, Nishizawa Y, Owaribe K (1996) Demonstration of the molecular shape of BP180, a 180-kDa bullous pemphigoid antigen and its potential for trimer formation. J Biol Chem 271:13739–13745
Hogervorst F, Kuikman I, von dem Borne AEG Kr, Sonnenberg A (1990) Cloning and sequence analysis of $\beta 4$ cDNA: an integrin subunit that contains a unique 118 kd cytoplasmic domain. EMBO J 9:765–770
Hogervorst F, Kuikman I, van Kessel AG, Sonnenberg A (1991) Molecular cloning of the human $\alpha 6$ integrin subunit. Alternative splicing of $\alpha 6$ mRNA and chromosomal localization of the $\alpha 6$ and $\beta 4$ genes. Eur J Biochem 199:425–433
Hogervorst F, Admiraal LG, Niessen C, Kuikman I, Janssen H, Daams H, Sonnenberg A (1993) Biochemical characterization and tissue distribution of the A and B variants of the integrin $\alpha 6$ subunit. J Cell Biol 121:179–191
Hopkinson SB, Jones JC (2000) The N terminus of the transmembrane protein BP180 interacts with the N- terminal domain of BP230, thereby mediating keratin cytoskeleton anchorage to the cell surface at the site of the hemidesmosome. Mol Biol Cell 11:277–286
Hopkinson SB, Riddelle KS, Jones JC (1992) Cytoplasmic domain of the 180-kD bullous pemphigoid antigen, a hemidesmosomal component: molecular and cell biologic characterization. J Invest Dermatol 99:264–270
Hopkinson SB, Baker SE, Jones JC (1995) Molecular genetic studies of a human epidermal autoantigen (the 180-kD bullous pemphigoid antigen/BP180): identification of functionally important sequences within the BP180 molecule and evidence for an interaction between BP180 and $\alpha 6$ integrin. J Cell Biol 130:117–125
Imai T, Yoshie O (1993) C33 antigen and M38 antigen recognized by monoclonal antibodies inhibitory to syncytium formation by human T cell leukemia virus type 1 are both members of the transmembrane 4 superfamily and associate with each other and with CD4 or CD8 in T cells. J Immunol 151:6470–6481
Jarrett HW, Foster JL (1995) Alternate binding of actin and calmodulin to multiple sites on dystrophin. J Biol Chem 270:5578–5586
Jonkman MF, de Jong MC, Heeres K, Pas HH, van der Meer JB, Owaribe K, Martinez d, V, Niessen CM, Sonnenberg A (1995) 180-kD bullous pemphigoid antigen (BP180) is deficient in generalized atrophic benign epidermolysis bullosa. J Clin Invest 95:1345–1352
Kajiji S, Tamura RN, Quaranta V (1989) A novel integrin ($\alpha E\beta 4$) from human epithelial cells suggests a fourth family of integrin adhesion receptors. EMBO J 8:673–680
Karakesisoglou I, Yang Y, Fuchs E (2000) An epidermal plakin that integrates actin and microtubule networks at cellular junctions. J Cell Biol 149:195–208
Keep NH, Winder SJ, Moores CA, Walke S, Norwood FL, Kendrick-Jones J (1999) Crystal structure of the actin-binding region of utrophin reveals a head-to-tail dimer. Structure Fold Des 7:1539–1546

Kennel SJ, Godfrey V, Ch'ang LY, Lankford TK, Foote LJ, Makkinje A (1992) The β4 subunit of the integrin family is displayed on a restricted subset of endothelium in mice. J Cell Sci 101:145–150

Kennel SJ, Foote LJ, Cimino L, Rizzo MG, Chang LY, Sacchi A (1993) Sequence of a cDNA encoding the β4 subunit of murine integrin. Gene 130:209–216

Kere J, Srivastava AK, Montonen O, Zonana J, Thomas N, Ferguson B, Munoz F, Morgan D, Clarke A, Baybayan P, Chen EY, Ezer S, Saarialho-Kere U, de la Chappelle A, Schlessinger D (1996) X-linked anhidrotic (hypohidrotic) ectodermal dysplasia is caused by mutation in a novel transmembrane protein. Nat Genet 13:409–416

Kitadokoro K, Bordo D, Galli G, Petracca R, Falugi F, Abrignani S, Grandi G, Bolognesi M (2001) CD81 extracellular domain 3D structure: insight into the tetraspanin superfamily structural motifs. EMBO J 20:12–18

Koster J, Kuikman I, Kreft M, Sonnenberg A (2001) Two different mutations in the cytoplasmic domain of the integrin β4 subunit in nonlethal forms of epidermolysis bullosa prevent interaction of β4 with plectin. J Invest Dermatol 117:1405–1411

Koster J, Geerts D, Favre B, Borradori L, Sonnenberg A (2003) Analysis of the interactions between BP180, BP230, plectin and the integrin α6β4 important for hemidesmosome assembly. J Cell Sci 116:387–399

Kouklis PD, Hutton E, Fuchs E (1994) Making a connection: direct binding between keratin intermediate filaments and desmosomal proteins. J Cell Biol 127:1049–1060

Kowalczyk AP, Bornslaeger EA, Borgwardt JE, Palka HL, Dhaliwal AS, Corcoran CM, Denning MF, Green KJ (1997) The amino-terminal domain of desmoplakin binds to plakoglobin and clusters desmosomal cadherin-plakoglobin complexes. J Cell Biol 139:773–784

Krieger M (1992) Molecular flypaper and atherosclerosis: structure of the macrophage scavenger receptor. Trends Biochem Sci 17:141–146

Lagaudriere-Gesbert C, Le Naour F, Lebel-Binay S, Billard M, Lemichez E, Boquet P, Boucheix C, Conjeaud H, Rubinstein E (1997) Functional analysis of four tetraspans, CD9, CD53, CD81, and CD82, suggests a common role in costimulation, cell adhesion, and migration: only CD9 upregulates HB-EGF activity. Cell Immunol 182:105–112

Letai A, Coulombe PA, Fuchs E (1992) Do the ends justify the mean? Proline mutations at the ends of the keratin coiled-coil rod segment are more disruptive than internal mutations. J Cell Biol 116:1181–1195

Leung CL, Sun D, Zheng M, Knowles DR, Liem RK (1999) Microtubule actin cross-linking factor (MACF): a hybrid of dystonin and dystrophin that can interact with the actin and microtubule cytoskeletons. J Cell Biol 147:1275–1286

Leung CL, Zheng M, Prater SM, Liem RK (2001) The BPAG1 locus: Alternative splicing produces multiple isoforms with distinct cytoskeletal linker domains, including predominant isoforms in neurons and muscles. J Cell Biol 154:691–697

Leung CL, Green KJ, Liem RK (2002) Plakins: a family of versatile cytolinker proteins. Trends Cell Biol 12:37–45

Li K, Tamai K, Tan EM, Uitto J (1993) Cloning of type XVII collagen. Complementary and genomic DNA sequences of mouse 180-kilodalton bullous pemphigoid antigen (BPAG2) predict an interrupted collagenous domain, a transmembrane segment, and unusual features in the 5′-end of the gene and the 3′-untranslated region of the mRNA. J Biol Chem 268:8825–8834

Liu CG, Maercker C, Castanon MJ, Hauptmann R, Wiche G (1996) Human plectin: organization of the gene, sequence analysis, and chromosome localization (8q24). Proc Natl Acad Sci U S A 93:4278–4283

Liu Z, Diaz LA, Troy JL, Taylor AF, Emery DJ, Fairley JA, Giudice GJ (1993) A passive transfer model of the organ-specific autoimmune disease, bullous pemphigoid, using antibodies generated against the hemidesmosomal antigen, BP180. J Clin Invest 92:2480–2488

Lloyd C, Yu QC, Cheng J, Turksen K, Degenstein L, Hutton E, Fuchs E (1995) The basal keratin network of stratified squamous epithelia: defining K15 function in the absence of K14. J Cell Biol 129:1329–44

Lo SH, Janmey PA, Hartwig JH, Chen LB (1994) Interactions of tensin with actin and identification of its three distinct actin-binding domains. J Cell Biol 125:1067–1075

Lohi J (2001) Laminin-5 in the progression of carcinomas. Int J Cancer 94:763–767

Maatta A, DiColandrea T, Groot K, Watt FM (2001) Gene targeting of envoplakin, a cytoskeletal linker protein and precursor of the epidermal cornified envelope. Mol Cell Biol 21:7047–7053

Maecker HT, Todd SC, Levy S (1997) The tetraspanin superfamily: molecular facilitators. FASEB J 11:428–442

Magin TM, Schroder R, Leitgeb S, Wanninger F, Zatloukal K, Grund C, Melton DW (1998) Lessons from keratin 18 knockout mice: formation of novel keratin filaments, secondary loss of keratin 7 and accumulation of liver- specific keratin 8-positive aggregates. J Cell Biol 140:1441–1451

Mainiero F, Pepe A, Wary KK, Spinardi L, Mohammadi M, Schlessinger J, Giancotti FG (1995) Signal transduction by the $\alpha 6\beta 4$ integrin: distinct $\beta 4$ subunit sites mediate recruitment of Shc/Grb2 and association with the cytoskeleton of hemidesmosomes. EMBO J 14:4470–4481

Mainiero F, Pepe A, Yeon M, Ren Y, Giancotti FG (1996) The intracellular functions of $\alpha 6\beta 4$ integrin are regulated by EGF. J Cell Biol 134:241–253

Mannion BA, Berditchevski F, Kraeft SK, Chen LB, Hemler ME (1996) Transmembrane-4 superfamily proteins CD81 (TAPA-1), CD82, CD63, and CD53 specifically associated with integrin $\alpha 4\beta 1$ (CD49d/CD29). J Immunol 157:2039–2047

Marinkovich MP, Lunstrum GP, Keene DR, Burgeson RE (1992) The dermal-epidermal junction of human skin contains a novel laminin variant. J Cell Biol 119:695–703

McGowan KA, Marinkovich MP (2000) Laminins and human disease. Microsc Res Tech 51:262–279

McGrath JA, Gatalica B, Christiano AM, Li K, Owaribe K, McMillan JR, Eady RA, Uitto J (1995) Mutations in the 180-kD bullous pemphigoid antigen (BPAG2), a hemidesmosomal transmembrane collagen (COL17A1), in generalized atrophic benign epidermolysis bullosa. Nat Genet 11:83–86

McLean WH, Pulkkinen L, Smith FJ, Rugg EL, Lane EB, Bullrich F, Burgeson RE, Amano S, Hudson DL, Owaribe K, McGrath JA, McMillan JR, Eady RA, Leigh IM, Christiano AM, Uitto J (1996) Loss of plectin causes epidermolysis bullosa with muscular dystrophy: cDNA cloning and genomic organization. Genes Dev 10:1724–1735

McMillan JR, McGrath JA, Tidman MJ, Eady RA (1998) Hemidesmosomes show abnormal association with the keratin filament network in junctional forms of epidermolysis bullosa. J Invest Dermatol 110:132–137

Mellerio JE, Smith FJ, McMillan JR, McLean WH, McGrath JA, Morrison GA, Tierney P, Albert DM, Wiche G, Leigh IM, Geddes JF, Lane EB, Uitto J, Eady RA (1997) Recessive epidermolysis bullosa simplex associated with plectin mutations: infantile respiratory complications in two unrelated cases. Br J Dermatol 137:898–906

Meng JJ, Bornslaeger EA, Green KJ, Steinert PM, Ip W (1997) Two-hybrid analysis reveals fundamental differences in direct interactions between desmoplakin and cell type-specific intermediate filaments. J Biol Chem 272:21495–21503

Meng X, Klement JF, Leperi DA, Birk DE, Sasaki T, Timpl R, Uitto J, Pulkkinen L (2003) Targeted inactivation of murine laminin gamma2-chain gene recapitulates human junctional epidermolysis bullosa. J Invest Dermatol 121:720–731

Nakano A, Pulkkinen L, Murrell D, Rico J, Lucky AW, Garzon M, Stevens CA, Robertson S, Pfendner E, Uitto J (2001) Epidermolysis bullosa with congenital pyloric atresia: novel mutations in the β4 integrin gene (ITGB4) and genotype/phenotype correlations. Pediatr Res 49:618–626

Niessen CM, Cremona O, Daams H, Ferraresi S, Sonnenberg A, Marchisio PC (1994a) Expression of the integrin α6β4 in peripheral nerves: localization in Schwann and perineural cells and different variants of the β4 subunit. J Cell Sci 107:543–552

Niessen CM, Hogervorst F, Jaspars LH, de Melker AA, Delwel GO, Hulsman EH, Kuikman I, Sonnenberg A (1994b) The α6β4 integrin is a receptor for both laminin and kalinin. Exp Cell Res 211:360–367

Niessen CM, Raaij-Helmer MH, Hulsman EH, van der Neut, R, Jonkman MF, Sonnenberg A (1996) Deficiency of the integrin β4 subunit in junctional epidermolysis bullosa with pyloric atresia: consequences for hemidesmosome formation and adhesion properties. J Cell Sci 109:1695–1706

Niessen CM, Hulsman EH, Rots ES, Sanchez-Aparicio P, Sonnenberg A (1997a) Integrin α6β4 forms a complex with the cytoskeletal protein HD1 and induces its redistribution in transfected COS-7 cells. Mol Biol Cell 8:555–566

Niessen CM, Hulsman EH, Oomen LC, Kuikman I, Sonnenberg A (1997b) A minimal region on the integrin β4 subunit that is critical to its localization in hemidesmosomes regulates the distribution of HD1/plectin in COS-7 cells. J Cell Sci 110:1705–1716

Nievers MG, Schaapveld RQ, Oomen LC, Fontao L, Geerts D, Sonnenberg A (1998) Ligand-independent role of the β4 integrin subunit in the formation of hemidesmosomes. J Cell Sci 111:1659–1672

Nievers MG, Schaapveld RQ, Sonnenberg A (1999) Biology and function of hemidesmosomes. Matrix Biol 18:5–17

Nievers MG, Kuikman I, Geerts D, Leigh IM, Sonnenberg A (2000) Formation of hemidesmosome-like structures in the absence of ligand binding by the α6β4 integrin requires binding of HD1/plectin to the cytoplasmic domain of the β4 integrin subunit. J Cell Sci 113:963–973

Nikolic B, Mac NE, Mir B, Wiche G (1996) Basic amino acid residue cluster within nuclear targeting sequence motif is essential for cytoplasmic plectin-vimentin network junctions. J Cell Biol 134:1455–1467

Norgett EE, Hatsell SJ, Carvajal-Huerta L, Cabezas JC, Common J, Purkis PE, Whittock N, Leigh IM, Stevens HP, Kelsell DP (2000) Recessive mutation in desmoplakin disrupts desmoplakin-intermediate filament interactions and causes dilated cardiomyopathy, woolly hair and keratoderma. Hum Mol Genet 9:2761–2766

Norwood FL, Sutherland-Smith AJ, Keep NH, Kendrick-Jones J (2000) The structure of the N-terminal actin-binding domain of human dystrophin and how mutations in this domain may cause Duchenne or Becker muscular dystrophy. Structure Fold Des 8:481–491

O'Keefe EJ, Erickson HP, Bennett V (1989) Desmoplakin I and desmoplakin II. Purification and characterization. J Biol Chem 264:8310–8318

Okuda T, Matsuda S, Nakatsugawa S, Ichigotani Y, Iwahashi N, Takahashi M, Ishigaski T, Hamaguchi M (1999) Molecular cloning of macrophin, a human homologue of Drosophila kakapo with a close structural similarity to plectin and dystrophin. Biochem Biophys Res Commun 264:568–574

Okumura M, Uematsu J, Hirako Y, Nishizawa Y, Shimizu H, Kido N, Owaribe K (1999) Identification of the hemidesmosomal 500 kDa protein (HD1) as plectin. J Biochem (Tokyo) 126:1144–1150

Orian-Rousseau V, Aberdam D, Fontao L, Chevalier L, Meneguzzi G, Kedinger M, Simon-Assmann P (1996) Developmental expression of laminin-5 and HD1 in the intestine: epithelial to mesenchymal shift for the laminin $\gamma2$ chain subunit deposition. Dev Dyn 206:12–23

Peters B, Kirfel J, Bussow H, Vidal M, Magin TM (2001) Complete cytolysis and neonatal lethality in keratin 5 knockout mice reveal its fundamental role in skin integrity and in epidermolysis bullosa simplex. Mol Biol Cell 12:1775–1789

Pihlajaniemi T, Rehn M (1995) Two new collagen subgroups: membrane-associated collagens and types XV and XVII. Prog Nucleic Acid Res Mol Biol 50:225–262

Pulkkinen L, Christiano AM, Airenne T, Haakana H, Tryggvason K, Uitto J (1994) Mutations in the $\gamma2$ chain gene (LAMC2) of kalinin/laminin 5 in the junctional forms of epidermolysis bullosa. Nat Genet 6:293–297

Pulkkinen L, Smith FJ, Shimizu H, Murata S, Yaoita H, Hachisuka H, Nishikawa T, McLean WH, Uitto J (1996) Homozygous deletion mutations in the plectin gene (PLEC1) in patients with epidermolysis bullosa simplex associated with late-onset muscular dystrophy. Hum Mol Genet 5:1539–1546

Pulkkinen L, Kimonis VE, Xu Y, Spanou EN, McLean WH, Uitto J (1997a) Homozygous $\alpha6$ integrin mutation in junctional epidermolysis bullosa with congenital duodenal atresia. Hum Mol Genet 6:669–674

Pulkkinen L, McGrath J, Airenne T, Haakana H, Tryggvason K, Kivirikko S, Meneguzzi G, Ortonne JP, Christiano AM, Uitto J (1997b) Detection of novel LAMC2 mutations in Herlitz junctional epidermolysis Bullosa. Mol Med 3:124–135

Pulkkinen L, Kim DU, Uitto J (1998) Epidermolysis bullosa with pyloric atresia: novel mutations in the $\beta4$ integrin gene (ITGB4). Am J Pathol 152:157–166

Pyke C, Romer J, Kallunki P, Lund LR, Ralfkiaer E, Dano K, Tryggvason K (1994) The $\gamma2$ chain of kalinin/laminin 5 is preferentially expressed in invading malignant cells in human cancers. Am J Pathol 145:782–791

Pytela R, Wiche G (1980) High molecular weight polypeptides (270,000–340,000) from cultured cells are related to hog brain microtubule-associated proteins but copurify with intermediate filaments. Proc Natl Acad Sci U S A 77:4808–4812

Rabinovitz I, Toker A, Mercurio AM (1999) Protein kinase C-dependent mobilization of the $\alpha6\beta4$ integrin from hemidesmosomes and its association with actin-rich cell protrusions drive the chemotactic migration of carcinoma cells. J Cell Biol 146:1147–1160

Reddy DP, Muller H, Tran N, Nguyn H, Schaecke L, Bruckner-Tuderman L, Marinkovich P (1998) The extracellular domain of BP180 binds laminin-5. J Invest Dermatol 110:593a

Rezniczek GA, de Pereda JM, Reipert S, Wiche G (1998) Linking integrin $\alpha6\beta4$-based cell adhesion to the intermediate filament cytoskeleton: direct interaction between the $\beta4$ subunit and plectin at multiple molecular sites. J Cell Biol 141:209–225

Roh JY, Yee C, Lazarova Z, Hall RP, Yancey KB (2000) The 120-kDa soluble ectodomain of type XVII collagen is recognized by autoantibodies in patients with pemphigoid and linear IgA dermatosis. Br J Dermatol 143:104–111

Roper K, Gregory SL, Brown NH (2002) The 'Spectraplakins': cytoskeletal giants with characteristics of both spectrin and plakin families. J Cell Sci 115:4215–4225

Ruhrberg C, Hajibagheri MA, Simon M, Dooley TP, Watt FM (1996) Envoplakin, a novel precursor of the cornified envelope that has homology to desmoplakin. J Cell Biol 134:715–729

Ruhrberg C, Hajibagheri MA, Parry DA, Watt FM (1997) Periplakin, a novel component of cornified envelopes and desmosomes that belongs to the plakin family and forms complexes with envoplakin. J Cell Biol 139:1835–1849

Ruzzi L, Gagnoux-Palacios L, Pinola M, Belli S, Meneguzzi G, D'Alessio M, Zambruno G (1997) A homozygous mutation in the integrin α6 gene in junctional epidermolysis bullosa with pyloric atresia. J Clin Invest 99:2826–2831

Ryan MC, Tizard R, Van Devanter DR, Carter WG (1994) Cloning of the LamA3 gene encoding the α3 chain of the adhesive ligand epiligrin. Expression in wound repair. J Biol Chem 269:22779–22787

Ryan MC, Lee K, Miyashita Y, Carter WG (1999) Targeted disruption of the LAMA3 gene in mice reveals abnormalities in survival and late stage differentiation of epithelial cells. J Cell Biol 145:1309–1323

Sanchez-Aparicio P, Martinez d, V, Niessen CM, Borradori L, Kuikman I, Hulsman EH, Fassler R, Owaribe K, Sonnenberg A (1997) The subcellular distribution of the high molecular mass protein, HD1, is determined by the cytoplasmic domain of the integrin β4 subunit. J Cell Sci 110:169–178

Schaapveld RQ, Borradori L, Geerts D, van Leusden MR, Kuikman I, Nievers MG, Niessen CM, Steenbergen RD, Snijders PJ, Sonnenberg A (1998) Hemidesmosome formation is initiated by the β4 integrin subunit, requires complex formation of β4 and HD1/plectin, and involves a direct interaction between β4 and the bullous pemphigoid antigen 180. J Cell Biol 142:271–284

Schwarz EM, Benzer S (1997) *Calx*, a Na-Ca exchanger gene of *Drosophila melanogaster*. Proc Natl Acad Sci U S A 94:10249–10254

Seifert GJ, Lawson D, Wiche G (1992) Immunolocalization of the intermediate filament-associated protein plectin at focal contacts and actin stress fibers. Eur J Cell Biol 59:138–147

Shang M, Koshikawa N, Schenk S, Quaranta V (2001) The LG3 module of laminin-5 harbors a binding site for integrin α3β1 that promotes cell adhesion, spreading, and migration. J Biol Chem 276:33045–33053

Sincock PM, Mayrhofer G, Ashman LK (1997) Localization of the transmembrane 4 superfamily (TM4SF) member PETA-3 (CD151) in normal human tissues: comparison with CD9, CD63, and α5β1 integrin. J Histochem Cytochem 45:515–525

Smith EA, Fuchs E (1998) Defining the interactions between intermediate filaments and desmosomes. J Cell Biol 141:1229–1241

Smith FJ, Eady RA, Leigh IM, McMillan JR, Rugg EL, Kelsell DP, Bryant SP, Spurr NK, Geddes JF, Kirtschig G, Milana G, de Bono AG, Owaribe K, Wiche G, Pulkkinen L, Uitto J, McLean WH, Lane EB (1996) Plectin deficiency results in muscular dystrophy with epidermolysis bullosa. Nat Genet 13:450–457

Snellman A, Tu H, Vaisanen T, Kvist AP, Huhtala P, Pihlajaniemi T (2000) A short sequence in the N-terminal region is required for the trimerization of type XIII collagen and is conserved in other collagenous transmembrane proteins. EMBO J 19:5051–5059

Sonnenberg A, Linders CJ, Daams JH, Kennel SJ (1990) The α6β1 (VLA-6) and α6β4 protein complexes: tissue distribution and biochemical properties. J Cell Sci 96:207–217

Sonnenberg A, Calafat J, Janssen H, Daams H, Raaij-Helmer LM, Falcioni R, Kennel SJ, Aplin JD, Baker J, Loizidou M, Garrod D (1991) Integrin α6β4 complex is located in hemidesmosomes, suggesting a major role in epidermal cell-basement membrane adhesion. J Cell Biol 113:907–917

Spinardi L, Ren YL, Sanders R, Giancotti FG (1993) The $\beta 4$ subunit cytoplasmic domain mediates the interaction of $\alpha 6\beta 4$ integrin with the cytoskeleton of hemidesmosomes. Mol Biol Cell 4:871–884

Spinardi L, Einheber S, Cullen T, Milner TA, Giancotti FG (1995) A recombinant tail-less integrin $\beta 4$ subunit disrupts hemidesmosomes, but does not suppress $\alpha 6\beta 4$-mediated cell adhesion to laminins. J Cell Biol 129:473–487

Stanley JR (1993) Cell adhesion molecules as targets of autoantibodies in pemphigus and pemphigoid, bullous diseases due to defective epidermal cell adhesion. Adv Immunol 53:291–325

Stappenbeck TS, Green KJ (1992) The desmoplakin carboxyl terminus coaligns with and specifically disrupts intermediate filament networks when expressed in cultured cells. J Cell Biol 116:1197–1209

Stappenbeck TS, Bornslaeger EA, Corcoran CM, Luu HH, Virata ML, Green KJ (1993) Functional analysis of desmoplakin domains: specification of the interaction with keratin versus vimentin intermediate filament networks. J Cell Biol 123:691–705

Steinbock FA, Nikolic B, Coulombe PA, Fuchs E, Traub P, Wiche G (2000) Dose-dependent linkage, assembly inhibition and disassembly of vimentin and cytokeratin 5/14 filaments through plectin's intermediate filament-binding domain. J Cell Sci 113:483–491

Steinert PM, North AC, Parry DA (1994) Structural features of keratin intermediate filaments. J Invest Dermatol 103:19S–24S

Stepp MA, Spurr-Michaud S, Tisdale A, Elwell J, Gipson IK (1990) $\alpha 6\beta 4$ integrin heterodimer is a component of hemidesmosomes. Proc Natl Acad Sci U S A 87:8970–8974

Sterk LM, Geuijen CA, Oomen LC, Calafat J, Janssen H, Sonnenberg A (2000) The tetraspan molecule CD151, a novel constituent of hemidesmosomes, associates with the integrin $\alpha 6\beta 4$ and may regulate the spatial organization of hemidesmosomes. J Cell Biol 149:969–982

Sun Y, Zhang J, Kraeft SK, Auclair D, Chang MS, Liu Y, Sutherland R, Salgia R, Griffin JD, Ferland LH, Chen LB (1999) Molecular cloning and characterization of human trabeculin-α, a giant protein defining a new family of actin-binding proteins. J Biol Chem 274:33522–33530

Suzuki S, Naitoh Y (1990) Amino acid sequence of a novel integrin $\beta 4$ subunit and primary expression of the mRNA in epithelial cells. EMBO J 9:757–763

Svitkina TM, Verkhovsky AB, Borisy GG (1996) Plectin sidearms mediate interaction of intermediate filaments with microtubules and other components of the cytoskeleton. J Cell Biol 135:991–1007

Takizawa Y, Shimizu H, Nishikawa T, Hatta N, Pulkkinen L, Uitto J (1997) Novel ITGB4 mutations in a patient with junctional epidermolysis bullosa-pyloric atresia syndrome and altered basement membrane zone immunofluorescence for the $\alpha 6\beta 4$ integrin. J Invest Dermatol 108:943–946

Tamura RN, Rozzo C, Starr L, Chambers J, Reichardt LF, Cooper HM, Quaranta V (1990) Epithelial integrin $\alpha 6\beta 4$: complete primary structure of $\alpha 6$ and variant forms of $\beta 4$. J Cell Biol 111:1593–1604

Tamura RN, Cooper HM, Collo G, Quaranta V (1991) Cell type-specific integrin variants with alternative α chain cytoplasmic domains. Proc Natl Acad Sci U S A 88:10183–10187

Tasanen K, Eble JA, Aumailley M, Schumann H, Baetge J, Tu H, Bruckner P, Bruckner-Tuderman L (2000) Collagen XVII is destabilized by a glycine substitution mutation in the cell adhesion domain Col15. J Biol Chem 275:3093–3099

Timpl R, Brown JC (1994) The laminins. Matrix Biol 14:275–281

Trusolino L, Bertotti A, Comoglio PM (2001) A signaling adapter function for $\alpha6\beta4$ integrin in the control of HGF-dependent invasive growth. Cell 107:643–654

Uematsu J, Nishizawa Y, Sonnenberg A, Owaribe K (1994) Demonstration of type II hemidesmosomes in a mammary gland epithelial cell line, BMGE-H. J Biochem (Tokyo) 115:469–476

Uitto J, Pulkkinen L, Smith FJ, McLean WH (1996) Plectin and human genetic disorders of the skin and muscle. The paradigm of epidermolysis bullosa with muscular dystrophy. Exp Dermatol 5:237–246

van den Heuvel AP, Vries-Smits AM, van Weeren PC, Dijkers PF, de Bruyn KM, Riedl JA, Burgering BM (2002) Binding of protein kinase B to the plakin family member periplakin. J Cell Sci 115:3957–3966

van der Neut R, Krimpenfort P, Calafat J, Niessen CM, Sonnenberg A (1996) Epithelial detachment due to absence of hemidesmosomes in integrin $\beta4$ null mice. Nat Genet 13:366–369

van Leusden MR, Kuikman I, Sonnenberg A (1997) The unique cytoplasmic domain of the human integrin variant $\beta4E$ is produced by partial retention of intronic sequences. Biochem Biophys Res Commun 235:826–830

Verrando P, Pisani A, Ortonne JP (1988) The new basement membrane antigen recognized by the monoclonal antibody GB3 is a large size glycoprotein: modulation of its expression by retinoic acid. Biochim Biophys Acta 942:45–56

Verrando P, Blanchet-Bardon C, Pisani A, Thomas L, Cambazard F, Eady RA, Schofield O, Ortonne JP (1991) Monoclonal antibody GB3 defines a widespread defect of several basement membranes and a keratinocyte dysfunction in patients with lethal junctional epidermolysis bullosa. Lab Invest 64:85–92

Vidal F, Baudoin C, Miquel C, Galliano MF, Christiano AM, Uitto J, Ortonne JP, Meneguzzi G (1995) Cloning of the laminin $\alpha3$ chain gene (LAMA3) and identification of a homozygous deletion in a patient with Herlitz junctional epidermolysis bullosa. Genomics 30:273–280

Virata ML, Wagner RM, Parry DA, Green KJ (1992) Molecular structure of the human desmoplakin I and II amino terminus. Proc Natl Acad Sci U S A 89:544–548

Wadsworth S, Halvorson MJ, Coligan JE (1992) Developmentally regulated expression of the $\beta4$ integrin on immature mouse thymocytes. J Immunol 149:421–428

Weaver VM, Lelievre S, Lakins JN, Chrenek MA, Jones JC, Giancotti F, Werb Z, Bissell MJ (2002) $\beta4$ integrin-dependent formation of polarized three-dimensional architecture confers resistance to apoptosis in normal and malignant mammary epithelium. Cancer Cell 2:205–216

Wiche G (1998) Role of plectin in cytoskeleton organization and dynamics. J Cell Sci 111:2477–2486

Wiche G, Baker MA (1982) Cytoplasmic network arrays demonstrated by immunolocalization using antibodies to a high molecular weight protein present in cytoskeletal preparations from cultured cells. Exp Cell Res 138:15–29

Wiche G, Krepler R, Artlieb U, Pytela R, Denk H (1983) Occurrence and immunolocalization of plectin in tissues. J Cell Biol 97:887–901

Wiche G, Krepler R, Artlieb U, Pytela R, Aberer W (1984) Identification of plectin in different human cell types and immunolocalization at epithelial basal cell surface membranes. Exp Cell Res 155:43–49

Wiche G, Becker B, Luber K, Weitzer G, Castanon MJ, Hauptmann R, Stratowa C, Stewart M (1991) Cloning and sequencing of rat plectin indicates a 466-kD polypeptide chain with a three-domain structure based on a central α-helical coiled coil. J Cell Biol 114:83–99

Wright MD, Tomlinson MG (1994) The ins and outs of the transmembrane 4 superfamily. Immunol Today 15:588–594

Xiong JP, Stehle T, Zhang R, Joachimiak A, Frech M, Goodman SL, Arnaout MA (2002) Crystal structure of the extracellular segment of integrin $\alpha V\beta 3$ in complex with an Arg-Gly-Asp ligand. Science 296:151–155

Yang Y, Dowling J, Yu QC, Kouklis P, Cleveland DW, Fuchs E (1996) An essential cytoskeletal linker protein connecting actin microfilaments to intermediate filaments. Cell 86:655–665

Yang Y, Bauer C, Strasser G, Wollman R, Julien JP, Fuchs E (1999) Integrators of the cytoskeleton that stabilize microtubules. Cell 98:229–238

Yauch RL, Berditchevski F, Harler MB, Reichner J, Hemler ME (1998) Highly stoichiometric, stable, and specific association of integrin $\alpha 3\beta 1$ with CD151 provides a major link to phosphatidylinositol 4- kinase, and may regulate cell migration. Mol Biol Cell 9:2751–2765

Yurchenco PD, O'Rear JJ (1994) Basal lamina assembly. Curr Opin Cell Biol 6:674–681

# Part III
Immunoglobulin Superfamily

CEA-Related CAMs

A. K. Horst · C. Wagener (✉)

Institut für Klinische Chemie, Universitätsklinikum Hamburg-Eppendorf,
Martinistr. 52, 20251 Hamburg, Germany
wagener@uke.uni-hamburg.de

1	Introduction	284
2	Genomic Organization and Regulation of Transcription	287
2.1	Gene Structure	287
2.2	Regulation of Transcription	288
3	Three-Dimensional Structure	288
4	Expression Pattern	291
4.1	Expression of CEACAMs in Healthy Adult Tissues	291
4.2	Expression of CEACAMs in Healthy Adult Colon	294
4.3	Expression of CEACAMs During Gestation and Embryonic Development	295
4.4	Dysregulation of the Expression of CEACAMs in Human Tumors	296
5	Biological Functions of CEACAMs	299
5.1	CEACAMs Are Cellular Adhesion Molecules	299
5.2	CEACAMs Display Versatile Signal Transduction Properties	301
5.2.1	Mechanisms of Signal Transduction by CEACAM1	301
5.2.2	CEACAM1 Interacts with Components of the Cytoskeleton	303
5.2.3	CEACAM1 Acts as a Tumor Suppressor	304
5.2.4	CEACAM1 Promotes Invasion	306
5.2.5	CEACAM1 Is a Substrate for the Insulin Receptor and Regulates Insulin Clearance	307
5.3	CEACAMs Are Modulators of the Innate and Adaptive Immune Response	309
5.3.1	CEACAM1 Is a Positive Regulator of Neutrophil Effector Function	309
5.3.2	CEACAM1 Regulates T and B Lymphocyte Function	311
5.3.3	CEACAM1 Is a Novel Co-inhibitory Receptor on Human Natural Killer Cells	312
5.4	CEACAMs Are Receptors for Microbes and Viruses	313
5.4.1	CEACAMs Are Receptors for *Salmonellae* and *Escherichia coli*	313
5.4.2	CEACAMs Are Receptors for Neisseria	314
5.4.3	CEACAMs Are Receptors for *Haemophilus Influenzae*	316
5.4.4	CEACAMs Are Receptors for Murine Hepatitis Virus	317
5.5	CEACAM1 Modulates Angiogenesis	319
5.5.1	CEACAM1 Is an Angiogenic Growth Factor	319
5.5.2	CEACAM1 Induces Secretion of Angiostatic Factors in CEACAM1-Transfected Prostate Carcinoma Cells	321
6	Perspectives	322
	References	324

Abstract The carcinoembryonic antigen (CEA) family comprises a large number of cellular surface molecules, the CEA-related cell adhesion molecules (CEACAMs), which belong to the Ig superfamily. CEACAMs exhibit a complex expression pattern in normal and malignant tissues. The majority of the CEACAMs are cellular adhesion molecules that are involved in a great variety of distinct cellular processes, for example in the integration of cellular responses through homo- and heterophilic adhesion and interaction with a broad selection of signal regulatory proteins, i.e., integrins or cytoskeletal components and tyrosine kinases. Moreover, expression of CEACAMs affects tumor growth, angiogenesis, cellular differentiation, immune responses, and they serve as receptors for commensal and pathogenic microbes. Recently, new insights into CEACAM structure and function became available, providing further elucidation of their kaleidoscopic functions.

Keywords CEA-related cellular adhesion molecules · Immunoglobulin superfamily member · Adhesion · Signal transduction · Microbial receptor

1
Introduction

Carcinoembryonic antigen (CEA)-related cellular adhesion molecules (CEACAMs; pronounced C-CAMs) belong to the Ig superfamily of cellular surface molecules. CEA was discovered by Gold and Freedman as a tumor-associated antigen in human colorectal carcinoma (Gold and Freedman 1965). Originally, CEA was considered an oncofetal protein that is re-expressed in adult tissues only during carcinogenesis. Later, it became evident that CEA is expressed both in embryonic and healthy adult tissues. CEA levels in serum are elevated during the progression of various malignant diseases, namely cancer of the colon, breast, and lung. CEA serves as a clinical tumor marker and is of important prognostic relevance in the evaluation of progressive colonic carcinoma: increasing serum levels indicate recurrence and residual disease (Thompson 1991). After the cDNA of CEA had been cloned and characterized independently by various groups, CEA could be assigned to the Ig superfamily by sequence homology studies (Beauchemin et al. 1987; Oikawa et al. 1987; Paxton 1987; Zimmermann et al. 1987; Schrewe et al. 1990).

Members of the Ig superfamily of cell surface proteins do not only exert immune regulatory functions, but they also mediate cellular recognition and adhesion (Williams and Barclay 1988; Brümmendorf and Lemmon 2001; Juliano 2002). In fact, out of 26,383 human genes with a known or predicted function, 577 (1.9%) encode proteins that are involved in cell adhesion, and 264 (0.9%) encode immunoglobulins. Proteins containing Ig domains are among the most frequently expressed in the human genome, with 930 Ig domains contained in 381 Ig family members (Venter et al. 2001). Ig domains can form rods when arranged in series, and their ability to specifically recognize self or non-self proteins, or participate in *cis*- and *trans*-interactions, and regulate signaling and adhesion, make them principal players in orchestrating complex cellular responses. Moreover, alternative splicing and extensive N-linked glycosylation are used as a versatile tool to generate different isoforms, creating an even broader platform for recognition processes and signal integration.

The CEA-related adhesion molecules were discovered upon their cross-reactivity with CEA antisera and based on sequence homology studies after cloning of novel cDNAs. The first CEACAMs to be identified were the NCAs (non-specific cross-reacting antigens), independently described by the group of von Kleist et al. (1972) and Mach et al. (1972). Biliary glycoprotein (BGP) was originally described by Svenberg (1976) as being expressed in normal human gall bladder mucosa and at the bile canalicular surface of hepatocytes. BGP, also referred to as NCA-160, C-CAM (cell–cell adhesion molecule), C-CAM105, H4A, or pp120, is now called CEACAM1. In the course of the discovery of new genes and proteins that belong to the CEA gene or protein family, their nomenclature became quite inconsistent. Novel proteins were also named CGMs (CEA-gene family members), and well-conserved homologs to CEA-like genes in rodents have been assigned a variety of names. To date, 29 CEA-like genes are known in the human genome, 19 in the murine genome and 7 in the rat. According to their structural similarities, they were classified into three subgroups: CEA-like-genes, the PSG-subgroup (pregnancy-specific glycoproteins), and the pseudogenes. An overview of the CEA-family subgroup is shown in Fig. 1.

In this review, the nomenclature used is based on its most recent revision (Beauchemin et al. 1999; http://cea.klinikum.uni-muenchen.de). For common historical names and abbreviations, the reader is referred to Fig. 1 and Beauchemin et al. (1999).

All members of the CEA-subgroup share the following common characteristics. They are members of the Ig superfamily, they are membrane bound, and they are heavily glycosylated adhesion molecules (Odin et al 1986; Beauchemin et al. 1999). CEACAMs are multifunctional. Their biological functions comprise cellular homophilic and heterophilic adhesion, suppression of tumor cell growth, regulation of cell growth and differentiation, binding of a variety of different pathogens, and acting as growth factors for endothelial cells, as well as binding to a variety of intracellular adaptors in central signal transduction cascades.

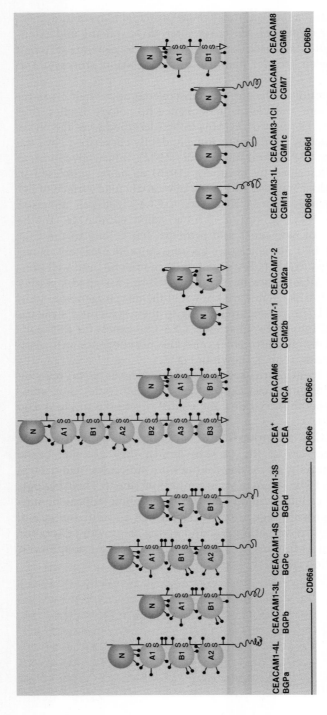

Fig. 1. Schematic presentation and nomenclature of the CEA-subgroup. The immunoglobulin variable-like domains are shown in *dark gray*. Note that they do not contain disulfide linkages as in the immunoglobulin constant-like domains. Positions of N-linked glycosylation sites are represented by *lollipops*. Linkage to the plasma membrane by glycosylphosphatidylinositol anchors is indicated by *arrows*. Different splice variants with either a long (*L*) or short (*S*) cytoplasmic domain are indicated. Names of the individual family members according to the latest revision of the CEA gene family nomenclature are shown in the *upper line* (Beauchemin 1999). In the *second line*, common historical names of the proteins and different splice variants are used. In the *third line*, CEACAMs are grouped according to their designation as leukocyte differentiation antigens. +, For historical reasons, the product of the *CEACAM5* gene, the CEA protein, is referred to as CEA. (Reproduced from Hammarström and Baranov 2001, with permission from Elsevier Science)

2
Genomic Organization and Regulation of Transcription

2.1
Gene Structure

The human CEA gene family comprises 29 genes that have been divided into three subgroups according to their DNA and amino acid sequence and structural homology: (1) 12 CEA genes, 7 of which are expressed (CEACAM1-12); (2) 11 PSG genes, 9 of which are expressed, and (3) a group of 6 pseudogenes (CEACAM13 through 18), for which no cDNA has yet been identified.

The CEA gene family emerged from a common ancestor gene that initially gave rise to block of three genes with the same transcriptional orientation. Repeated duplication and inversion of these gene blocks produced the CEA family gene cluster, located on the long arm of human chromosome 19 (q13.2, between CYP2A and D19S15). Within its expanse of 1.8 Mb, the CEA- and PSG-subgroup genes are organized in two clusters of 250 and 850 kb, respectively. The proximal half includes members of the CEA subgroup, and the distal half comprises the 11 PSG-subgroup genes with the pseudogenes interspersed. The two clusters are interrupted by a 700-kb region encoding 5 nonrelated proteins (Thompson et al. 1989; Kahn et al. 1992; Kahn et al. 1994; Olsen et al. 1994; Teglund et al. 1994). CEACAM1 is the most ancestral gene of the CEA-gene family. Whereas there is only one gene encoding the human and rat CEACAM1-protein, there are two closely related genes in the mouse (*Ceacam1* and *Ceacam2*).

The aminoterminal Ig variable-type domain (IgV) is highly homologous to that of other CEA-gene family members. At the same time, the IgV-like domain is more susceptible to variations in its amino acid sequence than the Ig constant-like (IgC) C2-like domains. CEACAMs are undergoing a dynamic evolution. All members of the CEA gene family share a common intron–exon organization. They contain an exon encoding 5′ UTR (untranslated region) and part of the leader peptide, followed by an exon encoding the rest of the signal-peptide (34 aa) for membrane targeting and the N-terminal IgV-like domain (108 aa). It is followed by exons encoding the IgC set-like domains of either the A- (93 aa) or the B-type (85 aa) that are arranged in A+B pairs. Each of these exons encodes a complete Ig domain. Downstream of this region are exons encoding the C-terminal portion and the 3′ UTR.

The number of Ig domains in CEACAM1 varies from one to four domains as a result of alternative mRNA splicing. This also gives rise to two isoforms that differ in the length of their cytoplasmic domains, a long form, referred to as CEACAM1-L, composed of 70–73 amino acids, or a short form, referred to as CEACAM1-S, and comprising 10–12 amino acids (McCuaig et al. 1992; Edlund et al. 1993; Najjar et al. 1993; McCuaig et al. 1993). In contrast to the human *CEACAM1* gene, there are two allelic variants in the mouse and in the rat, *Ceacam1a* and *Ceacam1b* (Edlund et al. 1993; McCuaig et al. 1993). These allelic variants exhibit their greatest genetic variation within the N-terminal domain. The most important isoforms of murine CEACAM1 contain either two or four extracellular Ig-like domains (the N-teminal domain and either the A2 or A1 and the B1 domain (Beauchemin 1999).

So far, the splice variants of human CEACAM1 are the best characterized (Barnett et al. 1989; Barnett et al. 1993). To date, there are at least 12 known splice variants. The most prominent isoforms are CEACAM1-4L and CEACAM1-4S, each containing four extracellular Ig-like domains and either a long or a short cytoplasmic tail.

2.2
Regulation of Transcription

The promoters of the CEA gene family member genes exhibit features of housekeeping genes. They lack classical TATA or CCAAT boxes, and they contain GC-rich regions and Sp1 sites. In the human and rat *CEACAM1*-gene, binding sites for activator protein (AP)1 and AP2-like factors, USF (upstream stimulatory factor), and HNF-4 (hepatic nuclear factor-4) could be identified (Hauck et al. 1994; Najjar et al. 1996).

However, in contrast to classical housekeeping genes, *CEACAM1*-genes are subjected to cell- and tissue-specific developmental and differential regulation (Schrewe et al. 1990; Hauck et al. 1991) Furthermore, they respond to changes in the hormonal status in certain tissues and cell lines: their expression can be induced by cyclic adenosine monophosphate (cAMP), retinoids, androgens, estrogens, glucocorticoids, or insulin (Svalander et al. 1990; Botling et al. 1995; Hsieh et al. 1995; Daniels et al. 1996; Najjar et al. 1996; Makarovskiy et al. 1999; Phan et al. 2001). CEA gene family members respond to inflammatory stimuli as well, such as interferons, tumor necrosis factors, and interleukins (Takahashi et al. 1993; Chen et al. 1996; Kammerer et al. 1998).

3
Three-Dimensional Structure

Modeling of the three-dimensional structure of members of the CEA gene family revealed the characteristic Ig fold (Bates et al. 1992; Boehm 1996; Tan et al. 2002). All human CEA-subgroup members contain one IgV-like domain and zero to six C-like domains. They are linked to the membrane by a transmembrane anchor, followed by a long or short cytoplasmic tail, a glycosylphosphatidylinositol (GPI)-anchor or, in case of the soluble PSG-subgroup, a short hydrophilic tail. In contrast to the human CEA family proteins, rodent CEACAMs may contain several IgV-like domains (Rudert et al. 1992). GPI-linked proteins of the CEA family are not expressed in rodents. CEA-related adhesion molecules are heavily glycosylated; they contain zero to six Asn-X-Ser/Thr-motifs for potential N-linked glycosylation per Ig domain. The glycosylation of CEACAMs comprises high-mannose and complex type oligosaccharides like lactosaminoglycans type I and type II chains that are terminated by fucosyl- and sialyl-residues (Odin et al. 1986; Mahrenholz et al. 1993; Stocks et al. 1993; Fukushima et al. 1995; Kannicht et al. 1999; Sanders and Kerr 1999). Type I and type II lactosaminoglycans constitute the Lewis blood group antigens. Prominent carbohydrate moieties of this group on CEACAM1 and CEA are Lewis^x and sialyl-Lewis^x antigens (Sanders and Kerr 1999). On CEACAM1, the presence of high-mannose residues seems to be restricted to its membrane proximal A2-domain (Mahrenholz et al. 1993; see also Fig. 7).

Modeling of the CEACAMs' three-dimensional structures, i.e., CEACAM1 and CEA, has been performed on the basis of structural comparison with other representative members of the Ig superfamily, namely CD2, CD4, CD8, and the Bence-Jones protein REI (Tan et al. 2002 and references therein). In a recent structural analysis performed with a soluble murine CEACAM1a, splice variant comprising the N-terminal IgV-like domain and the membrane proximal IgC-like domain (CEACAM1a[1,4]), the structure of the IgC-like domains in CEACAMs was found to resemble the I-like Ig fold rather than the C2-fold. Referring to data obtained earlier after structural modeling of human CEA, it can be hypothesized due to high inter-species structural conservation of the CEACAMs, that the A-type domains belong to the I1 set and the B-type domains belong to the I2 set fold. The variable-like N-terminal domains of the CEACAMs lack the characteristic disulfide linkage between the beta strands B and F but nonetheless meet the criteria for a V set Ig-like fold (Bates et al. 1992; Harpaz and Chothia 1994; Chothia et al. 1998; Wang and Springer 1998; Tan et al. 2002).

CEACAMs exhibit unique structural features when compared to other members of the Ig superfamily. Structurally, the N-terminal domain of CEACAM1 and other members of the CEA family is exceptional in that its CC' loop is convoluted and folds back onto the A'GFCC'C'' β-sheets, called CFG face hereafter (Fig. 2). Moreover, within the CEA family, the ABED face in the N-terminal domain is much more conserved than the CFG face. The CFG faces of variable-like domains are frequently used in cell surface recognition (Stuart et al. 1995; Wang and Springer 1998). This has been demonstrated for viral binding to host cell receptors, as specific sequences within this region confer the specificity for a cognate receptor. The variability in this region of the CEACAMs determines their distinctive binding properties in homophilic and heterophilic adhesion but also in their binding to extracellular ligands and as targets for microbes and viruses. Regarding the role of the N-terminal domain in *trans*-homophilic adhesion among CEACAM1

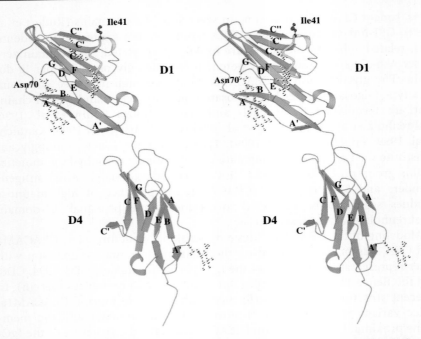

Fig. 2. Stereo view of the ribbon drawing of soluble murine CEACAM1a[1,4], which contains the amino-terminal IgV-like (*D1*) and the membrane proximal IgC-like domain (*D4*). The *ribbon drawing* reveals the anti-parallel β-sheets that constitute the immunoglobulin fold. The uniquely convoluted CC′ loop in the N-terminal domain which is involved in molecular recognition processes by CEACAMs, such as binding murine hepatitis virus and other ligands (see there in this review), is highlighted in *yellow*. The predicted key virus-binding residue Ile41 on the CC′ loop is shown in *red* in *ball-and-stick* representation. The N-linked glycan at Asn70 in the D1 domain that is conserved through the whole CEA family is labeled accordingly. (Reproduced from Tan et al. 2002, with permission)

and CEA, for example, it has been shown that interactions between the CGF face, more specifically the CC′ loop, are essential to promote homophilic binding (Teixeira et al. 1994; Watt et al. 2001).

This review will focus on functions of CEACAM1, as they are known to date. The data summarized here have been collected from various experimental in vitro-model systems, as well as from in vivo studies in mouse and rat. Additionally, clinical data are included.

4
Expression Pattern

4.1
Expression of CEACAMs in Healthy Adult Tissues

CEACAMs display a very heterogeneous expression pattern. Crossreactivity between several anti-CEACAM antibodies hampered precise characterization of their individual expression pattern until more specific antibodies became available. Summaries about reactivity of certain anti-CEACAM antibodies and antisera can be found in Nap et al. (1992) Schölzel et al. (2000) and Watt et al. (2001).

To date, the expression patterns of CEA (encoded by the *CEACAM5* gene), CEACAM1, CEACAM6, and CEACAM7 have been characterized best (Prall et al. 1996; Stanners 1998). Recently, more detailed information about the expression pattern of CEACAM6 and CEACAM7 have become available (Schölzel et al. 2000). However, there is relatively little information available on the expression pattern of CEACAM3, CEACAM4, and CEACAM8. Generally, CEACAMs are subjected to developmental and differential regulation in a spatiotemporal fashion.

Despite their different sites of expression, CEACAMs can be categorized into four major groups:

1. Selective epithelial expression pattern: CEA, CEACAM7—with pronounced apical expression
2. Expression on granulocytes: CEACAM3,8
3. Broad expression: CEACAM1,6
4. Predominant expression pattern in the syncytiotrophoblast: CEACAM-ps1–11 (formerly PSGs,)

CEACAM1, CEACAM3, and CEACAM4 contain a hydrophobic transmembrane domain followed by a long or short cytoplasmic domain, whereas CEACAM2, CEA, CEACAM6, and CEACAM7 are attached to the membrane via a glycosylphosphatidyl inositol anchor. CEA family members with GPI anchors are not expressed in rodents. The expression pattern of CEACAMs in mouse and human is very similar. However, in contrast to human, there are two CEACAM1 genes in mice, *Ceacam1* and *Ceacam2*, that display slightly different expression patterns; whereas gene products of the *Ceacam1* gene are predominantly found in the majority of epithelia, leucocytes, and endothelia, *Ceacam2* is mainly expressed in spleen, kidney, and testes (Robitaille et al. 1999; Han et al. 2001).

CEACAM1 displays the broadest expression pattern among CEACAMs that is conserved in humans and rodents. It is expressed on various epithelia, such as esophagus (glandular epithelial cells), stomach (pyloric mucous

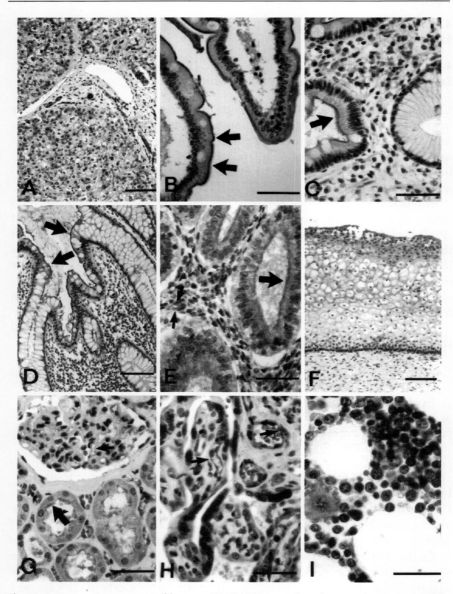

Fig. 3A–I. Expression pattern of human CEACAM1 in various human tissues in paraffin sections, as revealed by binding of the monoclonal anti-human CEACAM1-antibody 4D1/C2 (Drzeniek 1991) in: **A** liver, APAAP detection (alkaline phosphatase-anti alkaline phosphatase), **B** small intestine, ABC detection (avidin-biotin-complex), **C** glands from the gastroduodenal junction with enterocytes and mucosa glands, avidin biotin peroxidase complex (ABC) detection, **D** colon, alkaline phosphatase anti-alkaline phosphatase (APAAP) detection, **E** endometrium, APAAP detection, **F** uterine portio, APAAP detection, **G** kidney, APAAP detection, **H** placenta, APAAP detection, and **I** bone marrow,

cells, Brunner's gland cells), in epithelial cells of the duodenum, jejunum, and ileum; in colon (columnar epithelial cells, caveolated cells), pancreas (ductal epithelial cells), liver (bile canaliculi, bile duct epithelial cells), gall bladder (epithelia), in kidney epithelial cells in the proximal tubuli, the urinary bladder (transitional epithelial cells), prostate epithelial cells, cervix, squamous epithelial cells, endometrium, glandular epithelial cells, and in sweat and sebaceous glands (Fig. 3). Furthermore, CEACAM1 is expressed on granulocytes, leucocytes (T- and B-lymphocytes, monocytes), dendritic cells, and on endothelial cells in some organs (Hanenberg et al. 1994; Frängsmyr et al. 1995; Prall et al. 1996; Kammerer et al. 1998; Kammerer et al. 2001). Its two major isoforms, CEACAM1-L and CEACAM1-S, seem to be co-expressed in all tissues investigated so far (Baum et al. 1996; Turbide et al. 1997) with the exception of breast, endothelia, and T lymphocytes.

CEACAM6 also displays a fairly broad expression pattern, as it shares common sites of expression with CEACAM1: CEACAM6 is expressed in epithelia of different organs, and in granulocytes and monocytes (Kodera et al. 1993; Metze et al. 1996; Schölzel et al. 2000).

CEA exhibits a more restricted expression pattern in normal adult tissues. It is expressed on columnar epithelial cells in the colon, in mucous neck and goblet cells, in the pyrolic mucous cells of the stomach, and in squamous epithelial cells of the tongue, esophagus, prostate, and cervix as well as in secretory epithelia and duct cells of sweat glands. CEA expression has also been described on endothelia (Majuri et al. 1994). Interestingly, as shown in transgenic mouse models, the spatiotemporal expression pattern of CEA expressed under the transcriptional control of the *CEACAM5* promoter is preserved in transgenic animals (Eades-Perner et al. 1994). CEACAM7 is expressed in a pattern similar to CEA in colon, but is not found in granulocytes. CEACAM3 and CEACAM8 are expressed in granulocytes but have not been described in epithelial cells so far.

For the PSGs, the main expression site is the placenta, displaying high expression during the first trimester of pregnancy. Their expression is restricted to the syncytiotrophoblast (Rebstock et al. 1993; Zhou et al. 1997).

◄───

APAAP detection. CEACAM1 shows a pronounced apical location in B, C, D, E, and G (*large arrows*), and the staining of endothelia in E, G, and H (*small arrows*). Bars: A,D,F=100 µm; B, C, E, G, H=50 µm; I=25 µm. (Reproduced from Prall et al. 1996 with permission)

4.2
Expression of CEACAMs in Healthy Adult Colon

The expression of different members of the CEA-family has been extensively studied in the human colon, where they are a major component of the epithelial defense barrier, consisting of the glycocalyx and a supraepithelial mucus layer (Baranov et al. 1994; Frängsmyr et al. 1995; Frängsmyr et al. 1999; Hammarström and Baranov 2001). Four members of the CEA family

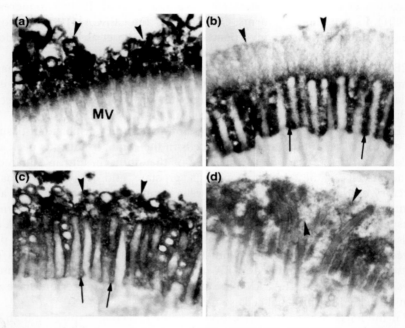

Fig. 4a–d. Immunoelectron microscopy of CEA, CEACAM1, CEACAM6, and CEACAM7 in normal human colon (indirect immunoperoxidase method). **a** Micrograph of the apical portion of a columnar cell. CEA-positive material is detected over the top of the microvilli (*MV*), and consists of thick long filaments and membrane vesicles (*arrowheads*). **b** Micrograph of the apical part of a mature columnar cell. Electron-dense CEACAM7-positive material is present between microvilli (*arrows*). Thin long matted filaments at the top of the microvilli are weakly stained. **c** Micrograph of the apical part of a mature columnar cell. CEACAM6-positive granular compact material, including membrane-bound vesicles, is seen between (*arrows*) and over the top of the microvilli (*arrowheads*). **d** Micrograph of the apical region of a mature columnar cell. A delicate CEACAM1-positive material that consists of very thin loose filaments is mainly located between the microvilli (*arrowheads*). Magnifications (a–d): ×20,000. The following murine monoclonal antibodies (mAb) were used: anti-CEA mAb Bu-103 (Baranov et al. 1994), anti-CEACAM6 mAbMox-36 (Baranov et al. 1994), anti-CEACAM1 mAb 4D1/C2 (Drzeniek et al. 1991), and anti-CEACAM7 mAb BAC2 (Frängsmyr et al. 1999). (Reproduced from Hammarström and Baranov 2001, with permission from Elsevier Science)

are expressed in the healthy colon, CEACAM1, CEA, CEACAM6, and CEACAM7. All four of these members of the CEA family are specifically localized on the apical surface of mature columnar epithelial cells (enterocytes), in microvesicles, filaments, and on highly differentiated epithelial cells at the crypt mouth. No expression can be detected on the basolateral side of the gut epithelium (Hansson et al. 1989; Frängsmyr et al. 1999). CEACAMs are a major component of the brush-border glycocalyx. However, regarding cellular differentiation, these CEACAMs display a compartmentalized expression pattern. Whereas CEA is exclusively located on the top of microvilli, and CEACAM7 is restricted to their lateral sides, CEACAM1 and CEACAM6 are expressed at the apical and lateral part of microvilli (Fig. 4). CEA and CEACAM6 are also expressed on goblet cells. Undifferentiated or maturing epithelial cells of mid and lower crypts are devoid of CEACAM1 and CEACAM7, and synthesize CEA and CEACAM6 only at low levels (Hammarström and Baranov 2001; Fig. 4).

4.3
Expression of CEACAMs During Gestation and Embryonic Development

Expression of CEA family members starts during the early fetal period and is maintained throughout life. Their expression can be detected within the first trimester of pregnancy in the embryo in humans and rodents and increases towards the end of embryonic development. Moreover, CEACAM1 and PSGs are expressed in the invasive trophoblast during gestation (Rebstock et al. 1993; Daniels et al. 1996; Bamberger et al. 2000).

CEA is present in normal human tissues in the fetus and in the adult in endoderm-derived tissues exclusively (Nap et al. 1988). The expression of CEA and other CEA-gene family members can be detected in early stages of pregnancy during the first trimester in human (Wagener et al. 1983; von Kleist et al. 1986). CEACAM1 expression also starts in the early phases of development and displays an interesting spatiotemporal regulation. In human and in mouse, it is expressed in the placenta on the maternal–fetal interface of the invasive extravillous trophoblast. CEACAM1 is also present on epithelial cells of pregnancy epithelium, but the decidua is devoid of CEACAM1. As in the human, CEACAM1 is also expressed during early stages in rodent development. Its expression can be detected as early as day 7.5 in the postimplantation conceptus, namely in the visceral yolk sac and in the invasive syncytiotrophoblast (Huang et al. 1990; Daniels et al. 1996; Sawa et al. 1997; Bamberger et al. 2000). Furthermore, CEACAM1 is expressed on maternal blood vessels, and a strong reactivity for anti-CEACAM1-specific antibodies is also found on embryonic capillaries. In the mouse embryo, the first sites of CEACAM1 expression are the primitive gut during early morphogenesis of the surface ectoderm, and in areas of epithelial–mesenchymal interactions, i.e., in the dermis, meninges, lung, kidney, salivary glands, and the pancreas. CEACAM1 is also detected during myogenesis and odontogenesis (Rass et al. 1994; Lüning et al. 1995; Daniels et al. 1996).

4.4
Dysregulation of the Expression of CEACAMs in Human Tumors

CEACAMs are expressed in various tumors of epithelial origin (colorectal carcinoma, lung adenocarcinoma, mucinous ovarian carcinoma, endometrial adenocarcinoma). Overall, the expression of CEA is more restricted in tumors than CEACAM1 and CEACAM6, which are also dysregulated in leukemias, hepatocellular carcinoma, and melanoma (summarized in Table 1).

In contrast to other CEA family members that are down- or upregulated in human tumors, CEACAM1 displays a rather conflicting expression pattern. In human colon, prostate, and hepatocellular carcinoma, for example, CEACAM1 is downregulated when compared to healthy tissue specimens. However, CEACAM1 is upregulated in gastric and squamous lung cell carcinomas or in malignant melanoma. In cases of CEACAM1 downregulation during early tumorigenesis, the loss of CEACAM1 expression is accompanied by a dramatic alteration in tissue architecture, caused by changes in cell polarity and adhesion. The effects observed after downregulation of CEACAM1 expression are phenotypically quite similar to those observed after the mutation of cadherins or loss of their expression during malignant progression (Birchmeier and Behrens 1994; Bracke et al. 1996). The impact of CEACAM1 expression on tissue architecture has been clearly demonstrated for hepatic, colorectal, and prostate carcinomas (Hixson et al. 1985; Nollau et al. 1997; Busch et al. 2002). In this context, it is noteworthy that restoration of CEACAM1-4S expression in mammary carcinoma cells induces lumen formation and reversion of the tumorigenic phenotype in a cell culture model (Kirshner et al. 2003a). In a marked contrast to these observations, CEACAM1 is re-expressed in invasive melanoma and lung adenocarcinoma (Laack et al. 2002; Thies et al. 2002). In the case of malignant melanoma, CEACAM1 expression was especially found at the invasive front of the tumors and maintained in their metastatic lesions. Moreover, CEACAM1 was identified as a *cis*-binding partner of integrin $\alpha_V\beta_3$ in a variety of epithelial cell lines, endothelial cells, and malignant melanoma (Brümmer et al. 2001). Hence, it is possible that *cis*-interaction of CEACAM1 with integrins can promote invasion. Downregulation of CEACAM1 during the progression of colonic carcinoma is an early event in tumorigenesis, indicating that loss of its expression is the result of a genetically based alteration (Neumaier et al. 1993). However, with regards to invasion, its invasive potential seems in part to depend on its binding partners in *cis* and its phosphorylation status (A.K. Horst, C. Wagener, and N. Beauchemin, manuscript in preparation).

Table 1. Expression of CEA family members in human tumors (Hammarström 1999). Overview of dysregulated expression of CEA gene family proteins in a variety of human tumors

Type of tumor	CEA	CEACAM6 (NCA)	CEACAM1 (BGP)	CEACAM8 (CGM6)	CEACAM3 (CGM1)	CEACAM7 (CGM2)	CEACAM4 CGM7	PSGs	Reference(s)
Epithelial									
Colorectal carcinoma	+	+↑	+↓			+↓			Thompson, 1994; Baranov, 1994; Kim et al. 1992; Jothy et al. 1993; Neumaier, 1993; Thompson et al. 1993; Cournoyer et al. 1988; Sheahan et al. 1990; Tsutsumi et al. 1990; Shi et al. 1994; Thompson et al. 1997
Gastric carcinoma	+↑	+↑	+↑			+↑			Kinugasa et al. 1998; Shi et al. 1994; Kodera et al. 1993
Lung adenocarcinoma	+	+	+						Kim et al. 1992; Cournoyer et al. 1988; Shi et al. 1994; Robbins et al. 1993; Laack et al. 2002; Sienel et al. 2003
Lung squamous cell carcinoma	–		+↑						Tsutsumi et al. 1990; Ohwada et al. 1994
Breast carcinomas	(+)	+	+			–			Thompson, 1994; Shi et al. 1994; Robbins et al. 1993; Cournoyer et al. 1988; Thompson et al. 1993; Bamberger et al. 2002
Pancreatic carcinoma	+	+	(+)	–	–	+	–		Shi et al. 1994
Gallbladder carcinoma	+	+	(+)	–	–				Shi et al. 1994
Mucinous ovarian carcinoma	+	(+)	(+)	–	–			–	Thompson 1994
Serous ovarian carcinoma	(+)	+	+						Thompson et al. 1993
Endometrial adenocarcinoma	+	+	+↓				–	–	Thompson et al. 1993
Hepatocellular carcinoma	–								Shi et al. 1994; Hinoda et al. 1990; Tanaka et al. 1997
Thyroid carcinoma	–								Shi et al. 1994
Nasopharyngeal carcinoma	–								Shi et al. 1994
Prostate			+↓						Busch et al. 2002

Table 1. (continued)

Type of tumor	CEA	CEACAM6 (NCA)	CEACAM1 (BGP)	CEACAM8 (CGM6)	CEACAM3 (CGM1)	CEACAM7 (CGM2)	CEACAM4 CGM7	PSGs	Reference(s)
Other									
Malignant mesothelioma	–								Dejmek and Hjerpe 1994
Small cell lung carcinoma	+		–						Kim et al. 1992; Ohwada et al. 1994
Acute lymphoblastic leukaemia	–	+	(+)		–		–		Hanenberg et al. 1994
Multiple myeloma			+						Satoh et al. 2002
Melanoma	–		+						Shi et al. 1994; Thies et al. 2002
Different sarcoma	–								Shi et al. 1994
Hydatidiform mole								+	Leslie et al. 1990
Choriocarcinoma								+	Leslie et al. 1990

Note that the expression of certain CEACAMs has not been determined completely in all malignant tissues summarized here. Also see Hammarström (1999) and references therein.
↓↑, Specific CEACAMs that are up- or downregulated in malignant versus healthy tissue specimens.
+, Greater than 50% of individual tumor samples were CEACAM positive.
(+), 10% to 50% CEACAM-positive tumor samples.

5
Biological Functions of CEACAMs

5.1
CEACAMs Are Cellular Adhesion Molecules

CEACAMs in human, mouse, and rat all act as homophilic cell adhesion molecules in vitro (Ocklind and Öbrink 1982; Rojas et al. 1990; McCuaig et al. 1992). Interestingly, despite its high homology to other CEACAMs, murine CEACAM2 does not function as a cell adhesion molecule (Robitaille et al. 1999).

The biological functions of the CEACAMs are largely determined by their ability to act as cellular and intercellular adhesion molecules. Adhesion of CEACAMs to each other or engagement with extracellular ligands induces specific signal transduction through CEACAMs, more specifically through CEACAM1-L and CEACAM3-L. Except for CEACAM3, CEACAM4, and CEACAM7, CEACAMs of the CEA-subgroup can interact with each other by homophilic or heterophilic binding. CEACAM1, CEA, and CEACAM6 exhibit homophilic and heterophilic adhesion among each other, whereas CEACAM8 seems to bind only to CEACAM6 (Oikawa et al. 1991). Binding of CEACAM1 to itself, CEA, or CEACAM6 was demonstrated by the use of recombinant CEACAM1 as well as in heterologous transfection. The adhesive properties of CEACAM1 are regulated by dimerization, the phosphorylation status of its cytoplasmic tail, its binding to calmodulin, and the activity of tissue transglutaminase (Öbrink 1997; Hunter et al. 1998). Recently, it was shown that adhesive properties of CEACAM1-4L can be modified after cleavage of its cytoplasmic tail by caspase-3 (Houde et al. 2003). Both the long-tail and short-tail isoform of CEACAM1 (CEACAM1-L and CEACAM1-S, respectively) undergo dimer formation in their cell-bound and soluble form, as shown using chemical crosslinkers (Hunter et al. 1996). Additionally, CEACAM1-L was identified as a substrate for tissue transglutaminase and it produces CEACAM1-L dimers by intracytoplasmic crosslinking (Hunter et al. 1998). CEACAMs preferentially form homodimers, i.e., CEACAM1-L binds to CEACAM1-L and CEACAM1-S binds to CEACAM1-S. Homodimerization of Ig superfamily members has been described for CD84, JAMs (junctional adhesion molecules), CD146, PECAM-1 (platelet endothelial cell adhesion molecule-1, CD31), ICAM-1 (intercellular adhesion molecule, CD54), N-CAM (neural cell adhesion molecule), and has also been established for cadherins (Rao et al. 1994; Shapiro et al. 1995; Nagar et al. 1996; Sun et al. 1996; Alais et al. 2001; Kostrewa et al. 2001; Martin et al. 2001). For cadherins and CEA, homodimer formation in *cis* is implicated in reinforcing adhesion of cadherins and CEA (Bates et al. 1992; Shapiro et al. 1995; Nagar et al. 1996).

In epithelial cells, a chemical equilibrium between dimers and monomers exists. This chemical equilibrium is influenced by the proliferative status

A

CEACAM1-L cytoplasmic domain

Rat 445-YFLYSRKSGGGSDHRDLTEHKPSTSSHNLGPS DDSPNKVDDVSYSVLNFNAQQSKRPTSASSSP--TETV YSVVKKK-519
Mouse 445-YFLYSRKSGGGSDQRDLTEHKPSTSNHNLAPS DNSPNKVDDVAYTVLNFNSQQPNRPTSAPSSPRATETV YSEVKKK-521
Human 446-CFLHFGKTGRASDQRDLTEHKPSVSNHTQDHSN DPPNKMNEVTYSTLNFEAQQPTQPTSASPSLTATEII YSEVKKQ-522

B

Mouse CEACAM1-L cytoplasmic domain

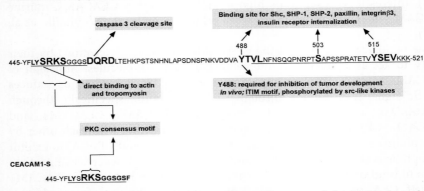

Fig. 5. A Comparison and amino acid sequence alignment of the highly homologous CEACAM1-L cytoplasmic domain in rat, mouse, and human. Conserved residues are *highlighted*. Conserved elements of signal transduction motifs are shown in *bold type*: the CEACAM1-L cytoplasmic domains contain conserved motifs containing tyrosine residues that are embedded into an imperfect ITAM (D/ExxxxxxD/ExxYxxLxxxxxxYxxL/I) or two perfect ITIMs (S/I/V/LxYxxL/V/I). Though the rat and mouse CEACAM1-L cytoplasmic domains are slightly shorter than the human CEACAM1-L cytoplasmic domain, Tyr513 in rat corresponds structurally and functionally to Tyr515 in mouse and Tyr516 in man. B Schematic summary of signal transduction and ligand binding properties of CEACAM1-S and CEACAM1-L cytoplasmic domains. Binding sites for calmodulin are *underlined*, binding sites for cytoskeletal components such as tropomyosin or actin, growth factor receptors or enzymes such as protein kinases or phosphatases and other intracellular adaptor proteins are *highlighted* appropriately. Residues known to be subject to phosphorylation and dephosphorylation are Tyr488, Ser503, and Tyr515. These residues are key mediators of CEACAM1-L signal transduction and adhesion

of the cell, the overall expression levels of the two isoforms, the ratio of CEACAM1-L and CEACAM1-S, and the levels of phosphorylation of the cytoplasmic tail on tyrosine and serine residues (Edlund et al. 1996; Öbrink 1997). CEACAM1-L and CEACAM1-S contain phosphorylation target sites for protein kinase C (PKC) that are located in proximity to calmodulin binding sites (Edlund et al. 1996). Calmodulin has two binding sites in CEACAM1-L and one in CEACAM1-S. The increase of intracellular calcium during cellular activation processes leads to the dissociation of CEACAM1 dimers in vitro (Öbrink 1997; Fig. 5).

In each case of *cis*- or *trans*-interaction between CEACAMs, their first Ig domain is essential (Cheung et al. 1993; Teixeira 1994; Watt et al. 2001). Homodimers interact through reciprocal bonds in their N-terminal domain (Wikström et al. 1996).

Examples for CEACAM function mediated by homophilic adhesion events include the establishment of tissue organization during embryonic development in the intestinal epithelium and in hepatocytes, in placental trophoblasts, during odontogenesis and myogenesis, during vascularization of the central nervous system, in neutrophil activation and extravasation during inflammatory responses, in regulation of T cell responses, in angiogenesis and regulation of cell proliferation (Öbrink 1997; Hammarström 1999; Zimmermann 2002). Examples of heterophilic interactions that involve CEACAMs are the adhesion to other CEACAM family members, adhesion to E-selectin, galectin-3, outer membrane proteins of *Neisseria meningitidis* and *Neisseria gonorrhoe*, *Haemophilus influenzae*, fimbriae of *Salmonella typhimurium* and *Escherichia coli*, and murine hepatitis virus, as described in detail below.

5.2
CEACAMs Display Versatile Signal Transduction Properties

5.2.1
Mechanisms of Signal Transduction by CEACAM1

The primary structure of the cytoplasmic domain displays a high conservation between humans and rodents. The most remarkable feature of the CEACAM1-L cytoplasmic domain lies in the two ITIM motifs (immunoreceptor tyrosine-based inhibition motif) that are fundamental in the regulation of signal transduction through CEACAM1-L. The CEACAM1-L cytoplasmic domain contains two tyrosine residues that have been shown to be phosphorylated. The first tyrosine, Tyr488, and the second tyrosine, Tyr515, are each located within an ITIM. Tyr515 in mouse corresponds to Tyr513 in rat and Tyr516 in human. ITIM motifs occur in a wide range of actual and potential coinhibitory receptors, in cytokine receptors or signaling kinases and intermediates. ITIMs are identified by the restricted consensus V/IxYxxL/V, but this sequence may be rather generally defined by the sequence V/I/L/SxYxxL/V/I (Sinclair 2000). When considering Tyr488 and Tyr515 together, these residues are located within an imperfect ITAM (immunoreceptor tyrosine-based activation motif; Fig. 5). In contrast to classic ITAMs, the spacing between the two tyrosine residues is not 10 amino acids, but rather 24–26 amino acids (Cambier 1995). The membrane proximal residue within the first ITIM, Tyr488, is phosphorylated by kinases of the src family, and by growth factor receptor tyrosine kinases such as the insulin receptor (IR) and the epidermal growth factor (EGF) receptor (Phillips et al.

1987). In epithelial cells, CEACAM1-L is phosphorylated by c-src, and in granulocytes by c-src, lyn, and hck (Skubitz et al. 1995). In hepatocytes, CEACAM1-L becomes tyrosine phosphorylated upon interaction with the IR tyrosine kinase (Margolis et al. 1990). In vitro, these tyrosine residues can be phosphorylated after treatment of cultured cells with the tyrosine phosphatase inhibitor pervanadate (Lin et al. 1995; Beauchemin et al. 1997). Regulated phosphorylation on Tyr488 of its cytoplasmic domain is of fundamental importance for signal transduction through CEACAM1-L. After phosphorylation on this residue, CEACAM1 interacts with components of the cytoskeleton, such as actin and paxillin, but also β_3 integrin (see below; Ebrahimnejad et al. 2000; Sadekova et al. 2000; Brümmer et al. 2001; Schumann et al. 2001).

Additionally, tyrosine phosphorylated CEACAM1-L becomes a substrate for the ubiquitously expressed phosphotyrosine phosphatases Src homology 2 domain-containing protein tyrosine phosphatase (SHP)-1 and SHP-2 (Beauchemin et al. 1997; Huber et al. 1999). Phosphorylation on both CEACAM1-L Tyr488 and Tyr515 is required for the recruitment of these SH2 domain-containing phosphatases. Binding of SHP-1 and SHP-2 has been described for coinhibitory receptors and other adhesion molecules that contain ITIM motifs in their cytoplasmic portion, such as a number of hematopoietic cell surface receptors: the Fcγ receptor IIB, CD22, and KIRs (killer inhibitory receptors) on natural killer cells (Long 1999; Sinclair 2000). Binding of SHP-2 to platelet endothelial cell adhesion molecule-1 (PECAM-1, CD31) is also well characterized. Strikingly, the cytoplasmic domains of PECAM-1 and CEACAM1-L share a high homology within their C-terminal region (Huber et al. 1999). Both the PECAM-1 and CEACAM1-L cytoplasmic domains are transiently phosphorylated on two tyrosine residues during adhesion processes, or, as for PECAM-1, during platelet aggregation (Famiglietti et al. 1997). Abrogation of tyrosine phosphorylation on PECAM-1 can modulate its homophilic and heterophilic binding properties. A very similar model for regulated CEACAM1 dimerization and its subsequent association with intracytoplasmic adaptor proteins like serine and tyrosine kinases or phosphatases has been proposed by Öbrink's group (Öbrink et al. 2002 and references therein). Clustering of CEACAM1-L through monoclonal antibodies induces transient dephosphorylation (Budt et al. 2002).

In addition to tyrosine phosphorylation, CEACAM1-L becomes phosphorylated on Ser503. Basal phosphorylation on Ser503 was first described on CEACAM1-L in context with its activity as a bile salt transporter in rat bile canalicular cells that is associated with a co-purifying ecto-ATPase (Sippel et al. 1994). Furthermore, Ser503 phosphorylation is essential during complex formation between CEACAM1-L and the IR, and for its tumor cell growth suppressive effects (see below). Ser503 phosphorylation, however, appears to be regulated in part through tyrosine phosphorylation, indicating that distinct signals of different signal transduction pathways can act through CEACAM1-L.

Although CEACAM1-L has been identified as the major active splice variant of CEACAM1 in signal transduction, CEACAM1-S also exerts interesting signal transduction properties, such as interaction with cytoskeletal components (see below), and phosphorylation on serine and threonine residues by protein kinase C isoenzymes (Edlund et al. 1998; cf. Fig. 5). However, its biological function is less well studied. In the mammary gland, CEACAM1-4S mediates apoptosis during mammary morphogenesis, and re-expression of CEACAM1-4S in breast cancer cell lines reverts their malignant phenotype to a normal phenotype when grown in 3D culture matrices (Kirshner et al. 2003a). Additionally, annexin II, a regulator of the secretory differentiation of mammary gland cells, has been identified as an intracellular ligand for CEACAM1-S (Kirshner et al. 2003b).

5.2.2
CEACAM1 Interacts with Components of the Cytoskeleton

In brush border cells of rat small intestine and in skeletal muscle cells, CEACAM1-L is associated with cortical actin, providing evidence for the implication of CEACAM1 in the organization and maintenance of tissue architecture (Hansson et al. 1989; Da Silva-Azevedo et al. 1999). It is assumed that this requires homophilic adhesion between CEACAM1 molecules on neighboring cells (Da Silva-Azevedo et al. 1999; Sadekova et al. 2000). Furthermore, CEACAM1-L also binds tropomyosin with its binding site in close proximity to the actin-binding site (Sadekova et al. 2000; Schumann et al. 2001). Using transfection studies in an adenocarcinoma cell line, it was discovered that the interaction with G-actin (globular actin) is dependent on tyrosine phosphorylation of the CEACAM1 long cytoplasmic tail. However, CEACAM1-S also contains actin-binding sites in its cytoplasmic portion. The individual sequences for actin binding in CEACAM1-S and CEACAM1-L are slightly different from each other. This might explain why tyrosine phosphorylation in CEACAM1-L can actually modulate its actin-binding properties (Schumann et al. 2001). The association with F-actin (fibrillar actin) can occur but is suspected to be indirect (Sadekova et al. 2000; Schumann et al. 2001). Importantly, CEACAM1-L localization in epithelial cells is restricted to their apical and lateral surface, i.e., cell–cell contacts (Sundberg and Öbrink 2002). This was also confirmed after microinjection of CEACAM1-L cDNA into fibroblasts and colon carcinoma cells (Sadekova et al. 2000). In contrast to CEACAM1-L, CEACAM1-S displays a rather diffuse localization in epithelial cells and fibroblasts after microinjection in a transient expression system (Sadekova et al. 2000). When stably transfected canine kidney epithelial cells (MDCK) are used, CEACAM1-S exhibits pronounced apical expression, but is not found in the lateral compartment. The explanation for this might be the fact that the MDCK cells were allowed to form layers and were grown on filters that encourage cellular polarization (Sundberg and Öbrink 2002). In the microinjection experiments, CEACAM1-L and CEACAM1-S were co-expressed with constitutively active Rho–guanosine triphosphatases (GTPases) on untreated culture dishes (Sadekova et al. 2000). Rho GTPases control a variety of cellular processes such as organization of the cytoskeleton, gene transcription, and adhesion (Mackay and Hall 1998). It was revealed that Rho–GTPase activity was required to target CEACAM1-L to the cell periphery in epithelial cells, more specifically to sites of cell–cell contacts. Targeting of CEACAM1-L to cell–cell contacts is induced by activated cdc42 and Rac1, and requires the CEACAM1-L transmembrane domain (Fournes et al. 2003). This targeting could be abolished by use of chaotropic agents such as cytochalasin D that disrupts F-actin. No such effects on CEACAM1-S localization could be detected. Shively's group proposed a regulatory mechanism for actin polymerization through CEACAM1, assuming CEACAM1-S interacts with G-actin, serving as a putative G-actin polymerization site, and with CEACAM1-L as an anchoring platform for polymerized actin filaments (Schumann et al. 2001).

5.2.3
CEACAM1 Acts as a Tumor Suppressor

First indications pointing toward an involvement of CEACAM1 in suppression of tumor cell growth were obtained after observing differential regulation of CEACAM1 expression in hepatoma (Hixson et al. 1985). In hepatocarcinoma cells, loss of CEACAM1 expression was observed in malignant cells when compared to normal specimens (Tanaka et al. 1997). Later, it was discovered that CEACAM1 is also downregulated in other tumors of epithelial origin in human, mouse and rat, such as colon (Neumaier et al. 1993; Rosenberg et al. 1993; Nollau et al. 1997), prostate (Kleinerman et al. 1995a; Pu et al. 1999; Busch et al. 2002), breast (Riethdorf et al. 1997; Huang et al. 1998; Bamberger et al. 1998), endometrium (Bamberger et al. 1998), and bladder (Kleinerman et al. 1996), leading to the hypothesis that CEACAM1 behaves as a tumor suppressor. CEACAM1 downregulation is an early step during malignant transformation in human tumors, for example during the development of colon carcinomas, where its expression is markedly decreased in the microadenoma and adenoma stages (Ilantzis et al. 1997; Nollau et al. 1997). Intriguingly, a 1-bp deletion has been identified within a microsatellite region of the 3′ UTR of the human *CEACAM1* gene, which is suspected to be involved in tumor onset and progression by dramatically decreasing CEACAM1 expression (Ruggiero et al. 2003).

The antitumoral potential of CEACAM1 has been extensively investigated in cell culture and orthotopic xenograft models that provided further evidence for its suppressive effects on tumor cell growth in colon and prostate cancer (Hsieh et al. 1995; Kleinerman et al. 1995b, 1996; Kunath et al. 1995; Luo et al. 1999; Estrera et al. 2001). Furthermore, it was revealed that the tumor suppressive effects exerted by CEACAM1-L in human, mouse, or rat are dependent on the presence of its long cytoplasmic tail and that physiological levels of CEACAM1 expression are required to sustain this function (Luo et al. 1997; Turbide et al. 1997; Izzi et al. 1999). The cytoplasmic tail of CEACAM1-L is necessary and sufficient to mediate growth inhibitory activity whereas CEACAM1-S does not display tumor cell growth inhibitory function (Turbide et al. 1997). These studies lead to the identification of amino acids that are crucial for the promotion of the reduction in tumor cell growth; as in other biological functions mediated by CEACAM1, phosphorylation on Tyr488, located within the ITIM, and on Ser503 play a key regulatory role in this context (Izzi et al. 1999; Estrera et al. 2001; Fournes et al. 2001).

Abrogation of tyrosine phosphorylation on Tyr488 or Ser503 by conversion of these amino acids to Phe or Ala, respectively, impairs the cell growth inhibitory effects. In contrast to Tyr488, tyrosine phosphorylation on Tyr515, as shown in mouse models, does not seem to influence the tumor suppressive potential of CEACAM1-L. The exact underlying molecular mechanisms for tumor suppression by CEACAM1 are still poorly understood. It is unclear at present whether dephosphorylation of the CEACAM1-L cytoplasmic domain by the tyrosine phosphatases SHP-1 and SHP-2 or a serine phosphatase is necessary in vivo to reduce tumoral cell proliferation.

Another aspect of tumor suppression by adhesion molecules is the phenomenon of contact inhibition leading to an arrest in cellular proliferation. In a cell culture model comparing a cell line derived from normal prostate epithelium and a bladder cancer cell line, homophilic cellular adhesion through CEACAM1 in *trans* and the expression of its isoform ratios was shown to be different in quiescent and proliferating cells (Singer et al. 2000). As mentioned above, the abilities of CEACAM1 to interact with components of the cytoskeleton and src-kinases or tyrosine phosphatases depend on the isoform ratio expressed by a particular cell type and engagement of these isoforms in dimer formation. The delicate equilibrium of CEACAM1-L monomers versus CEACAM1-L dimers regulates the balance between proliferative or anti-proliferative effects (Öbrink et al. 2002).

In general, few examples are known which reveal a causal context between the impact of signal transduction through adhesion molecules and malignant progression and dissemination of cells from a primary tumor. Well-studied examples in this context are E-cadherin and N-CAM (Birchmeier and Behrens 1994; Cavallaro et al. 2001; Hajra and Fearon 2002). Apart from the cadherin–catenin pathway, membrane association of the tyrosine phosphatases SHP-1 and SHP-2 can be observed in contact-inhibited cells. This association is required for E-cadherin-mediated growth arrest through $p27^{Kip}$, an effector of cyclin E-dependent kinase activity (Pallen and Tong 1991; St Croix et al. 1998). Interestingly, CEACAM1 expression on quiescent, confluent bladder carcinoma cells correlates with the expression of $p27^{Kip}$ (Singer et al. 2000), whereas $p27^{Kip}$ is downregulated in proliferating cells. Additionally, CEACAM1 expression in human breast carcinomas is associated with the expression of tumor suppressors such as Rb (retinoblastoma protein) and $p27^{Kip}$ (Bamberger et al. 2002). Furthermore, adhesion molecules like cadherins and N-CAM respond to signaling by growth factor receptors, such as EGF and fibroblast growth factor (FGF), respectively. During tumor progression, their adhesive properties are modulated by the influence of these growth factors or, as it has been described for E-cadherin, their expression is downregulated upon EGF receptor activation (Hazan and Norton 1998; Cavallaro et al. 2001; Al Moustafa et al. 2002). CEACAM1 is also a substrate for growth factor receptors or responds to growth factor stimuli (Phillips et al. 1987; Ergün et al. 2000).

CEACAM1 is not downregulated in all carcinomas; its expression can also be upregulated in certain tumors, as mentioned above. Intriguingly, CEACAM1 expression also correlates with β_3 integrin expression in regulated and dysregulated invasive processes such as trophoblast invasion and the progression of malignant melanoma (Bamberger et al. 2000; Thies et al. 2002). The influence of putative interacting partners for CEACAM1 in *cis* and *trans* on the invasive behavior of primary tumors has not been investigated in detail so far. β_3 Integrin is the only *cis*-interacting adhesion molecule for CEACAM1-L identified to date (Brümmer et al. 2001). Moreover, the expression pattern of the CEACAM1 isoforms CEACAM1-S and CEACAM1-L in the majority of malignant human tumors has not been defined yet. In invasive human lung adenocarcinomas, for example, CEACAM1-S expression is markedly upregulated when compared to normal tissue specimen that predominantly express CEACAM1-L (Wang et al. 2000).

5.2.4
CEACAM1 Promotes Invasion

Contradictory to the observations mentioned above, CEACAM1 is overexpressed in human invasive melanomas, primary lung tumors (squamous cell carcinoma, adenocarcinoma and small cell carcinoma), and stomach tumors (Kim et al. 1992; Ohwada et al. 1994; Kinugasa et al. 1998; Wang et al. 2000; Laack et al. 2002; Thies et al. 2002). In human malignant melanoma, CEACAM1 expression is strongest at the invasive front of the primary tumor, and is preserved in distant metastatic lesions (Thies et al. 2002). Furthermore, CEACAM1 expression correlates with the development of metastatic disease and is an independent prognostic parameter. A similar correlation between CEACAM1 expression and poor prognosis is found in adenocarcinoma of the lung (Laack et al. 2002). However, in metastatic lesions of lung tumors, *CEACAM1* mRNA expression appears to be decreased when compared to its expression levels in primary tumors (Ohwada et al. 1994). Moreover, CEACAM1 associates in *cis* with integrin $\alpha_v\beta_3$ at the invasive front of malignant human melanoma and at the apical surface of glandular cells of pregnancy endometrium (Brümmer et al. 2001). The concentrated colocalization at the tumor–stroma interface of invading melanoma and in the transitional region from proliferative to invasive extravillous trophoblast of the maternal–fetal interface indicates a role for CEACAM1/integrin β_3 complexes in cellular invasion (Brümmer et al. 2001).

5.2.5
CEACAM1 Is a Substrate for the Insulin Receptor and Regulates Insulin Clearance

Cellular sensitivity towards insulin is regulated by insulin-induced internalization and recycling of its receptor. Upon internalization of the receptor–ligand complex, the receptor is recycled, whereas insulin is targeted for degradation by passage through endosomal and lysosomal compartments. CEACAM1-L has been identified as an endogenous substrate for the tyrosine kinase activity of the insulin and EGF receptors in rat hepatoma cells and rat liver cell membranes (Rees-Jones and Taylor 1985; Phillips et al. 1987). In contrast to other substrates of the insulin receptor (IR), CEACAM1-L does not directly bind to insulin-like growth factor (IGF)-1 (Najjar et al. 1997).

The stimulation of insulin-induced endocytosis of the IR by CEACAM1-L requires a coordinated sequence of phosphorylation events in the CEACAM1-L and the IR cytoplasmic tails. Upon activation of the IR by ligand binding, it is autophosphorylated on diverse tyrosine residues and CEACAM1-L is recruited into the IR endocytosis complex (Choice et al. 1998; Najjar et al. 1998). This process is dependent on basal phosphorylation of CEACAM1-L on Ser503 by an unidentified cAMP-dependent kinase (Najjar et al. 1995). Two functionally distinct domains in the IR intracellular portion promote the phosphorylation of CEACAM1-L and its physical interaction with the receptor complex. Mutational analysis revealed that Tyr1316 in the C-terminal region of the β-chain of the IR is essential to promote phosphorylation of CEACAM1-L on Tyr488 (Najjar et al. 1997). Phosphorylation of CEACAM1-L on this tyrosine mediates the interaction with another, yet unidentified intracellular adapter that facilitates binding of the receptor-endocytosis complex to adaptor protein 2 adaptin proteins in clathrin vesicles and thus targets insulin for degradation (Najjar 2002; Fig. 6).

To initiate internalization, however, CEACAM1-L phosphorylation is required but not sufficient. Najjar et al. showed that phosphorylation on a conserved tyrosine residue in the IR juxtamembrane domain is fundamental in this process (Najjar et al. 1998). Transfection studies in heterologous systems confirmed that transfection with CEACAM1-L is sufficient to induce IR internalization (Soni et al. 2000).

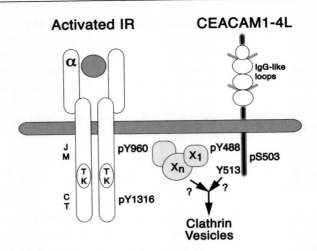

Fig. 6. Proposed model of receptor-mediated insulin endocytosis in the rat. Activation of the tyrosine kinase (*TK*) of the insulin receptor (*IR*) by insulin binding phosphorylates the receptor at many sites, including Tyr960 (*Y960*) in the juxtamembrane domain (*JM*) and Y1316 in the cytoplasmic portion (*CT*) of the β-subunit of the IR. Phosphorylation of Y1316 regulates phosphorylation of CEACAM1-4L on Y488. This causes CEACAM1-4L binding to an intracellular molecule (X_1) which mediates its indirect interaction with phosphorylated Y960 in the IR. Molecule X_i might function either alone or as a part of a complex of proteins (X_n). CEACAM1-4L, through Y513 and/or the mediator proteins, might then target the insulin endocytosis complex to AP2 adaptin proteins in clathrin vesicles to target insulin for degradation. Reproduced from Najjar 2002 with permission from Elsevier Science

CEACAM1-L downregulates the mitogenic effects of insulin by regulating insulin clearance but also through the interaction with Shc, an adaptor molecule implicated in inhibition of cellular proliferation. For the interaction with Shc, however, phosphorylation of CEACAM1-L by the IR is crucial (Poy et al. 2002a).

The in vivo regulatory effects of CEACAM1-L on insulin clearance were shown in a transgenic mouse model consisting of L-SACC1-transgenic mice that express a mutant form of rat CEACAM1-L (Ser503Ala) which cannot be phosphorylated on Ser503. The liver-targeted expression of this mutant under the ApoAI lipoprotein promoter caused secondary insulin resistance resulting from impaired insulin clearance by IR internalization in the mice (Poy et al. 2002b). Moreover, the CEACAM1-L Ser503Ala mutant exerts dominant-negative effects on the naturally expressed wildtype form of CEACAM1-L, emphasizing the immediate impact of signal transduction through CEACAM1-L in IR endocytosis and insulin clearance. The transgenic animals develop visceral adiposity with elevated plasma free-fatty acids and plasma and hepatic triglyceride levels, resembling the phenotype of type II diabetes.

5.3
CEACAMs Are Modulators of the Innate and Adaptive Immune Response

5.3.1
CEACAM1 Is a Positive Regulator of Neutrophil Effector Function

CEA family members expressed on neutrophilic granulocytes include CEACAM1-L (CD66a), CEACAM3 (CD66d), CEACAM6 (CD66c), and CEACAM8 (CD66b). CEACAM1-L was initially described as a phosphoprotein with a relative molecular weight of 160 kDa on human neutrophilic granulocytes with crossreactive antibodies recognizing various members of the CD66 cluster of differentiation antigens (Skubitz et al. 1992). Its identity as CEACAM1 or CD66a could be revealed after a more detailed study by Wagener's group using monoclonal antibodies directed against CEA family members on human neutrophils (Drzeniek et al. 1991; Stoffel et al. 1993). CEACAM1 is an activation antigen that is upregulated rapidly from intracellular storage components after stimulation (Watt et al. 1991; Kuroki et al. 1995). In neutrophil development, CEACAM1 is expressed from the myelocyte stage and maintained throughout neutrophil differentiation (Elghetany 2002). It is detected in low levels on resting cells, but its expression is rapidly upregulated after stimulation with chemotactic peptides such as formyl-methionyl-leucyl-phenylalanine (fMLP), or calcium ionophores and phorbol esters (Skubitz et al. 1992). Upregulation of CEACAM1 in inflammatory disease has also been reported (Honig et al. 1999).

Upon neutrophil activation, CEACAM1-L becomes tyrosine phosphorylated and binds to protein kinases such as hck, lyn, and src (Brümmer et al. 1995; Skubitz et al. 1995). After phosphorylation, enhanced association of these kinases with the cytoskeletal components of neutrophils can be observed. Ligation of CEACAM1-L with monoclonal antibodies or F(ab')$_2$ fragments leads to upregulation of β_2 integrin (CD11b/CD18) expression and activation, production of cytotoxic oxygen species, downregulation of L-selectin, and increase of neutrophil adhesion to endothelial cells and fibrinogen and fibronectin (Stocks et al. 1995; Skubitz et al. 1996; Stocks et al. 1996; Klein 1996; Ruchaud-Sparagano et al. 1997; Nair and Zingde 2001). Homotypic adhesion of neutrophils to HUVECs (human umbilical vein endothelial cells) and subsequent activation of integrins can be enhanced by synthetic peptides containing sequences of the N-terminal domain of CEACAM1-L (Skubitz et al. 2001). Since CEACAM1-L is also expressed on activated HUVECs, homotypic adhesion via CEACAM1-L could further enhance adhesion to endothelia and trigger signal transduction pathways that may modulate integrin-binding activity.

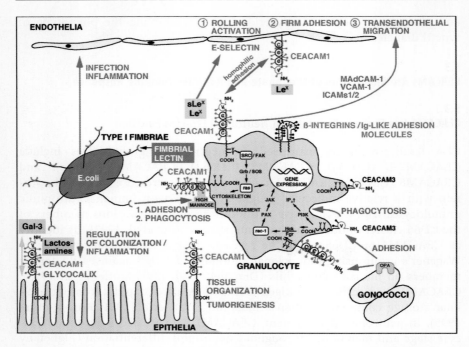

Fig. 7. Involvement of CEACAMs in the regulation of tissue colonization and infection by microbes through protein–carbohydrate and protein–protein interactions, as shown representatively for type 1-fimbriated *E. coli* and Opa$^+$ gonococci. Carbohydrate moieties displayed by CEACAMs might be key players in fine-tuning interactions between individual CEACAMs or CEACAMs and their physiological ligands. CEACAM1 and CEACAM3 signal transduction capacities upon microbe binding are exemplified for neutrophilic granulocytes; please note that signal transduction conveyed by CEACAMs through specific signal transduction adaptor molecules such as src family kinases or GTPases of the Rho family also accounts for epithelia and endothelia. *Gal-3*, galectin-3; *Le^x*, Lewis^x; *sLe^x*, sialyl Lewis^x; *OPA*, opacity protein

On human neutrophils, CEACAM1 is the main carrier of complex oligosaccharides, such as type I and type II lactosaminoglycan chains. Since it is the main carrier of Lewis^x and sialyl-Lewis^x-epitopes on human granulocytes, CEACAM1 has been suggested as a lectin receptor for cellular lectins, such as vascular selectins. Lectin–glycoconjugate interactions initiate tethering of lymphocytes and granulocytes to the vascular surface, allowing subsequent firm adhesion and facilitating extravasation into inflamed tissues at sites of bacterial assault (McEver 2002). Furthermore, CEACAM1 is a receptor for the soluble mammalian lectin galectin-3 that mediates protein binding to extracellular matrix components (Stocks et al. 1990; Stocks and Kerr 1993; Ochieng et al. 1998; Feuk-Lagerstedt et al. 1999; Fig. 7).

5.3.2
CEACAM1 Regulates T and B Lymphocyte Function

CEACAM1-L is expressed on human and murine T and B lymphocytes, dendritic cells, and natural killer (NK) cells (Khan et al. 1993; Möller et al. 1996; Kammerer et al. 2001; Boulton and Gray-Owen 2002). CEACAM1-L is the only member of the CEA family with a cytoplasmic domain that has been characterized on B and T lymphocytes (Singer et al. 2002). On NK cells and T lymphocytes, CEACAM1 expression is inducible after stimulation, whereas B lymphocytes express CEACAM1 constitutively. In primary mouse B lymphocytes, CEACAM1 displays a distinct isoform expression pattern in resting and activated B cells: in concert with surface IgM-crosslinking, CEACAM-L triggers B cell proliferation, Ig secretion and β_2 integrin-mediated homotypic adhesion (Greicius et al. 2003). In resting T cells, CEACAM1 is present in intracellular stores and only in very low abundance on the cell surface; but as an activation antigen, it is rapidly upregulated after ligation of the TCR/CD3 complex or stimulation with interleukin (IL)-2 (Möller et al. 1996; Kammerer et al. 1998). Dendritic cells express both CEACAM1-S and CEACAM1-L. Two splice variants of CEACAM1 are expressed on human T lymphocytes—CEACAM1-3L and CEACAM1-4L—that are glycosylated in a lymphocyte-specific manner (Kammerer et al. 1998). The mechanism for CEACAM1 release from intracellular compartments might be analogous to that revealed for CTLA4, a coinhibitory receptor that is also retained in the cytoplasm but becomes rapidly translocated to the cell surface upon T cell activation and tyrosine phosphorylation of its cytoplasmic domain. Retention of CTLA4 in the cytoplasm is regulated by complex formation with adaptor protein 1, also called AP47. AP47 regulates the sorting of transmembrane proteins between the trans-Golgi network and the endosomal compartment. Tyrosine phosphorylation is not compatible with AP47-complex formation. Interestingly, Blumberg et al. showed that CEACAM1-L in murine T cells can bind SHP-1 and AP47 in its phosphorylated and dephosphorylated state, respectively (Nakajima et al. 2002). Once exposed on the cell surface, CEACAM1-L ligation by monoclonal antibodies or F(ab) fragments stimulates IL secretion by intestinal T lymphocytes (IL-7 and IL-15) and release of specific chemokines and interleukins by dendritic cells, such as macrophage inflammatory protein 1α, and ~2, monocyte chemoattractant protein 1, and IL-6 and IL-12 (Donda et al. 2000; Kammerer et al. 2001). In addition, CEACAM1-L-induced signaling increases the surface expression of costimulatory molecules (CD40, CD54, CD80, CD86), indicating a regulatory role for CEACAM1 in dendritic cell maturation and activation (Kammerer et al. 2001). The release of these cytokines leads to priming of naïve MHCII-restricted CD4$^+$ T cells with a T helper 1 effector phenotype, favoring a humoral immune response (Kammerer et al. 2001). In intestinal T lymphocytes, cellular activation through CEACAM1-L induces an increase of nuclear factor (NF)-κB- and AP1-mediated transcription (Donda et al. 2000).

Contradictory to these positive regulatory effects on T cell function, CEACAM1-L can also act as a negative regulator of T cell responses. The above-mentioned reports describe regulatory influences of CEACAM1-L on the immune response as effects of either ligation of CEACAM1-L by specific antibodies or as secondary effects after stimulation with interleukins and chemokines or activation of the TCR/CD3 complex. However, a direct function for CEACAM1-L as a coinhibitory receptor on T lymphocytes was shown by Gray-Owen's group: The authors revealed that neisserial binding or binding of the neisserial Opa52 protein to its receptor CEACAM1-4L (see below) can arrest the activation and proliferation of $CD4^+$ T lymphocytes while the expression of the T cell activation marker CD69 is suppressed. Ligation of CEACAM1-L alone was shown to be sufficient for the suppression of $CD4^+$ T cell activation, an effect that could be enhanced after co-ligation of the ITAM-containing $CD3\epsilon$ chain. Furthermore, engagement of *Neisseria* with their receptor induces binding of the phosphatases SHP-1 and SHP-2 to the ITIM in the CEACAM1-L cytoplasmic domain. The reduction of T cell proliferation is a result of specific arrest in cell division and was not mediated through bacterial cytotoxicity (Bradbury 2002; Boulton and Gray-Owen 2002; Normark et al. 2002). Similarly, CEACAM1-L ligation with monoclonal antibodies can mediate negative regulation in T lymphocytes that results in inhibition of the delayed type hypersensitivity (DTH) reaction in early phases of T cell priming (Nakajima et al. 2002). Furthermore, CEACAM1-L ligation leads to downregulation of the cytolytic function of small intestinal intraepithelial lymphocytes ($CD8ab^+$/$TCRab^+$/CD28) and inhibits their CD3-directed and lymphokine–activated killer activity (Morales et al. 1999)

5.3.3
CEACAM1 Is a Novel Co-inhibitory Receptor on Human Natural Killer Cells

NK cells, such as neutrophils, belong to the innate immune defense and act by killing virus-infected or tumor cells in a MHC-I-dependent fashion. Their cytotoxicity is regulated by inhibitory, class I MHC-recognizing receptors that contain ITIM motifs in their cytosolic tail such as C-type lectins (CD94/NKG2A), killer Ig-related receptors, and the leukocyte Ig-like receptor. Among NK-specific receptors (NCRs), CD16 is implicated in mediating direct natural NK cell cytotoxicity (Long 1999; Mandelboim et al. 1999).

Natural killer cells that are freshly isolated express very low amounts of CEACAM1-L. However, CEACAM1-L-expression can be induced upon stimulation (Möller et al. 1996). CEACAM1-L is expressed on a subpopulation of activated human NK cells that are negative for CD16 ($CD16^-$) but positive for CD56 ($CD56^+$). Homotypic adhesion of CEACAM1-L on $CD16^-$/$CD56^+$ NK cells leads to inhibition of NK cell-mediated cytotoxicity in a novel

MHC-I-independent mechanism on a MHC-I-deficient melanoma cell line (Markel et al. 2002a). The degree of inhibition of CEACAM1-L on NK cytotoxicity correlates with the amount of CEACAM1-L expressed on the cell surface of target and effector cells. This leads to suppression of the adaptive and innate immune response to the tumor and provides a mechanism for tumoral evasion from immune surveillance. Recently, it was discovered that patients suffering from transporter associated with antigen processing 2 (TAP2) deficiency express unusually high levels of KIRs and CEACAM1, and that CEACAM1 exerts killer inhibitory function on these NK cells. This compensatory mechanism for class I MHC-dependent abrogation of killer activity leads to the inhibition of killing of tumor and autologous cells (Markel et al. 2003). Another situation that requires tight regulation of NK cell activity and suppression of their killer function is the invasion of the uterine endometrium by the extravillous trophoblast (Moffett-King 2002). $CD16^-/CD56^+$ NK cells constitute about 40% of decidual cells that upregulate CEACAM1-L expression upon stimulation with IL-2. Mandelboim's group showed that CEACAM1-L has a strong impact on the regulation of the local decidual immune response: CEACAM1-L-mediated homotypic interactions inhibit lysis, proliferation, and cytokine secretion of activated decidual NK, T, and NKT cells ($CD3^+/CD56^+$; Markel et al. 2002b). This allows maternal allorecognition of the fetus by uterine natural killer cells and invasion of the fetal extravillous trophoblast without challenging a cytotoxic immune reaction.

5.4
CEACAMs Are Receptors for Microbes and Viruses

5.4.1
CEACAMs Are Receptors for *Salmonellae* and *Escherichia coli*

CEACAM1, CEA, and CEACAM6 display high-mannose residues that are targeted by mannose-specific microbial lectins such as type 1 fimbriae of *E. coli* and *Salmonella typhimurium*. Type 1 fimbriae contain a lectin that specifically binds to terminally reducing α-mannosyl moieties (Manα1–3Man) present on glycans on cell surface receptors (Eshdat 1978).

Type 1 fimbriae bind to CEACAM1, CEA, and CEACAM6 on human intestinal epithelia and CEACAM1 and CEACAM6 on human neutrophilic granulocytes in vitro and in vivo in a mannose-dependent manner (Leusch et al. 1990, 1991a,b; Sauter et al. 1991; A.K. Horst, C. Wagener, unpublished data). Interestingly, the high-mannose glycan epitope for fimbrial binding to CEACAM1 and CEACAM6 seem to reside in different domains of these molecules. By mutational and biochemical analysis, Wagener's group demonstrated that CEACAM6 contains high-mannose moieties within its N-terminal domain. In contrast to this, high-mannose residues on CEACAM1 from human granulocytes are located within its membrane-proximal A2 domain

(Mahrenholz et al. 1993). On human granulocytes, CEACAM1 is the only member of the CEA family that contains an A2 domain, and mannose moieties within its N-terminal variable-like domain have not been demonstrated so far.

Activation of granulocytes is an important mechanism of the first-line defense: Binding of fimbriated bacteria triggers degranulation, production of cytotoxic oxygen species, activation of protein kinase C, and phospholipids turnover and bacterial uptake by lectinophagocytosis (Bar-Shavit et al. 1977; Sharon 1987; Gbarah et al. 1989). Furthermore, binding of fimbriated bacteria to intestinal epithelia is an essential function of CEACAMs in the regulation of the colonization of the commensal flora (Hammarström and Baranov 2001; Fig. 7).

5.4.2
CEACAMs Are Receptors for Neisseria

Initial attachment of gonococci to mucosal epithelia and neutrophils is mediated by bacterial pili (McGee et al. 1983). A firm secondary adhesion with subsequent ingestion and penetration into phagocytes or subepithelial layers, however, requires the expression of specific outer membrane proteins, the colony-associated Opa (opacity) proteins. Opa proteins trigger efficient opsonin-independent uptake of gonococci into phagocytic and non-phagocytic cells. Expression of gonococcal Opa proteins is phase variable (Stern et al. 1986; Meyer and van Putten 1989). Differential tropism of gonococci is determined by the expression of specific Opa variants (Kupsch et al. 1993). So far, two distinct classes of Opa proteins have been identified based on differential binding activity to their cellular receptors. The first class (Opa50-variants) binds to heparin sulfate proteoglycan-containing syndecan receptors (Chen et al. 1995), vitronectin, and fibronectin (Dehio et al. 1998). The second class of specific Opa-determinants (Opa52) targets CEACAMs. Heterologous transfection studies with *CEACAM* cDNAs revealed that meningococcal virulence-associated Opa-proteins, which are expressed by more than 95% of clinical and mucosal isolates of meningococci and gonococci, bind the N-terminal domain of CEACAM1-L. Additionally, meningococcal strains expressing capsule and sialylated lipopolysaccharide are also capable of binding to CEACAM1-L (Virji 1996). Besides CEACAM1-L, CEACAM3-L, CEA, and CEACAM6 are targeted by Opa proteins of *Neisseria*. However, different Opa variants display overlapping binding specificities for CEACAMs—some bind to CEACAM1, CEACAM3, CEA, and CEACAM6, whereas the binding of others is restricted to both CEACAM1 and CEA or CEA alone (Chen et al. 1997; Gray-Owen et al. 1997a,b; Bos et al. 1998). In contrast to these findings, Virji et al. demonstrated that CEACAM6 was not bound by certain Opa variants of different neisserial strains. The binding epitope within the N-terminal domain that is targeted by *Neisseria* is con-

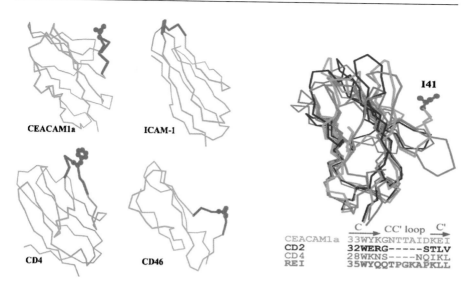

Fig. 8. Structural comparison of the immunoglobulin fold of the N-terminal domains of different members of the Ig superfamily which serve as receptors for various pathogens: soluble murine *CEACAM1a*, the receptor for murine coronavirus MHV; *ICAM-1*, the receptor for the major group of rhinoviruses; *CD4*, the primary receptor for HIV (human immunodeficiency virus); and *CD46*, the receptor for measles virus. The key virus binding residues are highlighted in *red*. Superposition (*right*) of the D1 domain of soluble murine CEACAM1a[1,4] (*cyan*), CD2 (*blue*), CD4 (*brown*) and REI (*green*). The unique convoluted conformation of the CC' loop in soluble murine CEACAM1a[1,4] is striking. The sequence alignment of the CC' loop regions of these four molecules are shown (*boted from Tan et al. 2002 with permission*)

fined to β-strand C in the non-glycosylated GFC face (cf. Fig. 8). Further studies revealed that differential tropism is mediated by a heterologous triplet of amino acids within this region (Popp et al. 1999).

Binding of Opa52$^+$ gonococci to CEACAM1-L provokes the respiratory burst in phagocytes and triggers tyrosine phosphorylation involved in CEACAM-mediated activation of the src-kinases Hck and Fgr (Hauck et al. 2000) and rapid activation of acid sphingomyelinase (Hauck et al. 2000). Acid sphingomyelinase activity is not only crucial for the activation of src-like kinases, but also for the small GTPase Rac1 (Grassme et al. 1997) and bacterial ingestion (Hauck 1998). Rac1 activates the stress-activated protein kinase pathway through PAK1 (p21-activated protein kinase) and JNK (Jun-N-terminal kinase; Hauck 1998; Fig. 7). Activation of src-like kinases and Rac1 drive cytoskeletal rearrangements that promote bacterial phagocytosis. This pathway is distinct from the opsonin-dependent Fcγ receptor-mediated signaling, since it does not involve Syk activation and is specific for Opa52$^+$ gonococci. Furthermore, the level of phosphorylation of the phosphotyrosine phosphatase SHP-1 and thus its activity is modulated by Opa-induced events. Increased tyrosine phosphorylation of SHP-1 by src-like kinases results in downregulation of its activity, thus reducing inhibitory pathways that might interfere with the bacterial uptake (Hauck 1998; Hauck et al. 1999). Neisserial binding to epithelia and endothelia triggers upregulation of CEACAM1 through inflammatory cytokines, such as tumor necrosis factor (TNF)-α, and through NF-κB-activated transcription (Muenzner et al. 2001, 2002). By this mechanism, *Neisseria* induce signal transduction pathways that allow efficient colonization of host tissue and encourage bacterial engulfment and transcellular passage.

The mechanism underlying CEACAM3-L-mediated uptake of *Neisseria* seems to employ different mechanisms when compared to CEACAM1-L: Bacterial uptake and intracellular survival was discovered to be associated with PI3 K (phosphatidylinsositiol-3-kinase) activity (Booth et al. 2003). In contrast to CEACAM1-L, CEACAM3-L induces IP_3 (phosphatidyl-inositol(3,4,5) trisphosphate) accumulation at sites of bacterial internalization. Furthermore, downstream products of the PI3 K pathway, such as phosphatidylinositol 3-phosphate, seem to be involved in the regulation of bacterial survival, due to their requirement for phagosomal maturation (Booth et al. 2003). CEACAM3 is expressed in granulocytes only and may support innate host defense. Though CEACAM3-L shares a high homology with CEACAM1-L, the CEACAM3-L cytoplasmic domain contains an ITAM instead of an ITIM that binds tyrosine kinases of the src-family and the calprotectin complex, and supports bacterial engulfment in a phosphotyrosine-dependent manner (Chen 2001; Streichert et al. 2001; McCaw et al. 2003).

5.4.3
CEACAMs Are Receptors for *Haemophilus Influenzae*

CEACAMs are also targeted by numerous strains of capsulate (typable) and acapsulate (non-typable) *Haemophilus influenzae*. Like gonococci, *Haemophilus* binds to CEACAMs through a variable outer membrane protein, the P5 protein, and interacts primarily with the N-terminal domain of CEACAMs. However, several strains seem to require more than the presence of the N-terminal domain for efficient binding of soluble CEACAMs, such as additional A or B domains (Virji et al. 2000). Non-adherent strains can reconstitute their ability to bind both to purified CEACAM1 and CEACAM1-transfected cells (CHO cells, Chinese hamster ovary cells) after transformation with the P5 protein. In contrast to the adhesin P5, *Haemophilus* expresses additional ligands that might target CEACAM1 on cell surfaces (Hill et al. 2001). Strikingly, site-directed mutagenesis with substitution of the surface-exposed amino acids revealed that the binding epitope for diverse *Haemophilus influenzae* strains lies within the same exposed region of the

CFG face that also harbors the sequence targeted by *Neisseria* (Bos et al. 1999; Virji et al. 1999; Fig. 8).

5.4.4
CEACAMs Are Receptors for Murine Hepatitis Virus

Murine hepatitis virus (MHV) strain A59 belongs to the enveloped positive-stranded RNA viruses in the coronaviridae family in the order nidovirales. Murine hepatitis viruses cause respiratory and enteric infections, hepatitis immune dysfunction, acute encephalitis, splenolysis, and chronic demyelinating disease (Compton et al. 1993). CEACAM1 was identified as a receptor for the murine hepatitis virus by Williams et al., and the cDNA was cloned by Holmes' group (Williams et al. 1990, 1991; Dveksler et al. 1991). In contrast to humans, mice express two allelic variants of the *Ceacam1*-gene, *Ceacam1a* and *Ceacam1b*, which each give rise to four distinct isoforms emerging from differential splicing. These isoforms consist of two or four extracellular domains and a long or short cytoplasmic tail (73 or 10 amino acids, respectively). All four murine CEACAM1a proteins and CEACAM2, consisting of the N-terminal domain and the A2 domain, are receptors for murine hepatitis virus strain A59 when recombinant proteins are expressed in a heterologous system (Baby Hamster Kidney cells; Nedellec et al. 1994). CEACAM1a is a high-affinity receptor for the 180-kDa viral spike glycoprotein of the murine hepatitis virus. Human HV seems to utilize a different receptor when compared to murine HV (Gagneten et al. 1996).

Most inbred laboratory mouse strains are susceptible to MHV infection and are homozygous for the CEACAM1a allele. However, mice homozygous for the CEACAM1b allelic variant display weaker MHV-A59 binding activity and are MHV resistant, such as SJL/J mice. To date, CEACAM1a proteins (including their splice variants) are the only receptors for MHV (Dveksler et al. 1993a). As shown in heterologous transfection studies, high levels of CEACAM2 expression also induce susceptibility towards MHV, although CEACAM2 displays a significantly lower affinity for the virus when compared to the CEACAM1a proteins (Nedellec et al. 1994). The differential expression of the CEACAM1-isoforms in different mouse tissues may explain the tissue tropism of different MHV strains. The target epitope for the viral spike protein resides within the N-terminal domain of CEACAM1, and glycosylation of the three potential N-linked glycosylation sites is not required for viral binding (Dveksler et al. 1995). Efficient binding, however, requires a CEACAM1 protein that contains more Ig domains than just the N-terminal domain (Dveksler et al. 1993b). In a partial mouse knockout system (p/p mice), targeted disruption of the murine *Ceacam1* gene leads to a marked reduction of susceptibility towards MHV (Blau et al. 2001). The p/p mice do indeed get infected by the MHVA59 virus, but interestingly, the liver produces only small foci compared to those observed in the wildtype mice, and these foci disappear 5 days post-infection.

In comparison to other members of the Ig superfamily, CEACAMs exhibit unique structural features which determine their distinctive homophilic and heterophilic adhesion properties (Fig. 8). This was demonstrated by Holmes' group in the analysis of the crystal structure of soluble murine CEACAM1a[1,4]: Superposition of the CEACAM1 N-terminal domain on the variable-like domain of other representative members of the IgSF, CD2, CD4, and the REI protein, reveals a protruding loop interconnecting the β-strands C and C' of CEACAM1-L that folds back onto the CFG face of the ABED β-sheet. The corresponding region in the REI protein, a typical variable domain of an antibody, however, depicts a hairpin-like structure (Tan et al. 2002; Fig. 8).

Comparison of the CEACAM1 N-terminal domain to the N-terminal domains of other virus receptors, like ICAM-I (intercellular cell adhesion molecule-I; CD54), a rhinovirus receptor, CD4, the primary receptor for HIV (human immunodeficiency virus), and CD46, the receptor for measles virus, reveals common structural features such as exposed hydrophobic residues (highlighted in red, left side of Fig. 8) that are crucial for viral binding and determine viral tropism (Wang and Springer 1998). In the CEACAM1 CC' loop, an Ile (Ile41) becomes exposed, which is a highly conserved residue in human and rodent CEACAMs. This Ile lies within a peptide sequence that has been defined by mutational analysis as the target for MHV binding. Transferring the structural data obtained from murine CEACAM1 to human CEACAM1, it is possible to assume that the N-terminal domain of human CEACAM1 and other CEACAMs exhibit very similar structural properties, and that the CC' loop also acquires a convoluted conformation (Tan et al. 2002). The importance of this conformational epitope also becomes evident in the context of neisserial interaction with CEACAMs, as well. The binding area on CEACAM1 and CEA targeted by *Neisseria* is constituted by amino acid residues on strand C, the CC' loop and strand F. Conserved amino acids within this region are responsible for stabilizing the convoluted conformation of the CC' loop (Fig. 8). Mutational analysis targeting these amino acids leads to abrogation of neisserial binding (Virji et al. 1999). In a different approach studying neutrophil adhesion to HUVECs, synthetic peptides containing amino acid sequences within the CC' loop and turns between the β-sheets were found to trigger neutrophil adhesion through β_2 integrin activation to HUVECs (Skubitz et al. 2000; Skubitz et al. 2001). These data emphasize that homophilic adhesion processes mediated by CEACAM1 or binding to other or yet unidentified ligands in *trans* mediate CEACAM-functions in vivo.

Moreover, homophilic interactions between CEACAMs is mediated by interactions of the CC' loop and the FG loops in the N-terminal domain. Interestingly, the ABED sheet is much more conserved in CEACAMs than the

CFG face, though the N-terminal domains of CEACAMs exhibit 70%–90% amino acid sequence identity. This might explain their selective binding properties and allows a certain flexibility for the interaction with different family members.

5.5
CEACAM1 Modulates Angiogenesis

5.5.1
CEACAM1 Is an Angiogenic Growth Factor

Angiogenesis, the sprouting outgrowth of blood vessels from pre-existing ones, is initiated by the angiogenic switch, a cascade triggered by a tip in the balance of pro-angiogenic and anti-angiogenic factors (Hanahan and Folkman 1996). Major regulatory stages in angiogenesis involve (1) the activation of endothelial cells by soluble growth factors that bind to their cognate receptor tyrosine kinases (Folkman 1996), (2) the initiation of endothelial cell migration and cellular interaction with extracellular matrix components, and (3) cellular proliferation and differentiation into endothelial tubes. Major growth factor families involved in angiogenesis are the vascular endothelial growth factor (VEGF)-receptor family, the tie-receptor family binding to angiopoietins, and the ephrins. Angiogenesis is initiated by signaling through one or several of these receptors either alone or in concert with adhesion molecules that are activated upon growth factor receptor signaling (Byzova et al. 2000; Carlson et al. 2001). In later stages of angiogenesis, however, adhesion molecules, i.e., immunoglobulins and integrins, are the major instruments orchestrating angiogenesis-related events such as cellular migration and the establishment of a vascular network on the basis of cellular adhesion and interaction with the extracellular matrix (Carmeliet and Jain 2000; Conway et al. 2001).

CEACAM1 is the only member of the CEA family with a cytoplasmic tail expressed on endothelia. Apart from CEACAM1, CEA is also expressed on endothelia (Majuri et al. 1994). Recently, CEACAM1 has been identified as an angiogenic growth factor (Ergün et al. 2000). In addition to growth factor receptors and their ligands, vascular adhesion molecules such as platelet/endothelial cell adhesion molecule (PECAM)1 (CD31) and the integrin $\alpha_V\beta_3$ also contribute to angiogenesis by triggering endothelial cell migration and mediating cell–cell and cell–extracellular matrix interactions (Ferrero et al. 1995; Varner et al. 1995). In this context, it is noteworthy that CEACAM1-4L interacts with integrin $\alpha_V\beta_3$ in human endothelial and epithelial cells (Brümmer et al. 2001). The first indication that CEACAM1 is involved in angiogenesis emerged from the findings that it is expressed on microvessels of proliferating tissue, such as microvessels in the developing central nervous system of the rat (Sawa et al. 1994) and in wound healing edges in granuloma tissue.

Furthermore, blood vessels on the fetal-maternal interface, i.e., in the decidua and the placenta, display CEACAM1 expression (Daniels et al. 1996; Prall et al. 1996; Bamberger et al. 2001). In the adult, CEACAM1 can be detected exclusively in small blood vessels, and no expression is found on mature, large blood vessels. In contrast to the distinct expression pattern in the human vasculature, CEACAM1 is expressed on the majority of large and small blood vessels in the mouse (A. Horst, N. Beauchemin, and C. Wagener, unpublished data). In human tumors, CEACAM1 is present in vessels of renal cell carcinoma, carcinomas of the urinary bladder and prostate, and Leydig cell tumors (Ergün et al. 2000).

CEACAM1 is present on capillary-like structures formed by endothelial progenitor cells when transplanted together with tumor cells into severe combined immunodeficiency (SCID) mice (Gehling et al. 2000). Purified native and recombinant CEACAM1 stimulates proliferation, chemotaxis, and tube formation of human microvascular endothelial cells and induces angiogenesis in the chorioallantois membrane (CAM) of the chicken. The angiogenic effects initiated by CEACAM1 were found to be additive in combination with VEGF or basic (b)FGF. Furthermore, stimulation of human dermal microvascular endothelial cells with VEGF and bFGF leads to induction of CEACAM1 expression, suggesting the existence of a functional link in a common signaling pathway (Ergün et al. 2000). As shown for the IR and the EGF receptor, CEACAM1 does act as a substrate for receptor tyrosine kinases (Phillips et al. 1987). The molecular link between vascular growth factor receptor signaling pathways and CEACAM1, however, still needs to be elucidated.

The fact that CEACAM1 is a major carrier of sialyl Lewis^x and Lewis^x residues on granulocytes suggests that CEACAM1 could function as a ligand for E-selectin (Kerr and Stocks 1992; Stocks and Kerr 1993; Sanders and Kerr 1999). Under physiological conditions, sialyl Lewis^x and Lewis^x-carrying glycoproteins support intercellular adhesion of endothelial cells (see also Fig. 7).

In summary, CEACAM1 exerts distinct functions in different stages of angiogenesis: during early stages, in its soluble form, it acts a chemoattractant, and its expression is upregulated in response to angiogenic stimuli (Ergün et al. 2000). Cleavage of cell-bound growth factor receptors and adhesion molecules has been described in the context of the regulation of their bioavailability and ligand-binding activities (Hornig et al. 2000; Carmeliet et al. 2001). In later stages of angiogenesis, as the nascent vasculature is stabilized, adhesion molecules are important tools for the recruitment of accessory cells, such as pericytes and the establishment of cell–cell junctions. CEACAM1 shares certain functional features with CD31 and junctional cell adhesion molecules (JAMs) that have been reported to support the organization of cell–cell junctions by homophilic interactions. These molecules do not only regulate vessel permeability and endothelial cell–cell interactions but also participate in cellular communication during neutrophil transmigration (Ferrero et al. 1995; Dejana et al. 2001). Additionally, CEACAM1 may facilitate the establishment of endothelial cell–cell contacts through binding of E-selectin and participate in the recruitment of accessory cells. By organizing cellular adhesion, CEACAM1 has been shown to promote tube formation in human endothelial cells and the lumen formation of epithelial cells in an extracellular matrix, providing evidence that it is indeed a key player in organizing luminal tissue architecture and regulating cellular differentiation (Huang et al. 1998; Kirshner et al. 2003a).

5.5.2
CEACAM1 Induces Secretion of Angiostatic Factors in CEACAM1-Transfected Prostate Carcinoma Cells

Unexpectedly, it was discovered that adenovirus-mediated CEACAM1-L expression in prostate cancer cells triggered the synthesis of a yet unidentified soluble angiostatic factor (Volpert et al. 2002). This study had originally been designed to further investigate the inhibitory effects of CEACAM1-L on prostate cancer cell growth. It was hypothesized that the tumor-suppressive effects of CEACAM1-L on prostate cancer cells is a result of the inhibition of tumor angiogenesis and that secretion of an angiostatic factor is specifically induced by re-expression of CEACAM1-L in these cells. Cell culture medium conditioned by wildtype adenovirus-CEACAM1-L-infected prostate carcinoma cells inhibited human endothelial cell migration and corneal neovascularization in vivo as well as an increase of endothelial cells. This effect could not be produced with media conditioned by a CEACAM1-L mutant (Ser503Ala) that is also incapable of conferring tumor-suppressive effects or endothelial cell apoptosis (Volpert et al. 2002).

In contrast to the above-mentioned model systems, purified or soluble CEACAM1-L per se has not been tested with regard to its angiogenic properties. Furthermore, it has not been mentioned whether CEACAM1 expression by endothelial cells co-cultured with the transfected prostate carcinoma cells was reduced or if secretion of soluble CEACAM1 by the endothelial cells was impaired. These two different approaches, however, provide important clues on how CEACAM1 can affect angiogenesis with respect to endothelial cell outgrowth and differentiation but also how changes in CEACAM1 expression during tumor progression in certain epithelia influence angiogenic events.

6
Perspectives

CEACAM1 is the most ancestral member of the CEA gene family. Expression of CEACAMs starts early during embryogenesis, and is maintained throughout life. CEACAMs are fundamental for establishing and maintaining tissue architecture. Although CEACAMs function as cellular adhesion molecules, they are not static, but rather involved in a great variety of dynamic processes, such as growth factor signaling, regulation of immune responses, and tumor cell growth suppression, as well as promotion of invasion. The signal transduction properties of CEACAM1 and its effects on cellular proliferation and differentiation are determined by its engagement in cellular adhesion and the expression level of different isoforms (Öbrink et al. 2002). As a signaling molecule, CEACAM1 displays dual functions that are exploited differently in different cellular contexts. On neutrophilic granulocytes, CEACAM1-L acts as a positive stimulator inducing activation and adhesion of neutrophils to endothelial cells through β_2 integrins, and it mediates opsonin-independent phagocytosis. An example for a negative regulatory effect through CEACAM1-L signal transduction can be observed after neisserial binding to CEACAM1-L on T lymphocytes and results in arrest of T cell proliferation (Boulton and Gray-Owen 2002). As reported by Mandelboim's group, CEACAM1-L expression on melanoma cells provides a mechanism for tumor cells to evade immune surveillance by downregulation of NK cell responses through homophilic interactions mediated by CEACAM1-L in *trans* (Markel et al. 2002a). It is worth noting that CEACAM1 is the only member of the CEA family with a long cytoplasmic domain that is capable of homophilic adhesion. Another important functional aspect of CEACAMs is that expression of membrane-bound forms with a GPI-anchor is predominantly—but not exclusively—found on epithelia, whereas CEACAMs that contain a long cytoplasmic domain and are capable of active signal transduction are also expressed on a variety of motile and circulating cells, such as lymphocytes, granulocytes, dendritic cells, or endothelial cells. Here, it is striking that especially CEACAM1-L expression is subject to upregulation after cellular activation by cytokines, chemokines, microbial proteins, or growth factors. However, it must not be neglected that other members of the CEA family, like CEA, CEACAM6, as well as CEACAM1, respond to inflammatory cytokines on intestinal epithelial cells (Takahashi 1993).

CEA is an ubiquitous clinical marker with prognostic relevance in the evaluation of various progressive malignant carcinomas. Its qualities as a diagnostic clinical marker, target for anticancer vaccines, and its therapeutic use has been reviewed elsewhere (Stanners 1998; Bunjes et al. 2001; Berinstein 2002; Pagel et al. 2002). Recently, CEACAM1 expression on invasive melanoma and adenocarcinoma of the lung was discovered as a novel clinical marker of high prognostic significance that is directly correlated with patient survival (Laack et al. 2002; Thies et al. 2002). In a multivariate analysis in lung cancer and melanoma patients, CEACAM1 was identified as an independent marker of poor prognosis and risk of metastasis (Laack et al. 2002; Thies et al. 2002; Sienel et al. 2003). In addition, the strongest CEACAM1 expression in malignant melanoma is observed on the invading front of the tumor. Its co-expression with β_3 integrin provides evidence for direct involvement of CEACAM1 in the regulation of invasive processes (Brümmer et al. 2001). In contrast to CEACAM1 upregulation in certain tumors, CEACAM1 expression is markedly reduced or even lost during early stages of tumor progression, such as in colonic carcinomas (Ilantzis et al. 1997; Nollau et al. 1997). In this case, tumor dissemination is encouraged by dramatic changes in tissue architecture and intercellular adhesion. Moreover, CEACAMs express Lewis blood group antigens that are associated with metastatic spread of primary tumors (Sanders and Kerr 1999). Besides its upregulation on tumoral tissues in a membrane bound form, CEACAM1 serum levels are elevated in jaundice and chronic liver diseases (Lucka et al. 1998; Draberova et al. 2000; Kondo et al. 2001). The effects of soluble CEACAM splice variants on the regulation of homophilic and heterophilic adhesion or their potential to act as soluble growth factors still need to be defined. It has not been investigated, however, if CEACAMs of the CEA family other than CEA and CEACAM6 are released from the cell membrane of normal or tumor cells and whether putative soluble forms have any implication in the regulation of adhesion or malignant progression. Furthermore, specific mutations associated with clinically relevant phenotypes have not been described yet except for CEA: patients with primary tumors and exceptionally high serum levels of CEA display a mutated form of CEA that is associated with increased hepatic metastasis (Zimmer and Thomas 2001).

With the perspective of developing new gene therapies that revert the malignant phenotype of primary tumors, a preclinical trial for the treatment of androgen-independent prostate cancer is currently under investigation. In this approach, adenovirus-mediated gene transfer of *CEACAM1-L* cDNA into PC-3 prostate carcinoma cells in a nude mouse xenograft model led to a marked reduction of tumor cell growth (Kleinerman et al. 1995b). In addition, the combination of *CEACAM1-L* gene therapy and treatment with the angiogenic inhibitor TNP-470 in this model resulted in enhanced suppression of tumor cell growth in vivo compared to restoration of CEACAM1-L expression alone (Pu et al. 2002). This study provides a novel therapeutic approach that takes advantage of the tumor suppressive effects of a cellular adhesion molecule and anti-angiogenic agents. It provides evidence that CEACAM1-L is a novel putative therapeutic agent for the treatment of malignant tumors.

Acknowledgements We would like to thank our colleagues Drs. Kathryn V. Holmes, Sonia M. Najjar, and Sten Hammarström for their kind permission to reproduce figures from their work. We would also like to thank Dr. Nicole Beauchemin for critical comments on the manuscript and Sabine Wuttke for her excellent assistance with the illustration of this review.

References

Al Moustafa AE, Yen L, Benlimame N, Alaoui-Jamali MA (2002) Regulation of E-cadherin/catenin complex patterns by epidermal growth factor receptor modulation in human lung cancer cells. Lung Cancer 37:49–56

Alais S, Allioli N, Pujades C, Duband JL, Vainio O, Imhof BA, Dunon D (2001) HEMCAM/CD146 downregulates cell surface expression of beta1 integrins. J Cell Sci 114:1847–1859

Bamberger AM, Riethdorf L, Nollau P, Naumann M, Erdmann I, Götze J, Brümmer J, Schulte HM, Wagener C, Löning T (1998) Dysregulated expression of CD66a (BGP, C-CAM), an adhesion molecule of the CEA family, in endometrial cancer. Am J Pathol 152:1401–1406

Bamberger AM, Sudahl S, Löning T, Wagener C, Bamberger CM, Drakakis P, Coutifaris C, Makrigiannakis A (2000) The adhesion molecule CEACAM1 (CD66a, C-CAM, BGP) is specifically expressed by the extravillous intermediate trophoblast. Am J Pathol 156:1165–1170

Bamberger AM, Sudahl S, Wagener C, Löning T (2001) Expression pattern of the adhesion molecule CEACAM1 (C-CAM, CD66a, BGP) in gestational trophoblastic lesions. Int J Gynecol Pathol 20:160–165

Bamberger AM, Kappes H, Methner C, Rieck G, Brümmer J, Wagener C, Löning T, Milde-Langosch K (2002) Expression of the adhesion molecule CEACAM1 (CD66a, BGP, C-CAM) in breast cancer is associated with the expression of the tumor-suppressor genes Rb, Rb2, and p27. Virchows Arch 440:139–144

Bar-Shavit Z, Ofek N, Goldman R, Mirelman D, Sharon N (1977) Mannose residues of phagocytes as receptors for the attachment of *E. coli* and *Salmonella typhi*. Biochem Biophys Res Commun 78:455–460

Baranov V, Yeung MM-W, Hammarström S (1994) Expression of CEA and nonspecific cross-reacting 50 kD Antigen in human normal and cancerous colon mucosa: comparative ultrastructural study with monoclonal antibodies. Cancer Res 54:3305–3314

Barnett TR, Kretschmer A, Austen DA, Goebel SJ, Hart JT, Elting JJ, Kamarck ME (1989) Carcinoembryonic antigens: alternative splicing accounts for the multiple mRNAs that code for novel members of the CEA family. J Cell Biol 108:267–276

Barnett TR, Drake L, Pickle W 2nd (1993) Human biliary glycoprotein gene: characterization of a family of novel alternatively spliced RNAs and their expressed proteins. Mol Cell Biol 13:1273–1282

Bates PA, Luo J, Sternberg MJ (1992) A predicted three-dimensional structure for the carcinoembryonic antigen (CEA). FEBS Lett 301:207–214

Baum O, Troll S, Hixson DC (1996) The long and the short isoform of cell-CAM 105 show variant-specific modifications in adult rat organs. Biochem Biophys Res Commun 227:775–781

Beauchemin N, Benchimol S, Cournoyer D, Fuks A, Stanners CP (1987) Isolation and characterization of full-length functional cDNA clones for human carcinoembryonic antigen. Mol Cell Biol 7:3221–3230

Beauchemin N, Kunath T, Robitaille J, Chow B, Turbide C, Daniels E, Veillette A (1997) Association of biliary glycoprotein with protein tyrosine phosphatase SHP-1 in malignant colon epithelial cells. Oncogene 14:783–790

Beauchemin N, Draber P, Dveksler G, Gold P, Gray-Owen S, Grunert F, Hammarsröm S, Holmes KV, Karlsson A, kuroki M, Lin S-H, Lucka L, Najjar SM, Neumaier M, Öbrink B, Shively JE, Skubitz KM, Stanners CP, Thomas P, Thompson JA, Virji M, von Kleist S, Wagener C, Watts S, Zimmermann W (1999) Redefined nomenclature for members of the carcinoembryonic antigen family. Exp Cell Res 252:243–249

Berinstein NL (2002) Carcinoembryonic antigen as a target for therapeutic anticancer vaccines: a review. J Clin Oncol 20:2197–2207

Birchmeier W, Behrens J (1994) Cadherin expression in carcinomas: role in the formation of cell junctions and the prevention of invasiveness. Biochim Biophys Acta 1198:11–26

Blau DM, Turbide C, Tremblay M, Olson M, Letourneau S, Michaliszyn E, Jothy S, Holmes KV, Beauchemin N (2001) Targeted disruption of the ceacam1 (mhvr) gene leads to reduced susceptibility of mice to mouse hepatitis virus infection. J Virol 75:8173–8186

Boehm MK, Mayans MO, Thornton JD, Begent RHJ, Keep PA, Perkins SJ (1996) Extended glycoprotein structure of the seven domains in human carcinoembryonic antigen by X-ray and neutron solution scattering and an automated curve fitting procedure: implications for cellular adhesion. J Mol Biol 259:718–736

Booth JW, Telio D, Liao EH, McCaw SE, Matsuo T, Grinstein S, Gray-Owen SD (2003) Phosphatidylinositol 3-kinases in CEACAM-mediated internalization of Neisseria gonorrhoeae. J Biol Chem 278:14037–14045

Bos MP, Kuroki M, Krop-Watorek A, Hogan D, Belland RJ (1998) CD66 receptor specificity exhibited by neisserial Opa variants is controlled by protein determinants in CD66 N-domains. Proc Natl Acad Sci U S A 95:9584–9589

Botling J, Oberg F, Nilsson K (1995) CD49f (alpha 6 integrin) and CD66a (BGP) are specifically induced by retinoids during human monocytic differentiation. Leukemia 9:2034–2041

Boulton IC, Gray-Owen SD (2002) Neisserial binding to CEACAM1 arrests the activation and proliferation of CD4+ T lymphocytes. Nat Immunol 3:229–236

Bracke ME, Van Roy FM, Mareel MM (1996) The E-cadherin/catenin complex in invasion and metastasis. Curr Top Microbiol Immunol 213:123–161

Bradbury J (2002) Neisseria gonorrhoeae evades host immunity by switching off T lymphocytes. Lancet 359:681

Brümmendorf T, Lemmon V (2001) Immunoglobulin superfamily receptors: *cis*-interactions, intracellular adapters and alternative splicing regulate adhesion. Curr Opin Cell Biol 13:611–618

Brümmer J, Neumaier M, Göpfert C, Wagener C (1995) Association of pp60c-src with biliary glycoprotein (CD66a), an adhesion molecule of the carcinoembryonic antigen family downregulated in colorectal carcinomas. Oncogene 11:1649–1655

Brümmer J, Ebrahimnejad A, Flayeh R, Schumacher U, Löning T, Bamberger AM, Wagener C (2001) *cis* Interaction of the cell adhesion molecule CEACAM1 with integrin beta(3). Am J Pathol 159:537–546

Budt M, Cichocka I, Reutter W, Lucka L (2002) Clustering-induced signaling of CEACAM1 in PC12 cells. Biol Chem 383:803–812

Bunjes D, Buchmann I, Duncker C, Seitz U, Kotzerke J, Wiesneth M, Dohr D, Stefanic M, Buck A, Harsdorf SV, Glatting G, Grimminger W, Karakas T, Munzert G, Dohner H, Bergmann L, Reske SN (2001) Rhenium 188-labeled anti-CD66 (a, b, c, e) monoclonal antibody to intensify the conditioning regimen prior to stem cell transplantation for patients with high-risk acute myeloid leukemia or myelodysplastic syndrome: results of a phase I-II study. Blood 98:565–572

Busch C, Hanssen TA, Wagener C, Öbrink B (2002) Down-regulation of CEACAM1 in human prostate cancer: correlation with loss of cell polarity, increased proliferation rate, and Gleason grade 3 to 4 transition. Hum Pathol 33:290–298

Byzova TV, Goldman CK, Pampori N, Thomas KA, Bett A, Shattil SJ, Plow EF (2000) A mechanism for modulation of cellular responses to VEGF: activation of the integrins. Mol Cell 6:851–860

Cambier JC (1995) Antigen and Fc receptor signaling. The awesome power of the immunoreceptor tyrosine-based activation motif (ITAM). J Immunol 155:3281–3285

Carlson TR, Feng Y, Maisonpierre PC, Mrksich M, Morla AO (2001) Direct cell adhesion to the angiopoietins mediated by integrins. J Biol Chem 276:26516–26525

Carmeliet P, Jain RK (2000) Angiogenesis in cancer and other diseases. Nature 407:249–257

Carmeliet P, Moons L, Luttun A, Vincenti V, Compernolle V, De Mol M, Wu Y, Bono F, Devy L, Beck H, Scholz D, Acker T, DiPalma T, Dewerchin M, Noel A, Stalmans I, Barra A, Blacher S, Vandendriessche T, Ponten A, Eriksson U, Plate KH, Foidart JM, Schaper W, Charnock-Jones DS, Hicklin DJ, Herbert JM, Collen D, Persico MG (2001) Synergism between vascular endothelial growth factor and placental growth factor contributes to angiogenesis and plasma extravasation in pathological conditions. Nat Med 7:575–583

Cavallaro U, Niedermeyer J, Fuxa M, Christofori G (2001) N-CAM modulates tumour-cell adhesion to matrix by inducing FGF-receptor signalling. Nat Cell Biol 3:650–657

Chen CJ, Lin TT, Shively JE (1996) Role of interferon regulatory factor-1 in the induction of biliary glycoprotein (cell CAM-1) by interferon-gamma. J Biol Chem 271:28181–28188

Chen T, Belland RJ, Wilson J, Swanson J (1995) Adherence of pilus- Opa+ gonococci to epithelial cells in vitro involves heparan sulfate. J Exp Med 182:511–517

Chen T, Grunert F, Medina-Marino A, Gotschlich EC (1997) Several carcinoembryonic antigens (CD66) serve as receptors for gonococcal opacity proteins. J Exp Med 185:1557–1564

Chen T, Bolland S, Chen I, Parker J, Pantelic M, Grunert F, Zimmermann W (2001) The CGM1a (CEACAM3/CD66d)-mediated phagocytotic pathway of Neisseria gonorrhea expressing opacity proteins is also the pathway to cell death. J Biol Chem 276:17413–17419

Cheung PH, Luo W, Qiu Y, Zhang X, Earley K, Millirons P, Lin SH (1993) Structure and function of C-CAM1. The first immunoglobulin domain is required for intercellular adhesion. J Biol Chem 268:24303–24310

Choice CV, Howard MJ, Poy MN, Hankin MH, Najjar SM (1998) Insulin stimulates pp120 endocytosis in cells co-expressing insulin receptors. J Biol Chem 273:22194–22200

Chothia C, Gelfand I, Kister A (1998) Structural determinants in the sequences of immunoglobulin variable domain. J Mol Biol 278:457–479

Compton SR, Barthold SW, Smith AL (1993) The cellular and molecular pathogenesis of coronaviruses. Lab Anim Sci 43:15–28

Conway EM, Collen D, Carmeliet P (2001) Molecular mechanisms of blood vessel growth. Cardiovasc Res 49:507–521

Cournoyer D, Beauchemin N, Boucher D, Benchimol S, Fuks A, Stanners CP (1988) Transcription of genes of the carcinoembryonic antigen family in malignant and nonmalignant human tissues. Cancer Res 48:3153–3157

Da Silva-Azevedo L, Reutter W (1999) The long isoform of the cell adhesion molecule C-CAM binds to actin. Biochem Biophys Res Commun 256:404-408

Daniels E, Letourneau S, Turbide C, Kuprina N, Rudinskaya T, Yazova AC, Holmes KV, Dveksler GS, Beauchemin N (1996) Biliary glycoprotein 1 expression during embryogenesis: correlation with events of epithelial differentiation, mesenchymal-epithelial interactions, absorption, and myogenesis. Dev Dyn 206:272-290

Dehio C, Gray-Owen SD, Meyer TF (1998) The role of neisserial Opa proteins in interactions with host cells. Trends Microbiol 6:489-495

Dejana E, Spagnuolo R, Bazzoni G (2001) Interendothelial junctions and their role in the control of angiogenesis, vascular permeability and leukocyte transmigration. Thromb Haemost 86:308-315

Dejmek A, Hjerpe A (1994) Carcinoembryonic antigen-like reactivity in malignant mesothelioma. A comparison between different commercially available antibodies. Cancer 73:464-469

Donda A, Mori L, Shamshiev A, Carena I, Mottet C, Heim MH, Beglinger C, Grunert F, Rochlitz C, Terracciano L, Jantscheff P, De Libero G (2000) Locally inducible CD66a (CEACAM1) as an amplifier of the human intestinal T cell response. Eur J Immunol 30:2593-2603

Draberova L, Cerna H, Brodska H, Boubelik M, Watt SM, Stanners CP, Draber P (2000) Soluble isoforms of CEACAM1 containing the A2 domain: increased serum levels in patients with obstructive jaundice and differences in 3-fucosyl-N-acetyl-lactosamine moiety. Immunology 101:279-287

Drzeniek Z, Lamerz R, Fenger U, Wagener C, Haubeck H-D (1991) Identification of membrane antigens in granulocytes and colonic carcinoma cells by a monoclonal antibody specific for biliary glycoprotein, a member of the carcinoembryonic antigen family. Cancer Lett 56:173-179

Dveksler GS, Pensiero MN, Cardellichio CB, Williams RK, Jiang GS, Holmes KV, Dieffenbach CW (1991) Cloning of the mouse hepatitis virus (MHV) receptor: expression in human and hamster cell lines confers susceptibility to MHV. J Virol 65:6881-6891

Dveksler GS, Dieffenbach CW, Cardellichio CB, McCuaig K, Pensiero MN, Jiang GS, Beauchemin N, Holmes KV (1993a) Several members of the mouse carcinoembryonic antigen-related glycoprotein family are functional receptors for the coronavirus mouse hepatitis virus-A59. J Virol 67:1-8

Dveksler GS, Pensiero MN, Dieffenbach CW, Cardellichio CB, Basile AA, Elia PE, Holmes KV (1993b) Mouse hepatitis virus strain A59 and blocking antireceptor monoclonal antibody bind to the N-terminal domain of cellular receptor. Proc Natl Acad Sci U S A 90:1716-1720

Dveksler GS, Basile AA, Cardellichio CB, Holmes KV (1995) Mouse hepatitis virus receptor activities of an MHVR/mph chimera and MHVR mutants lacking N-linked glycosylation of the N-terminal domain. J Virol 69:543-546

Eades-Perner A-M, van der Putten H, Hirth A, Thompson J, Neumaier M, von Kleist S, Zimmermann W (1994) Mice transgenic for the human carcinoembryonic antigen gene maintain its spatiotemporal expression pattern. Cancer Res 54:4169-4176

Ebrahimnejad A, Flayeh R, Unteregger G, Wagener C, Brümmer J (2000) Cell adhesion molecule CEACAM1 associates with paxillin in granulocytes and epithelial and endothelial cells. Exp Cell Res 260:365-373

Edlund M, Gaardsvoll H, Bock E, Öbrink B (1993) Different isoforms of stock-specific variants of the cell adhesion molecule C-CAM (cell CAM 105) in rat liver. Eur J Biochem 213:1109-1116

Edlund M, Blikstad I, Öbrink B (1996) Calmodulin binds to specific sequences in the cytoplasmic domain of C-CAM and down-regulates C-CAM self-association. J Biol Chem 271:1393–1399

Edlund M, Wikström K, Toomik R, Ek P, Öbrink B (1998) Characterization of protein kinase C-mediated phosphorylation of the short cytoplasmic domain isoform of C-CAM. FEBS Lett 425:166–170

Elghetany MT (2002) Surface antigen changes during normal neutrophilic development: a critical review. Blood Cells Mol Dis 28:260–274

Ergün S, Kilic N, Ziegeler G, Hansen A, Nollau P, Götze J, Wurmbach JH, Horst A, Weil J, Fernando M, Wagener C (2000) CEA-related cell adhesion molecule 1: a potent angiogenic factor and a major effector of vascular endothelial growth factor. Mol Cell 5:311–320

Eshdat Y, Ofek I, Yashouv-Gan Y, Sharon N, Mirelman D (1978) Isolation of a mannose-specific lectin from *E. coli* and its rule in the adherence of the bacteria to epithelial cells. Biochem Biophys Res Commun 85:1551–1559

Estrera VT, Chen DT, Luo W, Hixson DC, Lin SH (2001) Signal transduction by the CEACAM1 tumor suppressor. Phosphorylation of serine 503 is required for growth-inhibitory activity. J Biol Chem 276:15547–15553

Famiglietti J, Sun J, DeLisser HM, Albelda SM (1997) Tyrosine residue in exon 14 of the cytoplasmic domain of platelet endothelial cell adhesion molecule-1 (PECAM-1/CD31) regulates ligand binding specificity. J Cell Biol 138:1425–1435

Ferrero E, Ferrero ME, Pardi R, Zocchi MR (1995) The platelet endothelial cell adhesion molecule-1 (PECAM1) contributes to endothelial barrier function. FEBS Lett 374:323–326

Feuk-Lagerstedt E, Jordan ET, Leffler H, Dahlgren C, Karlsson A (1999) Identification of CD66a and CD66b as the major galectin-3 receptor candidates in human neutrophils. J Immunol 163:5592–5598

Folkman J, DAmore PA (1996) Blood vessel formation: what is the molecular basis? Cell 87:1153–1155

Fournes B, Sadekova S, Turbide C, Letourneau S, Beauchemin N (2001) The CEACAM1-L Ser503 residue is crucial for inhibition of colon cancer cell tumorigenicity. Oncogene 20:219–230

Fournes B, Farrah J, Olson M, Lamarche-Vane N, Beauchemin N (2003) Distinct Rho GTPase activities regulate epithelial cell localization of the adhesion molecule CEACAM1: involvement of the CEACAM1 transmembrane domain. Mol Cell Biol 23:7291–7304

Frängsmyr L, Baranov V, Prall F, Yeung MM, Wagener C, Hammarström S (1995) Cell- and region-specific expression of biliary glycoprotein and its messenger RNA in normal human colonic mucosa. Cancer Res 55:2963–2967

Frängsmyr L, Baranov V, Hammarström S (1999) Four carcinoembryonic antigen subfamily members, CEA, NCA, BGP and CGM2, selectively expressed in the normal human colonic epithelium, are integral components of the fuzzy coat. Tumour Biol 20:277–292

Fukushima K, Ohkura T, Kanai M, Kuroki M, Matsuoka Y, Kobata A, Yamashita K (1995) Carbohydrate structures of a normal counterpart of the carcinoembryonic antigen produced by colon epithelial cells of normal adults. Glycobiology 5:105–115

Gagneten S, Scanga CA, Dveksler GS, Beauchemin N, Percy D, Holmes KV (1996) Attachment glycoproteins and receptor specificity of rat coronaviruses. Lab Anim Sci 46:159–166

Gbarah A, Mhashilkar AM, Boner G, Sharon N (1989) Involvement of protein kinase C in activation of human granulocytes and peritoneal macrophages by type 1 fimbriated (mannose specific) *Escherichia coli*. Biochem Biophys Res Commun 165:1243–1249

Gehling UM, Ergün S, Schumacher U, Wagener C, Pantel K, Otte M, Schuch G, Schafhausen P, Mende T, Kilic N, Kluge K, Schäfer B, Hossfeld DK, Fiedler W (2000) In vitro differentiation of endothelial cells from AC133-positive progenitor cells. Blood 95:3106–3112

Gold P, Freedman SO (1965) Specific carcinoembryonic antigens of the human digestive system. J Exp Med 122:467–481

Grassme H, Gulbins E, Brenner B, Ferlinz K, Sandhoff K, Harzer K, Lang F, Meyer TF (1997) Acidic sphingomyelinase mediates entry of N. gonorrhoeae into nonphagocytic cells. Cell 91:605–615

Gray-Owen SD, Dehio C, Haude A, Grunert F, Meyer TF (1997a) CD66 carcinoembryonic antigens mediate interactions between Opa-expressing Neisseria gonorrhoeae and human polymorphonuclear phagocytes. EMBO J 16:3435–3445

Gray-Owen SD, Lorenzen DR, Haude A, Meyer TF, Dehio C (1997b) Differential Opa specificities for CD66 receptors influence tissue interactions and cellular response to Neisseria gonorrhoeae. Mol Microbiol 26:971–980

Greicius G, Severinson E, Beauchemin N, Öbrink B, Singer BB (2003) CEACAM1 is a potent regulator of B cell receptor complex-induced activation. J Leukoc Biol 74:126–134

Hajra KM, Fearon ER (2002) Cadherin and catenin alterations in human cancer. Genes Chromosomes Cancer 34:255–268

Hammarström S (1999) The carcinoembryonic antigen (CEA) family: structures, suggested functions and expression in normal and malignant tissues. Semin Cancer Biol 9:67–81

Hammarström S, Baranov V (2001) Is there a role for CEA in innate immunity in the colon? Trends Microbiol 9:119–125

Han E, Phan D, Lo P, Poy MN, Behringer R, Najjar SM, Lin SH (2001) Differences in tissue-specific and embryonic expression of mouse Ceacam1 and Ceacam2 genes. Biochem J 355:417–423

Hanahan D, Folkman J (1996) Patterns and emerging mechanisms of the angiogenic switch during angiogenesis. Cell 86:353–364

Hanenberg H, Baumann M, Quentin I, Nagel G, Grosse-Wilde H, von Kleist S, Gobel U, Burdach S, Grunert F (1994) Expression of the CEA gene family members NCA-50/90 and NCA-160 (CD66) in childhood acute lymphoblastic leukemias (ALLs) and in cell lines of B-cell origin. Leukemia 8:2127–2133

Hansson M, Blikstad I, Öbrink B (1989) Cell-surface location and molecular properties of cell-CAM 105 in intestinal epithelial cells. Exp Cell Res 181:63–74

Harpaz Y, Chothia C (1994) Many of the immunoglobulin superfamily domains in cell adhesion molecules and surface receptors belong to a new structural set which is close to that containing variable domains. J Mol Biol 238:528–539

Hauck W, Stanners CP (1991) Control of carcinoembryonic antigen gene family expression in a differentiating colon carcinoma cell line, Caco-2. Cancer Res 51:3526–3533

Hauck CR, Meyer TF, Lang F, Gulbins E (1998) CD66-mediated phagocytosis of opa52 Neisseria gonorrhoeae requires a src-like tyrosine kinase- and rac-1-dependent signalling pathway. EMBO J 17:443–454

Hauck CR, Gulbins E, Lang F, Meyer TF (1999) Tyrosine phosphatase SHP-1 is involved in CD66-mediated phagocytosis of Opa52-expressing Neisseria gonorrhoeae. Infect Immun 67:5490–5494

Hauck CR, Grassme H, Bock J, Jendrossek V, Ferlinz K, Meyer TF, Gulbins E (2000) Acid sphingomyelinase is involved in CEACAM receptor-mediated phagocytosis of Neisseria gonorrhoeae. FEBS Lett 478:260–266

Hauck W, Nedellec P, Turbide C, Stanners CP, Barnett TR, Beauchemin N (1994) Transcriptional control of the human biliary glycoprotein gene, a CEA gene family member down-regulated in colorectal carcinomas. Eur J Biochem 223:529–541

Hazan RB, Norton L (1998) The epidermal growth factor receptor modulates the interaction of E-cadherin with the actin cytoskeleton. J Biol Chem 273:9078–9084

Hill DJ, Toleman MA, Evans DJ, Villullas S, Van Alphen L, Virji M (2001) The variable P5 proteins of typeable and non-typeable Haemophilus influenzae target human CEACAM1. Mol Microbiol 39:850–862

Hinoda Y, Imai K, Nakagawa N, Ibayashi Y, Nakano T, Paxton RJ, Shively JE, Yachi A (1990) Transcription of biliary glycoprotein I gene in malignant and non-malignant human liver tissues. Int J Cancer 45:875–878

Hixson DC, McEntire KD, Öbrink B (1985) Alterations in the expression of a hepatocyte cell adhesion molecule by transplantable rat hepatocellular carcinomas. Cancer Res 45:3742–3749

Honig M, Peter HH, Jantscheff P, Grunert F (1999) Synovial PMN show a coordinated up-regulation of CD66 molecules. J Leukoc Biol 66:429–436

Hornig C, Barleon B, Ahmad S, Vuorela P, Ahmed A, Weich HA (2000) Release and complex formation of soluble VEGFR-1 from endothelial cells and biological fluids. Lab Invest 80:443–454

Houde C, Roy S, Leung N, Nicholson DW, Beauchemin N (2003) The cell adhesion molecule CEACAM1-L is a substrate of caspase-3-mediated cleavage in apoptotic mouse intestinal cells. J Biol Chem 278:16929–16935

Hsieh JT, Luo W, Song W, Wang Y, Kleinerman DI, Van NT, Lin SH (1995) Tumor suppressive role of an androgen-regulated epithelial cell adhesion molecule (C-CAM) in prostate carcinoma cell revealed by sense and antisense approaches. Cancer Res 55:190–197

Huang J, Simpson JF, Glackin C, Riethorf L, Wagener C, Shively JE (1998) Expression of biliary glycoprotein (CD66a) in normal and malignant breast epithelial cells. Anticancer Res 18:3203–3212

Huang JQ, Turbide C, Daniels E, Jothy S, Beauchemin N (1990) Spatiotemporal expression of murine carcinoembryonic antigen (CEA) gene family members during mouse embryogenesis. Development 110:573–588

Huber M, Izzi L, Grondin P, Houde C, Kunath T, Veillette A, Beauchemin N (1999) The carboxyl-terminal region of biliary glycoprotein controls its tyrosine phosphorylation and association with protein-tyrosine phosphatases SHP-1 and SHP-2 in epithelial cells. J Biol Chem 274:335–344

Huber M, Izzi L, Grondin P, Houde C, Kunath T, Veillette A, Beauchemin N (1999) The carboxy-terminal region of biliary glycoprotein controls its tyrosine phosphorylation and association with protein-tyrosine phosphatases SHP-1 and SHP-2 in epithelial cells. J Biol Chem 274:335–344

Hunter I, Sawa H, Edlund M, Öbrink B (1996) Evidence for regulated dimerization of cell–cell adhesion molecule (C-CAM) in epithelial cells. Biochem J 320:847–853

Hunter I, Sigmundsson K, Beauchemin N, Öbrink B (1998) The cell adhesion molecule C-CAM is a substrate for tissue transglutaminase. FEBS Lett 425:141–144

Ilantzis C, Jothy S, Alpert LC, Draber P, Stanners CP (1997) Cell-surface levels of human carcinoembryonic antigen are inversely correlated with colonocyte differentiation in colon carcinogenesis. Lab Invest 76:703–716

Izzi L, Turbide C, Houde C, Kunath T, Beauchemin N (1999) *cis*-Determinants in the cytoplasmic domain of CEACAM1 responsible for its tumor inhibitory function. Oncogene 18:5563–5572

Jothy S, Yuan SY, Shirota K (1993) Transcription of carcinoembryonic antigen in normal colon and colon carcinoma. In situ hybridization study and implication for a new in vivo functional model. Am J Pathol 143:250–257

Juliano RL (2002) Signal transduction by cell adhesion receptors and the cytoskeleton: functions of integrins, cadherins, selectins, and immunoglobulin-superfamily members. Annu Rev Pharmacol Toxicol 42:283–323

Kammerer R, Hahn S, Singer BB, Luo JS, von Kleist S (1998) Biliary glycoprotein (CD66a), a cell adhesion molecule of the immunoglobulin superfamily, on human lymphocytes: structure, expression and involvement in T cell activation. Eur J Immunol 28:3664–3674

Kammerer R, Stober D, Singer BB, Öbrink B, Reimann J (2001) Carcinoembryonic antigen-related cell adhesion molecule 1 on murine dendritic cells is a potent regulator of T cell stimulation. J Immunol 166:6537–6544

Kannicht C, Lucka L, Nuck R, Reutter W, Gohlke M (1999) N-glycosylation of the carcinoembryonic antigen related cell adhesion molecule, C-CAM, from rat liver: detection of oversialylated bi- and triantennary structures. Glycobiology 9:897–906

Kerr MA, Stocks SC (1992) The role of CD15-(Le(X))-related carbohydrates in neutrophil adhesion. Histochem J 24:811–826

Khan WN, Hammarström S, Ramos T (1993) Expression of antigens of the carcinoembryonic antigen family on B cell lymphomas and Epstein-Barr virus immortalized B cell lines. Int Immunol 5:265–270

Kim J, Kaye FJ, Henslee JG, Shively JE, Park JG, Lai SL, Linnoila RI, Mulshine JL, Gazdar AF (1992) Expression of carcinoembryonic antigen and related genes in lung and gastrointestinal cancers. Int J Cancer 52:718–725

Kinugasa T, Kuroki M, Takeo H, Matsuo Y, Ohshima K, Yamashita Y, Shirakusa T, Matsuoka Y (1998) Expression of four CEA family antigens (CEA, NCA, BGP and CGM2) in normal and cancerous gastric epithelial cells: up-regulation of BGP and CGM2 in carcinomas. Int J Cancer 76:148–153

Kirshner J, Chen CJ, Liu P, Huang J, Shively JE (2003a) CEACAM1-4S, a cell–cell adhesion molecule, mediates apoptosis and reverts mammary carcinoma cells to a normal morphogenic phenotype in a 3D culture. Proc Natl Acad Sci U S A 100:521–526

Kirshner J, Schumann D, Shively JE (2003b) CEACAM1, a cell–cell adhesion molecule, directly associates with annexin II in a 3D model of mammary morphogenesis. J Biol Chem M309115200

Klein ML, McGhee SA, Baranian J, Stevens L, Hefta SA (1996) Role of nonspecific cross-reacting antigen, a CD66 cluster antigen, in activation of human granulocytes. Infect Immun 64:4574–4579

Kleinerman DI, Troncoso P, Lin SH, Pisters LL, Sherwood ER, Brooks T, von Eschenbach AC, Hsieh JT (1995a) Consistent expression of an epithelial cell adhesion molecule (C-CAM) during human prostate development and loss of expression in prostate cancer: implication as a tumor suppressor. Cancer Res 55:1215–1220

Kleinerman DI, Zhang WW, Lin SH, Nguyen TV, von Eschenbach AC, Hsieh JT (1995b) Application of a tumor suppressor (C-CAM1)-expressing recombinant adenovirus in androgen-independent human prostate cancer therapy: a preclinical study. Cancer Res 55:2831–2836

Kleinerman DI, Dinney CP, Zhang WW, Lin SH, Van NT, Hsieh JT (1996) Suppression of human bladder cancer growth by increased expression of C-CAM1 gene in an orthotopic model. Cancer Res 56:3431–3435

Kodera Y, Isobe K, Yamauchi M, Satta T, Hasegawa T, Oikawa S, Kondoh K, Akiyama S, Itoh K, Nakashima I, et al (1993) Expression of carcinoembryonic antigen (CEA) and nonspecific crossreacting antigen (NCA) in gastrointestinal cancer; the correlation with degree of differentiation. Br J Cancer 68:130–136

Kondo Y, Hinoda Y, Akashi H, Sakamoto H, Itoh F, Hirata K, Kuroki M, Imai K (2001) Measurement of circulating biliary glycoprotein (CD66a) in liver diseases. J Gastroenterol 36:470–475

Kostrewa D, Brockhaus M, D'Arcy A, Dale GE, Nelboeck P, Schmid G, Mueller F, Bazzoni G, Dejana E, Bartfai T, Winkler FK, Hennig M (2001) X-ray structure of junctional adhesion molecule: structural basis for homophilic adhesion via a novel dimerization motif. EMBO J 20:4391–4398

Kunath T, Ordonez-Garcia C, Turbide C, Beauchemin N (1995) Inhibition of colonic tumor cell growth by biliary glycoprotein. Oncogene 11:2375–2382

Kupsch EM, Knepper B, Kuroki T, Heuer I, Meyer TF (1993) Variable opacity (Opa) outer membrane proteins account for the cell tropisms displayed by Neisseria gonorrhoeae for human leukocytes and epithelial cells. EMBO J 12:641–650

Kuroki M, Yamanaka T, Matsuo Y, Oikawa S, Nakazato H, Matsuoka Y (1995) Immunochemical analysis of carcinoembryonic antigen (CEA)-related antigens differentially localized in intracellular granules of human neutrophils. Immunol Invest 24:829–843

Laack E, Nikbakht H, Peters A, Kugler C, Jasiewicz Y, Edler L, Brümmer J, Schumacher U, Hossfeld DK (2002) Expression of CEACAM1 in adenocarcinoma of the lung: a factor of independent prognostic significance. J Clin Oncol 20:4279–4284

Leslie KK, Watanabe S, Lei KJ, Chou DY, Plouzek CA, Deng HC, Torres J, Chou JY (1990) Linkage of two human pregnancy-specific beta 1-glycoprotein genes: one is associated with hydatidiform mole. Proc Natl Acad Sci U S A 87:5822–5826

Leusch H-G, Hefta SA, Drzeniek Z, Hummel K, Markos-Pusztai Z, Wagener C (1990) *Escherichia coli* of human origin binds to carcinoembryonic antigen (CEA) and nonspecific crossreacting antigen (NCA). FEBS Lett 261:405–409

Leusch H-G, Drzeniek Z, Hefta SA, Markos-Pusztai Z, Wagener C (1991a) The putative role of members of the CEA-gene family (CEA, NCA, and BGP) as ligands for the bacterial colonization of different human epithelial tissues. Zentralbl Bakteriol 275:118–122

Leusch H-G, Drzeniek Z, Markos-Pusztai Z, Wagener C (1991b) Binding of *E. coli* and Salmonella strains to members of the CEA family: differential binding inhibition by aromatic α-glycosides of mannose. Infect Immun 59:2051–2057

Lin SH, Luo W, Earley K, Cheung P, Hixson DC (1995) Structure and function of C-CAM1: effects of the cytoplasmic domain on cell aggregation. Biochem J 311:239–245

Long EO (1999) Regulation of immune responses through inhibitory receptors. Annu Rev Immunol 17:875–904

Lucka L, Sel S, Danker K, Horstkorte R, Reutter W (1998) Carcinoembryonic antigen-related cell–cell adhesion molecule C-CAM is greatly increased in serum and urine of rats with liver diseases. FEBS Lett 438:37–40

Lüning C, Wroblewski J, Öbrink B, Hammarström L, Rozell B (1995) C-CAM expression in odontogenesis and tooth eruption. Connect Tissue Res 32:201–207

Luo W, Wood CG, Earley K, Hung M-C, Lin S-H (1997) Suppression of tumorigenicity of breast cancer cells by an epithelial cell adhesion molecule (C-CAM-1): the adhesion and growth suppression are mediated by different domains. Oncogene 14:1697–1704

Luo W, Tapolsky M, Earley K, Wood CG, Wilson DR, Logothetis CJ, Lin SH (1999) Tumor-suppressive activity of CD66a in prostate cancer. Cancer Gene Ther 6:313–321

Mackay DJ, Hall A (1998) Rho GTPases. J Biol Chem 273:20685–20688

Mahrenholz AM, Yeh CH, Shively JE, Hefta SA (1993) Microsequence and mass spectral analysis of nonspecific cross-reacting antigen 160, a CD15-positive neutrophil membrane glycoprotein. Biol Chem 268:13015–13018

Majuri ML, Hakkarainen M, Paavonen T, Renkonen R (1994) Carcinoembryonic antigen is expressed on endothelial cells. A putative mediator of tumor cell extravasation and metastasis. Apmis 102:432–438

Makarovskiy AN, Pu YS, Lo P, Earley K, Paglia M, Hixson DC, Lin SH (1999) Expression and androgen regulation of C-CAM cell adhesion molecule isoforms in rat dorsal and ventral prostate. Oncogene 18:3252–3260

Mandelboim O, Malik P, Davis DM, Jo CH, Boyson JE, Strominger JL (1999) Human CD16 as a lysis receptor mediating direct natural killer cell cytotoxicity. Proc Natl Acad Sci U S A 96:5640–5644

Margolis RN, Schell MJ, Taylor SI, Hubbard AL (1990) Hepatocyte plasma membrane ECTO-ATPase (pp120/HA4) is a substrate for tyrosine kinase activity of the insulin receptor. Biochem Biophys Res Commun 166:562–566

Markel G, Lieberman N, Katz G, Arnon TI, Lotem M, Drize O, Blumberg RS, Bar-Haim E, Mader R, Eisenbach L, Mandelboim O (2002a) CD66a interactions between human melanoma and NK cells: a novel class I MHC-independent inhibitory mechanism of cytotoxicity. J Immunol 168:2803–2810

Markel G, Wolf D, Hanna J, Gazit R, Goldman-Wohl D, Lavy Y, Yagel S, Mandelboim O (2002b) Pivotal role of CEACAM1 protein in the inhibition of activated decidual lymphocyte functions. J Clin Invest 110:943–953

Markel G, Mussaffi H, Ling KL, Salio M, Gadola S, Steuer G, Blau H, Achdout H, De Miguel M, Gonen-Gross T, Hanna J, Arnon TI, Qimron U, Volovitz I, Eisenbach L, Blumberg RS, Porgador A, Cerundolo V, Mandelboim O (2003) The mechanisms controlling NK cell autoreactivity in TAP2-deficient patients. Blood 103:1770–1778

Martin M, Romero X, de la Fuente MA, Tovar V, Zapater N, Esplugues E, Pizcueta P, Bosch J, Engel P (2001) CD84 functions as a homophilic adhesion molecule and enhances IFN-gamma secretion: adhesion is mediated by Ig-like domain 1. J Immunol 167:3668–3676

McCaw SE, Schneider J, Liao EH, Zimmermann W, Gray-Owen SD (2003) Immunoreceptor tyrosine-based activation motif phosphorylation during engulfment of Neisseria gonorrhoeae by the neutrophil-restricted CEACAM3 (CD66d) receptor. Mol Microbiol 49:623–637

McCuaig K, Turbide C, Beauchemin N (1992) mmCGM1a: a mouse carcinoembryonic antigen gene family member, generated by alternative splicing, functions as an adhesion molecule. Cell Growth Differ 3:165–174

McCuaig K, Rosenberg M, Nedellec P, Turbide C, Beauchemin N (1993) Expression of the Bgp gene and characterization of mouse colon biliary glycoprotein isoforms. Gene 127:173–183

McEver RP (2002) Selectins: lectins that initiate cell adhesion under flow. Curr Opin Cell Biol 14:581–586

McGee ZA, Stephens DS, Hoffman LH, Schlech WF, 3rd Horn RG (1983) Mechanisms of mucosal invasion by pathogenic Neisseria. Rev Infect Dis 5 Suppl 4:S708–714

Metze D, Soyer H-P, Zelger B, Neumaier M, Grunert F, Hartig C, Amann U, Bhardwaj R, Wagener C, Luger TA (1996) Expression of a glycoprotein of the carcinoembryonic antigen family in normal and neoplastic sebaceous glands. J Am Acad Dermatol 134:735–744

Meyer TF, van Putten JP (1989) Genetic mechanisms and biological implications of phase variation in pathogenic neisseriae. Clin Microbiol Rev 2 Suppl:S139–145

Moffett-King A (2002) Natural killer cells and pregnancy. Nat Rev Immunol 2:656–663

Möller MJ, Kammerer R, Grunert F, von Kleist S (1996) Biliary glycoprotein (BGP) expression on T cells and on a natural-killer-cell sub-population. Int J Cancer 65:740–745

Morales VM, Christ A, Watt SM, Kim HS, Johnson KW, Utku N, Texieira AM, Mizoguchi A, Mizoguchi E, Russell GJ, Russell SE, Bhan AK, Freeman GJ, Blumberg RS (1999) Regulation of human intestinal intraepithelial lymphocyte cytolytic function by biliary glycoprotein (CD66a). J Immunol 163:1363–1370

Muenzner P, Naumann M, Meyer TF, Gray-Owen SD (2001) Pathogenic Neisseria trigger expression of their carcinoembryonic antigen-related cellular adhesion molecule 1 (CEACAM1; previously CD66a) receptor on primary endothelial cells by activating the immediate early response transcription factor, nuclear factor-kappaB. J Biol Chem 276:24331–24340

Muenzner P, Billker O, Meyer TF, Naumann M (2002) Nuclear factor-kappa B directs carcinoembryonic antigen-related cellular adhesion molecule 1 receptor expression in Neisseria gonorrhoeae-infected epithelial cells. J Biol Chem 277:7438–7446

Nagar B, Overduin M, Ikura M, Rini JM (1996) Structural basis of calcium-induced E-cadherin rigidification and dimerization. Nature 380:360–364

Nair KS, Zingde SM (2001) Adhesion of neutrophils to fibronectin: role of the cd66 antigens. Cell Immunol 208:96–106

Najjar SM (2002) Regulation of insulin action by CEACAM1. Trends Endocrinol Metab 13:240–245

Najjar SM, Accili D, Philippe N, Jernberg J, Margolis R, Taylor SI (1993) pp120/ecto-ATPase, an endogenous substrate of the insulin receptor tyrosine kinase, is expressed as two variably spliced isoforms. J Biol Chem 268:1201–1206

Najjar SM, Philippe N, Suzuki Y, Ignacio GA, Formisano P, Accili D, Taylor SI (1995) Insulin-stimulated phosphorylation of recombinant pp120/HA4, an endogenous substrate of the insulin receptor tyrosine kinase. Biochemistry 34:9341–9349

Najjar SM, Boisclair YR, Nabih ZT, Philippe N, Imai Y, Suzuki Y, Suh DS, Ooi GT (1996) Cloning and characterization of a functional promoter of the rat pp120 gene, encoding a substrate of the insulin receptor tyrosine kinase. J Biol Chem 271:8809–8817

Najjar SM, Blakesley VA, Li Calzi S, Kato H, LeRoith D, Choice CV (1997) Differential phosphorylation of pp120 by insulin and insulin-like growth factor-1 receptors: role for the C-terminal domain of the beta-subunit. Biochemistry 36:6827–6834

Najjar SM, Choice CV, Soni P, Whitman CM, Poy MN (1998) Effect of pp120 on receptor-mediated insulin endocytosis is regulated by the juxtamembrane domain of the insulin receptor. J Biol Chem 273:12923–12928

Nakajima A, Iijima H, Neurath MF, Nagaishi T, Nieuwenhuis EE, Raychowdhury R, Glickman J, Blau DM, Russell S, Holmes KV, Blumberg RS (2002) Activation-induced expression of carcinoembryonic antigen-cell adhesion molecule 1 regulates mouse T lymphocyte function. J Immunol 168:1028–1035

Nap M, Mollgard K, Burtin P, Fleuren GJ (1988) Immunohistochemistry of carcino-embryonic antigen in the embryo, fetus and adult. Tumour Biol 9:145–153

Nap M, Hammarström M-L, Börmer O, Hammarström S, Wagener C, Handt S, Schreyer M, Mach J-P, Buchegger F, von Kleist S, Grunert F, Seguin P, Fuks A, Holm R, Lamerz R (1992) Specificity and affinity of monoclonal antibodies against CEA. Cancer Res 52:2329–2339

Nedellec P, Dveksler GS, Daniels E, Turbide C, Chow B, Basile AA, Holmes KV, Beauchemin N (1994) Bgp2, a new member of the carcinoembryonic antigen-related gene family, encodes an alternative receptor for mouse hepatitis viruses. J Virol 68:4525–4537

Neumaier M, Paululat S, Chan A, Matthaes P, Wagener C (1993) Biliary glycoprotein, a potential human cell adhesion molecule, is down-regulated in colorectal carcinomas. Proc Natl Acad Sci U S A 90:10744–10748

Neumaier M, Paulutat S, Chan A, Mattaes P, Wagener C (1993) BGP, a potential human adhesion molecule, is down-regulated in colorectal carcinoma. Proc Natl Acad Sci U S A 90:10744–10748

Nollau P, Prall F, Helmchen U, Wagener C, Neumaier M (1997) Dysregulation of carcinoembryonic antigen group members CGM2, CD66a (biliary glycoprotein), and nonspecific cross-reacting antigen in colorectal carcinomas. Comparative analysis by northern blot and in situ hybridization. Am J Pathol 151:521–530

Nollau P, Scheller H, Kona-Horstmann M, Rhde S, Hagenmüller F, Wagener C, Neumaier M (1997) Expression of CD66a (human C-CAM) and other members of the carcinoembryonic antigen gene family of adhesion molecules in human colorectal adenomas. Cancer Res 57:2354–2357

Normark S, Albiger B, Jonsson AB (2002) Gonococci cause immunosuppression by engaging a coinhibitory receptor on T lymphocytes. Nat Immunol 3:210–211

Öbrink B (1997) CEA adhesion molecules: multifunctional proteins with signal-regulatory properties. Curr Opin Struct Biol 9:616–626

Öbrink B, Sawa H, Scheffrahn I, Singer BB, Sigmundsson K, Sundberg U, Heymann R, Beauchemin N, Weng G, Ram P, Iyengar R (2002) Computational analysis of isoform-specific signal regulation by CEACAM1-A cell adhesion molecule expressed in PC12 cells. Ann N Y Acad Sci 971:597–607

Ochieng J, Leite-Browning ML, Warfield P (1998) Regulation of cellular adhesion to extracellular matrix proteins by galectin-3. Biochem Biophys Res Commun 246:788–791

Ocklind C, Öbrink B (1982) Intercellular adhesion of rat hepatocytes. Identification of a cell surface glycoprotein involved in the initial adhesion process. J Biol Chem 257:6788–6795

Odin P, Tingström A, Öbrink B (1986) Chemical characterization of cell-CAM 105, a cell-adhesion molecule isolated from rat liver membranes. Biochem J 236:559–568

Ohwada A, Takahashi H, Nagaoka I, Kira S (1994) Biliary glycoprotein mRNA expression is increased in primary lung cancer, especially in squamous cell carcinoma. Am J Respir Cell Mol Biol 11:214–220

Oikawa S, Nakazato H, Kosaki G (1987) Primary structure of human carcinoembryonic antigen (CEA) deduced from cDNA sequence. Biochem Biophys Res Commun 142:511–518

Oikawa S, Inuzuka C, Kuroki M, Arakawa F, Matsuoka Y, Kosaki G, Nakazato H (1991) A specific heterotypic cell adhesion activity between members of carcinoembryonic antigen family, W272 and NCA, is mediated by N-domains. J Biol Chem 266:7995–8001

Olsen A, Teglund S, Nelson D, Gordon L, Copeland A, Georgescu A, Carrano A, Hammarström S (1994) Gene organization of the pregnancy-specific glycoprotein region on human chromosome 19: assembly and analysis of a 700-kb cosmid contig spanning the region. Genomics 23:659–668

Pagel JM, Matthews DC, Appelbaum FR, Bernstein ID, Press OW (2002) The use of radioimmunoconjugates in stem cell transplantation. Bone Marrow Transplant 29:807–816

Pallen CJ, Tong PH (1991) Elevation of membrane tyrosine phosphatase activity in density-dependent growth-arrested fibroblasts. Proc Natl Acad Sci U S A 88:6996–7000

Paxton RJ, Mooser G, Pande H, Lee TD, Shively JE (1987) Sequence analysis of CEA: identification of glycosylation sites and homology with the immunoglobulin supergene family. Proc Natl Acad Sci U S A 84:920–924

Phan D, Sui X, Chen DT, Najjar SM, Jenster G, Lin SH (2001) Androgen regulation of the cell–cell adhesion molecule-1 (Ceacam1) gene. Mol Cell Endocrinol 184:115–123

Phillips SA, Perrotti N, Taylor SI (1987) Rat liver membranes contain a 120 kDa glycoprotein which serves as a substrate for the tyrosine kinases of the receptors for insulin and epidermal growth factor. FEBS Lett 212:141–144

Popp A, Dehio C, Grunert F, Meyer TF, Gray-Owen SD (1999) Molecular analysis of neisserial Opa protein interactions with the CEA family of receptors: identification of determinants contributing to the differential specificities of binding. Cell Microbiol 1:169–181

Poy MN, Ruch RJ, Fernstrom MA, Okabayashi Y, Najjar SM (2002a) Shc and CEACAM1 interact to regulate the mitogenic action of insulin. J Biol Chem 277:1076–1084

Poy MN, Yang Y, Rezaei K, Fernstrom MA, Lee AD, Kido Y, Erickson SK, Najjar SM (2002b) CEACAM1 regulates insulin clearance in liver. Nat Genet 30:270–276

Prall F, Nollau P, Neumaier M, Haubeck H-D, Drzeniek Z, Helmchen U, Löning T, Wagener C (1996) CD66a (BGP), an adhesion molecule of the CEA family, is expressed in epithelium, endothelium and myeloid cells in a wide range of normal human tissues. J Histochem Cytochem 44:35–41

Pu YS, Luo W, Lu HH, Greenberg NM, Lin SH, Gingrich JR (1999) Differential expression of C-CAM cell adhesion molecule in prostate carcinogenesis in a transgenic mouse model. J Urol 162:892–896

Pu YS, Do KA, Luo W, Logothetis CJ, Lin SH (2002) Enhanced suppression of prostate tumor growth by combining C-CAM1 gene therapy and angiogenesis inhibitor TNP-470. Anticancer Drugs 13:743–749

Rao Y, Zhao X, Siu CH (1994) Mechanism of homophilic binding mediated by the neural cell adhesion molecule NCAM. Evidence for isologous interaction. J Biol Chem 269:27540–27548

Rass A, Lüning C, Wroblewski J, Öbrink B (1994) Distribution of C-CAM in developing oral tissues. Anat Embryol (Berl) 190:251–261

Rebstock S, Lucas K, Weiss M, Thompson J, Zimmermann W (1993) Spatiotemporal expression of pregnancy-specific glycoprotein gene rnCGM1 in rat placenta. Dev Dyn 198:171–181

Rees-Jones RW, Taylor SI (1985) An endogenous substrate for the insulin receptor-associated tyrosine kinase. J Biol Chem 260:4461–4467

Riethdorf L, Lisboa BW, Henkel U, Naumann M, Wagener C, Löning T (1997) Differential expression of CD66a (BGP), a cell adhesion molecule of the carcinoembryonic antigen family, in benign, premalignant, and malignant lesions of the human mammary gland. J Histochem Cytochem 45:957–963

Robbins PF, Eggensperger D, Qi CF, Schlom J (1993) Definition of the expression of the human carcinoembryonic antigen and non-specific cross-reacting antigen in human breast and lung carcinomas. Int J Cancer 53:892–897

Robitaille J, Izzi L, Daniels E, Zelus B, Holmes KV, Beauchemin N (1999) Comparison of expression patterns and cell adhesion properties of the mouse biliary glycoproteins Bbgp1 and Bbgp2. Eur J Biochem 264:534–544

Rojas M, Fuks A, Stanners CP (1990) Biliary glycoprotein, a member of the immunoglobulin supergene family functions in vitro as a Ca2+ dependent intercellular adhesion molecule. Cell Growth Differentiation 1:527–533

Rosenberg M, Nedellec P, Jothy S, Fleiszer D, Turbide C, Beauchemin N (1993) The expression of mouse biliary glycorprotein, a carcinoembryonic antigen-related gene, is down-regulated in malignant mouse tissues. Cancer Res 55:4938–4945

Ruchaud-Sparagano MH, Stocks SC, Turley H, Dransfield I (1997) Activation of neutrophil function via CD66: differential effects upon beta 2 integrin mediated adhesion. Br J Haematol 98:612–620

Rudert F, Saunders AM, Rebstock S, Thompson JA, Zimmermann W (1992) Characterization of murine carcinoembryonic antigen gene family members. Mamm Genome 3:262–273

Ruggiero T, Olivero M, Follenzi A, Naldini L, Calogero R, Di Renzo MF (2003) Deletion in a (T)8 microsatellite abrogates expression regulation by 3′-UTR. Nucleic Acids Res 31:6561–6569

Sadekova S, Lamarche-Vane N, Li X, Beauchemin N (2000) The CEACAM1-L glycoprotein associates with the actin cytoskeleton and localizes to cell–cell contact through activation of Rho-like GTPases. Mol Biol Cell 11:65–77

Sanders DS, Kerr MA (1999) Lewis blood group and CEA related antigens; coexpressed cell–cell adhesion molecules with roles in the biological progression and dissemination of tumours. Mol Pathol 52:174–178

Satoh Y, Hayashi T, Takahashi T, Itoh F, Adachi M, Fukui M, Kuroki M, Imai K, Hinoda Y (2002) Expression of CD66a in multiple myeloma. J Clin Lab Anal 16:79–85

Sauter SL, Rutherfurd SM, Wagener C, Shively JE, Hefta SA (1991) Binding of nonspecific cross reacting antigen, a granulocyte membrane glycoprotein, to E.coli expressing type I fimbriae. Infect Immun 59:2485–2493

Sawa H, Kamada K, Sato H, Sendo S, Kondo A, Saito I, Edlund M, Öbrink B (1994) C-CAM expression in the developing rat central nervous system. Brain Res Dev Brain Res 78:35–43

Sawa H, Ukita H, Fukuda M, Kamada H, Saito I, Öbrink B (1997) Spatiotemporal expression of C-CAM in the rat placenta. J Histochem Cytochem 45:1021–1034

Schölzel S, Zimmermann W, Schwarzkopf G, Grunert F, Rogaczewski B, Thompson J (2000) Carcinoembryonic antigen family members CEACAM6 and CEACAM7 are differentially expressed in normal tissues and oppositely deregulated in hyperplastic colorectal polyps and early adenomas. Am J Pathol 156:595–605

Schrewe H, Thompson J, Bona M, Hefta LJ, Maruya A, Hassauer M, Shively JE, von Kleist S, Zimmermann W (1990) Cloning of the complete gene for carcinoembryonic antigen: analysis of its promoter indicates a region conveying cell type-specific expression. Mol Cell Biol 10:2738–2748

Schumann D, Chen CJ, Kaplan B, Shively JE (2001) Carcinoembryonic antigen cell adhesion molecule 1 directly associates with cytoskeleton proteins actin and tropomyosin. J Biol Chem 276:47421–47433

Shapiro L, Fannon AM, Kwong PD, Thompson A, Lehmann MS, Grubel G, Legrand JF, Als-Nielsen J, Colman DR, Hendrickson WA (1995) Structural basis of cell–cell adhesion by cadherins. Nature 374:327–337

Sharon N (1987) Bacterial lectins, cell–cell-recognition and infectious disease. FEBS Lett 217:145–157

Sheahan K, O'Brien MJ, Burke B, Dervan PA, O'Keane JC, Gottlieb LS, Zamcheck N (1990) Differential reactivities of carcinoembryonic antigen (CEA) and CEA-related monoclonal and polyclonal antibodies in common epithelial malignancies. Am J Clin Pathol 94:157–164

Shi ZR, Tacha D, Itzkowitz SH (1994) Monoclonal antibody COL-1 reacts with restricted epitopes on carcinoembryonic antigen: an immunohistochemical study. J Histochem Cytochem 42:1215–1219

Sienel W, Dango S, Woelfle U, Morresi-Hauf A, Wagener C, Brümmer J, Mutschler W, Passlick B, Pantel K (2003) Elevated expression of carcinoembryonic antigen-related cell adhesion molecule 1 promotes progression of non-small cell lung cancer. Clin Cancer Res 9:2260–2266

Sinclair NR (2000) Immunoreceptor tyrosine-based inhibitory motifs on activating molecules. Crit Rev Immunol 20:89–102

Singer BB, Scheffrahn I, Öbrink B (2000) The tumor growth-inhibiting cell adhesion molecule CEACAM1 (C-CAM) is differently expressed in proliferating and quiescent epithelial cells and regulates cell proliferation. Cancer Res 60:1236–1244

Singer BB, Scheffrahn I, Heymann R, Sigmundsson K, Kammerer R, Öbrink B (2002) Carcinoembryonic antigen-related cell adhesion molecule 1 expression and signaling in human, mouse, and rat leukocytes: evidence for replacement of the short cytoplasmic domain isoform by glycosylphosphatidylinositol-linked proteins in human leukocytes. J Immunol 168:5139–5146

Sippel CJ, Fallon RJ, Perlmutter DH (1994) Bile acid efflux mediated by the rat liver canalicular bile acid transport/ecto-ATPase protein requires serine 503 phosphorylation and is regulated by tyrosine 488 phosphorylation. J Biol Chem 269:19539–19545

Skubitz KM, Ducker TP, Goueli SA (1992) CD66 monoclonal antibodies recognize a phosphotyrosine-containing protein bearing a CEA cross-reacting antigen on the surface of human neutrophils. J Immunol 148:852–860

Skubitz KM, Campbell KD, Ahmed K, Skubitz AP (1995) CD66 family members are associated with tyrosine kinase activity in human neutrophils. J Immunol 155:5382–5390

Skubitz KM, Campbell KD, Skubitz AP (1996) CD66a, CD66b, CD66c, and CD66d each independently stimulate neutrophils. J Leukoc Biol 60:106–117

Skubitz KM, Campbell KD, Skubitz AP (2000) Synthetic peptides of CD66a stimulate neutrophil adhesion to endothelial cells. J Immunol 164:4257–4264

Skubitz KM, Campbell KD, Skubitz AP (2001) Synthetic peptides from the N-domains of CEACAMs activate neutrophils. J Pept Res 58:515–526

Soni P, Lakkis M, Poy MN, Fernström MA, Najjar SM (2000) The differential effects of pp120 (Ceacam 1) on the mitogenic action of insulin and insulin-like growth factor 1 are regulated by the nonconserved tyrosine 1316 in the insulin receptor. Mol Cell Biol 20:3896–3905

St Croix B, Sheehan C, Rak JW, Florenes VA, Slingerland JM, Kerbel RS (1998) E-Cadherin-dependent growth suppression is mediated by the cyclin-dependent kinase inhibitor p27(KIP1). J Cell Biol 142:557–571

Stanners C (1998) Cell adhesion and communication mediated by the CEA family: basic and clinical principles. Harwood Academic, Amsterdam

Stern A, Brown M, Nickel P, Meyer TF (1986) Opacity genes in Neisseria gonorrhoeae: control of phase and antigenic variation. Cell 47:61–71

Stocks SC, Kerr MA (1993) Neutrophil NCA-160 (CD66) is the major protein carrier of selectin binding carbohydrate groups LewisX and sialyl lewisX. Biochem Biophys Res Commun 195:478–483

Stocks SC, Kerr MA, Haslett C, Dransfield I (1995) CD66-dependent neutrophil activation: a possible mechanism for vascular selectin-mediated regulation of neutrophil adhesion. J Leukocyte Biol 58:40–48

Stocks SC, Albrechtsen M, Kerr MA (1990) Expression of the CD15 differentiation antigen (3-fucosyl-N-acetyl-lactosamine, LeX) on putative neutrophil adhesion molecules CR3 and NCA-160. Biochem J 268:275–280

Stoffel A, Neumaier M, Gaida FJ, Fenger U, Drzeniek Z, Haubeck HD, Wagener C (1993) Monoclonal, anti-domain and anti-peptide antibodies assign the molecular weight 160,000 granulocyte membrane antigen of the CD66 cluster to a mRNA species encoded by the biliary glycoprotein gene, a member of the carcinoembryonic antigen gene family. J Immunol 150:4978–4984

Streichert T, Ebrahimnejad A, Ganzer S, Flayeh R, Wagener C, Brümmer J (2001) The microbial receptor CEACAM3 is linked to the calprotectin complex in granulocytes. Biochem Biophys Res Commun 289:191–197

Stuart DI, Jones EY (1995) Recognition at the cell surface: recent structural insights. Curr Opin Struct Biol 5:735–743

Sun QH, DeLisser HM, Zukowski MM, Paddock C, Albelda SM, Newman PJ (1996) Individually distinct Ig homology domains in PECAM-1 regulate homophilic binding and modulate receptor affinity. J Biol Chem 271:11090–11098

Sundberg U, Öbrink B (2002) CEACAM1 isoforms with different cytoplasmic domains show different localization, organization and adhesive properties in polarized epithelial cells. J Cell Sci 115:1273–1284

Svalander PC, Odin P, Nilsson BO, Öbrink B (1990) Expression of cellCAM-105 in the apical surface of rat uterine epithelium is controlled by ovarian steroid hormones. J Reprod Fertil 88:213–221

Takahashi H, Okai Y, Paxton RJ, Hefta LJF, Shively JE (1993) Differential regulation of CEA and BGP by γ-IFN. Cancer Res 53:1612–1619

Tan K, Zelus BD, Meijers R, Liu JH, Bergelson JM, Duke N, Zhang R, Joachimiak A, Holmes KV, Wang JH (2002) Crystal structure of murine sCEACAM1a[1,4]: a coronavirus receptor in the CEA family. EMBO J 21:2076–2086

Tanaka K, Hinoda Y, Takahashi H, Sakamoto H, Nakajima Y, Imai K (1997) Decreased expression of biliary glycoprotein in hepatocellular carcinomas. Int J Cancer 74:15–19

Teglund S, Olsen A, Khan WN, Frängsmyr L, Hammarström S (1994) The pregnancy-specific glycoprotein (PSG) gene cluster on human chromosome 19: fine structure of the 11 PSG genes and identification of 6 new genes forming a third subgroup within the carcinoembryonic antigen (CEA) family. Genomics 23:669–684

Teixeira AM, Fawcett J, Simmons DL, Watt SM (1994) The N-domain of the biliary glycoprotein adhesion molecule mediates homotypic binding: domain interactions and epitope analysis of BGPc. Blood 84:211–219

Thies A, Moll I, Berger J, Wagener C, Brümmer J, Schulze HJ, Brunner G, Schumacher U (2002) CEACAM1 expression in cutaneous malignant melanoma predicts the development of metastatic disease. J Clin Oncol 20:2530–2536

Thompson J, Zimmermann W, Nollau P, Neumaier M, Weber-Arden J, Schrewe H, Craig I, Willocks T (1994) CGM2, a member of the carcinoembryonic antigen gene family is down-regulated in colorectal carcinoma. J Biol Chem 269:32924–32931

Thompson J, Seitz M, Chastre E, Ditter M, Aldrian C, Gespach C, Zimmermann W (1997) Down-regulation of carcinoembryonic antigen family member 2 expression is an early event in colorectal tumorigenesis. Cancer Res 57:1776–1784

Thompson JA, Grunert F, Zimmermann W (1991) Carcinoembryonic antigen gene family: molecular biology and clinical perspectives. J Clin Lab Anal 5:344–366

Thompson JA, Mossinger S, Reichardt V, Engels U, Beauchemin N, Kommoss F, von Kleist S, Zimmermann W (1993) A polymerase-chain-reaction assay for the specific identification of transcripts encoded by individual carcinoembryonic antigen (CEA)-gene-family members. Int J Cancer 55:311–319

Tsutsumi Y, Onoda N, Misawa M, Kuroki M, Matsuoka Y (1990) Immunohistochemical demonstration of nonspecific cross-reacting antigen in normal and neoplastic human tissues using a monoclonal antibody. Comparison with carcinoembryonic antigen localization. Acta Pathol Jpn 40:85–97

Turbide C, Kunath T, Daniels E, Beauchemin N (1997) Optimal ratios of biliary glycoprotein isoforms required for inhibition of colonic tumor cell growth. Cancer Res 57:2781–2788

Varner JA, Brooks PC, Cheresh DA (1995) REVIEW: the integrin alpha V beta 3: angiogenesis and apoptosis. Cell Adhes Commun 3:367–374

Virji M, Watt SM, Barker S, Makepeace K, Doyonass R (1996) The N-domain of the human CD66a adhesion molecule is a target for Opa proteins of Neisseria meningitidis and Neisseria gonorrhoeae. Mol Microbiol 22:929–939

Virji M, Evans D, Hadfield A, Grunert F, Teixeira AM, Watt SM (1999) Critical determinants of host receptor targeting by Neisseria meningitidis and Neisseria gonorrhoeae: identification of Opa adhesiotopes on the N-domain of CD66 molecules. Mol Microbiol 34:538–551

Virji M, Evans D, Griffith J, Hill D, Serino L, Hadfield A, Watt SM (2000) Carcinoembryonic antigens are targeted by diverse strains of typable and non-typable Haemophilus influenzae. Mol Microbiol 36:784–795

Volpert O, Luo W, Liu TJ, Estrera VT, Logothetis C, Lin SH (2002) Inhibition of prostate tumor angiogenesis by the tumor suppressor CEACAM1. J Biol Chem 16:16

von Kleist S, Chavanel G, Burtin P (1972) Identification of an antigen from normal human tissue that crossreacts with the carcinoembryonic antigen. Proc Natl Acad Sci U S A 69:2492–2494

von Kleist S, Winkler J, Migule I, Bohm N (1986) Carcinoembryonic antigen (CEA) expression in early embryogenesis: a study of the first trimester of gestation. Anticancer Res 6:1265–1272

Wagener C, Hain F, Fodisch HJ, Breuer H (1983) Localization of carcinoembryonic antigen in embryonic and fetal human tissues. Histochemistry 78:1–9

Wang J, Springer TA (1998) Structural specializations of immunoglobulin superfamily members for adhesion to integrins and viruses. Immunol Rev 163:197–215

Wang L, Lin SH, Wu WG, Kemp BL, Walsh GL, Hong WK, Mao L (2000) C-CAM1, a candidate tumor suppressor gene, is abnormally expressed in primary lung cancers. Clin Cancer Res 6:2988–2993

Watt SM, Sala-Newby G, Hoang T, Gilmore DJ, Grunert F, Nagel G, Murdoch SJ, Tchilian E, Lennox ES, Waldmann H (1991) CD66 identifies a neutrophil-specific epitope within the hematopoietic system that is expressed by members of the carcinoembryonic antigen family of adhesion molecules. Blood 78:63–74

Watt SM, Teixeira AM, Zhou GQ, Doyonnas R, Zhang Y, Grunert F, Blumberg RS, Kuroki M, Skubitz KM, Bates PA (2001) Homophilic adhesion of human CEACAM1 involves N-terminal domain interactions: structural analysis of the binding site. Blood 98:1469–1479

Wikström K, Kjellström G, Öbrink B (1996) Homophilic intercellular adhesion mediated by C-CAM is due to a domain 1-domain 1 reciprocal binding. Exp Cell Res 227:360–366

Williams AF, Barclay AN (1988) The immunoglobulin superfamily—domains for cell surface recognition. Annu Rev Immunol 6:381–405

Williams RK, Jiang GS, Snyder SW, Frana MF, Holmes KV (1990) Purification of the 110-kilodalton glycoprotein receptor for mouse hepatitis virus (MHV)-A59 from mouse liver and identification of a nonfunctional, homologous protein in MHV-resistant SJL/J mice. J Virol 64:3817–3823

Williams RK, Jiang GS, Holmes KV (1991) Receptor for mouse hepatitis virus is a member of the carcinoembryonic antigen family of glycoproteins. Proc Natl Acad Sci U S A 88:5533–5536

Zhou GQ, Baranov V, Zimmermann W, Grunert F, Erhard B, Mincheva-Nilsson L, Hammarström S, Thompson J (1997) Highly specific monoclonal antibody demonstrates that pregnancy-specific glycoprotein (PSG) is limited to syncytiotrophoblast in human early and term placenta. Placenta 18:491–501

Zimmer R, Thomas P (2001) Mutations in the carcinoembryonic antigen gene in colorectal cancer patients: implications on liver metastasis. Cancer Res 61:2822–2826

Zimmermann W (2002) Carcinoembryonic antigen. In: Creighton T (ed) Wiley Encyclopedia of Molecular Medicine. John Wiley and Sons, New York, pp 459–462

Zimmermann W, Ortlieb B, Friedrich R, von Kleist S (1987) Isolation and characterization of cDNA clones encoding the human carcinoembryonic antigen reveal a highly conserved repeating structure. Proc Natl Acad Sci U S A 84:2960–2964

Roles of Nectins in Cell Adhesion, Signaling and Polarization

K. Irie · K. Shimizu · T. Sakisaka · W. Ikeda · Y. Takai (✉)

Department of Molecular Biology and Biochemistry,
Osaka University Graduate School of Medicine/Faculty of Medicine,
2-2 Yamada-oka, Suita, Osaka 565-0871, Japan
ytakai@molbio.med.osaka-u.ac.jp

1	Introduction	345
2	General Properties of Nectins as CAMs	347
2.1	Molecular Structures of Nectins	347
2.2	Cell–Cell Adhesion Activity of Nectins	348
2.3	Tissue Distribution and Subcellular Localization of Nectins	349
3	Roles of Nectins in Cell–Cell Adhesion	350
3.1	Cooperative Roles of Nectins with Cadherins	350
3.1.1	Role of Nectins in Organization of AJs and TJs	350
3.1.2	Role of Nectins in Formation of Synapses	352
3.1.3	Association Mechanisms of Nectins, Cadherins, and TJ Components	354
3.2	Cadherin-Independent Roles of Nectins	355
3.2.1	Possible Roles of Nectins in Axonal Pathfinding	355
3.2.2	Role of Nectins in Organization of Sertoli Cell–Spermatid Junctions	357
3.3	Association of Nectins with the Actin Cytoskeleton	359
4	Roles of Nectins in Cell Signaling	359
4.1	Activation of Cdc42 and Rac by Nectins	359
4.2	Activation of JNK by Nectins	361
5	Role of Nectins in Cell Polarization	361
6	Nectins as Viral Receptors	364
7	Conclusions and Perspectives	365
References		366

Abstract Nectins are Ca^{2+}-independent immunoglobulin-like cell–cell adhesion molecules which constitute a family of four members. Nectins homophilically and heterophilically *trans*-interact and cause cell–cell adhesion. This nectin-based cell–cell adhesion plays roles in the organization of adherens junctions in epithelial cells and fibroblasts and synaptic junctions in neurons in cooperation with cadherins. The nectin-based cell–cell adhesion plays roles in the contacts between commissural axons and floor plate cells and in the organization of Sertoli cell–spermatid junctions in the testis, independently of cadherins. Nectins furthermore regulate intracellular signaling through Cdc42 and Rac small G proteins and cell polarization through cell polarity proteins. Pathologically, nectins serve as entry and cell–cell spread mediators of herpes simplex viruses.

Keywords Nectin · Afadin · Adherens junctions · Tight junctions · Cell signaling · Cell polarity · Small G proteins

Abbreviations

ADIP	Afadin DIL domain-interacting protein
AJs	Adherens junctions
aPKC	Atypical protein kinase C
CAMs	Cell adhesion molecules
Crb	Crumbs
DIL	Dilute
Dlg	Discs large
Dlt	Discs lost
ERK	Extracellular signal-regulated kinase
ES	Ectoplasmic specialization
F-actin	Actin filament
FHA	Forkhead-associated
gD	Glycoprotein D
GAP	GTPase-activating protein
GDI	Guanine nucleotide dissociation inhibitor
GEP	GDP/GTP exchange protein
GTP-Cdc42	GTP-bound active form of Cdc42 small G protein
GTP-Rac	GTP-bound active form of Rac small G protein
HSVs	Herpes simplex viruses
Ig	Immunoglobulin
JAM	Junctional adhesion molecule
JNK	c-Jun N-terminal kinase
Lgl	Lethal giant larvae
MAP kinase	Mitogen-activated protein kinase
MDCK cells	Madin–Darby canine kidney cells
mLgl	Mammalian Lgl
Nef	A recombinant extracellular fragment of a nectin fused to the Fc domain of IgG
Pals1	Protein associated with Lin7
PATJ	Pals1-associated tight junction protein
PR	Proline-rich domain
PSDs	Postsynaptic densities
RA	Ras-associated
Scrib	Scribble
Std	Stardust
TJs	Tight junctions

1
Introduction

Cells in multi-cellular organisms recognize their neighboring cells, adhere to them, and form cell–cell junctions. Such junctions have essential roles in various cellular functions, including morphogenesis, differentiation, proliferation, and migration (for reviews: Takeichi 1991; Gumbiner 1996; Vlemincks and Kemler 1999; Angst et al. 2000; Tepass et al. 2000; Yagi and Takeichi 2000). In epithelial cells, cell–cell adhesion is mediated through a junctional complex comprising tight junctions (TJs), adherens junctions (AJs), and desmosomes. These junctional structures are typically aligned from the apical to basal sides, although desmosomes are independently distributed in other areas. These junctions are generally made between homotypic cells and mediated by homophilic interactions of cell adhesion molecules (CAMs). At AJs, the transmembrane protein E-cadherin functions as a Ca^{2+}-dependent CAM (Takeichi 1991; Gumbiner 1996; Vlemincks and Kemler 1999; Angst et al. 2000; Tepass et al. 2000; Yagi and Takeichi 2000). E-cadherin is a member of the cadherin superfamily that comprises over 80 members, each of which is expressed in non-epithelial cells as well as epithelial cells (Tepass et al. 2000; Yagi and Takeichi 2000). TJs are likely to serve as a barrier that prevents solutes and water from passing between neighboring cells and as a fence between the apical plasma membrane and the basolateral plasma membrane in epithelial cells (for reviews: Tsukita et al. 1999, 2001). At TJs, claudins function as a Ca^{2+}-independent CAM (Tsukita et al. 1999, 2001). Two other transmembrane proteins, occludin and junction adhesion molecules (JAMs), have been identified at TJs (Tsukita et al. 2001). Occludin is a Ca^{2+}-independent CAM, but its physiological function has not yet been established, whereas JAMs are Ca^{2+}-independent immunoglobulin (Ig)-like CAMs which are involved in concentration of cell polarity proteins including PAR-3/PAR-6/atypical protein kinase C (aPKC) (for a review: Ohno 2001). TJs also concentrate other cell polarity proteins, Crumb/protein associated with Lin7 (Pals1)/Pals1-associated tight junction protein (PATJ) complexes (Ohno 2001). The formation and maintenance of TJs are generally dependent on the formation and maintenance of the E-cadherin-based AJs (Takeichi 1991; Gumbiner 1996; Vlemincks and Kemler 1999). It remains, however, unknown how cells recognize their neighboring cells and initiate the recruitment of E-cadherin to cell–cell contact sites, finally organizing the junctional complex. The formation and disruption of the junctional complex are furthermore dynamically regulated by many extracellular and intracellular signals: the formation of AJs is enhanced by the guanosine triphosphate (GTP)-bound active forms of Rac and Cdc42 small G proteins (GTP-Rac and GTP-Cdc42, respectively), whereas AJs are disrupted by many other extracellular signals, such as scatter factor/hepatocyte growth factor, phorbol esters, and activated oncogenes, such as Ras and Src (for a review:

Gumbiner 2000). The mechanisms of such dynamic organization of the junctional complex are also unknown.

There are other types of cell–cell junctions in many organs and tissues. Synapses are interneuronal asymmetric junctions at which active zones and postsynaptic densities (PSDs) are present at the presynaptic side and the postsynaptic side, respectively. Moreover, there are heterotypic junctions, such as those formed between differentiating germ cells and their supporter Sertoli cells in seminiferous epithelium in the testis and between specialized sensory cells and their supporting cells in sensory epithelia. However, in the case of these junctions, their molecular architectures or their mechanisms of organization are poorly understood. Since cadherins take part in Ca^{2+}-dependent homophilic cell–cell adhesion, Ca^{2+}-independent and heterophilic cell–cell adhesion are supposed to be mediated by other CAMs, such as an Ig superfamily member. Members of this family are defined by the presence of one or more copies of the Ig domain, a compact structure with two cysteine residues separated by 55–75 amino acids arranged as two antiparallel beta sheets. Members of the Ig CAMs function in a wide variety of cell types and are involved in many different biological processes including cell adhesion, axon guidance, and signaling (for a review: Brummendorf and Lemmon 2001).

Nectins, which constitute a family of four members, have recently emerged as Ca^{2+}-independent Ig-like CAMs (for reviews: Takai and Nakanishi 2003; Takai et al. 2003a). They play roles in the organization of adherens junctions in epithelial cells and fibroblasts and synaptic junctions in neurons in cooperation with cadherins (Takai and Nakanishi 2003; Takai et al. 2003a). Nectins also play roles in the contacts between commissural axons and floor plate cells and in the organization of Sertoli cell–spermatid junctions in the testis, independently of cadherins (Takai and Nakanishi 2003; N. Okabe, K. Shimizu, K. Ozaki-Kuroda, H. Nakanishi, K. Morimoto, M. Takeuchi, H. Katsumaru, F. Murakami, and Y. Takai, personal communication). Nectins furthermore regulate intracellular signaling through activation of Cdc42 and Rac and cell polarization through cell polarity proteins (Kawakatsu et al. 2002; Honda et al. 2003b; Takekuni et al. 2003; Shimizu and Takai 2003). Pathologically, nectins serve as entry and cell–cell spread mediators of herpes simplex viruses (HSVs) (for reviews: Campadelli-Fiume et al. 2000; Spear et al. 2000; Spear 2002; Takai and Nakanishi 2003). In this article, we describe roles of nectins in cell adhesion, signaling, and polarization.

2
General Properties of Nectins as CAMs

2.1
Molecular Structures of Nectins

The nectin family consists of four members, nectin-1, nectin-2, nectin-3, and nectin-4 (Takai and Nakanishi 2003; Takai et al. 2003a). All the members of nectin have two or three splice variants so that nectin-1α, nectin-1β, nectin-1γ, nectin-2α, nectin-2δ, nectin-3α, nectin-3β, and nectin-3γ isoforms exist (Takahashi et al. 1999; Satoh-Horikawa et al. 2000). Nectin-4 also has two splice variants (Reymond et al. 2001). Nectin-1α and nectin-2α were originally isolated as the poliovirus receptor-related proteins and named PRR-1 and PRR-2, respectively, although neither of them has subsequently been reported to serve as a poliovirus receptor (see Sect. 6 for details). They were later shown to serve as receptors for α-herpes virus, facilitating its entry and cell–cell spread and so were renamed HveC and HveB, respectively. All the members, except nectin-1γ, have an extracellular region with three Ig-like loops, a single transmembrane region, and a cytoplasmic region (Fig. 1). Nectin-1γ is a secreted protein which lacks the transmembrane region. Furthermore, all of them, except nectin-1β, nectin-3γ, and nectin-4,

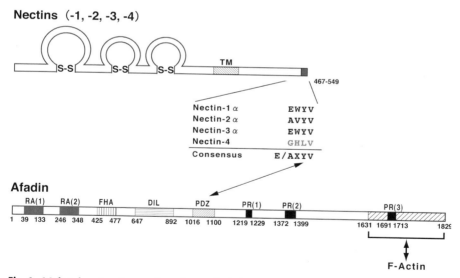

Fig. 1. Molecular structures of nectins and afadin. Nectins have a conserved motif of four amino acid residues (Glu/Ala-X-Tyr-Val) at their carboxy termini, and this binds the PDZ domain of afadin. *TM*, transmembrane region; *RA*, Ras associated domain; *FHA*, forkhead associated domain; *DIL*, dilute domain; *PDZ*, PDZ domain; and *PR*, proline-rich domain

have a conserved motif of four amino acid residues (Glu/Ala-X-Tyr-Val) at their carboxy termini, and this binds the PDZ domain of afadin, an actin filament (F-actin)-binding protein. This binding of afadin to nectins links nectins to the actin cytoskeleton. Although nectin-4 lacks the conserved motif, it binds the PDZ domain of afadin at its carboxy terminus.

Afadin has two splice variants, l-afadin and s-afadin (Mandai et al. 1997) (Fig. 1). l-Afadin, the larger variant, binds nectin through its PDZ domain and F-actin through its F-actin-binding domain. Afadin has two Ras-associated (RA) domains, a forkhead-associated (FHA) domain, a dilute (DIL) domain, and three proline-rich (PR) domains. Although l-afadin binds to F-actin, it does not have a cross-linking activity. s-Afadin, the smaller variant, has the two RA domains, the FHA domain, the DIL domain, the PDZ domain, and the two PR domains, but lacks the F-actin-binding domain and the third PR domain. Human s-afadin is identical to the gene product of *AF-6*, a gene that has been identified as an *ALL-1* fusion partner involved in acute myeloid leukemias (Prasad et al. 1993). Unless otherwise specified, afadin refers to l-afadin in this article.

2.2
Cell–Cell Adhesion Activity of Nectins

Cadherin-deficient L fibroblasts were originally used to analyze the adhesive activity of E-cadherin (Nagafuchi et al. 1987) and subsequently used to show

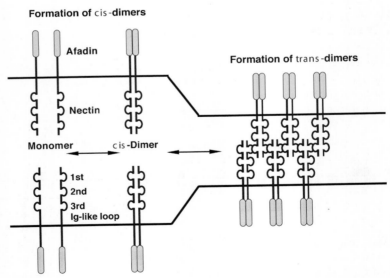

Fig. 2. A model for the intercellular adhesion activity of nectins. Nectins first form homo-*cis*-dimers and then homo- or hetero-*trans*-dimers, causing cell–cell adhesion

that nectins also have cell–cell adhesion activity (Takai and Nakanishi 2003). In contrast to E-cadherin, nectin's adhesion activity is Ca^{2+} independent. Each nectin forms *cis*-dimers, followed by formation of *trans*-dimers (Fig. 2). Each nectin forms homo-*cis*-dimers, but not hetero-*cis*-dimers. Each protein then forms homo-*trans*-dimers. Nectin-3, furthermore, forms hetero-*trans*-dimers with either nectin-1 or nectin-2. Nectin-4 hetero-*trans*-dimerizes with nectin-1, but neither nectin-1 nor nectin-2 hetero-*trans*-dimerizes in *trans*. These hetero-*trans*-dimers show much higher affinity than homo-*trans*-dimers. This property of nectin is different from that of cadherins, which form mainly homo-*trans*-dimers (Takeichi 1991). The second Ig-like loop of each nectin is necessary for the formation of the *cis*-dimers, whereas the first Ig-like loop is required for the formation of the *trans*-dimers, but not for the formation of the *cis*-dimer (Momose et al. 2002; Yasumi et al. 2003). The function of the third Ig-like loop is currently unknown. The interaction of each nectin with afadin is not essential for the formation of the *cis*-dimers or the *trans*-dimers. Whereas E-cadherin shows strong cell adhesion activity that is accompanied by compaction, the nectin-mediated adhesion does not show such a property.

2.3
Tissue Distribution and Subcellular Localization of Nectins

Nectin-1, nectin-2, and nectin-3 are ubiquitously expressed in a variety of cells including epithelial cells, neurons, and fibroblasts (Takai and Nakanishi 2003). Nectin-2 and nectin-3 are expressed in cells where cadherins are not expressed, such as blood cells and spermatids. Human nectin-4 is mainly expressed in placenta. l-Afadin is ubiquitously expressed, whereas s-afadin is specifically expressed in neurons. In the absorptive epithelial cells of mouse small intestine, nectin-2 and afadin are highly concentrated at the E-cadherin-based AJs, under which F-actin bundles lie (see Sect. 3.3). Both nectin-2 and afadin are absent from the claudin-based TJs and desmosomes. This distribution pattern is different from that of E-cadherin which, although E-cadherin is concentrated at AJs, is widely distributed from the apical to basal sides of the lateral membrane (Tsukita et al. 1992; Gumbiner 1996). Neither nectin-2 nor afadin is found at cell–matrix AJs, as shown by studies in costameres of heart and in cultured epithelial cells such as Madin–Darby canine kidney (MDCK) cells and mouse mammary tumor cells. In cultured fibroblasts that lack TJs, such as L cells stably expressing E-cadherin, nectin-2 and afadin colocalize with E-cadherin at AJs. Nectin-3 and afadin localize at cell–cell adhesion sites in NIH3T3 fibroblasts (Honda et al. 2003b). In epithelial cells and fibroblasts, each nectin localizes symmetrically at both sides of the plasma membranes of the AJs.

3
Roles of Nectins in Cell–Cell Adhesion

3.1
Cooperative Roles of Nectins with Cadherins

3.1.1
Role of Nectins in Organization of AJs and TJs

When AJs and TJs are formed in epithelial cells, primordial spot-like junctions are first formed at the tips of the cellular protrusions that radiate from adjacent cells (Yonemura et al. 1995; Adams et al. 1998; Vasioukhin et al. 2000) (Fig. 3). These primordial junctions fuse with each other to form short line-like junctions, which develop into more matured AJs. During and/or after this process, TJs are formed at the apical side of AJs to complete the junctional complex. E-cadherin, α-catenin, β-catenin, and ZO-1 are first assembled to form spot-like junctions, which are gradually matured to line-like junctions (AJs), followed by the assembly of claudins and occludin to form TJs (Ando-Akatsuka et al. 1999; Asakura et al. 1999; Sakisaka et al. 1999). Nectins play a role in formation of this junctional complex, especially at the initial stage (Takai and Nakanishi 2003; Takai et al. 2003a). Nectins and E-cadherin separately form *trans*-dimers that form micro-clusters at cell–cell contact sites, when the two migrating cells contact through protrusions, such as filopodia and lamellipodia (Fig. 3). Because nectins kinetically form micro-clusters more rapidly than E-cadherin, the nectin-based micro-clusters are mainly formed at the initial stage. The nectin-based micro-clusters then recruit E-cadherin, which results in formation of a mixture of the nectin- and E-cadherin-based micro-clusters. The nectin and E-cadherin molecules in these micro-clusters are associated through afadin and catenins that are linked to the actin cytoskeleton. The E-cadherin-based micro-clusters that form slowly and independently of the nectin-based micro-clusters rapidly recruit the nectin–afadin complex to form other primordial spot-like junctions. These primordi-

Fig. 3A–C. A model for the role and mode of action of nectins in formation of the junctional complex in epithelial cells. **A** Formation of AJs. When two migrating cells contact through their protrusions, nectins and E-cadherin separately form *trans*-dimers that form micro-clusters at cell–cell contact sites. The nectin-based micro-clusters are mainly formed at the initial stage. The nectin-based micro-clusters then recruit E-cadherin, which results in formation of a mixture of the nectin- and E-cadherin-based micro-clusters (primordial spot-like junctions). These primordial junctions fuse to form short line-like junctions, which develop into more matured AJs. Peripheral membrane proteins are not shown. **B** Formation of TJs. During and/or after the formation of AJs, JAMs are first assembled at the apical side of AJs, followed by the recruitment of claudins and occludin,

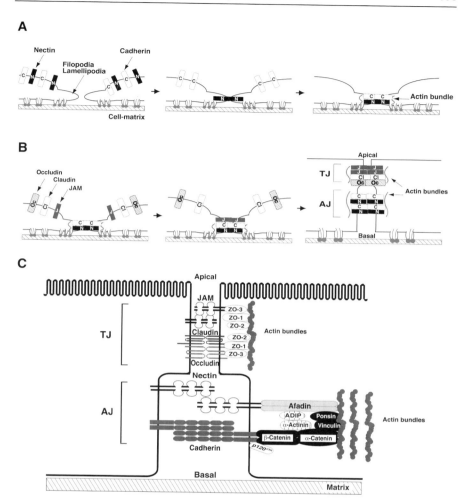

which eventually leads to the establishment of the claudin-based TJs. Peripheral membrane proteins are not shown. C The junctional complex of AJs and TJs, which is formed as described in A and B, in epithelial cells. At AJs, E-cadherin serves as an essential CAM. The cytoplasmic region binds β-catenin, which in turn binds α-catenin. α-Catenin is associated with the actin bundles directly and indirectly through vinculin and α-actinin. Nectin also functions as a CAM at AJs, but is more highly concentrated at AJs than E-cadherin. The cytoplasmic region binds afadin that is directly associated with the actin bundles. Afadin and α-catenin are associated with each other presumably through two connector units, a ponsin–vinculin unit and an afadin DIL domain-interacting protein (ADIP)-α-actinin unit. At TJs, claudins and JAMs function as CAMs. Occludin is another transmembrane protein at TJs. Their cytoplasmic regions bind ZO-1, ZO-2, and ZO-3. ZO-1 and ZO-2 are directly associated with F-actin and form a dimer with ZO-3. Actin bundles and peripheral membrane proteins shown in the *right side* cell are omitted in the *left side* cell

al junctions fuse with each other to form short line-like junctions, which develop into more matured AJs. During and/or after the formation of AJs, JAMs are first assembled at the apical side of AJs, followed by the recruitment of claudins and occludin, presumably through ZO-1, ZO-2, and ZO-3, which eventually leads to the establishment of the claudin-based TJs (Fig. 3).

Many lines of evidence are available indicating this role of nectins in the formation of AJs and TJs (Mandai et al. 1997, 1999; Asakura et al. 1999; Ikeda et al. 1999; Sakisaka et al. 1999; Takahashi et al. 1999; Miyahara et al. 2000; Satoh-Horikawa et al. 2000; Suzuki et al. 2000; Tachibana et al. 2000; Sozen et al. 2001; Yokoyama et al. 2001; Fukuhara et al. 2002a,b; Peng et al. 2002; Honda et al. 2003a,b,c; Tanaka et al. 2003). Typical examples are that (1) overexpression of nectin increases the velocity of the formation of AJs and TJs in epithelial cells and that of AJs in fibroblasts (Honda et al. 2003a,c); (2) inhibition of the nectin-based cell–cell adhesion by its specific inhibitors, such as glycoprotein D (gD) and a recombinant extracellular fragment of a nectin fused to the Fc domain of IgG (Nef), conversely reduces the velocity of the formation of AJs and TJs in epithelial cells and that of AJs in fibroblasts (Fukuhara et al. 2002a,b; Honda et al. 2003a,c). gD, an envelope protein of herpes simplex virus type 1, one of the α-herpes viruses (for reviews: Campadelli-Fiume et al. 2000; Spear et al. 2000), binds to nectin-1α, and inhibits not only the formation of homo-*trans*-dimers of nectin-1α but also the formation of hetero-*trans*-dimers between nectin-1α and nectin-3α (Sakisaka et al. 2001; see Sect. 6). Nef forms a *trans*-dimer with a cellular nectin and thereby inhibits the formation of the nectin-based cell–cell adhesion (Honda et al. 2003a); (3) in the ectoderm of afadin-deficient mice and embryoid bodies, organization of not only AJs but also TJs is highly impaired (Ikeda et al. 1999); and (4) mutations in the nectin-1 gene are responsible for cleft lip/palate-ectodermal dysplasia, Margarita island ectodermal dysplasia, and Zlotogora-Ogür syndrome, which is characterized by cleft lip/palate, syndactyly, and ectodermal dysplasia (Suzuki et al. 2000; Sozen et al. 2001). It is unknown whether the nectin-based cell–cell adhesion regulates only the velocity of the formation of AJs or is essential for the formation and/or maintenance of AJs.

3.1.2
Role of Nectins in Formation of Synapses

At the synapses between the mossy fiber terminals and the pyramidal cell dendrites in the CA3 area of the hippocampus, both synaptic junctions and puncta adherentia junctions are highly developed and actively remodeled in an activity-dependent manner (Amaral and Dent 1981). Synapses are formed by the meeting of axons and dendrites, but synaptic junctions and puncta adherentia junctions are not morphologically differentiated at primitive synapses. During the maturation of synapses, membrane domain specialization gradually occurs in the CA3 area of the hippocampus (Amaral

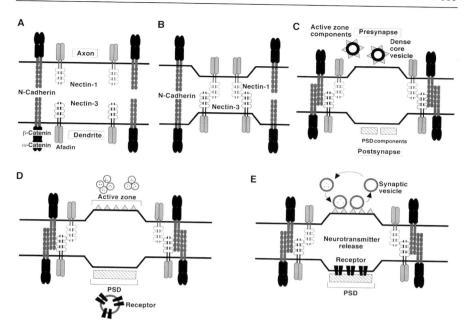

Fig. 4A–E. A model for the role and mode of action of nectins in formation of synapses. **A** Synapses are formed by the meeting of axons and dendrites during their maturation. **B** The nectin–afadin system first forms primordial junctions. **C** The N-cadherin–catenin system is recruited to the nectin-based cell–cell adhesion sites. **D** The components of active zones would then be recruited to the primordial junctions to form active zones at the presynaptic side. The components of PSDs would be assembled at the postsynaptic side. **E** The membrane receptors are transported to the postsynaptic side to form PSDs. The membrane domains, comprising synaptic junctions and puncta adherentia junctions, would then gradually become segregated, followed by maturation of synapses. At the puncta adherentia junctions, nectin-1 and nectin-3 localize asymmetrically at the presynaptic and postsynaptic sides, respectively. Afadin, N-cadherin, and catenins localize symmetrically at both sides. Synaptic junctions are associated with presynaptic active zones where synaptic vesicles, Ca^{2+} channels, and many other components, such as bassoon, localize, and with PSDs, where neurotransmitter receptors localize

and Dent 1981). It is postulated that primordial junctions form first, followed by the transport of the components of active zones on dense core vesicles and subsequent formation of active zones at the presynaptic side (for reviews: see Desbach et al. 2001; Ziv and Garner 2001). At the postsynaptic side, the components of PSDs are assembled and membrane receptors on vesicles are transported to this region.

Nectin-1 and nectin-3 localize asymmetrically at the presynaptic and postsynaptic sides, respectively, of the plasma membranes of the puncta adherentia junctions and form hetero-*trans*-dimers (Mizoguchi et al. 2002) (Fig. 4). Afadin, N-cadherin, and αN-catenin localize symmetrically at both sides (Nishioka et al. 2000; Mizoguchi et al. 2002). s-Afadin, but not l-afadin, localizes at PSDs (Buchert et al. 1999; Nishioka et al. 2000). In at least some brain regions other than the CA3 area of hippocampus, such as the optic tectum, N-cadherin localizes at both synaptic junctions and puncta adherentia junctions (Yamagata et al. 1995).

Inhibition of the nectin-1- and nectin-3-based adhesion by the inhibitor of nectin-1, gD, in cultured rat hippocampal neurons results in a decrease in size and a concomitant increase in number of synapses (Mizoguchi et al. 2002). This reflects partial inhibition of the formation of the hetero-*trans*-dimer between nectin-1 and nectin-3, which may affect the N-cadherin-mediated adhesion and eventually lead to formation of smaller synapses. Because N-cadherin also plays a role in the organization of synapses (Yamagata et al. 1995; Tang et al. 1998), it is likely that the formation of the hetero-*trans*-dimer between nectin-1 and nectin-3 plays an important role in the determination of the position and the size of synapses in cooperation with the N-cadherin–catenin system. This role of nectins is consistent with the finding that mutations in the nectin-1 gene are responsible for cleft lip/palate–ectodermal dysplasia, Zlotogora-Ogür syndrome, which is characterized by mental retardation in addition to ectodermal dysplasia (Suzuki et al. 2000).

On the basis of the evidence thus far available, we propose a following model for the role and mode of action of nectins in formation of synapses in the CA3 area of hippocampus (Takai et al. 2003b) (Fig. 4). During synaptogenesis, the nectin–afadin system first forms primordial junctions, and this is followed by the recruitment of the N-cadherin–catenin system to the primordial junctions. The components of active zones would then be recruited to the primordial junctions to form active zones at the presynaptic side. At the postsynaptic side, the components of PSDs would be assembled, and membrane receptors are transported to the primordial junctions. The membrane domains, comprising synaptic junctions and puncta adherentia junctions, would then gradually become segregated, followed by maturation of synapses.

3.1.3
Association Mechanisms of Nectins, Cadherins, and TJ Components

How nectins are physically associated with cadherins is not known, but both afadin and α-catenin are essential for this association. The cytoplasmic region of E-cadherin and α-catenin may be involved in, but are not essential for, this association. Afadin directly binds α-catenin in vitro, but this binding is not strong (Tachibana et al. 2000; Pokutta et al. 2002). The direct binding of these proteins may occur in vivo, but it is more likely that a posttranslational modification(s) of either or both proteins and/or an unidentified molecule(s) are required for the binding of α-catenin to afadin (Tachibana et al. 2000; Pokutta et al. 2002). Consistently, two connector units for nectins and cadherins have been identified (Fig. 3C). One is a ponsin–vinculin unit (Mandai et al. 1999). Ponsin is an afadin- and vinculin-binding protein and vinculin is an F-actin- and α-catenin-binding protein. The other is an afadin DIL domain-interacting protein (ADIP)-α-actinin unit (Asada et al. 2003). ADIP is an afadin- and α-actinin-binding protein and α-actinin is an α-catenin-binding protein. In fibroblasts, nectins and cadherins associate through these linkers and the actin cytoskeleton is not involved in this association (Honda et al. 2003b).

It is not known, either, how the nectin–afadin and cadherin–catenin systems regulate the formation of TJs in epithelial cells. ZO-1 indirectly associates with the nectin–afadin system in a manner independent of the E-cadherin–catenin system (Yokoyama et al. 2001). During the formation of the junctional complex of AJs and TJs, JAMs are first recruited to the apical side of the nectin-, E-cadherin-based junctions, where ZO-1 colocalizes with afadin, followed by recruitment of claudins and occludin to the apical side of the nectin-/E-cadherin-based junctions (Fukuhara et al. 2002a,b). ZO-1 is also translocated from the nectin-/E-cadherin-based cell–cell adhesion site to its apical side. Thus, the association of ZO-1 to the nectin–afadin system may play a role in recruiting the CAMs of TJs to the apical side of AJs. The precise order of recruitment of CAMs and ZO-1 remains unknown. One possibility is that JAMs are first recruited to the apical side of the nectin-/E-cadherin-based junctions, and then the ZO-1 associated with the nectin–afadin system is translocated to the apical side to bind to JAMs. Alternatively, the ZO-1 associated with the nectin–afadin system is first translocated to the apical side, and then JAMs are recruited to the apical side by binding to ZO-1.

3.2
Cadherin-Independent Roles of Nectins

3.2.1
Possible Roles of Nectins in Axonal Pathfinding

During development of the central nervous system, commissural axons grow toward the ventral midline (Fig. 5). After crossing the floor plate, they abruptly change their trajectory from the circumferential to the longitudinal axis. The contacts between the commissural axons and the floor plate cells are presumably involved in this axonal guidance, but their mechanisms or structures have not fully been understood. Nectin-1 and nectin-3 asymmetrically localize at the commissural axon side and the floor plate cell side, respectively, of the plasma membranes at their contact sites and heterophilically *trans*-interact there (N. Okabe, K. Shimizu, K. Ozaki-Kuroda, H. Nakanishi, K. Morimoto, M. Takeuchi, H. Katsumaru, F. Murakami, and Y. Takai, personal communication) (Fig. 5). The immunofluorescence signal or

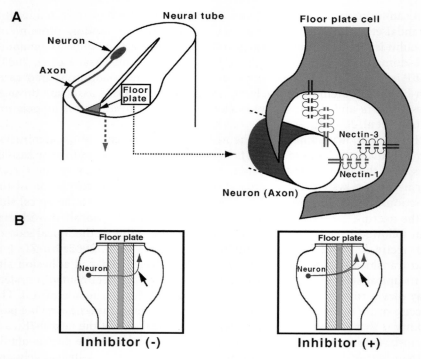

Fig. 5A, B. A model for the role and mode of action of nectins in the contacts between commissural axons and floor plate cells. **A** During development of the central nervous system, commissural axons grow toward the ventral midline. Nectin-1 and nectin-3 asymmetrically localize at the commissural axon side and the floor plate cell side, respectively, of the plasma membranes at their contact sites and heterophilically *trans*-interact there. **B** In vitro inhibition of the endogenous *trans*-interaction between nectin-1 and nectin-3 at the contacts between commissural axons and floor plate cells by the nectin inhibitors, gD and Nef-3. *Inhibitor (−)*: After crossing the floor plate, the commissural axons abruptly change their trajectory from the circumferential to the longitudinal axis. *Inhibitor (+)*: The nectin inhibitors impair the contacts and perturb the longitudinal turns of the commissural axons at the contralateral sites

the immunogold particle for N-cadherin or β-catenin is not concentrated at the contact sites, suggesting that N-cadherin and β-catenin play less important roles than nectins at this type of contact. The contacts between the commissural axons and the floor plate cells may not be associated with F-actin-binding proteins, including afadin, ZO-1, vinculin, and α-actinin. The contacts are not associated with the prominent actin cytoskeleton visualized by transmission electron microscopy (Yaginuma et al. 1991; N. Okabe, K. Shimizu, K. Ozaki-Kuroda, H. Nakanishi, K. Morimoto, M. Takeuchi, H. Katsumaru, F. Murakami, and Y. Takai, personal communication). Taken together, it is likely that the contacts between the commissural axons and the floor plate cells mediated by nectin-1 and nectin-3 is different from and not as strong as other junctions associated with the actin cytoskeleton.

In vitro inhibition of the endogenous *trans*-interaction between nectin-1 and nectin-3 by the nectin inhibitors, gD and Nef-3, results in impairment of the contacts and failure in longitudinal turns of the commissural axons at the contralateral sites of the rat hindbrain, suggesting that the contacts by the hetero-*trans*-interaction between nectin-1 and nectin-3 are involved in the trajectory of the commissural axons (N. Okabe, K. Shimizu, K. Ozaki-Kuroda, H. Nakanishi, K. Morimoto, M. Takeuchi, H. Katsumaru, F. Murakami and Y. Takai, personal communication). Because the commissural axons and the floor plate cells communicate or transfer signals through their contact sites (for a review: Stoeckli and Landmesser 1998 and Stoeckli, this volume), the impairment of the nectin-1 and nectin-3-based contacts by the nectin inhibitors may cause the axons unable to transfer signals insides the cells, resulting in abnormal turns and loss of the proper direction (Fig. 5) (see Sect. 4). Alternatively, the hetero-*trans*-interaction of nectins may play a role in mechanical fixation of the commissural axons to the floor plate cells and in maintenance of the contact sites for axonal pathfinding. Unidirectional signaling from nectin-3 to nectin-1 may regulate growth cones to determine their directions. Such signaling from nectin-3 to nectin-1 may regulate the up- or down-regulation of receptors for the guidance cues.

3.2.2
Role of Nectins in Organization of Sertoli Cell–Spermatid Junctions

Sertoli cells form a unique type of the F-actin-based junctional complex referred to as the ectoplasmic specialization (ES) (for reviews: Russell and Griswold 1993; Vogl et al. 2000). ES contains F-actin bundles, which are arranged at regular intervals beneath the plasma membrane, and a flattened cistern of the endoplasmic reticulum, which is connected to microtubules (Fig. 6). This is supposed to function as a scaffold that stabilizes an adhesive domain in the plasma membrane of Sertoli cells. Vinculin is concentrated at the Sertoli cell–Sertoli cell junctions and the Sertoli cell–spermatid junctions (Grove and Vogl 1989; Pfeiffer and Vogl 1991). Since the ES is formed only in Sertoli cells, the Sertoli cell–spermatid junctions are asymmetric. Moreover, unlike at typical AJs in epithelial cells, the existence of the cadherin–catenin system has been questioned at the Sertoli cell–spermatid junctions (Byers et al. 1991; Cyr et al. 1992; Anderson et al. 1994). Nectin-2, nectin-3, and afadin colocalize with the F-actin that underlies the Sertoli cell–spermatid junctions (Ozaki-Kuroda et al. 2002) (Fig. 6). Nectin-2 and nectin-3 asymmetrically localize at the Sertoli cell side and the spermatid side, respectively, and heterophilically *trans*-interact there. Afadin colocalizes with and binds to nectin-2 at the Sertoli cell side, but it remains unknown whether afadin is present and binds to nectin-3 at the spermatid side

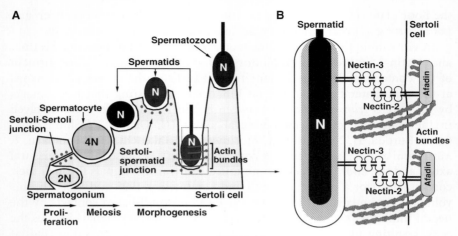

Fig. 6A, B. A model for the role and mode of action of nectins in formation of Sertoli cell–spermatid junctions. **A** Spermatogenesis. Sertoli cells, which constitute the single-layered seminiferous epithelium, support the differentiation of germ cells. Spermatogenic stem cells (spermatogonia) proliferate and differentiate into spermatocytes, which go through meiosis. The resultant haploid spermatids undergo morphological development (spermiogenesis) to be mature spermatozoa. **B** Sertoli cell–spermatid junctions. Nectin-2 in Sertoli cells and nectin-3 in spermatids form the hetero-*trans*-dimer at Sertoli cell–spermatid junctions. The actin bundles surround the spermatid heads like parallel rings. The nectin-based adhesive membrane microdomains show a one-to-one linkage with each F-actin bundle. Afadin colocalizes with and binds to nectin-2 at the Sertoli cell side, but it remains unknown whether afadin is present and binds to nectin-3 at the spermatid side. ES contains not only actin bundles but also a flattened cistern that is connected to microtubules. Microtubules are not shown

(Ozaki-Kuroda et al. 2002). The nectin-based adhesive membrane microdomains show one-to-one linkage with each F-actin bundle at the Sertoli cell–spermatid junctions (Ozaki-Kuroda et al. 2002) (see Sect. 3.3). The Sertoli cell–Sertoli cell junctions are similar to the junctional complex of epithelial cells, and the nectin–afadin system, the N-cadherin–catenin system, and the claudin-ZO-1 system form the junctional complex at the Sertoli cell–Sertoli cell junctions (Morita et al. 1999; Johnson and Boekelheide 2002; Ozaki-Kuroda et al. 2002). Nectin-2-deficient mice exhibit the male-specific infertility phenotype and have defects in the later steps of sperm morphogenesis, exhibiting distorted nuclei and abnormal distribution of mitochondria (Bouchard et al. 2000). In these mice, the structure of the Sertoli cell–spermatid junctions is severely impaired, and the localization of nectin-3 and afadin is disorganized (Ozaki-Kuroda et al. 2002). Thus, it is likely that the Sertoli cell–spermatid junctions rely largely on the nectin–afadin system. In addition, spermatozoa of nectin-2-deficient mice have a defect in binding to the zona pellucida and oocyte penetration (Mueller et al. 2003).

3.3
Association of Nectins with the Actin Cytoskeleton

Cortical F-actin underlying homotypic AJs in epithelial cells form a bundle around the apical end of the cell (Farquhar and Palade 1963; Hirano et al. 1987). As described in the Introduction and Sect. 3.1.1, the cadherin–catenin system is widely accepted as the major adhesion machinery of AJs, where they interact with the actin cytoskeleton directly via α-catenin, or indirectly via vinculin and α-actinin (for a review: Nagafuchi 2001). However, there are several reports showing that the depletion or significant reduction of the cadherin–catenin system does not hinder the formation of the apparently normal AJ structure (Tepass et al. 2000). Therefore, at least in some cases, an additional adhesive mechanism(s) is required to determine the precise size and localization of the AJ structure. The nectin–afadin system is an excellent candidate for the following reasons: (1) the nectin–afadin system concentrates even more strictly at epithelial AJs than the cadherin–catenin system (Mandai et al. 1997; Takahashi et al. 1999); (2) in afadin (–/–) mice, the organization of AJs is impaired in epithelial cells of embryos (Ikeda et al. 1999); and (3), afadin interacts directly with both nectin and F-actin and indirectly with the cadherin–catenin system (Tachibana et al. 2000). As described in Sect. 3.2.2, the nectin-based adhesive membrane microdomains show one-to-one linkage with each F-actin bundle at the Sertoli cell–spermatid junctions in testis (Ozaki-Kuroda et al. 2002). The structure of F-actin bundles at the Sertoli cell–spermatid junctions differs from that at typical AJs: F-actin is bundled tightly in parallel at the Sertoli cell–spermatid junctions and does not seem contractile (Vogl et al. 1991), while it is bundled loosely in antiparallel at typical AJs and exhibit the contractile property (Hirokawa and Tilney 1982). Nevertheless, their side-to-side association to the plasma membrane and their function, to scaffold the contact sites, are comparable. The nectin–afadin system may couple adhesion sites and F-actin bundles at AJs in cooperation with the E-cadherin–catenin system in epithelial cells.

4
Roles of Nectins in Cell Signaling

4.1
Activation of Cdc42 and Rac by Nectins

The *trans*-interaction of E-cadherin induces activation of Rac, but not that of Cdc42, in epithelial cells and fibroblasts (Nakagawa et al. 2001; Noren et al. 2001; Betson et al. 2002; Kovacs et al. 2002). The *trans*-interaction of E-cadherin does not induce activation of Rho in epithelial cells (Noren et al. 2001). The E-cadherin-induced activation of Rac is mediated through activation of phosphoinositide (PI)3 kinase (Kovacs et al. 2002), but it remains unknown how E-cadherin induces the activation of PI3 kinase or how PI3 kinase induces the activation of Rac. On the other hand, the *trans*-interactions of nectins induce not only activation of Rac but also that of Cdc42 in a PI3 kinase-independent manner in fibroblasts and epithelial cells (Kawakatsu et al. 2002; Honda et al. 2003b). It is unclear whether the *trans*-interactions of nectins induce activation of Rho. The nectin-induced activation of Rac requires activation of Cdc42, whereas the nectin-induced activation of Cdc42 does not require the activation of Rac. The C-terminal four amino acids of nectins, which are necessary for its binding of afadin, are not essential for the activation of Cdc42 and Rac. Cyclical activation and inactivation of these small G proteins are regulated by three types of regulators: Rho GDP/GTP exchange protein (GEP), Rho GTPase-activating protein (GAP), and Rho GDP dissociation inhibitor (GDI) (Takai et al. 2001). GEP enhances activation of the GDP-bound inactive form of the small G protein to the GTP-bound active form; GAP enhances inactivation of the GTP-bound active form of the small G protein to the GDP-bound inactive form; GDI binds the GDP-bound inactive form of the small G protein and keeps it in the cytosol and retrieves it from the membranes. E-Cadherin and nectins may induce either activation of GEP, inactivation of GAP, or both, but their precise molecular mechanism remains to be elucidated.

The formation of AJs is enhanced by GTP–Rac and GTP–Cdc42 in epithelial cells (Braga et al. 1997; Takaishi et al. 1997; Jou and Nelson 1998; Kodama et al. 1999; Ehrlich et al. 2002), whereas the formation of AJs is suppressed by inhibition of activation of Rac or the action of GTP–Cdc42 in epithelial cells (Ehrlich et al. 2002; Fukuhara et al. 2003). The formation of AJs is suppressed by inhibition of Rho (Braga et al. 1997, 1999). The modes of action of these small G proteins appear to be different: expression of a dominant-active mutant of Rac causes tight adhesion of the lateral plasma membrane from the apical to the basal area with interdigitation of the lateral plasma membrane in MDCK cells, whereas expression of a dominant-active mutant of Cdc42 causes similar tight adhesion of the lateral plasma membrane from the apical to the basal area, but without the interdigitation of the lateral plasma membrane (Takaishi et al. 1997; Jou and Nelson 1998; Kodama et al. 1999). The precise modes of action of Rac and Cdc42 remain unknown, but a plausible mechanism is that once cell–cell contacts are formed, mainly by the *trans*-interactions of nectins and partly by the *trans*-interaction of E-cadherin at the initial stage, Cdc42 and Rac are activated. GTP–Cdc42 formed in this way then induces formation of filopodia, which contact each other to form multiple cell–cell contact sites between the lateral plasma membranes of the adjacent cells (primordial spot-like junctions), whereas GTP–Rac formed in this way induces formation of lamellipodia which expands the cell–cell contact sites formed by the action of GTP–Cdc42 to form line-like junctions (Ehrlich et al. 2002; Fukuhara et al. 2003) (see Sect. 3.1.1 and Fig. 3).

Many downstream effectors have been identified for Rac and Cdc42. Some of these effectors are F-actin-binding proteins: IQGAP1 and IRSp53/WAVE for Rac and IQGAP1, NWASP and WASP for Cdc42 (for reviews: Takenawa and Miki 2001; Takai et al. 2001). Therefore, the nectin–afadin and E-cadherin–catenin systems regulate organization of the peripheral actin cytoskeleton in two different ways: one is organized by F-actin-binding proteins associated with nectins and cadherins, such as afadin, α-catenin, α-actinin, and vinculin; the other is organized by Cdc42 and Rac through their downstream effector F-actin-binding proteins. The actin cytoskeleton reorganized in these ways facilitates the formation of AJs and strengthen their cell–cell adhesion.

4.2
Activation of JNK by Nectins

GTP–Cdc42 and GTP–Rac induce activation of mitogen-activated protein (MAP) kinase cascades (for reviews: Vojtek and Cooper 1995; Denhardt 1996). Three major cascades have been identified: extracellular signal-regulated kinase (ERK), p38 MAP kinase, and c-Jun N-terminal kinase (JNK) pathways. Dominant-active mutants of Cdc42 and Rac induce activation of the p38 MAP kinase and JNK cascades, but not the ERK cascade. These small G proteins induce activation of JNK through activation of two JNK kinases, MAP kinase kinase (MKK)4 and MKK7 (Holland et al. 1997), and p38 MAP kinase through activation of two p38 MAP kinase kinases, MKK3 and MKK6 (Lee et al. 2001). It may be noted that GTP–Cdc42 and GTP–Rac formed by the *trans*-interactions of nectins selectively induce activation of JNK, but not p38 MAP kinase or ERK (Honda et al. 2003b). It remains unknown why GTP–Cdc42 or GTP–Rac formed by the *trans*-interactions of nectins does not induce the activation of p38 MAP kinase, but it may be due to the intracellular compartmentalization of the MAP kinase pathways. An overexpressed dominant-active mutant of Cdc42 or Rac may be distributed randomly to many compartments—including the nectin-based micro-domains—in cells and induce the activation of p38 MAP kinase as well as that of JNK. However, the nectin-based micro-domain may be linked only to the JNK pathway. Since the JNK signaling pathway is involved in regulation of many cellular events, including growth control, transformation, and programmed cell death (for reviews: Johnson and Lapadat 2002; Lin 2003), it is important to clarify the role of the nectin-induced activation of JNK.

5
Role of Nectins in Cell Polarization

Cell polarity is fundamental not only for cell functions but also for development and tissue maintenance (for reviews: Drubin and Nelson 1996; Yeaman et al. 1999). Mechanisms of establishment of cell polarity have been studied most extensively in *Caenorhabditis elegans* and *Drosophila* (for a review: Knoblich 2001). PAR proteins, PAR-1 to PAR-6, were first identified in *C. elegans* as indispensable proteins that are involved in the establishment of the anterior–posterior cell polarity of the one-cell embryo (for a review: Guo and Kemphues 1996). Thereafter, homologous proteins have been identified in *Drosophila* and mammals (for reviews: Knoblich 2001; Ohno 2001). In mammals, PAR-3, a mammalian homolog of the *par-3* gene product, was first identified as an aPKCζ-binding protein (Ohno 2001). PAR-3 localizes at TJs and forms a ternary complex with PAR-6 and aPKC in mammalian epithelial cells (Ohno 2001). These three proteins directly interact with each other and play a critical role in the apico-basal polarization of mammalian epithelial cells (Ohno 2001). This PAR-3/PAR-6/aPKC complex is an evolutionarily conserved cell polarization machinery that functions ubiquitously in a variety of biological context from warm embryos to differentiated mammalian cells (Ohno 2001). There are, furthermore, other sets of proteins that establish both AJs and membrane polarity in *Drosophila* (for reviews: Muller 2000; Tepass et al. 2001): these include Crumbs (Crb), Stardust (Std), and Discs lost (Dlt), which localize to the subapical region of epithelial cells, and Scribble (Scrib), Discs large (Dlg), and Lethal giant larvae (Lgl), which localize to the basolateral region. Mammalian homologs of Crb, Std, Dlt, Crumb, Pals1 and PATJ localize at TJs (Hurd at al. 2003; Matter and Balda 2003), whereas homologs of Dlg and Lgl localize at the basolateral region in epithelial cells (Musch et al. 2002).

A model for the mode of action of these polarity proteins in formation of the junctional complex of AJs and TJs in epithelial cells has recently been proposed (Plant et al. 2003; Yamanaka et al. 2003). Cell–cell contact initially stimulates the localization of the protein complex containing PAR-6, aPKC, and mammalian Lgl (mLgl) at the cell–cell contact sites. The complex is inactive for the formation of TJs. Once aPKC is activated, mLgl dissociates from the PAR-6/aPKC complex. This triggers the formation of the active PAR-3/PAR-6/aPKC complex that promotes the formation of TJs. The dissociated mLgl molecule remains in the lateral region of the plasma membrane and seems to contribute to the establishment of the basolateral plasma membrane. Although the mechanism for activation of aPKC remains to be clarified, binding of GTP–Cdc42 to PAR-6 is likely to trigger the activation. It has recently been shown that the Crumb/Pals1/PATJ complex can interact with the PDZ domain of PAR-6, and that this interaction is enhanced by GTP–Cdc42 (Hurd et al. 2003), suggesting that the Crumbs/Pals1 complex is also involved in the regulation of the interaction between mLgl and the PAR-6/aPKC complex, and that the PAR-6/aPKC complex, together with PAR-3, involves the Crumbs/Pals1/PATJ complex to promote the formation of TJs (Plant et al. 2003; Yamanaka et al. 2003). Thus, the dissociation of mLgl from the PAR-6/aPKC complex is likely to trigger the interaction of the

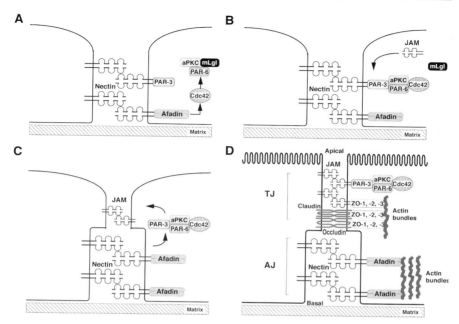

Fig. 7A–D. A model for the role and mode of action of nectins in formation of cell polarity in epithelial cells. **A** When two migrating cells contact through their protrusions, nectins form *trans*-dimers that form micro-clusters at cell–cell contact sites. The *trans*-interactions of nectins induce the activation of Cdc42, which binds to PAR-6 of the GTP–Cdc42/PAR-6/aPKC/mLgl complex. Nectins bind both afadin and PAR-3. **B** GTP–Cdc42/PAR-6 activates aPKC, which then phosphorylates mLgl, resulting in release of mLgl from the GTP–Cdc42/PAR-6/aPKC complex. The GTP–Cdc42/PAR-6/aPKC complex then forms a complex with PAR-3. The *trans*-interactions of nectins recruit JAMs to the apical side of AJs. **C** The GTP–Cdc42/PAR-3/PAR-6/aPKC complex translocates from nectins and binds to JAMs. **D** Claudins and occludin are recruited and eventually lead to the establishment of TJs. The cytoplasmic regions of JAMs, claudins, and occludin bind ZO-1, ZO-2, and ZO-3. Actin bundles and peripheral membrane proteins shown in the *right-side* cell are omitted in the *left-side* cell. E-cadherin or the Crumb/Pals1/PATJ complex is not shown (see Fig. 3C and text)

PAR-6/aPKC complex with the Crumbs/Pals1/PATJ complex in addition to its interaction with PAR-3 (Plant et al. 2003; Yamanaka et al. 2003).

In this proposed model, Cdc42 functions as an initial trigger for the formation of TJs through these polarity proteins. GTP–Cdc42 formed by the action of nectins is likely to play this role. The roles of nectins in recruiting JAMs to the apical side of AJs and in activation of Cdc42 suggest that nectins play a role in formation of cell polarity in a following way (Fig. 7): the *trans*-interactions of nectins recruit E-cadherin to form AJs and then recruit JAMs to the apical side of AJs. On the other hand, they induce activation of Cdc42, which binds to PAR-6. GTP–Cdc42/PAR-6 then activates aPKC, which then phosphorylates mLgl, resulting in the release of mLgl from the GTP–Cdc42/PAR-6/aPKC complex. The GTP–Cdc42/PAR-6/aPKC complex then forms a complex with PAR-3, resulting in association of this multiple complex to JAMs through PAR-3. It remains unknown where PAR-3 localizes before it binds to JAMs. Recent analysis indicates that PAR-3 binds to nectin-1 and nectin-3, but not nectin-2 (Takekuni et al. 2003). The first PDZ domain of PAR-3 binds to the C-terminal, four amino acids of nectins-1 and nectin-3. Thus, nectin-1 and nectin-3 bind both afadin and PAR-3. The precise role of this binding of PAR-3 to nectins remains unknown, but PAR-3 may bind to nectins before it binds to JAMs, and the complex formation of PAR-3 with GTP–Cdc42/PAR-6/aPKC releases PAR-3 from nectins and translocates the complex to JAMs.

6
Nectins as Viral Receptors

The interaction of a virus with its cellular receptor initiates dynamic events that enable the virus to enter the cells (Campadelli-Fiume et al. 2000; Schneider-Schaulies 2000; Spear et al. 2000). In infected tissues, the virus then spreads from cell to cell. Various membrane molecules, including CAMs, membrane receptors, and proteoglycans, have been identified as receptors for a variety of viruses (Schneider-Schaulies 2000). For example, intercellular adhesion molecule (ICAM) (Greve et al. 1989), coxsackievirus and adenovirus receptor (Bergelson 1997), and JAM (Barton et al. 2001), all of which physiologically function as CAMs, serve as receptors for adenovirus, reovirus, and rhinovirus, respectively. As described in Sect. 2.1, nectin-1α and nectin-2α were originally isolated as PRR-1 and PRR-2 (Morrison et al. 1992; Lopez et al. 1995; Eberlé et al. 1995), later shown to be receptors for α-herpes viruses, facilitating entry and intracellular spread, and named HveC and HveB, respectively (Cocchi et al. 1998, 2000; Geraghty et al. 1998; Warner et al. 1998; Lopez et al. 2001; Sakisaka et al. 2001). It remains to be seen whether nectin-3 and nectin-4 serve as receptors for viruses. Human nectin-1 can facilitate the entry of all α-herpes viruses tested so far, including HSV-1, HSV-2, and pseudorabies virus (Campadelli-Fiume et al. 2000; Spear et al. 2000). Human nectin-2 is active for a restricted number of α-herpes viruses (Campadelli-Fiume et al. 2000; Spear et al. 2000). The usual manifestations of HSV disease are mucocutaneous lesions. HSV disease establishes latent infection of neurons in sensory ganglia and causes recurrent lesions at the sites of primary infection. In HSV disease, the cell–cell spread significantly contributes to the pathogenesis. The underlying mechanisms for entry and cell–cell spread of HSVs through nectin-1 are not fully understood, but at least four viral glycoproteins, gD, gB, and the heterodimer of gH and gL, participate in the entry of HSV-1 (Campadelli-Fiume et al. 2000; Spear et al. 2000). Recombinant gD binds to nectin-1 of host cells and inhibits the HSV-1 infection, indicating that gD serves as a viral component that specifically interacts with the cellular receptor, nectin-1 (Campadelli-Fiume et al. 2000; Spear et al. 2000). The first Ig-like loop of nectin-1 is sufficient for the binding of gD and entry of HSV-1 (Cocchi et al. 1998), and the second and third Ig-like loops increase efficiency of the entry (Cocchi et al. 1998). The first Ig-like loop of nectin-2 is also critical for HSV entry (Martinez and Spear 2001). Accumulation of gD to nectin-1 may contribute to the disruption of cell–cell junctions as described for adenovirus infection (Spear 2002). This would be a mechanism for the virus to exit across airway epithelia. The interaction of nectin-1 with afadin increases the efficiency of cell–cell spread, but not the entry, of HSV-1 (Sakisaka et al. 2001). The E-cadherin–catenin system increases efficiency of both entry and cell–cell spread of HSV-1, providing an additional line of evidence that nectin and E-cadherin might be physically and functionally associated (Sakisaka et al. 2001).

7
Conclusions and Perspectives

We have described here that nectins play diverse roles: they organize a variety of cell–cell junctions, including homotypic symmetrical, homotypic asymmetrical, and heterotypic junctions in cooperation with, or independently of, cadherins; nectins induce activation of intracellular signaling pathways through Cdc42 and Rac and thereby regulate gene expression and formation of cell polarity; and pathogenically, nectins serve as entry and cell–cell spread mediators of HSVs. However, the detailed molecular mechanisms underlying the connections between nectins and CAMs, other than cadherins, signaling molecules, or polarity proteins, have not yet been fully elucidated. On the other hand, the elucidated roles of nectins have raised other new roles of nectins: for instance, nectins may be involved in disruption of various types of cell–cell junctions; and nectins may be involved in concentration of many biologically active molecules at cell–cell junctions. It may be fascinating that nectins may play a role in cell recognition, since the nectin-based junctions form more rapidly than the cadherin-based junctions. Elucidation of these unresolved issues will give us deeper insights into the molecular linkage between cell–cell junctions and various cell functions, such as morphogenesis, differentiation, proliferation, and migration, and also into the molecular mechanisms that underlie many human diseases arising from junctional disorders, such as cancer and vascular and mental diseases.

References

Adams CL, Chen YT, Smith SJ, Nelson WJ (1998) Mechanisms of epithelial cell–cell adhesion and cell compaction revealed by high resolution tracking of E-cadherin-green fluorescent protein. J Cell Biol 142:1105–1119

Amaral DG, Dent JA (1981) Development of the mossy fibers of the dentate gyrus. I. A light and electron microscopic study of the mossy fibers and their expansions. J Comp Neurol 195:51–86

Anderson AM, Edvardsen K, Skakkebaek NE (1994) Expression and localization of N- and E-cadherin in the human testis and epididymis. Int J Androl 17:174–180

Ando-Akatsuka Y, Yonemura S, Itoh M, Furuse M, Tsukita S (1996) Differential behavior of E-cadherin and occludin in their colocalization with ZO-1 during the establishment of epithelial cell polarity. J Cell Physiol 179:115–125

Angst BD, Marcozzi C, Magee AI (2000) The cadherin superfamily: diversity in form and function. J Cell Sci 114:629–641

Asada M, Irie K, Morimoto K, Yamada A, Ikeda W, Takeuchi M, Takai Y (2003) ADIP: A novel afadin- and α-actinin-binding protein localized at cell–cell adherens junctions. J Biol Chem 278:4103–4111

Asakura T, Nakanishi H, Sakisaka T, Takahashi K, Mandai K, Nishimura M, Sasaki T, Takai Y (1999) Similar and differential behavior between the nectin–afadin–ponsin and cadherin–catenin systems during the formation and disruption of polarized junctional alignment in epithelial cells. Genes Cells 4:573–581

Barton ES, Forrest JC, Connolly JL, Chappell JD, Liu Y, Schnell FJ, Nusrat A, Parkos CA, Dermody TS (2001) Junction adhesion molecule is a receptor for reovirus. Cell 104:441–451

Bergelson JM, Cunningham JA, Droguett G, Kurt-Jones EA, Krithivas A, Hong JS, Horwitz MS, Crowell RL, Finberg RW (1997) Isolation of a common receptor for Coxsackie B viruses and adenoviruses 2 and 5. Science 275:1320–1323

Betson M, Lozano E, Zhang J Braga VM (2002) Rac activation upon cell–cell contact formation is dependent on signaling from the epidermal growth factor receptor. J Biol Chem 277:36962–36969

Bouchard MJ, Dong Y, McDermott BM Jr Lam DH, Brown KR, Shelanski M, Bellve AR, Racaniello VR (2000) Defects in nuclear and cytoskeletal morphology and mitochondrial localization in spermatozoa of mice lacking nectin-2, a component of cell–cell adherens junctions. Mol Cell Biol 20:2865–2873

Braga VM, Machesky LM, Hall A Hotchin NA (1997) The small GTPases Rho and Rac are required for the establishment of cadherin-dependent cell–cell contacts. J Cell Biol 137:1421–1431

Braga VM, Del Maschio A, Machesky L, Dejana E (1999) Regulation of cadherin function by Rho and Rac: modulation by junction maturation and cellular context. Mol Biol Cell 10:9–22

Brummendorf T, Lemmon V (2001) Immunoglobulin superfamily receptors: cis-interactions, intracellular adaptors and alternative splicing regulate adhesion. Curr Opin Cell Biol 13:611–618

Buchert M, Schneider S, Meskenaite V, Adams MT, Canaani E, Baechi T, Moelling K, Hovens CM (1999) The junction-associated protein AF-6 interacts and clusters with specific Eph receptor tyrosine kinases at specialized sites of cell–cell contact in the brain. J Cell Biol 144:361–371

Byers S, Graham R, Dai HN, Hoxter B (1991) Development of Sertoli cell junctional specializations and the distribution of the tight-junction-associated protein ZO-1 in the mouse testis. Am J Anat 191:35–47

Campadelli-Fiume G, Cocchi F, Menotti L, Lopez M (2000) The novel receptors that mediate the entry of herpes simplex viruses and animal alphaherpesviruses into cells. Rev Med Virol 10:305–319

Cocchi F, Lopez M, Menotti L, Aoubala M, Dubreuil P, Campadelli-Fiume G (1998) The V domain of herpesvirus Ig-like receptor (HIgR) contains a major functional region in herpes simplex virus-1 entry into cells and interacts physically with the viral glycoprotein D. Proc Natl Acad Sci USA 95:15700–15705

Cocchi F, Menotti L, Dubreuil P, Lopez M, Campadelli-Fiume G (2000) Cell-to-cell spread of wild-type herpes simplex virus type 1, but not of syncytial strains, is mediated by the immunoglobulin-like receptors that mediate virion entry, nectin1 (PRR1/HveC/HIgR) and nectin2 (PRR2/HveB). J Virol 74:3909–3917

Cyr DG, Blaschuk OW, Robaire B (1992) Identification and developmental regulation of cadherin messenger ribonucleic acids in the rat testis. Endocrinology 131:139–145

Denhardt DT (1996) Signal-transducing protein phosphorylation cascades mediated by Ras/Rho proteins in the mammalian cell: the potential for multiplex signalling. Biochem J 318:729–747

Desbach T, Qualmann B, Kessels MM, Garner CC, Gundelfinger ED (2001) The presynaptic cytomatorix of brain synapses. Cell Mol Life Sci 58:94–116

Drubin GD Nelson WJ (1996) Origins of cell polarity. Cell 84:335–344

Eberlé F, Dubreuil P, Mattei MG, Devilard E, Lopez M (1995) The human PRR2 gene, related to the human poliovirus receptor gene (PVR), is the true homolog of the murine MPH gene. Gene 159:267–272

Ehrlich JS, Hansen MDH, Nelson WJ (2002) Spatio-temporal regulation of Rac1 localization and lamellipodia dynamics during epithelial cell–cell adhesion. Dev Cell 3:259–270

Farquhar MG, GE Palade (1963) Junctional complexes in various epithelia. J Cell Biol 17:375–412

Fukuhara A, Irie K, Nakanishi H, Takekuni K, Kawakatsu T, Yamada A, Katata T, Honda T, Sato T, Shimizu K, Ozaki H, Horiuchi H, Kita T, Takai Y (2002a) Involvement of nectin in the localization of junctional adhesion molecule at tight junctions. Oncogene 21:7642–7655

Fukuhara A, Irie K, Yamada A, Katata T, Honda T, Shimizu K, Nakanishi H, Takai Y (2002b) Roles of nectin in organization of tight junctions in epithelial cells. Genes Cells 7:1059–1072

Fukuhara A, Shimizu K, Kawakatsu T, Fukuhara T, Takai Y (2003) Involvement of nectin-activated Cdc42 small G protein in organization of adherens and tight junctions in Madin-Darby canine kidney cells. J Biol Chem 278:51885–51893

Geraghty RJ, Krummenacher C, Cohen GH, Eisenberg RJ, Spear PG (1998) Entry of alphaherpesviruses mediated by poliovirus receptor-related protein 1 and poliovirus receptor. Science 280:1618–1620

Greve JM, Davis G, Meyer AM, Forte CP, Yost SC, Marlor CW, Kamarck ME, McClelland A (1989) The major human rhinovirus receptor is ICAM-1. Cell 56:839–847

Grove BD, Vogl AW (1989) Sertoli cell ectoplasmic specializations: a type of actin-associated adhesion junction? J Cell Sci 93:309–323

Gumbiner BM (1996) Cell adhesion: the molecular basis of tissue architecture and morphogenesis. Cell 84:345–357

Gumbiner BM (2000) Regulation of cadherin adhesive activity. J Cell Biol 148:399–403

Guo S, Kemphues KJ (1996) Molecular genetics of asymmetric cleavage in the early Caenorhabditis elegans embryo. Curr Opin Genet Dev 6:408–415

Hirano S, Nose A, Hatta K, Kawakami A, Takeichi M (1987) Calcium-dependent cell–cell adhesion molecules (cadherins): subclass specificities and possible involvement of actin bundles. J Cell Biol 105:2501–2510

Hirokawa N, Tilney LG (1982) Interactions between actin filaments and between actin filaments and membranes in quick-frozen and deeply etched hair cells of the chick ear. J Cell Biol 95:249–261

Holland PM, Suzanne M, Campbell JS, Noselli S, Cooper JA (1997) MKK7 is a stress-activated mitogen-activated protein kinase kinase functionally related to hemipterous. J Biol Chem 272:24994–24998

Honda T, Shimizu K, Kawakatsu T, Yasumi M, Shingai T, Fukuhara A, Ozaki-Kuroda K, Irie K, Nakanishi H, Takai Y (2003a) Antagonistic and agonistic effects of an extracellular fragment of nectin of formation of E-cadherin-based cell–cell adhesion. Genes Cells 8:51–63

Honda T, Shimizu K, Kawakatsu T, Fukuhara A, Irie K, Nakamura T, Matsuda M, Takai Y (2003b) Cdc42 and Rac small G proteins activated by trans-interactions of nectins are involved in activation of c-Jun N-terminal kinase, but not in association of nectins and cadherin to form adherens junctions, in fibroblasts. Genes Cells 8:481–491

Honda T, Shimizu K, Fukuhara A, Irie K, Takai Y (2003c) Regulation by nectin of the velocity of the formation of adherens junctions and tight junctions. Biochem Biophys Res Commun 306:104–109

Hurd TW, Gao L, Roh MH, Macara IG, Margolis B (2003) Direct interaction of two polarity complexes implicated in epithelial tight junction assembly. Nat Cell Biol 5:137–142

Ikeda W, Nakanishi H, Miyoshi J, Mandai K, Ishizaki H, Tanaka M, Togawa A, Takahashi K, Nishioka H, Yoshida H, Mizoguchi A, Nishikawa S, Takai Y (1999) Afadin: a key molecule essential for structural organization of cell–cell junctions of polarized epithelia during embryogenesis. J Cell Biol 146:1117–1132

Johnson GL, Lapadat R (2002) Mitogen-activated protein kinase pathways mediated by ERK, JNK, and p38 protein kinases. Science 298:1911–1912

Johnson KJ, Boekelheide K (2002) Dynamic testicular adhesion junctions are immunologically unique. II. Localization of classic cadherins in rat testis. Biol Reprod 66:992–1000

Jou TS, Nelson WJ (1998) Effects of regulated expression of mutant RhoA and Rac1 small GTPases on the development of epithelial (MDCK) cell polarity. J Cell Biol 142:85–100

Kawakatsu T, Shimizu K, Honda T, Fukuhara T, Hoshino T, Takai Y (2002) trans-Interactions of nectins induce formation of filopodia and lamellipodia through the respective activation of Cdc42 and Rac small G proteins. J Biol Chem 277:50749–50755

Knoblich JA (2001) Asymmetric cell division during animal development. Nat Rev Mol Cell Biol 2:11–20

Kodama A, Takaishi K, Nakano K, Nishioka H, Takai Y (1999) Involvement of Cdc42 small G protein in cell–cell adhesion, migration and morphology of MDCK cells. Oncogene 18:3996–4006

Kovacs EM, Ali RG, McCormack AJ, Yap AS (2002) E-cadherin homophilic ligation directly signals through rac and phosphatidylionsitol 3-kinase to regulate adhesive contacts. J Biol Chem 277:6708–6718

Lee SH, Eom M, Lee SJ, Kim S, Park HJ, Park D (2001) BetaPix-enhanced p38 activation by Cdc42/Rac/PAK/MKK3/6-mediated pathway. Implication in the regulation of membrane ruffling. J Biol Chem 276:25066–25072

Lin A (2003) Activation of the JNK signaling pathway: breaking the brake on apoptosis. Bioessays. 25:17–24

Lopez M, Eberlé F, Mattei MG, Gabert J, Birg F, Bardin F, Maroc C, Dubreuil P (1995) Complementary DNA characterization and chromosomal localization of a human gene related to the poliovirus receptor-encoding gene. Gene 155:261–265

Lopez M, Cocchi F, Avitabile E, Leclerc A, Adelaide J, Campadelli-Fiume G, Dubreuil P (2001) Novel, soluble isoform of the herpes simplex virus (HSV) receptor nectin1 (or PRR1-HIgR-HveC) modulates positively and negatively susceptibility to HSV infection. J Virol 75:5684–5691

Mandai K, Nakanishi H, Satoh A, Obaishi H, Wada M, Nishioka H, Itoh M, Mizoguchi A, Aoki T, Fujimoto T, Matsuda Y, Tsukita S, Takai Y (1997) Afadin: A novel actin filament-binding protein with one PDZ domain localized at cadherin-based cell-to-cell adherens junction. J Cell Biol 139:517–528

Mandai K, Nakanishi H, Satoh A, Takahashi K, Satoh K, Nishioka H, Mizoguchi A, Takai Y (1999) Ponsin/SH3P12: an l-afadin- and vinculin-binding protein localized at cell-cell and cell-matrix adherens junctions. J Cell Biol 144:1001–1017

Martinez WM, Spear PG (2001) Structural features of nectin-2 (HveB) required for herpes simplex virus entry. J Virol 75:11185–11195

Matter K, Balda MS (2003) Signalling to and from tight junctions. Nat Rev Mol Cell Biol 4:225–236

Miyahara M, Nakanishi H, Takahashi K, Satoh-Horikawa K, Tachibana K, Takai Y (2000) Interaction of nectin with afadin is necessary for its clustering at cell-cell contact sites but not for its cis dimerization or trans interaction. J Biol Chem 275:613–618

Mizoguchi A, Nakanishi H, Kimura K, Matsubara K, Ozaki-Kuroda K, Katata T, Honda T, Kiyohara Y, Heo K, Higashi M, Tsutsumi T, Sonoda S, Ide C, Takai Y (2002) Nectin: an adhesion molecule involved in formation of synapses. J Cell Biol 156:555–565

Momose Y, Honda T, Inagaki M, Shimizu K, Irie K, Nakanishi H, Takai Y (2002) Role of the second immunoglobulin-like loop of nectin in cell–cell adhesion. Biochem Biophys Res Commun 293:45–49

Morita K, Sasaki H, Fujimoto K, Furuse M, Tsukita S (1999) Claudin-11/OSP-based tight junctions of myelin sheaths in brain and Sertoli cells in testis. J Cell Biol 145:579–588

Morrison ME, Racaniello VR (1992) Molecular cloning and expression of a murine homolog of the human poliovirus receptor gene. J Virol 66:2807–2813

Mueller S, Rosenquist TA, Takai Y, Bronson RA, Wimmer E (2003) Loss of nectin-2 at Sertoli-spermatid junctions leads to male infertility and correlates with severe spermatozoan head and midpiece malformation, impaired binding to the zona pellucida, and oocyte penetration. Biol Repro 69:1330–1340

Muller HA (2000) Genetic control of epithelial cell polarity: lessons from Drosophila. Dev Dyn 218:52–67

Musch A, Cohen D, Yeaman C, Nelson WJ, Rodriguez-Boulan E, Brennwald PJ (2002) Mammalian homolog of Drosophila tumor suppressor lethal (2) giant larvae interacts with basolateral exocytic machinery in Madin-Darby canine kidney cells. Mol Biol Cell 13:158–168

Nagafuchi A (2001) Molecular architecture of adherens junctions. Curr Opin Cell Biol 13:600–603

Nagafuchi A, Shirayoshi Y, Okazaki K, Yasuda K, Takeichi M (1987) Transformation of cell adhesion properties by exogenously introduced E-cadherin cDNA. Nature 329:341–343

Nakagawa M, Fukata M, Yamaga M, Itoh N, Kaibuchi K (2001) Recruitment and activation of Rac1 by the formation of E-cadherin-mediated cell–cell adhesion sites. J Cell Sci 114:1829–1838

Nishioka H, Mizoguchi A, Nakanishi H, Mandai K, Takahashi K, Kimura K, Satoh-Moriya A, Takai Y (2000) Localization of l-afadin at puncta adhaerentia-like junctions between the mossy fiber terminals and the dendritic trunks of pyramidal cells in the adult mouse hippocampus. J Comp Neurol 424:297–306

Noren NK, Niessen CM, Gumbiner BM, Burridge K (2001) Cadherin engagement regulates Rho family GTPases. J Biol Chem 276:33305–33308

Ohno S (2001) Intercellular junctions and cellular polarity: the PAR-aPKC complex, a conserved core cassette playing fundamental roles in cell polarity. Curr Opin Cell Biol 13:641–648

Ozaki-Kuroda K, Nakanishi H, Ohta H, Tanaka H, Kurihara H, Mueller S, Irie K, Ikeda W, Sasaki T, Wimmer E, Nishimune Y, Takai Y (2002) Nectin couples cell–cell adhesion and the actin scaffold at heterotypic testicular junctions. Curr Biol 12:1145–1150

Peng Y-F, Mandai K, Nakanishi H, Ikeda W, Asada M, Momose Y, Shibamoto S, Yanagihara K, Shiozaki H, Monden M, Takeichi M, Takai Y (2002) Restoration of E-cadherin-based cell–cell adhesion by overexpression of nectin in HSC-39 cells, a human signet ring cell gastric cancer cell line. Oncogene 21:4108–4119

Pfeiffer DC, Vogl AW (1991) Evidence that vinculin is co-distributed with actin bundles in ectoplasmic ("junctional") specializations of mammalian Sertoli cells. Anat Rec 231:89–100

Plant PJ, Fawcett JP, Lin DC, Holdorf AD, Binns K, Kulkarni S, Pawson T (2003) A polarity complex of mPar-6 and atypical PKC binds, phosphorylates and regulates mammalian Lgl. Nat Cell Biol 5:301–308

Pokutta S, Drees F, Takai Y, Nelson WJ, Weis WI (2002) Biochemical and structural definition of the l-afadin- and actin-binding sites of α-catenin. J Biol Chem 277:18817–18826

Prasad R, Gu Y, Alder H, Nakamura T, Canaani O, Saito H, Huebner K, Gale RP, Nowell PC, Kuriyama K (1993) Cloning of the ALL-1 fusion partner, the AF-6 gene, involved in acute myeloid leukemias with the t(6;11) chromosome translocation. Cancer Res 53:5624–5628

Reymond N, Fabre S, Lecocq E, Adelaide J, Dubreuil P, Lopez M (2001) Nectin4/PRR4: a new afadin-associated member of the nectin family that trans-interacts with nectin1/PRR1 through V domain interaction. J Biol Chem 276:43205–43215

Russell LD, Griswold MD (1993) The Sertoli cell. Cache River Press, Vienna

Sakisaka T, Nakanishi H, Takahashi K, Mandai K, Miyahara M, Satoh A, Takaishi K, Takai Y (1999) Different behavior of l-afadin and neurabin-II during the formation and destruction of cell–cell adherens junctions. Oncogene 18:1609–1617

Sakisaka T, Taniguchi T, Nakanishi H, Takahashi K, Miyahara M, Ikeda W, Yokoyama S, Peng YF, Yamanishi K, Takai Y (2001) Requirement of interaction of nectin-1alpha/HveC with afadin for efficient cell–cell spread of herpes simplex virus type 1. J Virol 75:4734–4743

Satoh-Horikawa K, Nakanishi H, Takahashi K, Miyahara M, Nishimura M, Tachibana K, Mizoguchi A, Takai Y (2000) Nectin-3, a new member of immunoglobulin-like cell adhesion molecules that shows homophilic and heterophilic cell–cell adhesion activities. J Biol Chem 275:10291–10299

Schneider-Schaulies J (2000) Cellular receptors for viruses: links to tropism and pathogenesis. J Gen Virol 81:1413–1429

Shimizu K, Takai Y (2003) Roles of the intercellular adhesion molecule nectin in intracellular signaling. J Biochem (Tokyo) 134:631–636

Sozen MA, Suzuki K, Tolarova MM, Bustos T, Fernandez Iglesias JE, Spritz RA (2001) Mutation of PVRL1 is associated with sporadic, non-syndromic cleft lip/palate in northern Venezuela. Nat Genet 29:141–142

Spear PG (2002) Viral interactions with receptors in cell junctions and effects on junctional stability. Dev Cell 3:462–464

Spear PG, Eisenberg RJ, Cohen GH (2000) Three classes of cell surface receptors for alphaherpesvirus entry. Virology 275:1–8

Stoeckli ET, Landmesser LT (1998) Axon guidance at choice points. Curr Opin Neurobiol 8:73–79

Suzuki K, Hu D, Bustos T, Zlotogora J, Richieri-Costa A, Helms JA, Spritz RA (2000) Mutations of PVRL1, encoding a cell–cell adhesion molecule/herpesvirus receptor, in cleft lip/palate-ectodermal dysplasia. Nat Genet 25:427–430

Tachibana K, Nakanishi H, Mandai K, Ozaki K, Ikeda W, Yamamoto Y, Nagafuchi A, Tsukita S, Takai Y (2000) Two cell adhesion molecules, nectin and cadherin, interact through their cytoplasmic domain-associated proteins. J Cell Biol 150:1161–1176

Takahashi K, Nakanishi H, Miyahara M, Mandai K, Satoh K, Satoh A, Nishioka H, Aoki J, Nomoto A, Mizoguchi A, Takai Y (1999) Nectin/PRR: an immunoglobulin-like cell adhesion molecule recruited to cadherin-based adherens junctions through interaction with afadin, a PDZ domain-containing protein. J Cell Biol 145:539–549

Takai Y, Nakanishi H (2003) Nectin and afadin: novel organizers of intercellular junctions. J Cell Sci 116:17–27

Takai Y, Sasaki T, Matozaki T (2001) Small GTP-binding proteins. Physiol Rev 81:153–208

Takai Y, Irie K, Shimizu K, Sakisaka T, Ikeda W (2003a) Nectins and nectin-like molecules: roles in cell adhesion, migration, and polarization. Cancer Sci 94:655–667

Takai Y, Shimizu K, Ohtsuka T (2003b) The roles of cadherins and nectins in interneuronal synapse formation. Curr Opin Neurobiol 13:520–526

Takaishi K, Sasaki T, Kotani H Nishioka H, Takai Y (1997) Regulation of cell–cell adhesion by rac and rho small G proteins in MDCK cells. J Cell Biol 139:1047–1059

Takeichi M (1991) Cadherin cell adhesion receptor as a morphogenetic regulator. Science 251:1451–1455

Takekuni K, Ikeda W, Fujito T, Morimoto K, Takeuchi M, Monden M, Takai Y (2003) Direct binding of cell polarity protein PAR-3 to cell–cell adhesion molecule nectin at neuroepithelial cells of developing mouse. J Biol Chem 278:5497–5500

Takenawa T, Miki H (2001) WASP and WAVE family proteins: key molecules for rapid rearrangement of cortical actin filaments and cell movement. J Cell Sci 114:1801–1809

Tanaka Y, Nakanishi H, Kakunaga S, Okabe N, Kawakatsu T, Shimizu K, Takai Y (2003) Role of nectin in formation of E-cadherin-based adherens junctions in keratinocytes: analysis with the N-cadherin dominant negative mutant. Mol Biol Cell 14:1597–1609

Tang L, Hung CP, Schuman EM (1998) A role for the cadherin family of cell adhesion molecules in hippocampal long-term potentiation. Neuron 20:1165–1175

Tepass U, Truong K, Godt D, Ikura M, Peifer M (2000) Cadherins in embryonic and neural morphogenesis. Nat Rev Mol Cell Biol 1:91–100

Tepass U, Tanentzapf G, Ward R, Fehon R (2001) Epithelial cell polarity and cell junctions in Drosophila. Annu Rev Genet 35:747–784

Tsukita S, Tsukita S, Nagafuchi A, Tonemura S (1992) Molecular linkage between cadherins and actin filaments in cell–cell adherens junctions. Curr Opin Cell Biol 4:834–839

Tsukita S, Furuse M, Itoh M (1999) Structural and signalling molecules come together at tight junctions. Curr Opin Cell Biol 11:628–633

Tsukita S, Furuse M, Itoh M (2001) Multifunctional strands in tight junctions. Nat Rev Mol Cell Biol 4:285–293

Vasioukhin V, Bauer C, Yin M, Fuchs E (2000) Direct actin polymerization is the driving force for epithelial cell–cell adhesion. Cell 100:209–219

Vlemincks K, Kemler R (1999) Cadherin and tissue formation: integrating adhesion and signaling. Bioessays 21:211–220

Vogl AW, Pfeiffer DC, Redenbach DM (1991) Ectoplasmic ("junctional") specializations in mammalian Sertoli cells: influence on spermatogenic cells. Ann NY Acad Sci 637:175–202

Vogl AW, Pfeiffer DC, Mulholland D, Kimel G, Guttman J (2000) Unique and multifunctional adhesion junctions in the testis: ectoplasmic specializations. Arch Histol Cytol 63:1–15

Vojtek AB, Cooper JA (1995) Rho family members: activators of MAP kinase cascades. Cell 82:527–529

Warner MS, Geraghty RJ, Martinez WM, Montgomery RI, Whitbeck JC, Xu R, Eisenberg RJ, Cohen GH, Spear PG (1998) A cell surface protein with herpesvirus entry activity (HveB) confers susceptibility to infection by mutants of herpes simplex virus type 1, herpes simplex virus type 2, and pseudorabies virus. Virology 246:179–189

Yagi T, Takeichi M (2000) Cadherin superfamily genes: functions, genomic organization, and neurologic diversity. Genes Dev 14:1169–1180

Yaginuma H, Homma S, Kunzi R, Oppenheim RW (1991) Pathfinding by growth cones of commissural interneurons in the chick embryo spinal cord: a light and electron microscopic study. J Comp Neurol 304:78–102

Yamagata M, Herman JP, Sanes JR (1995) Lamina-specific expression of adhesion molecules in developing chick optic tectum. J Neurosci 15:4556–4571

Yamanaka T, Horikoshi Y, Sugiyama Y, Ishiyama C, Suzuki A, Hirose T, Iwamatsu A, Shinohara A, Ohno S (2003) Mammalian Lgl forms a protein complex with PAR-6 and aPKC independently of PAR-3 to regulate epithelial cell polarity. Curr Biol 13:734–743

Yasumi M, Shimizu K, Honda T, Takeuchi M, Takai Y (2003) Role of each immunoglobulin-like loop of nectin for its cell–cell adhesion activity. Biochem Biophys Res Commun 302:61–66

Yeaman C, Grindstaff KK, Nelson WJ (1999) New perspectives on mechanisms involved in generating epithelial cell polarity. Physiol Rev 9:73–98

Yokoyama S, Tachibana K, Nakanishi H, Yamamoto Y, Irie K, Mandai K, Nagafuchi A, Monden M, Takai Y (2001) α-Catenin-independent recruitment of ZO-1 to nectin-based cell–cell adhesion sites through afadin. Mol Biol Cell 12:1595–1609

Yonemura S, Itoh M, Nagafuchi A, Tsukita S (1995) Cell-to-cell adherens junction formation and actin filament organization: similar and differences between non-polarized fibroblasts and polarized epithelial cells. J Cell Sci 108:127–142

Ziv NE, Garner CC (2001) Principles of glutamatergic synapse formation: seeing the forest for the tree. Curr Opin Neurobiol 11:536–543

Ig Superfamily Cell Adhesion Molecules in the Brain

E. T. Stoeckli

Institute of Zoology, University of Zurich, Winterthurerstrasse 190, 8057 Zurich, Switzerland
Esther.Stoeckli@zool.unizh.ch

1	Introduction	374
2	The Role of IgSF CAMs in Axon Guidance	376
2.1	IgSF CAMs, the Vanguard of Axon Guidance Cues	376
2.2	Axon Guidance Across the Midline of the Nervous System	377
2.3	Midline Contact Elicits a Switch in Growth Cones' Response from Attraction to Repulsion	380
2.4	The Interaction Between Dcc and Unc-5 Induces a Switch from Short- to Long-Range Effects	381
2.5	IgSF CAMs Play a Role in Subpopulation-Specific Guidance of Sensory Afferents	381
3	IgSF CAMs in Cell Migration	382
4	IgSF CAMs in Cerebellar Development	384
5	IgSF CAMS in Synaptogenesis	385
6	IgSF CAMs in Synaptic Maintenance and Plasticity	386
7	Intracellular Signaling Associated with IgSF CAMs	387
7.1	Intracellular Signaling Associated with Transmembrane IgSF CAMs	387
7.2	Intracellular Signaling Associated with Glycosylphosphatidyl-Inositol Anchored IgSF CAMs	388
7.3	Conformational Changes Are Sufficient to Induce a Switch in Intracellular Signaling Associated with IgSF CAMs	389
7.4	Changes in Signaling by Removal of IgSF CAMs from the Membrane	389
7.5	Silencing of IgSF CAM-Derived Signals	391
8	Outlook	392
References		392

Abstract Cell adhesion molecules of the immunoglobulin superfamily (IgSF CAMs) were discovered 25 years ago based on their role in cell–cell adhesion. Ever since, they have played a major role in developmental neuroscience research. The elucidation of IgSF CAM structure and function has been tightly linked to the establishment of new areas of research. Over the years, our view of the role of the IgSF CAMs has changed. First, they were thought to provide "specific glue" segregating subtypes of cells in the nervous system. Soon it became clear that IgSF CAMs can do much more. The focus shifted from simple adhesion to CAM-associated signaling that was shown to be involved in the promotion of axon growth and the regulation of cell migration. From there it was a small step to axon guidance, a field that has been given a lot of attention during the last decade. More recently, the involvement of IgSF CAMs in synapse formation and maturation has been discovered, although this last step in the formation of neural circuits was thought to be the domain of other families of cell adhesion molecules, such as the neuroligins, the neurexins, and the cadherins. Certainly, the most striking discovery in the context of IgSF CAMs has been the diversity of signaling mechanisms that are associated with them. The versatility of signals and their complexity make IgSF CAMs a perfect tool for brain development.

Keywords Axon guidance · Cell migration · Intracellular signaling · Clustering · Cell–cell recognition

1
Introduction

Making the right choices is one of the most important aspects of life. Even very early aspects of life depend on making choices; for instance, when cells have to connect to their own kind during organogenesis, and later, when cells have to migrate to find their appropriate position, or during the formation of neural circuits when the processes of neurons, the axons, have to find their target cells to form synapses. Making the right choices is essential for brain development. In the nervous system cells are mostly born in proliferative areas far away from their final location, and hence, they have to migrate considerable distances to reach their final position. Even more impressive than cell migration in the developing nervous system is the establishment of the intricate network of axonal and dendritic connections between neurons and their target cells. Axons have to navigate through the preexisting tissue to contact their partners and to establish functional synapses. All these steps require specific recognition processes, mediated by recognition molecules expressed on cell surfaces. It was based on this concept of specific cell–cell recognition that the first cell adhesion molecules (CAMs) of the immunoglobulin superfamily (IgSF) were discovered some 25 years ago (Brackenbury et al. 1977; Thiery et al. 1977). Despite intensive efforts to identify additional family members and to characterize their structure and function, they still have not revealed all their secrets. Nowadays, new subfamilies are no longer discovered at the high pace of the early 1990s. Instead, the results from structural and functional analyses, and discoveries of new interactions between the known family members and between the IgSF CAMs and other protein families have caught our attention, allowing us to assemble the pieces of evidence to a model of how IgSF CAMs are involved in development and function of the brain.

Based on the background of their discovery, cell adhesion molecules of the Ig superfamily were always suggested to play a role in cell–cell recognition. However, our view of how they do this has changed. Initially, IgSF

CAMs were thought to provide "specific glue," allowing some cells to adhere to selected partner cells (reviewed in Jessell 1988; Goodman and Shatz 1993). After the neurite outgrowth-promoting activity of some IgSF members had been demonstrated they were implicated in axon guidance. First, the hypothesis predicted that they would guide axons by providing differential adhesion along specific pathways. Soon, it became clear that this was not the case. In fact, no correlation between the strength of adhesion and the growth rate or the growth preference of neurites was found (Lemmon et al. 1992). Over time, the focus shifted from adhesion to signaling, starting a series of studies that concentrated on the elucidation of intracellular signals generated in response to the complex extracellular interactions of IgSF CAMs. The transformation of the extracellular interaction into an intracellular signal was especially intriguing for the subfamily of glycosylphosphatidyl–inositol-anchored IgSF CAMs, like axonin-1/transiently associated glycoprotein (TAG)-1 and F11/F3/contactin. As a result of these studies, the concept of clustering as a means to modulate the signal derived from CAM–CAM interactions was put forth. Functional evidence for clustering in the plane of the neuronal membrane, called *cis*- as opposed to *trans*-interac-

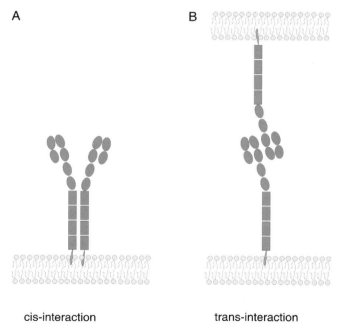

Fig. 1A, B. IgSF CAMs can interact with each other in two different modes: two molecules interacting with each other in the plane of the same membrane form a *cis*-interaction (**A**). Molecules from two different cell membranes bind each other in a *trans*-interaction (**B**)

tion, was demonstrated for axonin-1/TAG-1 and neuron–glia (Ng)CAM/L1 (Buchstaller et al. 1996; Stoeckli et al. 1996). Later, a switch in the association of axonin-1/TAG-1 and NgCAM/L1 with non-receptor tyrosine kinases of the src family to a casein–kinase II-related kinase was demonstrated depending on the involvement in *cis*- versus *trans*-interactions (Fig. 1; Kunz et al. 1996). Still, intracellular signaling in response to CAM–CAM interactions is poorly understood. In particular, it is not known how the enormous complexity of CAM–CAM interactions is transformed into specific intracellular signals. This is even more evident in light of recent findings that expand the range of interaction partners for IgSF CAMs (Castellani et al. 2000).

In line with the initial concept of cell–cell recognition, and in agreement with the requirements for guidance cues not just for axonal pathfinding (see Sect. 2) but also for cell migration (Sect. 3), is the involvement of IgSF CAMs in early steps of nervous system development. Several recent studies concentrate on this aspect of CAM function, for instance, in cerebellar development (Sect. 4). Given this wide spectrum of functions during development of the nervous system, it does not come as a surprise that malfunction or absence of IgSF CAMs can be linked to human disease (Fransen et al. 1995; Kenwrick and Doherty 1998; Weller and Gartner 2001).

Interestingly, the initial concept that IgSF CAMs might provide a specific glue that helps to segregate different cell types has gained new momentum lately. On the one hand, a role in adhesion without any effect on axon growth has been shown for the IgLON subfamily (McNamee et al. 2002). On the other hand, a role of IgSF CAMs in synapse formation, maintenance, and plasticity has been demonstrated (Sect. 5; and Sect. 6).

2
The Role of IgSF CAMs in Axon Guidance

2.1
IgSF CAMs, the Vanguard of Axon Guidance Cues

The precise formation of neuronal connections is the basis for the function of the nervous system. This seems to be an unsolvable problem in view of the overwhelming number of neurons that have to navigate to their target areas and recognize their target cells. Although there was no molecular background at the time, Santiago Ramón y Cajal postulated more than 100 years ago the existence of attractants that would guide growth cones, the tip of elongating axons, toward their targets. This idea was taken up by Sperry and formulated in his famous chemoaffinity hypothesis (Sperry 1963) that inspired generations of neuroscientists to look for such guidance cues. Goodman and colleagues who took the central nervous system of the grasshopper embryo as a model for axon guidance studies found evidence for the selective fasciculation of growing axons with preexisting fiber bundles. The "labeled pathway hypothesis" explained at least the pathway choices of some of the fibers following the so-called pioneer fibers, based on selective axon–axon recognition (Goodman et al. 1984). Studies by Bentley and colleagues using again the grasshopper embryo, but this time the limb bud, established the concept of guidepost cells in axon guidance (reviewed in Bentley and O'Connor 1992). This concept was not restricted to follower fibers but could explain pathway choices of pioneer fibers as well. Based on this background and based on the expression patterns of IgSF CAMs in the developing nervous system, a link of the IgSF CAMs to axonal pathfinding was obvious. Thus, newly identified IgSF CAMs were tested for their neurite outgrowth-promoting activity in vitro (e.g., Doherty et al. 1990; Furley et al. 1990; Gennarini et al. 1991; Stoeckli et al. 1991). However, neurite outgrowth promotion in a culture dish is a poor model for axon guidance in a developing organism. One of the first studies addressing the role of IgSF CAMs in axon guidance in vivo was carried out by Landmesser and colleagues (1988). They looked at the role of NgCAM/L1 and neural cell N(CAM) in hindlimb innervation in the chicken embryo. The intramuscular nerve branching pattern was thought to be regulated by CAM–CAM interactions, and, in particular, to be modulated by the polysialic acid modification of NCAM. Perturbation studies decreasing the axon–axon interactions by interfering with NgCAM interactions were shown to enhance the number of side branches. In contrast, the injection of anti-NCAM antibodies interfering with axon–muscle interactions produced the expected opposite result, a reduction in nerve branching and a tendency to more fasciculated axon growth (reviewed in Stoeckli and Landmesser 1998a).

2.2
Axon Guidance Across the Midline of the Nervous System

One of the best-studied pathway choices is the crossing of the ventral midline of the spinal cord by commissural axons (Stoeckli and Landmesser 1998b). Commissural neurons which are localized in the dorsolateral spinal cord, close to the dorsal root entry zone, project their axons ventromedially toward the floor plate, a triangular structure forming the ventral midline of the spinal cord. After crossing the midline, commissural axons turn rostrally into the longitudinal axis of the spinal cord, still keeping contact with the floor-plate border (Fig. 2). A number of in vitro and in vivo studies have contributed to our current view of commissural axon pathfinding (Kaprielian et al. 2000, 2001).

From in vitro studies, using spinal cord explants, we know that bone morphogenetic proteins (BMPs) secreted by the roof plate contribute to the initial ventral extension of commissural axons (Augsburger et al. 1999). Netrin-1, the first chemoattractant that was characterized at the molecular level

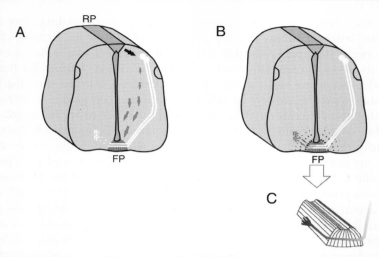

Fig. 2A–C. A cooperation of long-range (A) and short-range (B) guidance cues is necessary for commissural axon guidance. A long-range repellent, BMP7 (*black double-headed arrow*) is keeping commissural axons from approaching the roof plate (*RP*). Netrin-1 (*gray arrows*) is derived from the floor plate (*FP*) and attracts commissural axons toward the ventral midline. The specific pathway taken by the commissural axons is determined by short-range guidance cues (B). On their way to the floor plate, they are fasciculated due to the interactions of axonin-1 and NgCAM that are expressed on the axonal surface. Commissural axons enter the floor plate due to the interaction between axonin-1 on the growth cone and NrCAM expressed by the floor plate. The growth cone's contact with the floor plate results in the up-regulation of Robo (C) on the growth cone surface. The presence of Robo allows the growth cone to detect Slit, the repellent activity associated with the floor plate that is deposited in the extracellular matrix (*gradient indicated by black dots*) and, at the same time, it silences the attraction to netrin. After midline crossing, commissural axons turn rostrally into the longitudinal axis still keeping contact with the floor-plate surface

(Kennedy et al. 1994; Serafini et al. 1996), is secreted by the floor plate and attracts commissural axons ventrally (Serafini et al. 1996; Leonardo et al. 1997). In addition to these long-range guidance cues, commissural axons are guided by a variety of short-range guidance cues belonging to the IgSF CAMs. By in vivo studies, axonin-1 and NgCAM were shown to be involved in the fasciculation of commissural axons on their way to the floor plate (Stoeckli and Landmesser 1995). The interaction between axonin-1 and NgCAM-related cell adhesion molecule (NrCAM) that is expressed by floor-plate cells is required for commissural axons' entry into the floor plate, and thus, for midline crossing (Stoeckli and Landmesser 1995). A more detailed analysis of the events following the initial contact between commissural growth cones and floor-plate cells with time-lapse video microscopy revealed a negative signal derived from floor-plate cells (Stoeckli et al. 1997).

Such a negative signal derived from the midline of the nervous system had been identified also in genetic studies in *Drosophila* (reviewed in Tear et al. 1993). Clearly, axon guidance across the midline is determined by a balance between positive and negative signals derived from growth cone–floor plate interactions (reviewed in Stoeckli and Landmesser 1998b; Kaprielian et al. 2000, 2001). The molecular nature of the negative cue(s) associated with the midline remained elusive for quite some time even after Robo (Roundabout), the receptor mediating repellent signals derived from the midline, had been identified in a screen for guidance defects in the *Drosophila* ventral nerve cord (Seeger et al. 1993). Finally, the extracellular matrix protein Slit was identified as a ligand for Robo (Kidd et al. 1999). Due to the progress in genomics, both Robo and Slit were soon identified in vertebrates (Kidd et al. 1998a; Yuan et al. 1999) and additional family members were found in *Drosophila*. Two Slits and three Robos have been identified (reviewed by Wong et al. 2002). A direct interaction of Robo and Slit was shown for their vertebrate orthologs (Brose et al. 1999).

Genetic analyses in *Drosophila* indicated a link between the expression levels of Robo and Comm (Commissureless). Overexpression of *comm* resulted in a *robo* loss-of-function phenotype (Kidd et al. 1998b). Studies in two labs have recently led to the conclusion that Comm is regulating the expression of Robo at the posttranslational level (Myat et al. 2002; Keleman et al. 2002). In the presence of Comm, Robo was not inserted into the plasma membrane but shuttled directly to the late endosomes or lysosomes where it was degraded. Thus, axons expressing Comm could cross the midline because Robo levels were kept low during the time window of Comm expression, which was shown to coincide with midline crossing (Keleman et al. 2002). After midline crossing, Comm is down-regulated, resulting in the observed high expression levels of Robo in distal axon segments (Kidd et al. 1998a,b). In a yeast two-hybrid screen Nedd4, a ubiquitin ligase, was found to bind to the intracellular domain of Comm (Myat et al. 2002). The expression of mutant forms of Nedd4 revealed a requirement for an interaction of Comm and Nedd4 for the regulation of Robo levels.

Because Comm has not been found in vertebrates, the expression of Robo has to be regulated by a different mechanism. One possibility would be the regulation of Robo insertion via regulation of vesicle fusion. Evidence for a role of regulated vesicle fusion in commissural axon guidance has been found by in vivo studies using the chicken embryo as a model system (Pekarik et al. 2003). In these studies, a Rab GDI (Rab guanine nucleotide dissociation inhibitor) was found to be up-regulated when commissural axons crossed the midline. In the absence of Rab GDI, commissural axons stalled within the floor plate and failed to reach the contralateral border. This observation would be in agreement with the hypothesis that the expression of Robo is required to prevent commissural axons from lingering in the floor plate and to make them exit the floor plate on the contralateral side

(Fig. 2). Further support for the model that requires insertion of new receptors for guidance cues during floor-plate crossing was provided by the finding that postcommissural but not precommissural axons responded to Slit and Semaphorins (Zou et al. 2000) and that Ephrins and their receptors played a role in commissural axon guidance after midline crossing (Imondi et al. 2000).

Detailed analyses in *Drosophila* have refined our view about the roles of Slits and Robos in commissural axon guidance. Conclusions drawn from experiments carried out in the labs of Dickson and Goodman indicated a combinatorial code of Robo family members that determined the longitudinal fascicle to be followed by subpopulations of commissural axons in the *Drosophila* ventral nerve cord (Rajagopalan et al. 2000a,b; Simpson et al. 2000a,b). Whereas Robo was found to be important for the regulation of midline crossing, combinations of Robo2 and Robo3 in addition to Robo were determining the distance of the longitudinal fascicle that was joined after midline crossing (reviewed by Guthrie 2001).

The last chapter in the analysis of Robo function may not be written, however. In a recent study, evidence for a homophilic interaction of Robo family members that could promote neurite outgrowth was presented (Hivert et al. 2002). Although this may seem surprising in view of Robo's role as a receptor for the repellent activity of Slit, it is compatible with the results described in these studies. It remains to be tested in vivo whether Robo has a dual function at the midline.

2.3
Midline Contact Elicits a Switch in Growth Cones' Response from Attraction to Repulsion

The shift in the balance between positive and negative cues underlying pathfinding of commissural axons at the ventral midline was also observed in an in vitro preparation of the rat metencephalon. Murakami and colleagues (Shirasaki et al. 1998) showed that after contact with the floor plate, commissural axons lost responsiveness to netrin, although there was no change in the expression level of the IgSF CAM Dcc (Deleted in colorectal cancer), the receptor mediating attractive responses to netrin (Keino-Masu et al. 1996; Fazeli et al. 1997). An explanation for this observation was found later in experiments using an in vitro turning assay with *Xenopus* spinal neurons (Stein and Tessier-Lavigne 2001). A direct physical interaction between the cytoplasmic domains of the Slit receptor Robo and the netrin receptor Dcc was found to silence the attractive response to netrin. Thus, signals derived from molecular interactions between growth cone receptors and midline-derived guidance cues are not just added up. Rather some molecular interactions can actively switch off other signaling pathways (Sect. 7).

A similar switch in Dcc-mediated response to netrin from attraction to repulsion has been found when Unc-5 was coexpressed with Dcc (Hong et al. 1999). In both cases, a conformational change induced by ligand binding was suggested as a possible mechanism for alterations in the molecular interactions which in turn resulted in changes in the associated signaling pathway. Thus, conformational changes do not only occur in the extracellular domains of IgSF CAMs in response to *cis*-interactions (see Sect. 7 and Sonderegger et al. 2000) but seem to be a recurrent means to change CAM-associated signaling.

2.4
The Interaction Between Dcc and Unc-5 Induces a Switch from Short- to Long-Range Effects

A change in clustering, i.e., *cis*-interactions between the cytoplasmic domains of Unc-5 and Dcc were suggested as the underlying mechanism for the induction of short-range versus long-range repulsive effects of Unc-5 (Keleman and Dickson 2001; see also Sect. 7.5). According to this model, the formation of Dcc homodimers would mediate attraction, the formation of Unc5 homodimers would elicit short-range repulsion, whereas the formation of Dcc/Unc5 heterodimers would be responsible for long-range repulsion in response to netrin.

2.5
IgSF CAMs Play a Role in Subpopulation-Specific Guidance of Sensory Afferents

Using function-blocking antibodies to interfere with IgSF CAM function in the embryonic chicken spinal cord in vivo revealed a role of axonin-1 and F11/contactin in the guidance of sensory afferents to and along the longitudinal axis of the spinal cord (Perrin et al. 2001). Blocking the interactions of F11/ contactin resulted in an aberrant morphology of the dorsal root and the dorsal root entry zone (DREZ). The latter was shown to be caused by an erroneous ventral turn upon contact with the spinal cord. In addition, some fibers failed to extend along the longitudinal axis of the spinal cord. In the absence of axonin-1 function the dorsal gray matter was innervated prematurely. In contrast to later stages (Fitzgerald et al. 1993; Messersmith et al. 1995), the repellent sema3A was not restricted to the ventral spinal cord during early stages but was found throughout the gray matter, and thus, to prevent premature innervation of the dorsal horn (Fu et al. 2000). Neuropilin-1, a component of the sema3A receptor, was shown to bind to L1/NgCAM (Castellani et al. 2000; reviewed by Castellani and Rougon 2002), which in turn was also shown to bind to axonin-1 in *cis* (Buchstaller et al. 1996; Kunz et al. 1996; Stoeckli et al. 1996; Kunz et al. 1998). Thus, it remains to be tested whether a change in the *cis*-clustering of these components may be linked to the observed premature entry of sensory axons into the dorsal gray matter.

Both axonin-1 and F11 were demonstrated to play a role also at later stages of development of sensory afferents when collaterals of the primary axons were navigating to their specific target layers in the gray matter of the spinal cord (Perrin et al. 2001). In agreement with its expression pattern, axonin-1 was found to be required for nociceptive fibers to correctly target the dorsal-most layers. F11, which is predominantly expressed by proprioceptive collaterals, was shown to be required for their extension to the ventral horn, where they contact motoneurons. Surprisingly, both NrCAM and NgCAM were found to have very specific effects on subpopulation-specific sensory axon pathfinding (Perrin et al. 2001), although they can bind to both axonin-1 and F11 in addition to their homophilic binding capacity (Kuhn et al. 1991; Brümmendorf et al. 1993; Morales et al. 1993; Suter et al. 1995; Fitzli et al. 2000) and although NrCAM and NgCAM/L1 are widely expressed in both the gray and the white matter of the spinal cord during the time of collateral growth. NrCAM loss-of-function phenotypes were similar to those seen after blocking F11 interactions, resulting in the failure of proprioceptive collateral extension to the ventral spinal cord. The loss of NgCAM function mimicked the effects of axonin-1 perturbation, resulting in aberrant growth of nociceptive collaterals to more medial and more ventral layers of the gray matter.

3
IgSF CAMs in Cell Migration

Cell migration is a crucial step in the development of the nervous system. Neurons are born in germinal zones, far away from their final position in the mature nervous system (Hatten 1999; Marín and Rubenstein 2003). Two different modes of cell migration have been discriminated: radial and tangential migration. Whereas radial migration depends on a close association of the migrating cell with radial glia cells, tangential migration is either independent of mechanical guidance support or loosely follows axon tracts (Marín and Rubenstein 2001; Hatten 2002). Still, both types of migration share the requirement for cell–cell contact and molecular guidance cues for the navigation of migrating cells to their target. Recent findings suggest that the difference between radial and tangential migration may not be as fundamental as initially thought (Marín and Rubenstein 2003). In support of this, a recent study demonstrated a role of Reelin in both radial and chain migration in the rostral migratory stream (Frotscher 1998; Hack et al. 2002). Furthermore, cells can use both modes of migration during different phases of development. For instance, granule cells of the cerebellum get to their final position by tangential migration of their precursor cells from the rhombic lip to the cerebellar anlage. After differentiation to postmitotic granule cells, they undergo radial migration along processes of Bergmann glia cells to reach their final position in the internal granule cell layer (Hatten 1999).

Cell migration has been studied mostly in the cerebral cortex and in the cerebellum. However, evidence for the involvement of IgSF CAMs in cell migration comes also from other systems. In the nematode *Caenorhabditis elegans* mutations in unc-5, unc-6, and unc-40 were shown to result in cell migration defects in addition to the axon guidance defects that are well known from the mutation of their vertebrate homologs Unc-5, netrin, and Dcc, respectively (Leonardo et al. 1997; Kennedy 2000; reviewed by Wadsworth 2002). Similarly, the Slits were identified as repellents for axons expressing receptors of the Robo family (Brose et al. 1999; Li et al. 1999; see above) and they were shown to influence cell migration (reviewed in Park et al. 2002; Wong et al. 2002) providing evidence for the fact that cues involved in axon guidance can also be used as guidance cues by migrating cells (Tsai and Miller 2002). Slit derived from the septum was implicated as a repellent for migrating neuronal precursor cells from the anterior subventricular zone (SVZa; Hu and Rutishauser 1996; Wu et al. 1999). Furthermore, GABAergic neurons were shown to use Slit for their migration from the striatum to the neocortex (Zhu et al. 1999). However, in contrast to the effect of Slit on axon guidance, evidence for a repellent activity of Slit on migrating vertebrate cells has not yet been demonstrated directly in vivo (Bagri et al. 2002; Plump et al. 2002). This may be a problem of redundancy. In *C. elegans* with only one Slit gene, a cell migration defect is found in loss-of-function mutants (Hao et al. 2001).

An effect of the axon guidance molecule netrin-1 in cell migration has been demonstrated in the cerebellum (Alcántara et al. 2000). An attractive effect on netrin on precerebellar neurons appears to be mediated by Dcc (Fazeli et al. 1997; Alcántara et al. 2000). A repulsive effect on cell migration in cerebellar development is in line with the phenotype of mice lacking unc-5 genes (reviewed by Leonardo et al. 1997).

Another similarity between axon guidance and cell migration is the contribution of both repulsive and attractive cues. For instance, neuroblasts migrating from the subventricular zone to the olfactory bulb in the rostral migratory stream (RMS) were shown to depend on both attractive cues and the repulsive effect of Slit (see above). Netrin-1 is expressed in cells of the olfactory bulb whereas its receptors Dcc and neogenin are expressed in migrating cells (Murase and Horwitz 2002). Evidence from NCAM-deficient mice suggested that the polysialic acid (PSA) moiety of NCAM played a role in chain migration in the RMS (Chazal et al. 2000). A role for netrin and Dcc in the migration of pontine cells from their dorsal origin toward the ventral midline were demonstrated both in vitro and in vivo, providing evidence for the common role of these molecules in axon guidance and cell migration (Yee et al. 1999).

Tangential migration was described in the developing cortex and suggested to depend to some degree on axons (O'Rourke et al. 1995). In support of this, blocking TAG-1 function in cortical slices by the presence of function-blocking antibodies was shown to interfere with GABAergic interneuron migration along corticofugal fibers (Denaxa et al. 2001).

4
IgSF CAMs in Cerebellar Development

The analysis of mouse mutants lacking both NrCAM and L1/NgCAM revealed changes in cerebellar development that were not detectable in mice lacking only one of the two genes (Sakurai et al. 2001). The thickness of both the external granule cell layer (EGL) and the internal granule cell layer (IGL) were reduced in double-knockout mice compared to littermates that were lacking NrCAM but heterozygous for L1. As the IGL was affected more strongly than the EGL, a defect in granule cell migration was suggested as a possible explanation. Further support for this explanation was provided by the ectopic localization of Zic2-positive granule cells in the inner-most layer of the EGL suggesting that cell migration rather than differentiation was affected. In agreement with this, the expression of TAG-1 was not changed in double knockouts. TAG-1 was generally accepted as a marker for young postmitotic granule cells (Kuhar et al. 1993). However, recent studies comparing the expression of TAG-1 and markers for cell proliferation indicated that TAG-1 was expressed already before granule cells were postmitotic (Bizzoca et al. 2003).

A reduction in cerebellar size was also found in mice expressing F3/contactin under the control of the TAG-1 promoter (Bizzoca et al. 2003). As mentioned above, TAG-1 is a marker for very early stages of granule cell differentiation. Therefore, the TAG-1 promoter was used to drive expression of F3/contactin prematurely in the EGL. Interestingly, the decrease in cerebellar size found in these transgenic mice was only transient. The difference was most pronounced between postnatal day (P)6 and P11. By P30, the size of the cerebellum in transgenic and wild-type mice could no longer be distinguished. Due to a detailed quantitative analysis of cell proliferation and programmed cell death, the authors could demonstrate that the effect on cerebellar size was due to a transient decrease in cell proliferation in the EGL and a period of enhanced cell death in the IGL. Interestingly, the transient nature of the effect on cerebellar size correlated with the regulation of F3/contactin protein levels that were not reflected in the expression level of F3/contactin mRNA. Thus, future studies will be required to address the mechanism underlying the regulation of F3 protein.

A role for the IgSF CAM F3/F11/contactin in granule cell development had been described earlier in vitro (Buttiglione et al. 1998) and in vivo (Berglund et al. 1999). Mice lacking F3/contactin were severely ataxic and died during the first 3 weeks of life. In addition to the defects in cerebellar organization, defects in proprioceptive neural circuits in the spinal cord as detected in the chicken spinal cord may contribute to the ataxia (Perrin et al. 2001). Furthermore, F3/contactin was found to be crucial for the assembly of the node of Ranvier (Boyle et al. 2001) and to play a role in the clustering of Na^+ channels at this site (Kazarinova-Noyes et al. 2001). For its function as organizer, F3/contactin was shown to undergo *cis*-interactions with Caspr (Rios et al. 2000) and *trans*-interactions with neurofascin (Tait et al. 2000; Charles et al. 2002; reviewed by Falk et al. 2002).

5
IgSF CAMS in Synaptogenesis

Synapses are the business center of communication, the site where information is transferred from one neuron to the next. Synapses are asymmetric sites of cell–cell contact with a release site, the presynaptic bouton, on one side of the synaptic cleft and the acceptor site, the postsynaptic specialization, on the receiving cell on the other side of the cleft. For many years, synapse formation was mainly studied at the neuromuscular junction, the special peripheral synapse formed between motoneurons and muscle cells (reviewed by Sanes and Lichtman 2001). More recently, studies in the CNS have been carried out at the cellular and molecular level (reviewed by Garner et al. 2002; see references therein).

Synapse formation is a multi-step process that is initiated by the contact of a growth cone with a target cell and followed by the differentiation of the pre- and postsynaptic specialization. Underneath these specializations at the active zone are scaffolds that organize the release site presynaptically and the receptor assembly in the postsynaptic density (Phillips et al. 2001; Sheng 2001). Thus, as seen for axon guidance and cell migration, finding and contacting the right partner is the recurrent theme in synapse formation. Not surprisingly, cell adhesion molecules play an essential role in synapse formation. However, so far, the members of the Ig superfamily have not been found to play a major part in this process, in contrast to cadherins (see Part 1, this volume), neurexins, and neuroligins (Garner et al. 2002).

Only recently, evidence for an involvement of IgSF CAMs in synapse formation was presented. The Sidekicks, a subgroup of IgSF CAMs expressed at synapses in the retina, are at the interface between synapse formation and axonal pathfinding (Yamagata et al. 2002). They can undergo homophilic interactions. Ectopic expression of *sidekicks* was sufficient to drive synapse formation in specific sublaminae of the inner plexiform layer, where retinal ganglion cell dendrites and axons of bipolar and amacrine cells form synaptic contacts (Masland 2001). So far, loss-of-function experiments have not been done for the Sidekicks, therefore it is not known whether they are necessary for sublaminae-specific synapse formation. As described for other synaptic proteins, they have a PDZ-binding domain at their C-terminus. PDZ-binding domains are essential for the connection to the synaptic scaffold that organizes protein localization both pre- and postsynaptically (Sheng and Sala 2001; Garner et al. 2002).

A different approach was used for the identification of the SynCAMs. Vertebrate homologs of IgSF CAMs with PDZ domains were identified in a database search in analogy to the invertebrate system, where IgSF CAMs with PDZ domains, such as fasciclin II (Davis et al. 1997) and apCAM (Bailey et al. 1992) were shown to be involved in synapse formation. The SynCAMs (for synaptic cell adhesion molecules) were identified as multiple protein bands exclusively in lysates of brain tissue although their mRNA seemed to be distributed more widely (Biederer et al. 2002). The peak of SynCAM expression corresponds to the major period of synapse formation in the rat brain. Ectopic expression of SynCAM in non-neuronal cells was sufficient to induce synapse formation. When hippocampal neurons were co-cultured with 293 cells expressing SynCAM and the glutamate receptor GluR2, close to 30% of the cells exhibited spontaneous synaptic currents.

Synapse formation in vertebrates was described as a cooperation of the pre- and the postsynaptic cell (Sanes and Lichtman 1999). Recently, an additional player in synapse formation was identified by Shen and Bargmann in the nematode *C. elegans*. In a genetic study, they identified a signal from epithelial cells that was capable of inducing synapse formation, thus involving a signal from a third party in synapse formation (Shen and Bargmann 2003). Syg-1, a member of a new subgroup of the IgSF CAMs with four immunoglobulin domains, a transmembrane, and a cytoplasmic domain with a consensus sequence for PDZ binding, was thought to specify the location of the synapse.

6
IgSF CAMs in Synaptic Maintenance and Plasticity

Neuromuscular junctions in adult mice lacking all three isoforms of NCAM were found to be morphologically normal (Moscoso et al. 1998). Still, the analysis of their development revealed a difference in size and maturation, i.e., a delay in the elimination of the polyneuronal innervation (Moscoso et al. 1998; Rafuse et al. 2000). In detailed functional analyses, Landmesser and colleagues found that NCAM was essential for vesicle fusion at the presynaptic terminal, and thus for synaptic function (Polo-Parada et al. 2001). In knockout animals, the regulation of endo- and exocytosis of synaptic vesicles remained immature, resulting in continued vesicle fusion events outside of the presynaptic area. This in turn caused a lack in paired pulse facilitation and a profound depression in response to repetitive stimulation (Rafuse et al. 2000). Thus, NCAM appears to be much more important for synaptic function than for synapse formation.

These findings at the neuromuscular junction are in agreement with findings implicating NCAM in processes underlying synaptic plasticity in the CNS, such as memory and learning (reviewed by Kiss et al. 2001). In hippocampal slices, a requirement for NCAM and L1 was shown for the induction of long-term potentiation (LTP; Lüthi et al. 1994). In fact, removal of the polysialic acid modification of NCAM was sufficient to interfere with LTP induction (Muller et al. 1996). Contrary to expectations, a role of L1 in LTP could not be confirmed in vivo with two different protocols (Bliss et al. 2000). Still, a link of L1 to hippocampal LTP has been confirmed recently. Proteolytic cleavage of L1 was shown to be triggered by the synaptic activity-dependent activation of neuropsin, an extracellular serine protease expressed in the hippocampus (Chen et al. 1995; Matsumoto-Miyai et al. 2003). Experiments with NCAM knockout mice confirmed the findings from the slice cultures looking at the acute effect of loss of NCAM function (Cremer et al. 1994; Muller et al. 1996), although there were also reports about conflicting results (Holst et al. 1998).

7
Intracellular Signaling Associated with IgSF CAMs

A great number of studies have addressed the role of IgSF CAMs in cell–cell adhesion, neurite extension, axon guidance, cell migration, myelination, and disease. However, the majority of these studies looked at extracellular interactions of these proteins. A lot less is known about the nature of the intracellular signals that are initiated by these extracellular interactions.

7.1
Intracellular Signaling Associated with Transmembrane IgSF CAMs

The most straightforward mechanism of signal transduction for transmembrane IgSF CAMs is the direct activation of intracellular signaling pathways by their cytoplasmic tail (reviewed by Crossin and Krushel 2000). For instance, activation of the MAP kinase pathway was linked to NCAM (Schmid et al. 1999; Kolkova et al. 2000) and L1 (Schaefer et al. 1999; Schmid et al. 2000). Alternatively, indirect signaling was suggested by several studies that provided evidence for an association of NCAM and L1 with the FGFR in the plane of the plasma membrane (cis-interaction; Williams et al. 1994a,b; Saffell et al. 1997; reviewed by Dunican and Doherty 2000). The production of the lipid arachidonic acid in response to fibroblast growth factor (FGF) or CAM stimulation was shown to be involved in Ca^{2+}-dependent neurite outgrowth (Doherty et al. 2000; Dunican and Doherty 2000).

Obviously, signaling involves changes of the cytoskeleton that are directly involved in growth and guidance decisions. The actin cytoskeleton was given particular attention. A large number of proteins regulating polymerization/ depolymerization, stabilization, and branching of actin filaments have been described. The evidence linking extracellular cues to changes in actin cytoskeleton has been summarized in several specialized reviews (Mueller 1999; Korey and Van Vactor 2000; Dickson 2001; Luo 2002; Patel and Van Vactor 2002). In addition to affecting the cytoskeleton via small guanosine triphosphatase (GTPase)-dependent mechanisms, transmembrane IgSF CAMs, such as L1, neurofascin, and NrCAM can associate with the actin cytoskeleton via binding of ankyrin (Lustig et al. 2001; reviewed by Kamiguchi and Lemmon 2000a).

7.2
Intracellular Signaling Associated with Glycosylphosphatidyl-Inositol Anchored IgSF CAMs

More difficult to understand is the signal transduction of the glycosylphosphatidyl-inositol (GPI)-linked IgSF CAMs, such as TAG-1/axonin-1 and F3/F11/contactin. These molecules do not have direct access to proteins on the cytoplasmic side of the plasma membrane; still there are several reports about the association of GPI-linked molecules with the non-receptor tyrosine kinase fyn (Olive et al. 1995; Zisch et al. 1995; Kunz et al. 1996; Kunz et al. 1998; Kramer et al. 1999; reviewed by Crossin and Krushel 2000). Both axonin-1 and F3 could be co-immunoprecipitated with fyn. Furthermore, it was shown that the association between fyn and axonin-1 was dependent on the presence of a *cis*-interacting partner of axonin-1 (Kunz et al. 1996). Under experimental conditions that favored *cis*-interactions between NgCAM and axonin-1, axonin-1 was no longer associated with fyn, instead a caseinkinase II-related kinase co-immunoprecipitated with the axonin-1/NgCAM complexes. These findings raised several important issues. On the one hand, they point out that intracellular signaling associated with IgSF CAMs can only be assessed in the context of a specific situation when the full set of possible binding partners is available. Therefore, the ectopic expression of IgSF CAMs in cell lines is most likely not reflecting the full potential and the specificity of the signaling pathways in the in vivo situation. On the other hand, the results suggest that changes in signaling can occur rapidly and without the need for removal of a particular CAM from the cell surface. It is sufficient to provide IgSF CAMs with new binding partners in *cis*, and the choice of binding partners for *trans*-interactions and/or the intracellular signaling associated with a particular CAM can change dramatically (Sect. 7.3).

Along these lines, it is important to note that several reports indicate the importance of *cis*-interactions for the specificity of CAM signaling (e.g., Buchstaller et al. 1996; Stoeckli et al. 1996; Fitzli et al. 2000; Rios et al. 2000; Boyle et al. 2001; Charles et al. 2002; Falk et al. 2002).

7.3
Conformational Changes Are Sufficient to Induce a Switch in Intracellular Signaling Associated with IgSF CAMs

In an in vitro growth assay, commissural axons chose to grow on the mixed substratum when they were given a choice between stripes coated with a mixture of NrCAM and NgCAM or stripes coated with NgCAM only (Fitzli et al. 2000). Interestingly, growth cone morphology and size changed dramatically when growth cones chose to grow on the mixed substrate. However, this change was not induced by the nature of the substrate but by the fact that the growth cones had made a choice, as growth on a homogeneous substrate of mixed NgCAM/NrCAM did not result in large, complex growth cones. Similarly, when the choice was abolished by the addition of anti-axonin-1 antibodies to the medium, growth cone morphology did not change even though cells were exposed to alternating stripes of mixed NgCAM/NrCAM and NgCAM only. In vivo, changes in growth cone morphology can be observed along the trajectory of commissural axons. While growth cones advancing along fiber tracts exhibit a simple morphology, they assume a complex morphology at choice points, such as the floor plate or the optic chiasm (Bovolenta and Dodd 1990; Mason and Erskine 2000). Even though the choice between the two types of stripes in the in vitro assay was a rather artificial situation, it served as a model for the encounter of a choice point by growth cones. The sudden change in growth cone morphology at choice points is unlikely to be due to a change in surface expression of IgSF CAMs but rather reflects a change in signaling associated with the preexisting CAMs. Possible mechanisms for the switch in signaling are changes in *cis*-interacting partners in response to new *trans*-interacting binding partners encountered at the choice point, and, associated with this, conformational changes that are suggested based on structural studies of CAM–CAM interactions (Rader et al. 1996; Freigang et al. 2000; Kunz et al. 2002; reviewed by Sonderegger et al. 2000). Of particular interest are observations that *cis*- and *trans*-interactions of axonin-1 may cooperate in the enhancement of cell–cell contact areas (Kunz et al. 2002).

7.4
Changes in Signaling by Removal of IgSF CAMs from the Membrane

Although signaling associated with IgSF CAMs can be altered based on conformational changes without the need for protein removal from the membrane, internalization of CAMs was identified as a mechanism to regulate adhesion. The surface expression of L1 was shown to be dynamically regulated by intracellular trafficking (Kamiguchi et al. 1998; Kamiguchi and Lemmon 2000b; Kamiguchi and Yoshihara 2001). For this purpose, the cytoplasmic domain of L1 was shown to contain a sorting motif for clathrin-mediated internalization. Sophisticated labeling techniques were used to visualize the internalization of L1 in the central domain of growth cones advancing in an L1-dependent manner, its transport to the periphery, and its reinsertion into the plasma membrane at the leading edge.

More recently, the removal of another IgSF CAM, Robo, was shown to play a role in axon guidance across the midline (Sect. 2.2). Binding of the ubiquitin ligase DNedd4 to Commissureless (Comm) was required for Comm's ability to down-regulate Robo (Myat et al. 2002).

Another way to remove IgSF CAMs from the cell surface is their cleavage by metalloproteases (reviewed by McFarlane 2003). A disintegrin and metalloproteinase (ADAM)10, a member of the disintegrin metalloproteinase family, was shown to release L1/NgCAM from the cell surface (Gutwein et al. 2000, 2003), and thereby to influence L1-dependent cell migration (Mechtersheimer et al. 2001). Similarly, DCC, the receptor mediating the attractive response of commissural axons toward netrin (see below) was shown to be sensitive to metalloproteases (Galko and Tessier-Lavigne 2000). Genetic evidence for a role in metalloproteases in netrin function was found in *C. elegans* (Nishiwaki et al. 2000). Migration of the distal tip cells was affected in mig-17 mutants. MIG-17 is a member of the ADAM family. Furthermore, the *Drosophila* ortholog of ADAM10, kuzbanian, was suggested to play a role in the down-regulation of Robo that is required for commissural axons to cross the midline (Schimmelpfeng et al. 2001).

Biochemical studies indicated that axonin-1/TAG-1 could be released from the membrane by a glycosylphosphatidyl-inositol-specific phospholipase D (Lierheimer et al. 1997). Interestingly, the closely related glycosylphosphatidyl-anchored IgSF CAM F11/F3/contactin, was released with a much lower efficiency, reflecting the fact that soluble axonin-1 is the predominant form of axonin-1 (Stoeckli et al. 1989), whereas F11 is predominantly membrane-bound (Lierheimer et al. 1997). The biological significance of this difference is not known. Although soluble axonin-1 was competing for receptor binding with membrane-bound axonin-1 in vitro and in vivo (Stoeckli et al. 1991; Stoeckli and Landmesser 1995), it is not clear whether the specific release of axonin-1 is used as a mechanism to regulate axonin-1 function under physiological conditions in vivo, or whether soluble axonin-1 has some additional functions.

7.5
Silencing of IgSF CAM-Derived Signals

In explant cultures of the metencephalon, commissural axons lost their attractive response to the midline after they had been in contact with the floor plate (Shirasaki et al. 1998). At the time, this observation could not be explained at the molecular level, as no down-regulation of Dcc, the receptor mediating attraction in response to netrin, could be measured. However, more recently, an explanation was found using *Xenopus* spinal cord neurons for an in vitro guidance assay (Stein and Tessier-Lavigne 2001). Obviously, signals derived from guidance cues and their receptors are not simply integrated but perceived in a hierarchical order. The simultaneous exposure of growth cones to netrin and Slit completely abolished (silenced) attraction in response to netrin but not to other attractive cues, such as brain-derived neurotrophic factor (BDNF). An association between the cytoplasmic domains of Robo, the receptor for Slit, and Dcc, the receptor for netrin, was postulated as the mechanism underlying the silencing of attraction. In agreement with the observed behavior of growth cones, results from biochemical studies suggested that the interaction between the cytoplasmic domains of Robo and Dcc were only possible after a conformational change of Robo induced by Slit binding. Thus, the silencing of the attractive response in midline crossing represents another situation where the conformational change of an IgSF molecule is linked to a switch in signaling (Sect. 7.3).

A phenomenon related to silencing is the switch from attraction to repulsion that could be elicited in growth cones from *Xenopus* spinal cord neurons (Hong et al. 1999). Depending on the type of receptor expressed by the neuron, the growth cone was attracted to netrin or repelled by it. The expression of Dcc alone was responsible for the attractive response, whereas co-expression of Dcc and Unc-5 resulted in repulsion. Detailed analysis revealed a *cis*-interaction between Dcc and Unc-5 as the mechanims for the switch from attraction to repulsion (Hong et al. 1999; but see Keleman and Dickson 2001; Sect. 2.4).

A similar switch in responsiveness from attraction to repulsion in response to netrin was observed in vitro when levels of intracellular cyclic adenosine monophosphate (cAMP) were elevated (Ming et al. 1997; Song et al. 1997). The difference in response to a particular guidance cue by different cell types could thus be explained by a difference in the intracellular concentration of second messengers and does not necessarily depend on different surface receptors. In addition, the difference in cAMP or cyclic guanosine monophosphate (cGMP) levels could explain the observed difference in response to Sema3A by dendrites versus axon of the same neuron (Polleux et al. 2000; reviewed by Song and Poo 2001).

8
Outlook

Progress in our understanding of intracellular signaling mechanisms along with the elucidation of intra- and extracellular interactions of IgSF CAMs has provided us with a wealth of information about possible functions of IgSF CAMs in development and function of the brain. Still, many open questions remain to be addressed. Clearly, more in vivo studies will be required to understand the function of IgSF CAMs as the complexity and the versatility of their interactions and the associated changes in signaling cannot be mimicked with in vitro approaches. The availability of new strategies to produce regionally and temporally inducible knockout mice and alternative in vivo systems will allow us to tackle remaining issues and learn more about the function of IgSF CAMs.

References

Alcantara S, Ruiz M, De Castro F, Soriano E, Sotelo C (2000) Netrin 1 acts as an attractive or as a repulsive cue for distinct migrating neurons during the development of the cerebellar system. Development 127:1359–1372

Augsburger A, Schuchardt A, Hoskins S, Dodd J, Butler S (1999) BMPs as mediators of roof plate repulsion of commissural neurons. Neuron 24:127–141

Bagri A, Marin O, Plump AS, Mak J, Pleasure SJ, Rubenstein JL, Tessier-Lavigne M (2002) Slit proteins prevent midline crossing and determine the dorsoventral position of major axonal pathways in the mammalian forebrain. Neuron 33:233–248

Bailey CH, Chen M, Keller F, Kandel ER (1992) Serotonin-mediated endocytosis of apCAM: an early step of learning-related synaptic growth in Aplysia. Science 256:645–649

Bentley D, O'Connor TP (1992) Guidance and steering of peripheral pioneer growth cones in grasshopper embryos. In: Letourneau PC, Kater SB, Macagno ER (eds) The nerve growth cone. Raven Press, New York, pp 265–282

Berglund EO, Murai KK, Fredette B, Sekerkova G, Marturano B, Weber L, Mugnaini E, Ranscht B (1999) Ataxia and abnormal cerebellar microorganization in mice with ablated contactin gene expression. Neuron 24:739–750

Biederer T, Sara Y, Mozhayeva M, Atasoy D, Liu X, Kavalali ET, Sudhof TC (2002) SynCAM, a synaptic adhesion molecule that drives synapse assembly. Science 297:1525–1531

Bizzoca A, Virgintino D, Lorusso L, Buttiglione M, Yoshida L, Polizzi A, Tattoli M, Cagiano R, Rossi F, Kozlov S, Furley A, Gennarini G (2003) Transgenic mice expressing F3/contactin from the TAG-1 promoter exhibit developmentally regulated changes in the differentiation of cerebellar neurons. Development 130:29–43

Bliss T, Errington M, Fransen E, Godfraind JM, Kauer JA, Kooy RF, Maness PF, Furley AJ (2000) Long-term potentiation in mice lacking the neural cell adhesion molecule L1. Curr Biol 10:1607–1610

Bovolenta P, Dodd J (1990) Guidance of commissural growth cones at the floor plate in embryonic rat spinal cord. Development 109:435–447

Boyle ME, Berglund EO, Murai KK, Weber L, Peles E, Ranscht B (2001) Contactin orchestrates assembly of the septate-like junctions at the paranode in myelinated peripheral nerve. Neuron 30:385–397

Brackenbury R, Thiery JP, Rutishauser U, Edelman GM (1977) Adhesion among neural cells of the chick embryo. I. An immunological assay for molecules involved in cell-cell binding. J Biol Chem 252:6835–6840

Brose K, Bland KS, Wang KH, Arnott D, Henzel W, Goodman CS, Tessier-Lavigne M, Kidd T (1999) Slit proteins bind Robo receptors and have an evolutionarily conserved role in repulsive axon guidance. Cell 96:795–806

Brummendorf T, Hubert M, Treubert U, Leuschner R, Tarnok A, Rathjen FG (1993) The axonal recognition molecule F11 is a multifunctional protein: specific domains mediate interactions with Ng-CAM and restrictin. Neuron 10:711–727

Buchstaller A, Kunz S, Berger P, Kunz B, Ziegler U, Rader C, Sonderegger P (1996) Cell adhesion molecules NgCAM and axonin-1 form heterodimers in the neuronal membrane and cooperate in neurite outgrowth promotion. J Cell Biol 135:1593–1607

Buttiglione M, Revest JM, Pavlou O, Karagogeos D, Furley A, Rougon G, Faivre-Sarrailh C (1998) A functional interaction between the neuronal adhesion molecules TAG-1 and F3 modulates neurite outgrowth and fasciculation of cerebellar granule cells. J Neurosci 18:6853–6870

Castellani V, Rougon G (2002) Control of semaphorin signaling. Curr Opin Neurobiol 12:532–541

Castellani V, Chedotal A, Schachner M, Faivre-Sarrailh C, Rougon G (2000) Analysis of the L1-deficient mouse phenotype reveals cross-talk between Sema3A and L1 signaling pathways in axonal guidance. Neuron 27:237–249

Charles P, Tait S, Faivre-Sarrailh C, Barbin G, Gunn-Moore F, Denisenko-Nehrbass N, Guennoc AM, Girault JA, Brophy PJ, Lubetzki C (2002) Neurofascin is a glial receptor for the paranodin/Caspr-contactin axonal complex at the axoglial junction. Curr Biol 12:217–220

Chazal G, Durbec P, Jankovski A, Rougon G, Cremer H (2000) Consequences of neural cell adhesion molecule deficiency on cell migration in the rostral migratory stream of the mouse. J Neurosci 20:1446–1457

Chen ZL, Yoshida S, Kato K, Momota Y, Suzuki J, Tanaka T, Ito J, Nishino H, Aimoto S, Kiyama H, et al (1995) Expression and activity-dependent changes of a novel limbic-serine protease gene in the hippocampus. J Neurosci 15:5088–5097

Cremer H, Lange R, Christoph A, Plomann M, Vopper G, Roes J, Brown R, Baldwin S, Kraemer P, Scheff S, et al (1994) Inactivation of the N-CAM gene in mice results in size reduction of the olfactory bulb and deficits in spatial learning. Nature 367:455–459

Crossin KL, Krushel LA (2000) Cellular signaling by neural cell adhesion molecules of the immunoglobulin superfamily. Dev Dyn 218:260–279

Davis GW, Schuster CM, Goodman CS (1997) Genetic analysis of the mechanisms controlling target selection: target-derived Fasciclin II regulates the pattern of synapse formation. Neuron 19:561–573

Denaxa M, Chan CH, Schachner M, Parnavelas JG, Karagogeos D (2001) The adhesion molecule TAG-1 mediates the migration of cortical interneurons from the ganglionic eminence along the corticofugal fiber system. Development 128:4635–4644

Dickson BJ (2001) Rho GTPases in growth cone guidance. Curr Opin Neurobiol 11:103–110

Doherty P, Fruns M, Seaton P, Dickson G, Barton CH, Sears TA, Walsh FS (1990) A threshold effect of the major isoforms of NCAM on neurite outgrowth. Nature 343:464–466

Doherty P, Williams G, Williams EJ (2000) CAMs and axonal growth: a critical evaluation of the role of calcium and the MAPK cascade. Mol Cell Neurosci 16:283–295

Dunican DJ, Doherty P (2000) The generation of localized calcium rises mediated by cell adhesion molecules and their role in neuronal growth cone motility. Mol Cell Biol Res Commun 3:255–263

Falk J, Bonnon C, Girault JA, Faivre-Sarrailh C (2002) F3/contactin, a neuronal cell adhesion molecule implicated in axogenesis and myelination. Biol Cell 94:327–334

Fazeli A, Dickinson SL, Hermiston ML, Tighe RV, Steen RG, Small CG, Stoeckli ET, Keino-Masu K, Masu M, Rayburn H, Simons J, Bronson RT, Gordon JI, Tessier-Lavigne M, Weinberg RA (1997) Phenotype of mice lacking functional Deleted in colorectal cancer (Dcc) gene. Nature 386:796–804

Fitzgerald M, Kwiat GC, Middleton J, Pini A (1993) Ventral spinal cord inhibition of neurite outgrowth from embryonic rat dorsal root ganglia. Development 117:1377–1384

Fitzli D, Stoeckli ET, Kunz S, Siribour K, Rader C, Kunz B, Kozlov SV, Buchstaller A, Lane RP, Suter DM, Dreyer WJ, Sonderegger P (2000) A direct interaction of axonin-1 with NgCAM-related cell adhesion molecule (NrCAM) results in guidance, but not growth of commissural axons. J Cell Biol 149:951–968

Fransen E, Lemmon V, Van Camp G, Vits L, Coucke P, Willems PJ (1995) CRASH syndrome: clinical spectrum of corpus callosum hypoplasia, retardation, adducted thumbs, spastic paraparesis and hydrocephalus due to mutations in one single gene, L1. Eur J Hum Genet 3:273–284

Freigang J, Proba K, Leder L, Diederichs K, Sonderegger P, Welte W (2000) The crystal structure of the ligand binding module of axonin-1/TAG-1 suggests a zipper mechanism for neural cell adhesion. Cell 101:425–433

Frotscher M (1998) Cajal-Retzius cells, Reelin, and the formation of layers. Curr Opin Neurobiol 8:570–575

Fu SY, Sharma K, Luo Y, Raper JA, Frank E (2000) SEMA3A regulates developing sensory projections in the chicken spinal cord. J Neurobiol 45:227–236

Furley AJ, Morton SB, Manalo D, Karagogeos D, Dodd J, Jessell TM (1990) The axonal glycoprotein TAG-1 is an immunoglobulin superfamily member with neurite outgrowth-promoting activity. Cell 61:157–170

Galko MJ, Tessier-Lavigne M (2000) Function of an axonal chemoattractant modulated by metalloprotease activity. Science 289:1365–1367

Garner CC, Zhai RG, Gundelfinger ED, Ziv NE (2002) Molecular mechanisms of CNS synaptogenesis. Trends Neurosci 25:243–251

Gennarini G, Durbec P, Boned A, Rougon G, Goridis C (1991) Transfected F3/F11 neuronal cell surface protein mediates intercellular adhesion and promotes neurite outgrowth. Neuron 6:595–606

Goodman CS, Shatz CJ (1993) Developmental mechanisms that generate precise patterns of neuronal connectivity. Cell 72 Suppl: 77–98

Goodman CS, Bastiani MJ, Doe CQ, du Lac S, Helfand SL, Kuwada JY, Thomas JB (1984) Cell recognition during neuronal development. Science 225:1271–1279

Guthrie S (2001) Axon guidance: Robos make the rules. Curr Biol 11: R300–303

Gutwein P, Oleszewski M, Mechtersheimer S, Agmon-Levin N, Krauss K, Altevogt P (2000) Role of Src kinases in the ADAM-mediated release of L1 adhesion molecule from human tumor cells. J Biol Chem 275:15490–15497

Gutwein P, Mechtersheimer S, Riedle S, Stoeck A, Gast D, Joumaa S, Zentgraf H, Fogel M, Altevogt DP (2003) ADAM10-mediated cleavage of L1 adhesion molecule at the cell surface and in released membrane vesicles. Faseb J 17:292–294

Hack I, Bancila M, Loulier K, Carroll P, Cremer H (2002) Reelin is a detachment signal in tangential chain-migration during postnatal neurogenesis. Nat Neurosci 5:939–945

Hao JC, Yu TW, Fujisawa K, Culotti JG, Gengyo-Ando K, Mitani S, Moulder G, Barstead R, Tessier-Lavigne M, Bargmann CI (2001) C. elegans slit acts in midline, dorsal-ventral, and anterior-posterior guidance via the SAX-3/Robo receptor. Neuron 32:25–38

Hatten ME (1999) Central nervous system neuronal migration. Annu Rev Neurosci 22:511–539

Hatten ME (2002) New directions in neuronal migration. Science 297:1660–1663

Hivert B, Liu Z, Chuang CY, Doherty P, Sundaresan V (2002) Robo1 and Robo2 are homophilic binding molecules that promote axonal growth. Mol Cell Neurosci 21:534–545

Holst BD, Vanderklish PW, Krushel LA, Zhou W, Langdon RB, McWhirter JR, Edelman GM, Crossin KL (1998) Allosteric modulation of AMPA-type glutamate receptors increases activity of the promoter for the neural cell adhesion molecule, N-CAM. Proc Natl Acad Sci U S A 95:2597–2602

Hong K, Hinck L, Nishiyama M, Poo MM, Tessier-Lavigne M, Stein E (1999) A ligand-gated association between cytoplasmic domains of UNC5 and DCC family receptors converts netrin-induced growth cone attraction to repulsion. Cell 97:927–941

Hu H, Rutishauser U (1996) A septum-derived chemorepulsive factor for migrating olfactory interneuron precursors. Neuron 16:933–940

Imondi R, Wideman C, Kaprielian Z (2000) Complementary expression of transmembrane ephrins and their receptors in the mouse spinal cord: a possible role in constraining the orientation of longitudinally projecting axons. Development 127:1397–1410

Jessell TM (1988) Adhesion molecules and the hierarchy of neural development. Neuron 1:3–13

Kamiguchi H, Lemmon V (2000a) IgCAMs: bidirectional signals underlying neurite growth. Curr Opin Cell Biol 12:598–605

Kamiguchi H, Lemmon V (2000b) Recycling of the cell adhesion molecule L1 in axonal growth cones. J Neurosci 20:3676–3686

Kamiguchi H, Yoshihara F (2001) The role of endocytic l1 trafficking in polarized adhesion and migration of nerve growth cones. J Neurosci 21:9194–9203

Kamiguchi H, Long KE, Pendergast M, Schaefer AW, Rapoport I, Kirchhausen T, Lemmon V (1998) The neural cell adhesion molecule L1 interacts with the AP-2 adaptor and is endocytosed via the clathrin-mediated pathway. J Neurosci 18:5311–5321

Kaprielian Z, Imondi R, Runko E (2000) Axon guidance at the midline of the developing CNS. Anat Rec 261:176–197

Kaprielian Z, Runko E, Imondi R (2001) Axon guidance at the midline choice point. Dev Dyn 221:154–181

Kazarinova-Noyes K, Malhotra JD, McEwen DP, Mattei LN, Berglund EO, Ranscht B, Levinson SR, Schachner M, Shrager P, Isom LL, Xiao ZC (2001) Contactin associates with Na+ channels and increases their functional expression. J Neurosci 21:7517–7525

Keino-Masu K, Masu M, Hinck L, Leonardo ED, Chan SS, Culotti JG, Tessier-Lavigne M (1996) Deleted in Colorectal Cancer (DCC) encodes a netrin receptor. Cell 87:175–185

Keleman K, Dickson BJ (2001) Short- and long-range repulsion by the *Drosophila* Unc5 netrin receptor. Neuron 32:605–617

Keleman K, Rajagopalan S, Cleppien D, Teis D, Paiha K, Huber LA, Technau GM, Dickson BJ (2002) Comm sorts robo to control axon guidance at the *Drosophila* midline. Cell 110:415–427

Kennedy TE (2000) Cellular mechanisms of netrin function: long-range and short-range actions. Biochem Cell Biol 78:569–575

Kennedy TE, Serafini T, de la Torre JR, Tessier-Lavigne M (1994) Netrins are diffusible chemotropic factors for commissural axons in the embryonic spinal cord. Cell 78:425–435

Kenwrick S, Doherty P (1998) Neural cell adhesion molecule L1: relating disease to function. Bioessays 20:668–675

Kidd T, Brose K, Mitchell KJ, Fetter RD, Tessier-Lavigne M, Goodman CS, Tear G (1998a) Roundabout controls axon crossing of the CNS midline and defines a novel subfamily of evolutionarily conserved guidance receptors. Cell 92:205–215

Kidd T, Russell C, Goodman CS, Tear G (1998b) Dosage-sensitive and complementary functions of roundabout and commissureless control axon crossing of the CNS midline. Neuron 20:25–33

Kidd T, Bland KS, Goodman CS (1999) Slit is the midline repellent for the robo receptor in *Drosophila*. Cell 96:785–794

Kiss JZ, Troncoso E, Djebbara Z, Vutskits L, Muller D (2001) The role of neural cell adhesion molecules in plasticity and repair. Brain Res Brain Res Rev 36:175–184

Kolkova K, Novitskaya V, Pedersen N, Berezin V, Bock E (2000) Neural cell adhesion molecule-stimulated neurite outgrowth depends on activation of protein kinase C and the Ras-mitogen-activated protein kinase pathway. J Neurosci 20:2238–2246

Korey CA, Van Vactor D (2000) From the growth cone surface to the cytoskeleton: one journey, many paths. J Neurobiol 44:184–193

Kramer EM, Klein C, Koch T, Boytinck M, Trotter J (1999) Compartmentation of Fyn kinase with glycosylphosphatidylinositol-anchored molecules in oligodendrocytes facilitates kinase activation during myelination. J Biol Chem 274:29042–29049

Kuhar SG, Feng L, Vidan S, Ross ME, Hatten ME, Heintz N (1993) Changing patterns of gene expression define four stages of cerebellar granule neuron differentiation. Development 117:97–104

Kuhn TB, Stoeckli ET, Condrau MA, Rathjen FG, Sonderegger P (1991) Neurite outgrowth on immobilized axonin-1 is mediated by a heterophilic interaction with L1(G4). J Cell Biol 115:1113–1126

Kunz B, Lierheimer R, Rader C, Spirig M, Ziegler U, Sonderegger P (2002) Axonin-1/TAG-1 mediates cell–cell adhesion by a cis-assisted trans-interaction. J Biol Chem 277:4551–4557

Kunz S, Ziegler U, Kunz B, Sonderegger P (1996) Intracellular signaling is changed after clustering of the neural cell adhesion molecules axonin-1 and NgCAM during neurite fasciculation. J Cell Biol 135:253–267

Kunz S, Spirig M, Ginsburg C, Buchstaller A, Berger P, Lanz R, Rader C, Vogt L, Kunz B, Sonderegger P (1998) Neurite fasciculation mediated by complexes of axonin-1 and Ng cell adhesion molecule. J Cell Biol 143:1673–1690

Landmesser L, Dahm L, Schultz K, Rutishauser U (1988) Distinct roles for adhesion molecules during innervation of embryonic chick muscle. Dev Biol 130:645–670

Lemmon V, Burden SM, Payne HR, Elmslie GJ, Hlavin ML (1992) Neurite growth on different substrates: permissive versus instructive influences and the role of adhesive strength. J Neurosci 12:818–826

Leonardo ED, Hinck L, Masu M, Keino-Masu K, Fazeli A, Stoeckli ET, Ackerman SL, Weinberg RA, Tessier-Lavigne M (1997) Guidance of developing axons by netrin-1 and its receptors. Cold Spring Harb Symp Quant Biol 62:467–478

Li HS, Chen JH, Wu W, Fagaly T, Zhou L, Yuan W, Dupuis S, Jiang ZH, Nash W, Gick C, Ornitz DM, Wu JY, Rao Y (1999) Vertebrate slit, a secreted ligand for the transmembrane protein roundabout, is a repellent for olfactory bulb axons. Cell 96:807–818

Lierheimer R, Kunz B, Vogt L, Savoca R, Brodbeck U, Sonderegger P (1997) The neuronal cell-adhesion molecule axonin-1 is specifically released by an endogenous glycosylphosphatidylinositol-specific phospholipase. Eur J Biochem 243:502–510

Luo L (2002) Actin cytoskeleton regulation in neuronal morphogenesis and structural plasticity. Annu Rev Cell Dev Biol 18:601–635

Lustig M, Zanazzi G, Sakurai T, Blanco C, Levinson SR, Lambert S, Grumet M, Salzer JL (2001) Nr-CAM and neurofascin interactions regulate ankyrin G and sodium channel clustering at the node of Ranvier. Curr Biol 11:1864–1869

Luthi A, Laurent JP, Figurov A, Muller D, Schachner M (1994) Hippocampal long-term potentiation and neural cell adhesion molecules L1 and NCAM. Nature 372:777–779

Marin O, Rubenstein JL (2001) A long, remarkable journey: tangential migration in the telencephalon. Nat Rev Neurosci 2:780–790

Marin O, Rubenstein JL (2003) Cell migration in the forebrain. Annu Rev Neurosci 26:441–483

Masland RH (2001) The fundamental plan of the retina. Nat Neurosci 4:877–886

Mason C, Erskine L (2000) Growth cone form, behavior, and interactions in vivo: retinal axon pathfinding as a model. J Neurobiol 44:260–270

Matsumoto-Miyai K, Ninomiya A, Yamasaki H, Tamura H, Nakamura Y, Shiosaka S (2003) NMDA-dependent proteolysis of the presynaptic adhesion molecule L1 in the hippocampus by neuropsin. J Neurosci 23:7723–7736

McFarlane S (2003) Metalloproteases: carving out a role in axon guidance. Neuron 37:559–562

McNamee CJ, Reed JE, Howard MR, Lodge AP, Moss DJ (2002) Promotion of neuronal cell adhesion by members of the IgLON family occurs in the absence of either support or modification of neurite outgrowth. J Neurochem 80:941–948

Mechtersheimer S, Gutwein P, Agmon-Levin N, Stoeck A, Oleszewski M, Riedle S, Postina R, Fahrenholz F, Fogel M, Lemmon V, Altevogt P (2001) Ectodomain shedding of L1 adhesion molecule promotes cell migration by autocrine binding to integrins. J Cell Biol 155:661–673

Messersmith EK, Leonardo ED, Shatz CJ, Tessier-Lavigne M, Goodman CS, Kolodkin AL (1995) Semaphorin III can function as a selective chemorepellent to pattern sensory projections in the spinal cord. Neuron 14:949–959

Ming GL, Song HJ, Berninger B, Holt CE, Tessier-Lavigne M, Poo MM (1997) cAMP-dependent growth cone guidance by netrin-1. Neuron 19:1225–1235

Morales G, Hubert M, Brummendorf T, Treubert U, Tarnok A, Schwarz U, Rathjen FG (1993) Induction of axonal growth by heterophilic interactions between the cell surface recognition proteins F11 and Nr-CAM/Bravo. Neuron 11:1113–1122

Moscoso LM, Cremer H, Sanes JR (1998) Organization and reorganization of neuromuscular junctions in mice lacking neural cell adhesion molecule, tenascin-C, or fibroblast growth factor-5. J Neurosci 18:1465–1477

Mueller BK (1999) Growth cone guidance: first steps towards a deeper understanding. Annu Rev Neurosci 22:351–388

Muller D, Wang C, Skibo G, Toni N, Cremer H, Calaora V, Rougon G, Kiss JZ (1996) PSA-NCAM is required for activity-induced synaptic plasticity. Neuron 17:413–422

Murase S, Horwitz AF (2002) Deleted in colorectal carcinoma and differentially expressed integrins mediate the directional migration of neural precursors in the rostral migratory stream. J Neurosci 22:3568–3579

Myat A, Henry P, McCabe V, Flintoft L, Rotin D, Tear G (2002) *Drosophila* Nedd4, a ubiquitin ligase, is recruited by Commissureless to control cell surface levels of the roundabout receptor. Neuron 35:447–459

Nishiwaki K, Hisamoto N, Matsumoto K (2000) A metalloprotease disintegrin that controls cell migration in *Caenorhabditis elegans*. Science 288:2205–2208

O'Rourke NA, Sullivan DP, Kaznowski CE, Jacobs AA, McConnell SK (1995) Tangential migration of neurons in the developing cerebral cortex. Development 121:2165–2176

Olive S, Dubois C, Schachner M, Rougon G (1995) The F3 neuronal glycosylphosphatidylinositol-linked molecule is localized to glycolipid-enriched membrane subdomains and interacts with L1 and fyn kinase in cerebellum. J Neurochem 65:2307–2317

Park HT, Wu J, Rao Y (2002) Molecular control of neuronal migration. Bioessays 24:821–827

Patel BN, Van Vactor DL (2002) Axon guidance: the cytoplasmic tail. Curr Opin Cell Biol 14:221–229

Pekarik V, Bourikas D, Miglino N, Joset P, Preiswerk S, Stoeckli ET (2003) Screening for gene function in chicken embryo using RNAi and electroporation. Nat Biotechnol 21:93–96

Perrin FE, Rathjen FG, Stoeckli ET (2001) Distinct subpopulations of sensory afferents require F11 or axonin-1 for growth to their target layers within the spinal cord of the chick. Neuron 30:707–723

Phillips GR, Huang JK, Wang Y, Tanaka H, Shapiro L, Zhang W, Shan WS, Arndt K, Frank M, Gordon RE, Gawinowicz MA, Zhao Y, Colman DR (2001) The presynaptic particle web: ultrastructure, composition, dissolution, and reconstitution. Neuron 32:63–77

Plump AS, Erskine L, Sabatier C, Brose K, Epstein CJ, Goodman CS, Mason CA, Tessier-Lavigne M (2002) Slit1 and Slit2 cooperate to prevent premature midline crossing of retinal axons in the mouse visual system. Neuron 33:219–232

Polleux F, Morrow T, Ghosh A (2000) Semaphorin 3A is a chemoattractant for cortical apical dendrites. Nature 404:567–573

Polo-Parada L, Bose CM, Landmesser LT (2001) Alterations in transmission, vesicle dynamics, and transmitter release machinery at NCAM-deficient neuromuscular junctions. Neuron 32:815–828

Rader C, Kunz B, Lierheimer R, Giger RJ, Berger P, Tittmann P, Gross H, Sonderegger P (1996) Implications for the domain arrangement of axonin-1 derived from the mapping of its NgCAM binding site. Embo J 15:2056–2068

Rafuse VF, Polo-Parada L, Landmesser LT (2000) Structural and functional alterations of neuromuscular junctions in NCAM-deficient mice. J Neurosci 20:6529–6539

Rajagopalan S, Nicolas E, Vivancos V, Berger J, Dickson BJ (2000a) Crossing the midline: roles and regulation of Robo receptors. Neuron 28:767–777

Rajagopalan S, Vivancos V, Nicolas E, Dickson BJ (2000b) Selecting a longitudinal pathway: Robo receptors specify the lateral position of axons in the *Drosophila* CNS. Cell 103:1033–1045

Rios JC, Melendez-Vasquez CV, Einheber S, Lustig M, Grumet M, Hemperly J, Peles E, Salzer JL (2000) Contactin-associated protein (Caspr) and contactin form a complex that is targeted to the paranodal junctions during myelination. J Neurosci 20:8354–8364

Saffell JL, Williams EJ, Mason IJ, Walsh FS, Doherty P (1997) Expression of a dominant negative FGF receptor inhibits axonal growth and FGF receptor phosphorylation stimulated by CAMs. Neuron 18:231–242

Sakurai T, Lustig M, Babiarz J, Furley AJ, Tait S, Brophy PJ, Brown SA, Brown LY, Mason CA, Grumet M (2001) Overlapping functions of the cell adhesion molecules Nr-CAM and L1 in cerebellar granule cell development. J Cell Biol 154:1259–1273

Sanes JR, Lichtman JW (1999) Development of the vertebrate neuromuscular junction. Annu Rev Neurosci 22:389–442

Sanes JR, Lichtman JW (2001) Induction, assembly, maturation and maintenance of a postsynaptic apparatus. Nat Rev Neurosci 2:791–805

Schaefer AW, Kamiguchi H, Wong EV, Beach CM, Landreth G, Lemmon V (1999) Activation of the MAPK signal cascade by the neural cell adhesion molecule L1 requires L1 internalization. J Biol Chem 274:37965–37973

Schimmelpfeng K, Gogel S, Klambt C (2001) The function of leak and kuzbanian during growth cone and cell migration. Mech Dev 106:25–36

Schmid RS, Graff RD, Schaller MD, Chen S, Schachner M, Hemperly JJ, Maness PF (1999) NCAM stimulates the Ras-MAPK pathway and CREB phosphorylation in neuronal cells. J Neurobiol 38:542–558

Schmid RS, Pruitt WM, Maness PF (2000) A MAP kinase-signaling pathway mediates neurite outgrowth on L1 and requires Src-dependent endocytosis. J Neurosci 20:4177–4188

Seeger M, Tear G, Ferres-Marco D, Goodman CS (1993) Mutations affecting growth cone guidance in *Drosophila*: genes necessary for guidance toward or away from the midline. Neuron 10:409–426

Serafini T, Colamarino SA, Leonardo ED, Wang H, Beddington R, Skarnes WC, Tessier-Lavigne M (1996) Netrin-1 is required for commissural axon guidance in the developing vertebrate nervous system. Cell 87:1001–1014

Shen K, Bargmann CI (2003) The immunoglobulin superfamily protein SYG-1 determines the location of specific synapses in *C. elegans*. Cell 112:619–630

Sheng M (2001) Molecular organization of the postsynaptic specialization. Proc Natl Acad Sci U S A 98:7058–7061

Sheng M, Sala C (2001) PDZ domains and the organization of supramolecular complexes. Annu Rev Neurosci 24:1–29

Shirasaki R, Katsumata R, Murakami F (1998) Change in chemoattractant responsiveness of developing axons at an intermediate target. Science 279:105–107

Simpson JH, Bland KS, Fetter RD, Goodman CS (2000a) Short-range and long-range guidance by Slit and its Robo receptors: a combinatorial code of Robo receptors controls lateral position. Cell 103:1019–1032

Simpson JH, Kidd T, Bland KS, Goodman CS (2000b) Short-range and long-range guidance by slit and its Robo receptors. Robo and Robo2 play distinct roles in midline guidance. Neuron 28:753–766

Sonderegger P, Welte W, Stoeckli ET (2000) Sensing cues for axon guidance—from extracellular protein conformation to intracellular signalling. In: The ELSO Gazette: e-magazine of the European Life Scientist Organization (http://www.the-elso-gazette.org/magazines/reviews/review1.asp) Issue 1 (1 September, 2000)

Song H, Poo M (2001) The cell biology of neuronal navigation. Nat Cell Biol 3: E81–88

Song HJ, Ming GL, Poo MM (1997) cAMP-induced switching in turning direction of nerve growth cones. Nature 388:275–279

Sperry RW (1963) Chemoaffinity in the orderly growth of nerve fiber patterns and connections. Proc Natl Acad Sci USA 50:703–710

Stein E, Tessier-Lavigne M (2001) Hierarchical organization of guidance receptors: silencing of netrin attraction by slit through a Robo/DCC receptor complex. Science 291:1928–1938

Stoeckli ET, Landmesser LT (1995) Axonin-1, Nr-CAM, and Ng-CAM play different roles in the in vivo guidance of chick commissural neurons. Neuron 14:1165–1179

Stoeckli ET, Landmesser LT (1998a) Molecular mechanisms of growth cone guidance in the vertebrate nervous system. In: Sonderegger P (ed) Ig superfamily molecules in the nervous system, vol 6. Harwood Academic, Amsterdam, pp 161–181

Stoeckli ET, Landmesser LT (1998b) Axon guidance at choice points. Curr Opin Neurobiol 8:73–79

Stoeckli ET, Lemkin PF, Kuhn TB, Ruegg MA, Heller M, Sonderegger P (1989) Identification of proteins secreted from axons of embryonic dorsal-root-ganglia neurons. Eur J Biochem 180:249–258

Stoeckli ET, Kuhn TB, Duc CO, Ruegg MA, Sonderegger P (1991) The axonally secreted protein axonin-1 is a potent substratum for neurite growth. J Cell Biol 112:449–455

Stoeckli ET, Ziegler U, Bleiker AJ, Groscurth P, Sonderegger P (1996) Clustering and functional cooperation of Ng-CAM and axonin-1 in the substratum-contact area of growth cones. Dev Biol 177:15–29

Stoeckli ET, Sonderegger P, Pollerberg GE, Landmesser LT (1997) Interference with axonin-1 and NrCAM interactions unmasks a floor-plate activity inhibitory for commissural axons. Neuron 18:209–221

Suter DM, Pollerberg GE, Buchstaller A, Giger RJ, Dreyer WJ, Sonderegger P (1995) Binding between the neural cell adhesion molecules axonin-1 and Nr-CAM/Bravo is involved in neuron-glia interaction. J Cell Biol 131:1067–1081

Tait S, Gunn-Moore F, Collinson JM, Huang J, Lubetzki C, Pedraza L, Sherman DL, Colman DR, Brophy PJ (2000) An oligodendrocyte cell adhesion molecule at the site of assembly of the paranodal axo-glial junction. J Cell Biol 150:657–666

Tear G, Seeger M, Goodman CS (1993) To cross or not to cross: a genetic analysis of guidance at the midline. Perspect Dev Neurobiol 1:183–194

Thiery JP, Brackenbury R, Rutishauser U, Edelman GM (1977) Adhesion among neural cells of the chick embryo. II. Purification and characterization of a cell adhesion molecule from neural retina. J Biol Chem 252:6841–6845

Tsai HH, Miller RH (2002) Glial cell migration directed by axon guidance cues. Trends Neurosci 25:173–175; discussion 175–176

Wadsworth WG (2002) Moving around in a worm: netrin UNC-6 and circumferential axon guidance in C. elegans. Trends Neurosci 25:423–429

Weller S, Gartner J (2001) Genetic and clinical aspects of X-linked hydrocephalus (L1 disease): mutations in the L1CAM gene. Hum Mutat 18:1–12

Williams EJ, Furness J, Walsh FS, Doherty P (1994a) Activation of the FGF receptor underlies neurite outgrowth stimulated by L1, N-CAM, and N-cadherin. Neuron 13:583–594

Williams EJ, Furness J, Walsh FS, Doherty P (1994b) Characterisation of the second messenger pathway underlying neurite outgrowth stimulated by FGF. Development 120:1685–1693

Wong K, Park HT, Wu JY, Rao Y (2002) Slit proteins: molecular guidance cues for cells ranging from neurons to leukocytes. Curr Opin Genet Dev 12:583–591

Wu W, Wong K, Chen J, Jiang Z, Dupuis S, Wu JY, Rao Y (1999) Directional guidance of neuronal migration in the olfactory system by the protein Slit. Nature 400:331–336

Yamagata M, Weiner JA, Sanes JR (2002) Sidekicks: synaptic adhesion molecules that promote lamina-specific connectivity in the retina. Cell 110:649–660

Yee KT, Simon HH, Tessier-Lavigne M, O'Leary DM (1999) Extension of long leading processes and neuronal migration in the mammalian brain directed by the chemoattractant netrin-1. Neuron 24:607–622

Yuan W, Zhou L, Chen JH, Wu JY, Rao Y, Ornitz DM (1999) The mouse SLIT family: secreted ligands for ROBO expressed in patterns that suggest a role in morphogenesis and axon guidance. Dev Biol 212:290–306

Zhu Y, Li H, Zhou L, Wu JY, Rao Y (1999) Cellular and molecular guidance of GABAergic neuronal migration from an extracortical origin to the neocortex. Neuron 23:473–485

Zisch AH, D'Alessandri L, Amrein K, Ranscht B, Winterhalter KH, Vaughan L (1995) The glypiated neuronal cell adhesion molecule contactin/F11 complexes with src-family protein tyrosine kinase Fyn. Mol Cell Neurosci 6:263–279

Zou Y, Stoeckli E, Chen H, Tessier-Lavigne M (2000) Squeezing axons out of the gray matter: a role for slit and semaphorin proteins from midline and ventral spinal cord. Cell 102:363–375

Part IV
Physiology and Pathology of Cell Adhesion

Part IV

Adhesion Mechanisms of Endothelial Cells

P. F. Bradfield (✉) · B. A. Imhof

Department of Pathology, Faculty of Medicine, CMU, University of Geneva,
1 rue Michel Servet, Geneva 4, CH1211 Switzerland
paul.bradfield@medecine.unige.ch

1	Immune Surveillance and Homeostasis: The Role of the Endothelium. . . .	406
2	Cell Adhesion Molecules	406
3	Targeting Leukocyte Adhesion	408
3.1	Primary Adhesion	409
3.2	Activation.	410
3.3	Secondary Adhesion	410
3.4	Diapedesis	410
4	Regulation of Cell Trafficking by Distinct Populations of Endothelial Cells .	413
4.1	Recirculation of Leukocytes: Deciphering the Role of the Lymphatic Endothelium	415
5	Inflammation and Wound Healing: Synergy Between the Endothelium and the Extra-vasculature.	416
6	Cell Adhesion Molecules, the Players in Mediating Endothelial Function . .	418
6.1	Selectins and Their Ligands	418
6.1.1	Selectins.	418
6.1.2	Selectin Ligands: Sialomucins and Carbohydrate Moieties	419
6.1.3	Function and Patterns of Expression	420
6.2	Integrins	421
6.3	The Immunoglobulin Superfamily	424
6.3.1	ICAM and VCAM	425
6.3.2	Mucosal Adressin Cell Adhesion Molecule 1	426
6.3.3	CD31	426
6.3.4	Junctional Adhesion Molecules	427
6.4	CD99	428
6.5	CD44	428
6.6	Vascular Adhesion Protein-1	429
References		429

Abstract Endothelial cells express a diverse and exquisite array of adhesion molecules and cell surface receptors. Adhesion molecules expressed on endothelial cells not only maintain structural integrity of the vasculature, but also mediate more dynamic processes such as the highly regulated movement of leukocytes from free flow into different tissue compartments. Recent studies have focused on the molecular processes that mediate endothelial cell function and their ability to respond rapidly to changes in their immediate microenvironment, as well as maintaining routine cell trafficking through specialist tissue compartments. Adhesion molecules expressed on the endothelium mediate the movement of leukocytes into the underlying extravasculature to mediate a diverse array of functions including immune effector responses, cellular interactions in specialist lymphatic microenvironments and recirculation back into the vasculature. The true diversity and capacity of adhesion molecules capable of being expressed on the endothelium is now beginning to emerge, demonstrating new levels of complexity as specialist subsets of endothelium are characterised that define specific, yet diverse functions. In this chapter, the role of cell adhesion molecules in mediating endothelial cell function is discussed, from how their different physiochemical properties contribute to function, to how specific ligand interactions expressed on leukocyte cell populations contribute to functions ranging from constitutive cell trafficking to inflammation.

Keywords Endothelium · Leukocyte · Migration · Vascular junctions · Adhesion molecules · Trafficking · Inflammation

1
Immune Surveillance and Homeostasis: The Role of the Endothelium

Immune responses are optimised by constant monitoring and tight control of cell positioning to ensure a rapid and effective immune response when required. Integral to immune surveillance is cell trafficking, a process that confines cells to different routes of cell movement, dependent on cell type and function. This also allows cells with specific functions to move from the extravasculature and exist within defined compartments, optimising relevant interactions and effector responses, but minimising encounters with other cell types. Cell trafficking also ensures that cells involved with immunosurveillance have a much higher degree of contact with a potential antigen, and are therefore able to initiate a more rapid effector response.

Leukocytes as well as other cell types express a number of cell surface molecules that are able to bind to a wide range of ligands expressed on endothelial cell and stromal cell populations which provide an 'address code', allowing a cell to determine its position as well as direction and function. These include cell adhesion molecules that can physically anchor cells and allow migration through tissue, and chemokines, which provide cells with information about direction leading to recruitment, proliferation, apoptosis and retention (Imhof and Dunon 1995; Butcher and Picker 1996; Campbell and Butcher 2000).

2
Cell Adhesion Molecules

The ability of an individual cell to assimilate information about its immediate environment is critical in maintaining the integrity of multicellular organisms. Co-ordinating information about adjacent cells and extracellular matrix protein (ECM) provides an important cue that regulates cell proliferation, apoptosis, differentiation and migration (Gumbiner and Yamada 1995). Such stimuli are therefore essential for the growth and development of complex multicellular organisms.

Cell adhesion in multicellular organisms has been known for many years. The extravasation of leukocytes from the vasculature into the surrounding tissue was observed by intravital microscopy over a century ago. The molecular nature of cell adhesion remained unclear for many decades. It was not until more recent times where the development of techniques in immunology, biochemistry and molecular biology allowed molecules involved in cell adhesion to be cloned and characterised. These studies showed that the interaction of a cell with adjacent cells or extracellular matrix is mediated through families of cell surface proteins, which were named cell adhesion molecules (CAMs).

CAMs are sub-divided into different families dependent on molecular structure as well as their physical nature of adhesion. Higher affinity interactions tend to be less dynamic and are involved with maintenance of tissue structure, whereas less avid interactions are more transient and are involved with leukocyte and endothelial cell biology. Classically, adhesion molecules have been defined as a cell surface receptor whose interaction with a ligand is capable of attaching a cell to another cell, or matrix substrate. This working definition was reinforced by observations that CAMs are attached to the cytoskeleton and engagement of ligand results in multimerisation and redistribution within the cell membrane.

The physical properties vary considerably between different adhesion molecules (Bruinsma 1997). Although affinity constants are low in comparison to growth factors and their receptors, it is generally considered more useful to measure on/off rates in defining specific characteristics of CAMs. CAMs were originally analysed by constructing recombinant chimeric proteins of the external domains to study the intricacies behind functional adhesion. Adhesion events were broadly categorised into homophilic events (binding to self) and heterophilic events (binding to non-self). CAMs were also capable of *cis*-interactions (binding on the same cell surface) and *trans*-interactions (binding onto adjacent cell membranes or to extracellular matrix).

Recently however, it has been shown that CAMs are capable of transmitting both mechanical and biochemical signals across the cell membrane (Buckley et al. 1998). Such signalling events demonstrate how engagement of CAMs by their ligand can trigger intracellular signalling cascades leading to proliferation, apoptosis, differentiation and migration. *Cis*-interactions in CAMs may therefore have a role in imposing order for other cell surface receptors by organising clustering and co-localisation, thus reinforcing cell adhesion and signalling events (Yap et al. 1997).

Engagement of different CAMs by their ligands may have synergistic effects on common intracellular pathways. This may explain why such apparently distinct CAMs as well as chemokine and growth factor receptors have to be engaged for a particular cell function to occur. This may occur by direct interactions, as with glycosylphosphatidyl inositol (GPI)-anchored proteins CD14, CD16b and CD87 associating with β_2 integrin Mac-1 to modulate signalling. Examples of more distal 'crosstalk' between different adhesion molecules are platelet/endothelial CAM (PECAM)-1 (CD31) modulating the rate of integrin-mediated adhesion and speed of migration (haptotaxis) of neutrophils on cultured endothelial monolayers by 'inside out' signalling (Rainger et al. 1997; Reedquist et al. 2000).

CAMs are ideally suited to behaving as mediators of intracellular signalling because of their role in mechanical movement and attachment. Such associations allow for directly physical and intimate interactions with the immediate environment of the cell. This combined synergistic effect on intracellular pathways from other CAMs as well as to chemokine and growth factors receptors, allows a cell to participate in a series of synchronised interactions leading to distinct changes in cell adhesion properties, enabling migration through the endothelial barrier. The distinct physiochemical and biological properties of these CAM families and how they participate in leukocyte recruitment from the vasculature will be considered in detail in a future section (Sect. 12).

3
Targeting Leukocyte Adhesion

Targeting leukocyte homing and migration from the vasculature into underlying tissue requires a process that can overcome the high shear stresses experienced in the vasculature and permit migration into a specific tissue or sites of inflammation. These adhesive interactions mediate the initial interactions with leukocytes in free flow, through to stabilisation and diapedesis, and into underlying tissues by a series of specialised adhesion pathways. This multistep cascade involves the sequential engagement of adhesion and signalling receptors in a combinatorial process providing a series of progressive 'decision-making' processes for extravasation, or return to circulation. Therefore co-ordinated expression of these molecules provides an 'area code' capable of discriminating against leukocyte sub-sets and mediating homing to particular tissues or sites of inflammation (Figs. 1 and 3) (Campbell et al. 1999b).

In this section, a brief overview is given of the multistep paradigm describing adhesion and signalling receptor cascades involved in leukocyte recruitment and the role of the endothelium, with more detailed descriptions given in later sections. The multistep adhesion cascade of leukocyte and en-

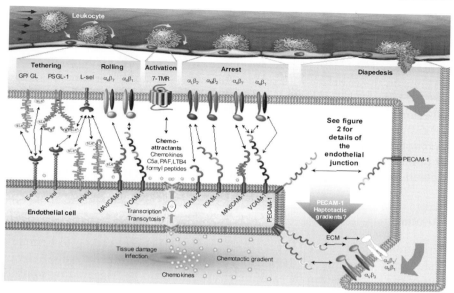

Fig. 1. The five steps of leukocyte recruitment from the vasculature. Adhesion molecules for leukocytes are detailed at the *top* and endothelial cells *below*. The *arrows* in the *middle portion* indicate the various interactions possible between molecules. Endothelial cells are responsible for the expression, or transportation of chemokines from the abluminal side to display on lumenal surfaces (transcytosis). Exposure to chemotactic agent on lumenal surfaces leads to activation of integrins expressed on leukocytes allowing firm adhesion (arrest) and bipolar migration (diapedesis) into the tissue. (Adapted from von Andrian and Mackay 2000)

dothelium interactions can be roughly divided into four or five overlapping steps; primary adhesion (tethering and rolling), activation, secondary adhesion (arrest) and diapedesis (Fig. 1).

3.1
Primary Adhesion

Primary adhesion describes the capturing of leukocytes from free flow and tethering them to the endothelium. This interaction is mediated by selectins and their ligands resulting in 'rolling' of the cell in the direction of flow along the vascular wall within minutes of an inflammatory stimulus. Interestingly other molecules such as α_4 integrins, CD44 and vascular adhesion protein (VAP)-1 have been observed to mediate rolling, but at much slower speeds (von Andrian and Mackay 2000). These molecules may participate in the transition from rolling to firm adhesion and provide additional cues for cell homing as part of the multistep cascade (Bargatze et al. 1995). Endothelium and platelets constitutively expresses P-selectin, which is stored within

Weibel–Palade bodies of endothelial cells and α granules of platelets. An inflammatory stimulus triggers rapid fusion with the lumenal cell membrane, allowing for the immediate capture and recruitment of leukocytes to the affected tissue (Sect. 18). Leukocytes in the circulation possess long thin protrusions on their surface with preferential expression of L-selectin and α_4 integrins, but exclude adhesion molecules not involved in the initial primary adhesion events (Sect. 18). The high on/off rates of selectin binding and spatial distribution on specialised membrane structures facilitates transient interactions with endothelium that reduces cell velocity from 4,000 μm/s to 40 μm/s, allowing interactions with immobilised chemokines or other chemotactic factors bound to the endothelial cell surface (Sect. 13).

3.2
Activation

Activation describes the triggering of integrin function, leading to firm adhesion and 'flattening' of the leukocyte against the endothelial wall. Activation of the integrins can be extremely rapid and can be triggered by soluble factors such as formyl-methionyl-leucyl-phenylalanine (fMLP), platelet-activating factor (PAF) and C5A, and chemokines (Sect. 4). Chemokines can be expressed by endothelium or transcytosed onto endothelial lumenal surfaces (Sect. 4) where binding of chemokines to chemokine receptors on the surface of the leukocyte leads to activation of G proteins and the activation of integrins (Sect. 17).

3.3
Secondary Adhesion

The secondary adhesion phase requires the activation of integrins leading to stable adhesion, and the arrest of the leukocyte. Activation of integrins on the leukocyte leads to adhesion to endothelial ligands such as intercellular adhesion molecule (ICAM)-1, ICAM-2, vascular (V)CAM-1 and mucosal endothelial addressin (Mad)CAM-1 (Sect. 18) (Fig. 1). Leukocytes are then able to migrate over the endothelial lumenal surface in order to encounter an intercellular junction, whereupon endothelial junction proteins undergo rearrangement and leukocytes are able to undergo migration into the underlying tissue (Oppenheimer-Marks et al. 1994).

3.4
Diapedesis

Although emigration of neutrophils by a transcellular route through endothelium (transcytosis) have been documented (Feng et al. 1998), the most commonly described mechanism involves the migration, or diapedesis, of leukocytes between adjoining endothelial cells (bipolar migration). This involves multiple interactions, as endothelium possess a complex network of proteins that serve as barriers preventing solutes leaking from the vasculature, and preserving cell polarity by maintaining a molecular boundary between the basolateral and apical plasma membrane domains (Johnson-Leger et al. 2000). These barriers, or junctions, are highly localised structures that are concerned with different functional aspects of endothelial function, with differential expression levels defining specialised endothelial structures.

Barriers known as tight junctions are formed from discrete clusters of the molecules occludin, claudins and junctional adhesion molecules (JAM) expressed close to the lumenal surface. Intercellular homotypic adhesion between these molecules maintains very close contact between endothelial cells, and appears to be primarily concerned with permeability control. CAMs that are associated with tight junctions are often associated with restricting cell trafficking, and may well have a role in exclusion of leukocytes from specific sites. Presently there are 16 members of the claudin family, although specific roles for each member are still unclear (Tsukita et al. 1999). Claudin-3, -4 and -8 have been shown to be preferentially expressed in the liver and kidney (Morita et al. 1999a), with claudin-1 and -11 expression being implicated in regulating endothelial cell trafficking at sites of immune privilege such as the brain (Imhof et al. 2001) or the testes (Morita et al. 1999b).

Whilst research so far has succeeded in characterising the molecular mechanisms mediating the capture and firm adhesion of leukocytes onto the endothelial surface, mechanisms that mediate the lumenal to ablumenal passage of leukocytes between endothelial cells (diapedesis) have only just started to be elaborated.

Molecules that can mediate such a complex and heavily regulated process often have the capacity to redistribute rapidly at or from the endothelial junction during an inflammatory response. As discussed, molecules such as CD31 (Sect. 21), vascular endothelial (VE)-cadherin, the family of junctional adhesion molecules (Sect. 22) and CD99 (Sect. 23) are known to play a critical role in regulating diapedesis. One critical factor is their capacity for homophilic binding, allowing maintenance of endothelial barriers and the capacity to switch to heterophilic binding on leukocytes during diapedesis (Fig. 2). What has become increasingly evident is that these molecules also have the capacity for *cis* and *trans* heterophilic interactions, and these complexes may also make a significant contribution to the multistep combinatorial decision-making process regulating leukocyte trafficking and homing.

Diapedesis of migrating leukocytes is precipitated by disruption of adherens junctions, which are barriers comprised of homotypic adhesion across intercellular junctions, between VE-cadherin expressed along endothelial cell borders (Fig. 2). Leukocytes migrating on the endothelium surface can redistribute membrane-bound proteases onto their leading edge of migration and disrupt VE-cadherin complexes, leading to their diffusion

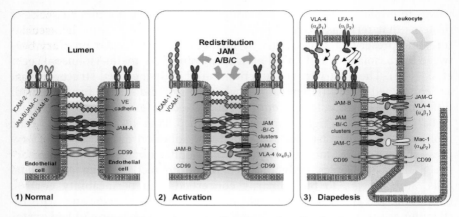

Fig. 2. The endothelial cell junction. Localisation and redistribution of CD99, VE-cadherin, JAM-A, -B and -C at endothelial cell junctions during activation, and their role in mediating movement of leukocytes through inter-endothelial junctions from the lumen into the extravasculature

over the endothelial cell surface (Cepinskas et al. 1999). This leads to redistribution of associated cytoplasmic proteins such as β-catenin, reorganisation of the actin cytoskeleton and increased endothelial permeability and diapedesis (Behrens et al. 1996; Buckley and Simmons 1997; Johnson-Leger et al. 2000). However, experimental observations indicate that leukocyte migration occurs preferentially at tricellular corners, where tight and adherens junctions become discontinuous, and therefore can 'bypass' any potential disruption of endothelial junctions (Burns et al. 1997). Diapedesis and movement out of the vasculature may be further facilitated by the shedding of specialist adhesion molecules such as L-selectin (Walcheck et al. 1996) and other CAMs (Gearing et al. 1992). CD31 (PECAM-1) has also proved to be an attractive candidate for mediating diapedesis, as expression is concentrated around endothelial junctions. CD31 may direct cells through endothelial layers by the establishment of haptotactic gradients (Sect. 21), as expression is concentrated at the basolateral plasma membrane domains of endothelium (Dejana and Del Maschio 1995) (Fig. 1). More recent findings have indicated that CD31 in endothelial cells is constitutively recycled between the cell membrane and subcellular compartments proximal to intercellular junctions. In diapedesis, recycled CD31 is targeted to specific regions of the endothelial junction engaged in homophilic adhesion with the transmigrating monocyte and assists the movement of the cell through the junction (Mamdouh et al. 2003).

Molecules such as CD31 and integrin-associated proteins (IAPs) have shown a capacity to regulate diapedesis as they are expressed on both leukocytes and endothelial cells, and can regulate integrin function in response to chemotactic and haptotactic gradients (Cooper et al. 1995; Rainger et al. 1999). The identification of the integrin $\alpha_v\beta_3$, also expressed on endothelial and leukocyte populations, as a ligand for CD31 (Buckley et al. 1996) may demonstrate that integrins, CD31, and IAP may modulate diapedesis by formation of junctional complexes involved in homotypic and heterotypic interactions.

The mechanism regulated by CD99 is quite separate from that of CD31, since blocking both molecules gives an accumulative and complete abolishment of diapedesis, demonstrating that there are individual molecular processes that participate in and regulate diapedesis (Schenkel et al. 2002; Muller 2003).

Diapedesis therefore represents a series of barriers that must be overcome, involving a diverse array of molecular interactions from delocalisation and proteolytic degradation of junctional proteins to haptotactic migration along gradients of junctional molecules across endothelial barriers. The physical engagement of these junctional molecules may well serve to ensure that the leukocyte acts as a 'plug' as it proceeds through the endothelium and does not compromise the integrity of the endothelium (Johnson-Leger et al. 2000). Exciting new progress is being made in this field and what is becoming evident is, rather than a series of combinatorial decision-making steps mediated by distinct families of cell molecules, diapedesis may represent the non-sequential interplay and feedback mechanisms of many families of molecules (Johnson-Leger and Imhof 2003).

Such complexity may allow the vasculature to intrinsically regulate the fine balance of maintaining structural integrity and homeostasis yet alter vascular permeability to cell traffic and solutes during inflammation and constitutive cell trafficking.

4
Regulation of Cell Trafficking by Distinct Populations of Endothelial Cells

Classically, descriptions of the underlying mechanisms of leukocytes recruitment from the vasculature focus on effector cell populations recruited by the activated endothelium during inflammation. The constitutive expression of ligands for naïve T and B lymphocyte populations by endothelial cell populations found within secondary lymph nodes, known as high endothelial venules (HEVs), are able to regulate the routine trafficking into the lymph node. Naïve T cells express L-selectin and the chemokines receptor CCR7 and HEV constitutively express GlyCAM-1, CD34 and MAdCAM-1 that present mucin-like glycoprotein ligands for L-selectin. This, in combination with lumenal presentation of CCR7 chemokines ligand CCL19 and CCL21, provide the 'area code' that induces the rolling and firm adhesion processes

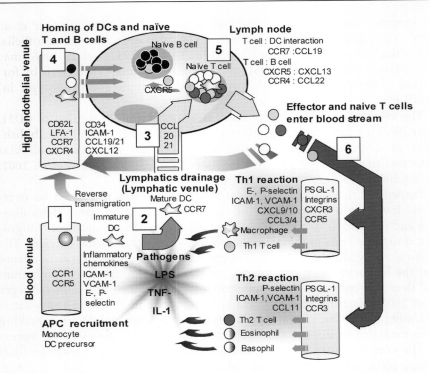

Fig. 3. Chemokine, chemokine receptor and adhesion molecule expression determines cell trafficking in the afferent and effector phases during an immune response. (*1*) Dendritic cell (*DC*) precursors are recruited by inflammatory chemokine production at a site of infection. (*2*) Phagocytosis of pathogen induces DC maturation and a switch from inflammatory chemokine receptor expression to CCR7. (*3*) Dendritic cells reverse transmigrate into the vasculature, or enter the lymphatics, draining to the lymph node and migrating to the T cell areas. (*4*) Reverse-transmigrated DCs and naïve T and B cells enter the lymph node and migrate to T and B cell areas. (*5*) T cells interact with DC and become activated. Activated T cells migrate to B cell areas to stimulate antigen-specific B cells for antibody production. (*6*) T helper (*Th*)1 or Th2 cells enter the vasculature and migrate to peripheral sites of inflammation to initiate an effector response (Salusto and Lanzavecchia 1999)

preceded by migration into the secondary lymph node (Girard et al 1998; Salusto and Lanzavecchia 1999), (Fig. 3).

Memory T cells are recruited to extra-lymphoid sites in every part of the body, showing remarkable tissue selectivity. Homing of memory T cells to tissues where antigen was originally encountered enables a rapid response upon re-encounter, and is central to optimising effector responses associated with immunological memory. The best-studied memory lymphocyte populations that home to tertiary lymphoid sites are those found in the skin and gastrointestinal tract. Homing of T cells involves a series of combinatorial 'decision processes' involving multistep sequential engagement of adhesion molecules, chemokines and chemokine receptors on endothelial surfaces that are tissue-specific, allowing for recruitment and retention of very specific sub-sets of T cells (Butcher and Picker 1996).

The chemokine receptors CCR4 and CCR9 plus the CAM cutaneous lymphocyte antigen (CLA) are expressed at high levels on memory T cells found at cutaneous sites. The chemokine CCL17 is produced constitutively by keratinocytes and concentrated on endothelial surfaces, leading to engagement of CCR4 and firm adhesion of CLA$^+$ T cells rolling on endothelial surfaces expressing E-selectin (Campbell et al. 1999a). Homing of memory CD4 T cells can also be restricted to different segments of the gastrointestinal tract. The chemokine CCL25 is restricted to epithelial cells of the small intestine, leading to homing of intraepithelial lymphocyte and lamina propria memory T lymphocyte populations expressing the chemokine receptor CCR9 and high levels of $\alpha_4\beta_7$ (Zabel et al. 1999; Kunkel et al. 2000; Papadakis et al. 2000).

These findings suggest that many homing chemokines and chemokine receptors are not necessarily exclusive to particular tissues, or in initiating polarised T helper (Th)1 or Th2 responses. Therefore these chemokines, presented on endothelial lumenal surfaces, and chemokine receptors, along with other 'addressins' may play a more combined role in the proper and simultaneous positioning of all T cell subsets during inflammation and routine trafficking (D'Ambrosio et al. 1998; Campbell and Butcher 2000).

4.1
Recirculation of Leukocytes: Deciphering the Role of the Lymphatic Endothelium

Within the extravasculature, the lymphatics drain lymph and immune cells into afferent lymph vessels leading to the lymph nodes (Fig. 3). Lymph and immune cells that leave the lymph node exit through the efferent lymphatic vessel and collect in the thoracic duct for drainage into the left subclavian vein. Until recently, lymphatic vessels had to be classified by histological methods for the lack of red blood cells within the lumen, and little was known about the actual phenotype of the lymphatic endothelium.

This difficulty in categorising lymphatic endothelium illustrates the similarities with vascular endothelium. Lymphatic endothelium are morphologically very similar to vascular and high endothelial venules and exhibit classical adherens, tight and gap junctions They also express molecules such as CD31, VE-cadherin, catenins and many integrins, suggesting that leukocyte migration into the afferent lymphatics across lymphatic endothelium involves a combinatorial multistep process with similar levels of stringency observed in the vasculature (Kriehuber et al. 2001). The similarity in adhesion molecule profiles observed between lymphatic endothelium and vascular endothelium would suggest that the lymphatic endothelium must have evolved additional mechanisms to facilitate leukocyte trafficking to compensate for the lack of flow. Recently, the discovery and characterisation of novel adhesion molecules that are restricted to specific subsets of endothelium has demonstrated a level of complexity beyond selectins, chemokines and integrins (Vestweber 2003).

The expression of a recently characterised molecule called lymphatic and vascular endothelial molecule (CLEVER)-1 has been shown to be restricted to lymphatic and HEV endothelium (Irjala et al. 2003). Blocking of CLEVER-1 function in vivo using monoclonal antibodies restricts lymphocyte trafficking through lymph nodes and suggests that this molecule plays an important role in leukocyte homing to specialised lymphoid environments through the vasculature and across HEV, as well as the afferent lymphatics through lymphatic endothelium (Irjala et al. 2003; Vestweber 2003).

Recently, numerous molecules have also been identified that are expressed on lymphatic endothelial cells (LEC). These include the molecule LYVE-1, a receptor for the ECM mucopolysaccharide hyaluronan (HA), which appears to be involved with sequestering chemokines (bound to HA) that mediate leukocyte trafficking towards the lymph node, or be involved with HA metabolism (Jackson 2003). Other molecules include a glomerular podocyte membrane mucoprotein called podoplanin, although its function in LEC remains unclear (Breiteneder-Geleff et al. 1999). Vascular endothelial growth factor receptor (VEGFR)3 has also been identified on LEC (Makinen et al. 2001). VEGFR3 has two known ligands from the vascular endothelial growth factor (VEGF) family, VEGF C and VEGF D, and initial studies have demonstrated that this receptor plays a central role in mediating the growth and survival of LEC (Makinen et al. 2001).

Other cell surface molecules recently identified include D6, a chemokine receptor (Nibbs et al. 2001), and the macrophage mannose receptor, a novel receptor for L-selectin which can mediate adhesion between LEC and lymphocytes (Irjala et al. 2001). They also express the chemokines CCL19 and CCL21, which mediate chemotaxis of CCR6- and CCR7-positive cells into the lymph node (Fig. 3), but the function of these recently identified markers on LEC have yet to be defined (Jackson 2003).

5
Inflammation and Wound Healing: Synergy Between the Endothelium and the Extra-vasculature

The role of inflammation is to allow cells and proteins of the vasculature to enter damaged or infected tissues. It is the product of many localised and systemic responses to tissue injury and infection. Such responses allow the targeting of bodily immune resources from the vasculature by encouraging a greater turnover of cells and soluble factors at sites of infection and damage. They are mediated by the activation of the soluble precursors of the kinin, clotting and fibrinolytic systems (via Hageman factor) as a direct consequence of damage or infection to the endothelium and extravasculature.

These inflammatory mediators can rapidly precipitate inflammatory events, often by acting directly on the inactive components of the plasma enzyme system or by directly activating effector leukocyte populations. Inflammatory mediators are extremely diverse and can potentially act on nearly every aspect of the immune process. They are small soluble molecules that are either derived from exogenous factors such as bacteria (lipopolysaccharide from gram negative bacteria), or endogenous factors which often exist as precursors within the cytoplasm of cells and are secreted upon stimulation (e.g. histamine secreted by mast cells). Different inflammatory mediators are released during distinct phases of inflammation, depending on the nature and duration of the inflammatory insult (Austyn and Wood 1994; Kuby 1997).

Early phase mediators are released very early upon inflammation and often exist as precursors ready for rapid release or effect. Their effect is often localised and mostly mediates events in the acute phase response. They are generally vasoactive (e.g. histamine) and can trigger activation of the clotting or fibrinolytic pathways. Late phase mediators such as the prostaglandins and leukotrienes have to be synthesised upon inflammation from arachidonic acid as opposed to existing as proforms that can be immediately activated. They are often produced in response to activation of the kinin and coagulation pathways and have more systemic effects on blood vessel walls and on smooth muscle. Therefore, late phase proteins modulate the extent of inflammation and regulate the balance of inflammatory and non-inflammatory events. Despite the diversity of stimuli and inflammatory mediators that can trigger and protract the inflammatory process, it can be divided into four phases.

The acute vascular response occurs within seconds of tissue injury, leading to vasodilation and increased permeability of the capillary wall. This leads to increased blood flow (hyperaemia), increased redness (erythema) and the entry of protein-rich fluid into the tissue (oedema). The consequences of these effects leads to the localised accumulation of fluid (a wheal) and the characteristic swelling associated with inflammation at the site of injury or infection. The increased flow and permeability also contributes to an increase in cellular traffic of polymorphonuclear leukocytes.

The acute cellular response occurs over a period of minutes and hours and is characterised by the appearance of polymorphonuclear leukocytes, particularly neutrophils, at the site of injury where they participate in direct microbial killing. The deposition of fibrinogen and fibronectin around the lesion leads to the accumulation of platelets and red blood cells and eventual clotting, which physically seals damaged blood vessels. Another soluble mediator, bradykinin, a potent vasodilator, also activates pain receptors and contraction of smooth muscle cells, encouraging protection by the individual of the wounded or infected area.

The chronic cellular response describes the recruitment phase of different leukocyte cell populations such as macrophages and lymphocytes. Macrophages, as well as being involved in direct microbial killing, are responsible for clearance of tissue debris and remodelling of tissue. Effector lymphocyte populations are involved in antigen-specific responses and will 'home' to tissue where the antigen has been encountered.

The resolution phase is where tissue injury has been repaired and infectious agents eliminated. Mechanisms for attaining resolution and tissue homeostasis are processes involving the removal of effector cells and wound healing. This entails the removal of any blood clots (fibrinolysis), and the migration of fibroblasts and endothelial cells into damaged or missing tissue to restore tissue architecture. Tissue remodelling in wound healing can involve extensive migration of fibroblasts and often is associated with scarring (fibrosis) (Austyn and Wood 1994; Kuby 1997).

6
Cell Adhesion Molecules, the Players in Mediating Endothelial Function

6.1
Selectins and Their Ligands

6.1.1
Selectins

The selectins, a family of Ca^{2+}-dependent carbohydrate-binding proteins (C-type lectins) play an essential role in leukocyte homing to lymphoid tissues and accumulation at sites of inflammation. Leukocytes in flow become tethered onto the endothelium and roll due to the shear forces experienced at the lumenal wall. Such an interaction can occur due to the high bond dissociation/association rates and ensures the contact zone between the cell and lumen is rapidly translated along the vessel wall. The capture and rolling of leukocytes by the selectins on the endothelium prepare the leukocyte for tight adhesion to the endothelium and diapedesis into the extravasculature (Issekutz 1992; Shimizu et al. 1992).

The selectins consist of three members: P-selectin, expressed on platelets and activated endothelium; E-selectins also expressed on activated endothelium; and L-selectin, which is expressed on most circulating leukocytes. They are type I transmembrane glycoproteins consisting of an amino terminal calcium-type lectin domain (C-type), an epidermal growth factor domain (EGF), two to nine short consensus repeats (SCRs) and a short cytoplasmic tail. Homology between the three selectins varies between 60%–65% for the C-type lectin and EGF domains, to 40%–45% for the SCR domains. This is consistent with selectins being able to bind carbohydrate moieties that are similar, but not necessarily identical, that are presented on diverse ligands (Barclay et al. 1997). Evidence from crystallography and mutational studies showing the structural basis of selectin ligand binding supports this. All domains appear to be essential for adhesion, with the C-type lectin and EGF involved in direct engagement of the ligand (Tedder et al. 1995). The conjoint effect of these two domains in ligand binding has further demonstrated how the different selectins have similar yet distinct patterns of carbohydrate recognition. Each of these separate modules is encoded by a separate exon, suggesting that lectins evolved from a common primordial ancestor into a family of proteins capable of considerable diversity and cross-reactivity of binding.

Selectins are extremely important in acute inflammation and cell trafficking. This can be seen with clinical diseases such as leukocyte adhesion deficiency (LAD) II, where there is a complete lack of recruitment of neutrophils from the vasculature to sites of inflammation. This is caused by a defect in fucosylation during post-translational modification, leading to an absence of functional selectin ligands (von Andrian et al. 1993). This phenotype has been confirmed by individual selectin knockout mice, which often appear to have a normal phenotype, with combined selectin knockouts required for the complete abolishment of leukocyte rolling (Jung Ley 1999). Studies with knockout mice have confirmed that L- and P-selectin mediate the initial capturing of leukocytes, with the combined action of all three selectins required for optimal rolling (Tedder et al. 1995).

6.1.2
Selectin Ligands: Sialomucins and Carbohydrate Moieties

The nature of selectin binding is still an area of intense investigation. Selectins recognise fucosylated structures containing sialyl-Lewis X (sLe^x) (von Andrian et al. 1993). A number of ligands for selectins have been identified and, with the exception of E-selectin ligand (ESL)-1, belong to a family of cell surface molecules called sialomucins (Shimizu and Shaw 1993). Sialomucins are threonine- and serine-rich proteins that are heavily O-glycosylated, allowing optimal exposure of multiple terminal sugars. The best characterised ligand for P-selectin is P-selectin glycoprotein ligand (PSGL)-1, which can mediate adhesion under static and flow conditions (McEver et al. 1995) with functional binding dependent on tyrosine sulphation near the amino terminus of PSGL-1 (Pouyani and Seed 1995; Sako et al. 1995). E-selectin has a lower affinity for PSGL-1 requiring the amino terminus or sulphation binding to a separate site on the same molecule. E-selectin also binds ESL-1, a distinct transmembrane cell surface molecule (Steegmaier et al. 1997). Although not a sialomucin, it requires myeloid-specific fucosylation N-linked chains for E-selectin adhesion. E-selectin also has a number of low-affinity ligands such as an epitope of PSGL-1 known as CLA (Berg et al. 1991), L-selectin, CD66 and β_2 integrins (Carlos and Harlan 1994), although whether such low-affinity interactions can mediate any physiological effect is still unclear. L-selectin binds to the heavily glycosylated cell surface proteins CD34, GlyCAM-1 and MadCAM-1, as well as a spectrum of glycolipids expressed constitutively on HEVs.

It is evident that there is considerable cross-reactivity in selectin ligand binding, which is further compounded by carbohydrate recognition being reliant on three variables known as triad systems (Crocker and Feizi 1996). Triad adhesion requires optimal presentation of the carbohydrate moiety by a carrier protein or glycolipid for adhesion between counter-receptors of lectin and carbohydrate. This degree of flexibility allows cell adhesion to be additionally controlled at a post-translational level of the carrier protein. For example, PSGL-1 is expressed by all lymphocytes, but only a small number can bind P-selectin, demonstrating tightly regulated post-translational regulation (Moore et al. 1995). Triad interactions are difficult to investigate by more conventional technical procedures due to the heterogeneity of protein glycosylation on the cell surface. New strategies devised to investigate the role of carbohydrates in cell–cell interactions are being developed and are helping us define new families of carbohydrate CAMs that extend far beyond that of just the selectin family (Feizi 2000).

6.1.3
Function and Patterns of Expression

Endothelium constitutively expresses P-selectin, which is stored within Weibel–Palade bodies of endothelial cells and the α granules of platelets (McEver 1990; Wagner 1993). An inflammatory stimuli triggers rapid fusion with the lumenal cell membrane, allowing for the immediate capture and recruitment of leukocytes to the affected tissue. E-selectin expression is protein synthesis-dependent and peaks 4–6 h after the inflammatory stimuli, dropping to a basal level after 24 h. Expression is more prolonged than P-selectin and is thought to have a role in allowing for an extended recruitment period after stimulation. Myeloid cells throughout differentiation and nearly all neutrophils, eosinophils and monocytes express L-selectin. The majority of B lymphocytes and virgin T lymphocytes also express L-selectin, but it is only expressed by certain sub-populations of memory T lymphocytes and natural killer (NK) cells. The role of L-selectin in lymphocyte homing, by mediating adhesion to high endothelial venules in peripheral lymph nodes, has been well characterised (Gallatin et al. 1983) (Fig. 3), but the broad expression of L-selectin by effector cells suggests that it has a role to play in general leukocyte trafficking (Barclay et al. 1997). L-selectin is preferentially expressed in clusters on the tips of microvilli or membrane ruffles

(Erlandsen et al. 1993), which enhances contact initiation with ligand and dramatically enhances the initial tethering interactions out of flow (von Andrian et al. 1995; Finger et al. 1996; Stein et al. 1999). L-selectin is rapidly shed by proteolytic cleavage after coming into direct contact and 'rolling' on the endothelium. Such a transient interaction may be a means of allowing the rapid detachment of a rolling leukocyte 'scanning' the lumenal wall for further signals. If an appropriate 'decision-making' signal is not received within a certain time frame, the cell becomes detached from the endothelium. Self-limiting mechanisms for controlling neutrophil recruitment by endothelium may also be mediated by engagement of L-selectin ligand on adjacent neutrophils, leading to proteolytic cleavage and detachment, thus maintaining a 'critical mass' of rolling neutrophils (Chen et al. 1995).

Selectins are also able to mediate intracellular signal transduction events capable of altering cell behaviour and function. Although selectins have short cytoplasmic tails, selectin function and signal transduction is dependent on attachment to the cytoskeleton. Several signalling functions have been described that are dependent on selectin and cell type (Crockett-Torabi 1998). With L-selectin, signal transduction can induce a rapid flux in intracellular calcium, superoxide generation, and activation of ras and rac pathways, leading to structural changes in the cytoskeleton (Brenner et al. 1996, 1997; Crockett-Torabi 1998). Signalling pathways have been described for L-selectin that can lead to direct activation of β_2 integrins, mediating firm adhesion to the endothelial walls under flow (Gopalan et al. 1997). Lymphocyte binding to P-selectin also induces phosphorylation of a focal adhesion kinase (FAK) and paxillin (Haller et al. 1997).

6.2
Integrins

Integrins are a superfamily of cell surface heterodimers expressed on all nucleated cells. They consist of a non-covalently linked α and β subunit. So far, the number of α subunits characterised stands at 17, and β subunits at 8. Although some cell types express many of these different integrin subunits, subunit association occurs in a highly restricted manner. There are at least 23 different integrins with even further diversity generated at transcriptional and post-translational level. Integrin function is modulated by different activation states affecting ligand affinity and avidity. Integrins recognise a huge variety of ligands, which are present in extra-cellular matrix or on the cell surface. Their almost ubiquitous presence on cells demonstrates their diversity and flexibility in function, and they are known to have profound effects on cell behaviour and function.

The pivotal role that integrins play in modulating endothelial cell function was illustrated by the characterisation of clinical syndromes, where integrin gene defects were shown to lead to profound defects in recruitment of specific effector cell population to sites of infection and inflammation (von Andrian et al. 1993).

Both the α and β subunits have distinct domains with ligand specificity determined mostly by the α subunit and regulatory functions by the β subunit. Some α subunits have inserted regions called I domains (α_L, α_M, α_X, α_D, α_1 and α_2). These regions contain metal ion-dependent adhesion sites (MIDAS), which contain cation-binding sites required for ligand binding. Some α subunits do not contain I domains (α_3, α_5, α_6, α_7, α_8, α_v and α_{IIb}), but have a disulphide-linked heavy and light chain as a result of proteolytic cleavage. The cytoplasmic tail of the β subunit also contains motifs that are important for cell-signalling events and attachment to the cytoskeleton.

All integrin function and structure is cation dependent, although there is divergence in di-cation specificity and effect. Such differences suggest that di-cation binding can affect integrin conformation, and therefore regulate ligand binding. It is thought that di-cations may interact with integrins in such a way that substantially alters their molecular shape and 'expose' additional sites thus enhancing or inhibiting ligand binding (Bazzoni et al. 1995). The role of di-cations remains unclear, as some observations have been made with the di-cations well in excess of physiological concentrations, but may demonstrate the level of effect they have on integrin structural malleability, and the depth of control they have on function.

Levels of integrin expression are often considered to be constitutive, although this can be affected by factors such as cytokines. In some cell types, levels of integrin expression are controlled by transcription of the α subunit, since a reservoir of the β subunit associated with the chaperone calnexin exists within the cytoplasm (Lenter and Vestweber 1994). Integrin affinities for their ligand can be altered within a few minutes (Humphries Newham 1998). The kinetics of such transient changes is fundamental to many biological processes such as cell migration, which has been an area of intense interest for many years. The intracellular events leading to integrin activation and engagement of ligand are known as 'inside-out' signalling (Hughes and Pfaff 1998). This is associated with conformational changes of the integrins leading to an increase in ligand affinity, as well as events such as lateral diffusion and clustering of integrin molecules. Integrins can also regulate cell behaviour by 'outside-in' signalling. Simultaneous ligation of integrins and occupancy of growth factor receptors can lead to intracellular signalling events and changes in cell growth, migration, differentiation and proliferation (Buckley et al. 1998).

Cell adhesion and cell-signalling events directly associated with the integrin heterodimer have been well characterised, but it has become clear that integrins are also able to form dynamic *cis* interactions with other cell surface receptors to form distinct multi-receptor complexes. The breadth of integrin function is thought to be due to their capacity to associate with a wide range of cell receptors, intracellular signalling molecules and membrane proteins (Porter and Hogg 1998). These multi-receptor complexes can form independently of the integrin activation state or ligand engagement, and can lead to lateral diffusion of cell surface receptors that cluster into distinct focal adhesion complexes. Therefore the formation of such functional membrane units can lead to a switch between avidity and affinity as well the recruitment of a number of cell surface receptors that can participate in cell signalling (Petty and Todd 1996).

Integrins interact in 'cis' with many different families of cell surface receptors via extracellular domains. These include IAPs (CD47), which consists of an extracellular Ig superfamily (IgSF) C-type loop with five membranes spanning sequences and a short cytoplasmic tail. IAPs are one of four splice variants and are found on most cell types with the capacity to interact with more than one integrin, as well as integrin-independent functions. The function of IAPs when associated with β_3 integrin has been shown to be important in myeloid cell differentiation and cell migration. Human cells that lack IAP do not exhibit $\alpha_v\beta_3$-mediated binding, and transfection of the IAP—specifically the extracellular IgSF loop—reconstitutes $\alpha_v\beta_3$ function (Lindberg et al. 1996). Engagement of the $\alpha_v\beta_3$–IAP complex leads to intracellular signalling, which inactivates the integrin $\alpha_4\beta_1$, the ligand to VCAM, and thus increases the speed of cell migration (Imhof et al. 1997).

Other cell surface receptors include the transmembrane-4 superfamily (TM4SF) or tetraspans. The tetraspans are a family of more than 20 proteins and have four membrane-spanning domains with the N and C terminus contained within the cytoplasm (Berditchevski et al. 1996). Tetraspans also share many characteristics with CD47, as they are able to associate with more than one integrin as well as having functions that are integrin independent. Tetraspan interactions and stoichiometry are quite different from that of CD47. Tetraspans tend to be localised at specific sites on the cell membrane, such as microvilli and adherens junctions, and have been implicated in directing integrins to specific sites to increase their functional efficiency (Hemler et al. 1996). Tetraspans have been implicated in cell migration during development, providing spatial cues for processes such as cell polarisation, but a clear role for tetraspans have yet to be defined (Porter and Hogg 1998; Yanez-Mo et al. 2001).

Other cell surface receptors that associate with integrins include growth factor receptors (GFRs), which appear to modulate their function by clustering with integrins to enhance and augment cell signalling. Examples include the specific association of insulin receptor (IR) with the integrin $\alpha_v\beta_3$, leading to a more potent mitogenic effect of insulin upon engagement of $\alpha_v\beta_3$ with ligand (Vuori and Ruoslahti 1994). Physical and functional interactions of β_2 integrins with the GPI-linked membrane proteins CD14, CD16 and CD87 have also been described (Petty and Todd 1996). GPI proteins lack a transmembrane region and rely on association with transmembrane partners to mediate intracellular events. These interactions are extracellular and are mediated by carbohydrate moieties. When CD14 binds bacterial lipopolysaccharide (LPS), associated Mac-1 mediates intracellular signals that can lead to cell activation. This is similar to CD16, an F_c receptor (F_cRIIIB), where association with Mac-1 promotes intracellular events leading to phagocytosis (Todd and Petty 1997).

Integrins are known to interact with surface molecules such as caveolin, a molecule found in plasma membrane invaginations (caveolae). Caveolin acts by binding and clustering cell-signalling molecules such as the Ha-Ras and the Src-family tyrosine kinases, maintaining them in an inactivated state. It also associates with $\alpha_v\beta_3$, certain β_1 subunits and GPI-anchored urokinase-type plasminogen activation receptor (uPAR, CD87). Engagement of the $\alpha_v\beta_3$ or β_1 integrins by ligand can result in different outcomes that are dependent on interactions with caveolin. This can activate cell-signalling molecules such as H-Ras or Src-1 tyrosine kinases leading to cell proliferation or terminal differentiation (Wary et al. 1996).

In certain cell types, caveolin association can further regulate β_1 integrin function as it also associates with uPAR. The formation of such complexes can lead to a switch between integrin-mediated fibronectin binding to uPAR vitronectin binding (Wei et al. 1996). With cell types such as neutrophils, which do not express caveolin, the uPAR associates with Mac-1 directly. During chemotactic migration, uPAR dissociates from Mac-1 and associates with p150,95 (CD11c/CD18) at the leading edge of the cell (Bohuslav et al. 1995). Integrins are known to associate with many cell membrane molecules and can modulate the activity of proteases such as urokinase plasminogen activator (uPA), the ligand for uPAR, and metalloprotease (MMP)-2. Therefore integrin–protease complexes may have a key role in directing cell migration during diapedesis, angiogenesis and wound healing (Brooks et al. 1996; Yebra et al. 1996). Cell surface molecules such as CD98 can regulate the affinity of β_1 directly by interaction through cytoplasmic regions. Monoclonal antibodies to CD98 can induce $\alpha_3\beta_1$ activation leading to homotypic aggregation, and it is thought to have a role in regulating haemopoiesis (Warren et al. 1996). Although only a few integrin-associated proteins have been described, many more are thought to exist and are currently being investigated. Integrin function is reliant on the formation of clusters with other molecules to form an extremely diverse array of distinct functional units. The formation of these units is dynamic and is capable of forming microdomains or focal adhesions, leading to enhanced signalling and adhesion (Petty and Todd 1996; Buckley et al. 1998; Porter and Hogg 1998).

6.3
The Immunoglobulin Superfamily

The IgSF comprises a large family of glycoproteins mostly expressed at the cell surface. They are the most abundant cell surface molecules found on leukocytes, accounting for nearly 50% of all cell surface glycoproteins (Kreis Vale 1999). IgSF homologues are found throughout the animal kingdom, and functional adhesion of these molecules is thought to have evolved from single identical domains involved in simple *trans*-interactions on opposing cell membrane surfaces. The genes encoding these single ancestral domains evolved into more complex multidomain molecules capable of not only homophilic interactions, but also more complex heterophilic *cis*- and *trans*-interactions as well as signal transduction.

IgSF functions are quite diverse, with some molecules acting as receptors for growth factors and receptors for the F_c region of immunoglobulins, but the majority function as CAMs capable of intracellular signal transduction (Holness and Simmons 1994). IgSF structure consists of repeating units of Ig-like domains. The homology between different domains within the same protein, as well as between domains in different IgSF proteins is usually only 10%–30%. Functional diversity has been attributed to mutation and amplification of these domains from a single common ancestral gene. Despite this poor homology, these Ig-like modules have conformational similarities. The Ig modules consist of 7–9 antiparrallel β strands arranged in two layers that are stabilised by disulphide-linked cysteines separated by 55–75 amino acids. Further classification based on immunoglobulin structure has identified three types: V set, C1 set and C2 set domains. Further analysis has led to reclassification of C2 set domains into two distinct classes: a redefined C2 type and an I set type domain which is characterised by a shortened V-like domain (Harpaz and Chothia 1994; Chothia and Jones 1997). The characterisation of neuronal (N)CAM during development demonstrated dynamic patterns of expression as well as the capacity to alter cell behaviour and differentiation, through both signal transduction and mechanical adhesion (Doherty et al. 1991). This multiplicity of function is due to ligand redundancy and a capacity to mediate intracellular signalling events, and it explains why IgSF molecules have a role in development and the regulation of the immune system (Baldwin et al. 1996). Their role in leukocyte trafficking through endothelium and HEVs has been well characterised. These include ICAM-1 (CD54), ICAM-2 (CD102), ICAM-3 (CD50), VCAM-1 (CD106), PECAM-1 (CD31) and MadCAM-1. IgSF molecules mediate firm adhesion of rolling leukocytes to the endothelium, an essential prerequisite to extravasation into the extravasculature (Fig. 1).

More recently, research has focused on a new and exciting family of IgSF.

6.3.1
ICAM and VCAM

ICAM-1 and VCAM-1 both have wide tissue distributions with expression on many different cell types (Barclay et al. 1997) and have key roles in stabilising leukocyte–endothelium interactions during inflammation. Endothelial expression is upregulated by different inflammatory mediators, leading to firm adhesion of leukocytes expressing the corresponding activated integrin ligand (Figs. 1 and 3). Activation of integrin occurs by 'inside-out' signalling triggered by chemokine–chemokine receptor interactions during the initial leukocyte–endothelial interactions. ICAM-1 contributes to the extravasation of most leukocytes expressing the integrin ligands lymphocyte functional antigen (LFA)-1 ($\alpha_L\beta_2$), Mac-1 ($\alpha_M\beta_2$) and p150,93 ($\alpha_X\beta_2$); VCAM ligands are VLA-4 ($\alpha_4\beta_1$) and $\alpha_4\beta_7$. Soluble forms of ICAM and VCAM that are still biologically active can be found in serum, suggesting regulation by proteolytic cleavage from the cell surface (Gearing et al. 1992). ICAM-2 also binds the integrin LFA-1 and is constitutively expressed by endothelial cells and most leukocytes. The different structure and constitutive expression pattern of ICAM-2 suggests a role in trafficking of leukocyte populations, such as NK cells (Somersalo 1996) and DCs (Geijtenbeek et al. 2000), as well as a more diverse role in cell-mediated killing on NK cells (Helander et al. 1996). ICAM-3 also binds LFA-1 and the integrin $\alpha_D\beta_2$. It is expressed constitutively by all resting leukocytes and is thought to stabilise interactions between cells during immune responses (Kreis Vale 1999).

6.3.2
Mucosal Adressin Cell Adhesion Molecule 1

MadCAM-1 is expressed on HEVs. Ligands for MadCAM include L-selectin, which bind to the mucin-rich domain and is critical to the trafficking of naïve T cells through HEVs into lymph nodes (Fig. 1). The two IgSF domains can also mediate adhesion to the integrin $\alpha_4\beta_7$, which is expressed by a subset of memory T cells and is a prerequisite step for homing to the mucosal tissues of the mesenteric lymph nodes and Peyer's patches (Picker and Butcher 1992; Butcher and Picker 1996).

6.3.3
CD31

CD31, or platelet endothelial CAM 1 (PECAM-1), is found in abundance on endothelium, platelets and most leukocytes. The six different IgSF domains of CD31 have the capacity to interact with different ligands. Domains 1 and 6 are required for homophilic interactions, whereas only domains 1and 2 are required for heterophilic interactions with the integrin $\alpha_v\beta_3$ (Fawcett et al. 1995) (Fig. 1). CD31 has long been known to play a critical role in diapedesis (Muller and Randolph 1999); it has the capacity to directly transduce intracellular signal events that can modulate cell motility on the endothelial lumen by activating β_2 integrins (Rainger et al. 1997) and upregulate expression of the integrin $\alpha_6\beta_1$ (a receptor for laminin) on transmigrated neutrophils, facilitating their migration through the perivascular basement membrane (Dangerfield et al. 2002).

Such findings demonstrate the dual functionality of some adhesion molecules where weak adhesive interactions provide important microenvironmental cues and direct cells to areas of high adhesion, a process known s haptotaxis (Carter 1967; Bianchi et al. 1997).

6.3.4
Junctional Adhesion Molecules

JAM-A, -B and -C are a group of cell surface molecules belonging to the Ig superfamily. They have two extracellular Ig domains, a single transmembrane domain and a short cytoplasmic tail. JAM-A is expressed by epithelial and endothelial cells as well as leukocytes and platelets (Martin-Padura et al. 1998; Williams et al. 1999; Liu et al. 2000; Sobocka et al. 2000). JAM-B expression is expressed on endothelial cells, whereas JAM-C is expressed on endothelial cells as well as subpopulations of B and T cells, NK cells, dendritic cells, monocytes and platelets (Johnson-Leger et al. 2002; Santoso et al. 2002).

All JAM members are capable of forming homophilic interactions and heterophilic interactions (reviewed in Luscinskas et al. 2002; Muller 2003; Johnson-Leger and Imhof 2003). Preliminary studies on JAM-A demonstrated that it makes a major contribution to the structural integrity of the endothelial junction and plays a critical role in regulating leukocyte traffic from the lumen (Sect. 4). X-ray diffraction studies have shown that JAM-A is capable of forming *cis* and *trans* interactions (Kostrewa et al. 2001). In this model, JAM dimers orientated in *cis* on the endothelial cell surface are capable of forming *trans* interactions at their N terminus with identical JAM dimers on adjacent cells (Fig. 2) (Kostrewa et al. 2001). The complexity of such a two-dimensional macromolecular structure is further compounded if this model is extended to other JAM members, since JAM-B and JAM-C show preferential binding to each other (Cunningham et al. 2000; Arrate et al. 2001) (see Fig. 2).

More recently, other binding partners have been described for JAM members (see review Johnson-Leger and Imhof 2003) and it has been postulated that all are capable of forming complex *cis* and *trans* interactions with integrins (Ostermann et al. 2002) (Fig. 2). In vitro experiments have shown JAM-A to be a counter-receptor for the integrin LFA-1 and that it facilitates leukocyte diapedesis by redistributing onto lumenal apical surfaces from interendothelial junctions upon endothelial activation (Ozaki et al. 1999; Ostermann et al. 2002).

More complex interactions have been observed where JAM-B, expressed on endothelial cells, is able to interact with the integrin VLA-4 ($\alpha_4\beta_1$) on T cells, but only after prior engagement with JAM-C expressed on T cells (Cunningham et al. 2002).

JAM-B expressed on human platelets has been shown to be a counter-receptor for the leukocyte integrin Mac-1 ($\alpha_M\beta_2$) and plays a key role in mediating platelet–leukocyte interactions (Santoso et al. 2002). JAM-C also demonstrates a similar level of complexity to other JAM members, where interactions with JAM-B on adjacent endothelial cells can switch to the integrin Mac-1 expressed on adhered monocytic cell lines (C. Lamagna, G. Mandicourt, B.A. Imhof, M. Aurrand-Lions, submitted).

6.4
CD99

CD99 is a highly O-glycosylated, 32-kDa protein expressed by most leukocytes and endothelial cells. Originally characterised as a co-stimulatory molecule on T cells and thymocytes, a specific role in regulating diapedesis has only recently been elucidated (Bernard et al. 2000; Schenkel et al. 2002). CD99 exhibits *trans* homophilic adhesion, and expression on endothelial cells is concentrated around interendothelial junctions. Monoclonal antibodies directed against CD99 have been shown to block diapedesis of monocytes across activated endothelium (Schenkel et al. 2002).

Like CD31, CD99 has the capacity to modulate the activation state of integrins. Engagement of CD99 on T cells leads to activation of the integrin VLA-4, resulting in firm adhesion under flow conditions (Bernard et al. 2000).

6.5
CD44

CD44 represents a large family of heavily glycosylated proteins encoded by a single gene that contains 20 exons. The basic structure found in all CD44 molecules is encoded by 10 exons and represents the most common form of CD44 (CD44H). Binding of the glycosaminoglycan (GAG) hyaluronate is mediated by the membrane-distal link module region that is found on all CD44 isoforms, with additional binding and biological properties mediated by the variant isoforms (Barclay et al. 1997). Different isoforms are generated by insertion of combinations of the 10 additional exons (variable exons) into the mucin-like region contained within the centre of the molecule. Alternative splicing of these variable exons could potentially generate up to 100 different isoforms, but only 30 have been identified to date. The insertion or absence of these variable regions is known to contribute significantly to the function of CD44, as they encode additional GAG motifs such as chondroitin sulphate (CS) and heparin sulphate (HS) (Jackson et al. 1995). The multiplicity of function of CD44 may be to allow interactions with a range of tissue-specific ligands. The presence of such highly glycosylated negatively charged structures on the cell surface can have a profound effect on cell–cell and cell–ECM interactions, as they collectively impose a large electrostatic charge on the cell (Ruiz et al. 1995). A potential role for CD44 in lymphocyte trafficking can be postulated with the observation of CD44 expression on memory and effector lymphocyte populations, but not naïve populations (Butcher and Picker 1996). CD44 can also mediate interactions between different stromal cell populations in the extravasculature and has been implicated in haemopoiesis and development. The presence of these GAGs has also been shown to mediate *cis*-interactions, by presenting chemokines to chemokine receptors on the same cell surface (Ruiz et al. 1995).

6.6
Vascular Adhesion Protein-1

VAP-1 is a cell surface sialoglycoprotein originally identified by expression on endothelial cell structures in HEVs (Salmi and Jalkanen 1992; Jalkanen and Salmi 1993). Determining the exact role of VAP-1 has proved difficult, as functional expression in vitro has only been demonstrated on endothelial cells isolated form liver (Lalor et al. 2002). The expression of VAP-1 on endothelium can be observed in many tissues in vivo (Arvilommi et al. 1997) and is upregulated during inflammation (Salmi et al. 1993). Therefore, this discrepancy illustrates that expression of VAP-1 may be regulated in a tissue- and/or cell type-selective manner, and microenvironmental factors from the tissue-specific extravasculature regulate expression (Arvilommi et al. 1997).

Nevertheless, it has been clearly shown that VAP-1 is capable of mediating CD8 T cell adhesion and transmigration to hepatic endothelium and, in conjunction with other adhesion molecules, may have a role in differentially regulating the trafficking of leukocyte subsets (Lalor et al. 2002).

Funded by: The Swiss National Science Foundation 310A0-100697/1, Oncosuisse OCS-012335-02-2003.

References

Arrate MP, Rodriguez JM, Tran TM, Brock TA, Cunningham SA (2001) Cloning of human junctional adhesion molecule 3 (JAM3) and its identification as the JAM2 counter-receptor. J Biol Chem 276:45826–45832

Arvilommi AM, Salmi M, Jalkanen S (1997) Organ-selective regulation of vascular adhesion protein-1 expression in man. Eur J Immunol 27:1794–1800

Austyn JM, Wood KJ (1994) Principles of cellular and molecular immunology, 2nd edn. Oxford University Press, Oxford

Baldwin TJ, Fazeli MS, Doherty P, Walsh FS (1996) Elucidation of the molecular actions of NCAM and structurally related cell adhesion molecules. J Cell Biochem 61:502–513

Barclay AN, Brown MH, Law SK, McKnight A, Tomlinson MG, van der, Merwe PA (1997) The leukocyte antigen facts book, 2nd edn. Academic Press, London

Bargatze RF, Jutila MA, Butcher EC (1995) Distinct roles of L-selectin and integrins alpha 4 beta 7 and LFA-1 in lymphocyte homing to Peyer's patch-HEV in situ: the multistep model confirmed and refined. Immunity 3:99–108

Bazzoni G, Shih DT, Buck CA, Hemler ME (1995) Monoclonal antibody 9EG7 defines a novel beta 1 integrin epitope induced by soluble ligand and manganese, but inhibited by calcium. J Biol Chem 270:25570–25577

Behrens J, von Kries JP, Kuhl M, Bruhn L, Wedlich D, Grosschedl R, Birchmeier W (1996) Functional interaction of beta-catenin with the transcription factor LEF-1. Nature 382:638–642

Berditchevski F, Zutter MM, Hemler ME (1996) Characterization of novel complexes on the cell surface between integrins and proteins with 4 transmembrane domains (TM4 proteins). Mol Biol Cell 7:193–207

Berg EL, Yoshino T, Rott LS, Robinson MK, Warnock RA, Kishimoto TK, Picker LJ, Butcher EC (1991) The cutaneous lymphocyte antigen is a skin lymphocyte homing receptor for the vascular lectin endothelial cell-leukocyte adhesion molecule 1. J Exp Med 174:1461–1466

Bernard G, Raimondi V, Alberti I, Pourtein M, Widjenes J, Ticchioni M, Bernard A (2000) CD99 (E2) up-regulates alpha4beta1-dependent T cell adhesion to inflamed vascular endothelium under flow conditions. Eur J Immunol 30:3061–3065

Bianchi E, Bender JR, Blasi F, Pardi R (1997) Through and beyond the wall: late steps in leukocyte transendothelial migration. Immunol Today 18:586–591

Bohuslav J, Horejsi V, Hansmann C, Stockl J, Weidle UH, Majdic O, Bartke I, Knapp W, Stockinger H (1995) Urokinase plasminogen activator receptor, beta 2-integrins, and Src-kinases within a single receptor complex of human monocytes. J Exp Med 181:1381–1390

Breiteneder-Geleff S, Soleiman A, Kowalski H, Horvat R, Amann G, Kriehuber E, Diem K, Weninger W, Tschachler E, Alitalo K, Kerjaschki D (1999) Angiosarcomas express mixed endothelial phenotypes of blood and lymphatic capillaries: podoplanin as a specific marker for lymphatic endothelium. Am J Pathol 154:385–394

Brenner B, Gulbins E, Schlottmann K, Koppenhoefer U, Busch GL, Walzog B, Steinhausen M, Coggeshall KM, Linderkamp O, Lang F (1996) L-selectin activates the Ras pathway via the tyrosine kinase p56lck. Proc Natl Acad Sci U S A 93:15376–15381

Brenner B, Gulbins E, Busch GL, Koppenhoefer U, Lang F, Linderkamp O (1997) L-selectin regulates actin polymerisation via activation of the small G-protein Rac2. Biochem Biophys Res Commun 231:802–807

Brooks PC, Stromblad S, Sanders LC, von Schalscha TL, Aimes RT, Stetler-Stevenson WG, Quigley JP, Cheresh DA (1996) Localization of matrix metalloproteinase MMP-2 to the surface of invasive cells by interaction with integrin alpha v beta 3. Cell 85:683–693

Bruinsma R (1997) Les liaisons dangereuses: adhesion molecules do it statistically. Proc Natl Acad Sci U S A 94:375–376

Buckley CD, Simmons DL (1997) Cell adhesion: a new target for therapy. Mol Med Today 3:449–456

Buckley CD, Doyonnas R, Newton JP, Blystone SD, Brown EJ, Watt SM, Simmons DL (1996) Identification of alpha v beta 3 as a heterotypic ligand for CD31/PECAM-1. J Cell Sci 109:437–445

Buckley CD, Rainger GE, Bradfield PF, Nash GB, Simmons DL (1998) Cell adhesion: more than just glue (review). Mol Membr Biol 15:167–176

Burns AR, Walker DC, Brown ES, Thurmon LT, Bowden RA, Keese CR, Simon SI, Entman ML, Smith CW (1997) Neutrophil transendothelial migration is independent of tight junctions and occurs preferentially at tricellular corners. J Immunol 159:2893–2903

Butcher EC, Picker LJ (1996) Lymphocyte homing and homeostasis. Science 272:60–66

Campbell JJ, Butcher EC (2000) Chemokines in tissue-specific and microenvironment-specific lymphocyte homing. Curr Opin Immunol 12:336–341

Campbell JJ, Haraldsen G, Pan J, Rottman J, Qin S, Ponath P, Andrew DP, Warnke R, Ruffing N, Kassam N, Wu L, Butcher EC (1999a) The chemokine receptor CCR4 in vascular recognition by cutaneous but not intestinal memory T cells. Nature 400:776–780

Campbell JJ, Pan J, Butcher EC (1999b) Cutting edge: developmental switches in chemokine responses during T cell maturation. J Immunol 163:2353–2357

Carlos TM, Harlan JM (1994) Leukocyte-endothelial adhesion molecules. Blood 84:2068–2101

Carter SB (1967) Haptotaxis and the mechanism of cell motility. Nature 213:256–260

Cepinskas G, Sandig M, Kvietys PR (1999) PAF-induced elastase-dependent neutrophil transendothelial migration is associated with the mobilization of elastase to the neutrophil surface and localization to the migrating front. J Cell Sci 112 (Pt 12):1937–1945

Chen A, Engel P, Tedder TF (1995) Structural requirements regulate endoproteolytic release of the L-selectin (CD62L) adhesion receptor from the cell surface of leukocytes. J Exp Med 182:519–530

Chothia C, Jones EY (1997) The molecular structure of cell adhesion molecules. Annu Rev Biochem 66:823–862

Cooper D, Lindberg FP, Gamble JR, Brown EJ, Vadas MA (1995) Transendothelial migration of neutrophils involves integrin-associated protein (CD47). Proc Natl Acad Sci U S A 92:3978–3982

Crocker PR, Feizi T (1996) Carbohydrate recognition systems: functional triads in cell–cell interactions. Curr Opin Struct Biol 6:679–691

Crockett-Torabi E (1998) Selectins and mechanisms of signal transduction. J Leukoc Biol 63:1–14

Cunningham SA, Arrate MP, Rodriguez JM, Bjercke RJ, Vanderslice P, Morris AP, Brock TA (2000) A novel protein with homology to the junctional adhesion molecule. Characterization of leukocyte interactions. J Biol Chem 275:34750–34756

Cunningham SA, Rodriguez JM, Arrate MP, Tran TM, Brock TA (2002) JAM2 interacts with alpha4beta1. Facilitation by JAM3. J Biol Chem 277:27589–27592

D'Ambrosio D, Iellem A, Bonecchi R, Mazzeo D, Sozzani S, Mantovani A, Sinigaglia F (1998) Selective up-regulation of chemokine receptors CCR4 and CCR8 upon activation of polarized human type 2 Th cells. J Immunol 161:5111–5115

Dangerfield J, Larbi KY, Huang MT, Dewar A, Nourshargh S (2002) PECAM-1 (CD31) homophilic interaction up-regulates alpha6beta1 on transmigrated neutrophils in vivo and plays a functional role in the ability of alpha6 integrins to mediate leukocyte migration through the perivascular basement membrane. J Exp Med 196:1201–1211

Dejana E, Del Maschio A (1995) Molecular organization and functional regulation of cell to cell junctions in the endothelium. Thromb Haemost 74:309–312

Doherty P, Ashton SV, Moore SE, Walsh FS (1991) Morphoregulatory activities of NCAM and N-cadherin can be accounted for by G protein-dependent activation. Cell 67:21–33

Erlandsen SL, Hasslen SR, Nelson RD (1993) Detection and spatial distribution of the beta 2 integrin (Mac-1) and L-selectin (LECAM-1) adherence receptors on human neutrophils by high-resolution field emission SEM. J Histochem Cytochem 41:327–333

Fawcett J, Buckley C, Holness CL, Bird IN, Spragg JH, Saunders J, Harris A, Simmons DL (1995) Mapping the homotypic binding sites in CD31 and the role of CD31 adhesion in the formation of interendothelial cell contacts. J Cell Biol 128:1229–1241

Feizi T (2000) Carbohydrate-mediated recognition systems in innate immunity. Immunol Rev 173:79–88

Feng D, Nagy JA, Pyne K, Dvorak HF, Dvorak AM (1998) Neutrophils emigrate from venules by a transendothelial cell pathway in response to FMLP. J Exp Med 187:903–915

Finger EB, Bruehl RE, Bainton DF, Springer TA (1996) A differential role for cell shape in neutrophil tethering and rolling on endothelial selectins under flow. J Immunol 157:5085–5096

Gallatin WM, Weissman IL, Butcher EC (1983) A cell-surface molecule involved in organ-specific homing of lymphocytes. Nature 304:30–34

Gearing AJ, Hemingway I, Pigott R, Hughes J, Rees AJ, Cashman SJ (1992) Soluble forms of vascular adhesion molecules, E-selectin, ICAM-1, and VCAM-1: pathological significance. Ann N Y Acad Sci 667:324–331

Geijtenbeek TB, Krooshoop DJ, Bleijs DA, van Vliet SJ, van Duijnhoven GC, Grabovsky V, Alon R, Figdor CG, van Kooyk Y (2000) DC-SIGN-ICAM-2 interaction mediates dendritic cell trafficking. Nat Immunol 1:353–357

Gopalan PK, Smith CW, Lu H, Berg EL, McIntire LV, Simon SI (1997) Neutrophil CD18-dependent arrest on intercellular adhesion molecule 1 (ICAM-1) in shear flow can be activated through L-selectin. J Immunol 158:367–375

Gumbiner BM, Yamada KM (1995) Cell-to-cell contact and extracellular matrix [editorial]. Curr Opin Cell Biol 7:615–618

Haller H, Kunzendorf U, Sacherer K, Lindschau C, Walz G, Distler A, Luft FC (1997) T cell adhesion to P-selectin induces tyrosine phosphorylation of pp125 focal adhesion kinase and other substrates. J Immunol 158:1061–1067

Harpaz Y, Chothia C (1994) Many of the immunoglobulin superfamily domains in cell adhesion molecules and surface receptors belong to a new structural set which is close to that containing variable domains. J Mol Biol 238:528–539

Helander TS, Carpen O, Turunen O, Kovanen PE, Vaheri A, Timonen T (1996) ICAM-2 redistributed by ezrin as a target for killer cells. Nature 382:265–268

Hemler ME, Mannion BA, Berditchevski F (1996) Association of TM4SF proteins with integrins: relevance to cancer. Biochim Biophys Acta 1287:67–71

Holness CL, Simmons DL (1994) Structural motifs for recognition and adhesion in members of the immunoglobulin superfamily. J Cell Sci 107:2065–2070

Hughes PE, Pfaff M (1998) Integrin affinity modulation. Trends Cell Biol 8:359–364

Humphries MJ, Newham P (1998) The structure of cell-adhesion molecules. Trends Cell Biol 8:78–83

Imhof BA, Dunon D (1995) Leukocyte migration and adhesion. Adv Immunol 58:345–416

Imhof BA, Weerasinghe D, Brown EJ, Lindberg FP, Hammel P, Piali L, Dessing M, Gisler R (1997) Cross talk between alpha(v)beta3 and alpha4beta1 integrins regulates lymphocyte migration on vascular cell adhesion molecule 1. Eur J Immunol 27:3242–3252

Imhof BA, Engelhardt B, Vadas M (2001) Novel mechanisms of the transendothelial migration of leukocytes. Trends Immunol 22:411–414

Irjala H, Johansson EL, Grenman R, Alanen K, Salmi M, Jalkanen S (2001) Mannose receptor is a novel ligand for L-selectin and mediates lymphocyte binding to lymphatic endothelium. J Exp Med 194:1033–1042

Irjala H, Elima K, Johansson EL, Merinen M, Kontula K, Alanen K, Grenman R, Salmi M, Jalkanen S (2003) The same endothelial receptor controls lymphocyte traffic both in vascular and lymphatic vessels. Eur J Immunol 33:815–824

Issekutz TB (1992) Lymphocyte homing to sites of inflammation. Curr Opin Immunol 4:287–293

Jackson DG (2003) The lymphatics revisited: new perspectives from the hyaluronan receptor LYVE-1. Trends Cardiovasc Med 13:1–7

Jackson DG, Bell JI, Dickinson R, Timans J, Shields J, Whittle N (1995) Proteoglycan forms of the lymphocyte homing receptor CD44 are alternatively spliced variants containing the v3 exon. J Cell Biol 128:673–685

Jalkanen S, Salmi M (1993) Vascular adhesion protein-1 (VAP-1)—a new adhesion molecule recruiting lymphocytes to sites of inflammation. Res Immunol 144:746–749

Johnson-Leger C, Imhof BA (2003) Forging the endothelium during inflammation: pushing at a half-open door? Cell Tissue Res 314:93–105

Johnson-Leger C, Aurrand-Lions M, Imhof BA (2000) The parting of the endothelium: miracle, or simply a junctional affair? J Cell Sci 113:921–933

Johnson-Leger CA, Aurrand-Lions M, Beltraminelli N, Fasel N, Imhof BA (2002) Junctional adhesion molecule-2 (JAM-2) promotes lymphocyte transendothelial migration. Blood 100:2479–2486

Jung U, Ley K (1999) Mice lacking two or all three selectins demonstrate overlapping and distinct functions for each selectin. J Immunol 162:6755–6762

Kostrewa D, Brockhaus M, D'Arcy A, Dale GE, Nelboeck P, Schmid G, Mueller F, Bazzoni G, Dejana E, Bartfai T, Winkler FK, Hennig M (2001) X-ray structure of junctional adhesion molecule: structural basis for homophilic adhesion via a novel dimerization motif. EMBO J 20:4391–4398

Kreis T, Vale R (1999) Extracellular matrix, anchor, and adhesion proteins, 2nd edn. Oxford University Press, Oxford

Kriehuber E, Breiteneder-Geleff S, Groeger M, Soleiman A, Schoppmann SF, Stingl G, Kerjaschki D, Maurer D (2001) Isolation and characterization of dermal lymphatic and blood endothelial cells reveal stable and functionally specialized cell lineages. J Exp Med 194:797–808

Kuby J (1997) Immunology, 3rd edn. W.H. Freeman and Co, New York

Kunkel EJ, Campbell JJ, Haraldsen G, Pan J, Boisvert J, Roberts AI, Ebert EC, Vierra MA, Goodman SB, Genovese MC, Wardlaw AJ, Greenberg HB, Parker CM, Butcher EC, Andrew DP, Agace WW (2000) Lymphocyte CC chemokine receptor 9 and epithelial thymus-expressed chemokine (TECK) expression distinguish the small intestinal immune compartment: Epithelial expression of tissue-specific chemokines as an organizing principle in regional immunity. J Exp Med 192:761–768

Lalor PF, Edwards S, McNab G, Salmi M, Jalkanen S, Adams DH (2002) Vascular adhesion protein-1 mediates adhesion and transmigration of lymphocytes on human hepatic endothelial cells. J Immunol 169:983–992

Lenter M, Vestweber D (1994) The integrin chains beta 1 and alpha 6 associate with the chaperone calnexin prior to integrin assembly. J Biol Chem 269:12263–12268

Lindberg FP, Gresham HD, Reinhold MI, Brown EJ (1996) Integrin-associated protein immunoglobulin domain is necessary for efficient vitronectin bead binding. J Cell Biol 134:1313–1322

Liu Y, Nusrat A, Schnell FJ, Reaves TA, Walsh S, Pochet M, Parkos CA (2000) Human junction adhesion molecule regulates tight junction resealing in epithelia. J Cell Sci 113:2363–2374

Luscinskas FW, Ma S, Nusrat A, Parkos CA, Shaw SK (2002) Leukocyte transendothelial migration: a junctional affair. Semin Immunol 14:105–113

Makinen T, Veikkola T, Mustjoki S, Karpanen T, Catimel B, Nice EC, Wise L, Mercer A, Kowalski H, Kerjaschki D, Stacker SA, Achen MG, Alitalo K (2001) Isolated lymphatic endothelial cells transduce growth, survival and migratory signals via the VEGF-C/D receptor VEGFR-3. EMBO J 20:4762–4773

Mamdouh Z, Chen X, Pierini LM, Maxfield FR, Muller WA (2003) Targeted recycling of PECAM from endothelial surface-connected compartments during diapedesis. Nature 421:748–753

Martin-Padura I, Lostaglio S, Schneemann M, Williams L, Romano M, Fruscella P, Panzeri C, Stoppacciaro A, Ruco L, Villa A, Simmons D, Dejana E (1998) Junctional adhesion molecule, a novel member of the immunoglobulin superfamily that distributes at intercellular junctions and modulates monocyte transmigration. J Cell Biol 142:117–127

McEver RP (1990) Properties of GMP-140, an inducible granule membrane protein of platelets and endothelium. Blood Cells 16:73–80

McEver RP, Moore KL, Cummings RD (1995) Leukocyte trafficking mediated by selectin-carbohydrate interactions. J Biol Chem 270:11025–11028

Moore KL, Patel KD, Bruehl RE, Li F, Johnson DA, Lichenstein HS, Cummings RD, Bainton DF, McEver RP (1995) P-selectin glycoprotein ligand-1 mediates rolling of human neutrophils on P-selectin. J Cell Biol 128:661–671

Morita K, Furuse M, Fujimoto K, Tsukita S (1999a) Claudin multigene family encoding four-transmembrane domain protein components of tight junction strands. Proc Natl Acad Sci U S A 96:511–516

Morita K, Sasaki H, Fujimoto K, Furuse M, Tsukita S (1999b) Claudin-11/OSP-based tight junctions of myelin sheaths in brain and Sertoli cells in testis. J Cell Biol 145:579–588

Muller WA (2003) Leukocyte-endothelial-cell interactions in leukocyte transmigration and the inflammatory response. Trends Immunol 24:327–334

Muller WA, Randolph GJ (1999) Migration of leukocytes across endothelium and beyond: molecules involved in the transmigration and fate of monocytes. J Leukoc Biol 66:698–704

Nibbs RJ, Kriehuber E, Ponath PD, Parent D, Qin S, Campbell JD, Henderson A, Kerjaschki D, Maurer D, Graham GJ, Rot A (2001) The beta-chemokine receptor D6 is expressed by lymphatic endothelium and a subset of vascular tumors. Am J Pathol 158:867–877

Oppenheimer-Marks N, Kavanaugh AF, Lipsky PE (1994) Inhibition of the transendothelial migration of human T lymphocytes by prostaglandin E2. J Immunol 152:5703–5713

Ostermann G, Weber KS, Zernecke A, Schroder A, Weber C (2002) JAM-1 is a ligand of the beta(2) integrin LFA-1 involved in transendothelial migration of leukocytes. Nat Immunol 3:151–158

Ozaki H, Ishii K, Horiuchi H, Arai H, Kawamoto T, Okawa K, Iwamatsu A, Kita T (1999) Cutting edge: combined treatment of TNF-alpha and IFN-gamma causes redistribution of junctional adhesion molecule in human endothelial cells. J Immunol 163:553–557

Papadakis KA, Prehn J, Nelson V, Cheng L, Binder SW, Ponath PD, Andrew DP, Targan SR (2000) The role of thymus-expressed chemokine and its receptor CCR9 on lymphocytes in the regional specialization of the mucosal immune system. J Immunol 165:5069–5076

Petty HR, Todd RF (1996) Integrins as promiscuous signal transduction devices. Immunol Today 17:209–212

Picker LJ, Butcher EC (1992) Physiological and molecular mechanisms of lymphocyte homing. Annu Rev Immunol 10:561–591

Porter JC, Hogg N (1998) Integrins take partners: cross-talk between integrins and other membrane receptors. Trends Cell Biol 8:390–396

Pouyani T, Seed B (1995) PSGL-1 recognition of P-selectin is controlled by a tyrosine sulfation consensus at the PSGL-1 amino terminus. Cell 83:333–343

Rainger GE, Buckley C, Simmons DL, Nash GB (1997) Cross-talk between cell adhesion molecules regulates the migration velocity of neutrophils. Curr Biol 7:316–325

Rainger GE, Buckley CD, Simmons DL, Nash GB (1999) Neutrophils sense flow-generated stress and direct their migration through alphaVbeta3-integrin. Am J Physiol 276:H858–H864

Reedquist KA, Ross E, Koop EA, Wolthuis RM, Zwartkruis FJ, van Kooyk Y, Salmon M, Buckley CD, Bos JL (2000) The small GTPase, Rap1, mediates CD31-induced integrin adhesion. J Cell Biol 148:1151–1158

Ruiz P, Schwarzler C, Gunthert U (1995) CD44 isoforms during differentiation and development. Bioessays 17:17–24

Sako D, Comess KM, Barone KM, Camphausen RT, Cumming DA, Shaw GD (1995) A sulfated peptide segment at the amino terminus of PSGL-1 is critical for P-selectin binding. Cell 83:323–331

Salmi M, Jalkanen S (1992) A 90-kilodalton endothelial cell molecule mediating lymphocyte binding in humans. Science 257:1407–1409

Salmi M, Kalimo K, Jalkanen S (1993) Induction and function of vascular adhesion protein-1 at sites of inflammation. J Exp Med 178:2255–2260

Santoso S, Sachs UJ, Kroll H, Linder M, Ruf A, Preissner KT, Chavakis T (2002) The junctional adhesion molecule 3 (JAM-3) on human platelets is a counterreceptor for the leukocyte integrin Mac-1. J Exp Med 196:679–691

Schenkel AR, Mamdouh Z, Chen X, Liebman RM, Muller WA (2002) CD99 plays a major role in the migration of monocytes through endothelial junctions. Nat Immunol 3:143–150

Shimizu Y, Shaw S (1993) Cell adhesion. Mucins in the mainstream. Nature 366:630–631

Shimizu Y, Newman W, Tanaka Y, Shaw S (1992) Lymphocyte interactions with endothelial cells. Immunol Today 13:106–112

Sobocka MB, Sobocki T, Banerjee P, Weiss C, Rushbrook JI, Norin AJ, Hartwig J, Salifu MO, Markell MS, Babinska A, Ehrlich YH, Kornecki E (2000) Cloning of the human platelet F11 receptor: a cell adhesion molecule member of the immunoglobulin superfamily involved in platelet aggregation. Blood 95:2600–2609

Somersalo K (1996) Migratory functions of natural killer cells. Nat Immun 15:117–133

Steegmaier M, Blanks JE, Borges E, Vestweber D (1997) P-selectin glycoprotein ligand-1 mediates rolling of mouse bone marrow-derived mast cells on P-selectin but not efficiently on E-selectin. Eur J Immunol 27:1339–1345

Stein JV, Cheng G, Stockton BM, Fors BP, Butcher EC, von Andrian UH (1999) L-selectin-mediated leukocyte adhesion in vivo: microvillous distribution determines tethering efficiency, but not rolling velocity. J Exp Med 189:37–50

Tedder TF, Steeber DA, Chen A, Engel P (1995) The selectins: vascular adhesion molecules. FASEB J 9:866–873

Todd RF, Petty HR (1997) Beta 2 (CD11/CD18) integrins can serve as signaling partners for other leukocyte receptors. J Lab Clin Med 129:492–498

Tsukita S, Furuse M, Itoh M (1999) Structural and signalling molecules come together at tight junctions. Curr Opin Cell Biol 11:628–633

Vestweber D (2003) Lymphocyte trafficking through blood and lymphatic vessels: more than just selectins, chemokines and integrins. Eur J Immunol 33:1361–1364

von Andrian UH, Mackay CR (2000) T-cell function and migration. Two sides of the same coin. N Engl J Med 343:1020–1034

von Andrian UH, Berger EM, Ramezani L, Chambers JD, Ochs HD, Harlan JM, Paulson JC, Etzioni A, Arfors KE (1993) In vivo behavior of neutrophils from two patients with distinct inherited leukocyte adhesion deficiency syndromes. J Clin Invest 91:2893–2897

von Andrian UH, Hasslen SR, Nelson RD, Erlandsen SL, Butcher EC (1995) A central role for microvillous receptor presentation in leukocyte adhesion under flow. Cell 82:989–999

Vuori K, Ruoslahti E (1994) Association of insulin receptor substrate-1 with integrins. Science 266:1576–1578

Wagner DD (1993) The Weibel-Palade body: the storage granule for von Willebrand factor and P-selectin. Thromb Haemost 70:105–110

Walcheck B, Kahn J, Fisher JM, Wang BB, Fisk RS, Payan DG, Feehan C, Betageri R, Darlak K, Spatola AF, Kishimoto TK (1996) Neutrophil rolling altered by inhibition of L-selectin shedding in vitro. Nature 380:720–723

Warren AP, Patel K, McConkey DJ, Palacios R (1996) CD98: a type II transmembrane glycoprotein expressed from the beginning of primitive and definitive hematopoiesis may play a critical role in the development of hematopoietic cells. Blood 87:3676–3687

Wary KK, Mainiero F, Isakoff SJ, Marcantonio EE, Giancotti FG (1996) The adaptor protein Shc couples a class of integrins to the control of cell cycle progression. Cell 87:733–743

Wei Y, Lukashev M, Simon DI, Bodary SC, Rosenberg S, Doyle MV, Chapman HA (1996) Regulation of integrin function by the urokinase receptor. Science 273:1551–1555

Williams LA, Martin-Padura I, Dejana E, Hogg N, Simmons DL (1999) Identification and characterisation of human junctional adhesion molecule (JAM). Mol Immunol 36:1175–1188

Yanez-Mo M, Tejedor, R Rousselle, P Sanchez (2001) Tetraspanins in intercellular adhesion of polarized epithelial cells: spatial and functional relationship to integrins and cadherins. J Cell Sci 114:577–587

Yap AS, Brieher WM, Gumbiner BM (1997) Molecular and functional analysis of cadherin-based adherens junctions. Annu Rev Cell Dev Biol 13:119–146

Yebra M, Parry GC, Stromblad S, Mackman N, Rosenberg S, Mueller BM, Cheresh DA (1996) Requirement of receptor-bound urokinase-type plasminogen activator for integrin alphavbeta5-directed cell migration. J Biol Chem 271:29393–29399

Zabel BA, Agace WW, Campbell JJ, Heath HM, Parent D, Roberts AI, Ebert EC, Kassam N, Qin S, Zovko M, LaRosa GJ, Yang LL, Soler D, Butcher EC, Ponath PD, Parker CM, Andrew DP (1999) Human G protein-coupled receptor GPR-9-6/CC chemokine receptor 9 is selectively expressed on intestinal homing T lymphocytes, mucosal lymphocytes, and thymocytes and is required for thymus-expressed chemokine-mediated chemotaxis. J Exp Med 190:1241–1256

Pharmacology of Platelet Adhesion and Aggregation

B. Nieswandt[1] · S. Offermanns[2] (✉)

[1] Rudolf Virchow Center for Experimental Biomedicine, Vascular Biology, University of Würzburg, 97078 Würzburg, Germany
[2] Institute of Pharmacology, University of Heidelberg, Im Neuenheimer Feld 366, 69120 Heidelberg, Germany
Stefan.Offermanns@urz.uni-heidelberg.de

1	Introduction	438
2	Adhesion of Platelets to the Extracellular Matrix	439
2.1	von Willebrand Factor	440
2.1.1	Glycoprotein Ib-V-IX	441
2.2	Collagen	442
2.2.1	Glycoprotein VI	443
2.2.2	Integrin $\alpha 2\beta 1$	445
2.2.3	Integrins $\alpha 5\beta 1/\alpha 6\beta 1$	447
2.2.4	Integrin $\alpha IIb\beta 3$ (GPIIb/IIIa)	447
3	Platelet Aggregation	449
3.1	Integrin $\alpha IIb\beta 3$ in Aggregation	449
3.1.1	Inside-Out Activation	450
3.1.2	Outside-In Activation	454
3.2	Stabilization of Platelet Aggregates	454
4	Platelet-Endothelial Cell Interaction	455
5	Pharmacological Strategies to Interfere with Platelet Adhesive Functions	456
5.1	Inhibition of Platelet Adhesion	456
5.2	Inhibition of Integrin $\alpha IIb\beta 3$ (GPIIb/IIIa) Activation	458
5.2.1	Inhibition of TXA_2 Production or Action	458
5.2.2	Inhibition of ADP-Dependent Platelet Activation	459
5.3	Glycoprotein IIb/IIIa Antagonists	460
6	Conclusions	461
	References	462

Abstract At the injured vessel wall, blood platelets become activated and adhere to the subendothelial surface as well as to each other. These cellular adhesion processes are required for primary hemostasis, but can also lead to thrombosis. Considerable progress has been made during recent years in understanding the molecular mechanisms underlying platelet activation and adhesion. This knowledge will drive future efforts towards the development of new antiplatelet drugs for the prevention and treatment of cardiovascular diseases.

Keywords Platelets · Integrins · Collagen · Platelet adhesion · Antiplatelet drugs

1 Introduction

Platelets are small anuclear cell fragments which are generated from megakaryocytes in the bone marrow and then circulate for about 10 days in the blood before they are phagocytosed by tissue macrophages in the spleen and liver. During their normal life in the circulation, most platelets never undergo firm adhesion. The adhesive potential of platelets is only revealed when the endothelial cell layer which lines the blood vessels is damaged by injuries or alterations of the vessel wall such as in atherosclerosis. Under these conditions, components of the extracellular matrix are exposed on the luminal side of the vessel wall and are recognized by specific receptors on the platelet surface. Interaction between extracellular matrix and platelet re-

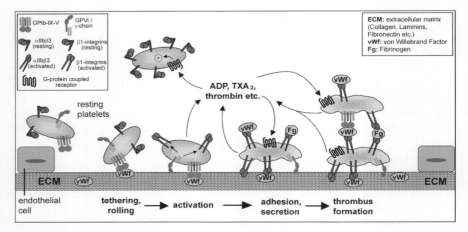

Fig. 1. Platelet adhesion and aggregation on the extracellular matrix (*ECM*). At sites of vascular injury the ECM becomes exposed to the flowing blood, allowing platelet adhesion and aggregation. Under high shear flow conditions, the initial contact (tethering) to the ECM is mediated predominantly by GP (GP)Ibα binding to von Willebrand factor (*vWF*) immobilized on collagen. The GPIbα-vWF interaction is essential at high shear rates (>500 s^{-1}), but may not be required at lower shear rates. In a second step, GPVI-collagen interactions initiate cellular activation followed by shifting of integrins to a high-affinity state and the release of second-wave agonists, most importantly adenosine diphosphate (*ADP*) and thromboxane A$_2$ (*TXA$_2$*). GPIb-mediated signalling may amplify GPVI-induced activation pathways. Cellular activation and upregulation of integrin affinity is a strict pre-requisite for adhesion. Firm adhesion of platelets to the ECM is mediated by activated β1 integrins and αIIbβ3. Integrins contribute to platelet activation directly via outside-in signalling and indirectly by stabilizing the GPVI-collagen interaction, resulting in sustained signalling and enhanced release. Released ADP and TXA$_2$ in turn amplify integrin activation on adherent platelets and mediate thrombus growth by activating additional platelets. This scheme does not exclude the involvement of other receptor-ligand interactions. *Fg*, fibrinogen

ceptors mediates the initial adhesion of platelets to the damaged vessel wall. In addition, various soluble stimuli are produced and released from platelets which strengthen platelet adhesion. These mediators also recruit more platelets into the growing platelet plug by exposing or activating receptors on the platelet surface which allow platelets to adhere to each other, a process called platelet aggregation (see Fig. 1). Platelet adhesion to the vessel wall and the subsequent formation of a platelet plug lead to the closure of smaller defects of the vessel wall and are required for primary hemostasis. Platelet plug formation is also the major pathomechanism underlying arterial thrombosis, since it occurs in myocardial infarction or ischemic stroke. The adhesive properties of platelets have to be regulated with high fidelity. This regulation has to ensure that platelets become activated under appropriate conditions in order to prevent blood loss in cases of vascular injury, while at the same time inappropriate platelet adhesion, which can lead to the clogging of blood vessels, has to be avoided. Platelets, which have evolved relatively late during evolution and which are only found in mammals, possess various adhesion receptors and a sophisticated regulatory machinery in order to adhere in response to a well-defined set of stimuli. Thus, platelets are a very instructive model for the cellular mechanisms underlying regulated adhesiveness. Their central role in arterial thrombosis has made the signalling mechanism and membrane proteins which are involved in regulating platelet adhesiveness prime targets for drugs used to treat and prevent arterial thrombosis.

2
Adhesion of Platelets to the Extracellular Matrix

At sites of vessel wall injury, the flowing blood comes in contact with the subendothelial extracellular matrix (ECM) which contains a large number of adhesive macromolecules, such as laminin, fibronectin, collagen, and von Willebrand factor (vWF). This triggers activation and adhesion of platelets, a multi-step process involving the concerted action of a variety of different surface receptors and their signalling pathways (see Fig. 1).

The mechanisms of platelet adhesion at sites of injury are to a large extent determined by the prevailing rheological conditions. Blood flows with a greater velocity in the center of the vessel than near the wall, thereby generating shear forces between adjacent layers of fluid that become maximal at the wall. The drag, which opposes platelet adhesion and aggregation, increases with the prevailing shear rates. Under conditions of high shear, such as is found in small arteries and arterioles, the initial tethering of platelets to the ECM is mediated by the interaction between the platelet receptor glycoprotein (GP)Ib and vWF bound to collagen. While mandatory at high shear, this interaction may not be relevant under conditions of low shear as found in veins and large arteries. The binding of GPIb to vWF has a fast off-rate and is therefore insufficient to mediate stable adhesion but rather maintains the platelet in close contact with the surface, although it continuously translocates in the direction of blood flow. During this 'rolling', platelets establish contacts with the thrombogenic ECM protein collagen through the immunoglobulin superfamily receptor, GPVI. While GPVI binds collagen with low affinity and thus is unable to mediate adhesion by itself, it triggers intracellular signals that shift platelet integrins to a high-affinity state and induce the release of the second-wave agonists adenosine diphosphate (ADP) and thromboxane A_2 (TXA_2), which in turn reinforce the activation process. Firm adhesion to the ECM is mediated by high-affinity β1-integrins which bind to collagen ($\alpha 2\beta 1$), fibronectin ($\alpha 5\beta 1$) and laminin ($\alpha 6\beta 1$), as well as the major platelet integrin, αIIbβ3, interacting with fibronectin and collagen-bound vWF.

2.1
von Willebrand Factor

vWF is a multimeric, adhesive glycoprotein that contains binding sites for collagen as well as for the two major platelet receptors, GPIb and integrin αIIbβ3, making it a central mediator in the adhesion process (Ruggeri 1999). vWF is found in the α-granules of platelets, in the Weibel-Palade bodies of endothelial cells, and in the plasma at a concentration of approximately 10 µg/ml in humans (Berndt et al. 2001). The mature subunit of vWF consists of 2,050 amino acids and is composed of four different repeating domains, designated A through D (Shelton-Inloes et al. 1986). The three homologous A domains span residues 497 to 1,111 and regulate interaction with different receptors and prothrombotic ligands of the subendothelial matrix. The A1 domain exclusively binds collagen type VI, whereas collagen I and III are bound via the A3 domain (Hoylaerts et al. 1997). The C1 domain contains the sequence Arg-Gly-Asp (RGD), which represents a binding motif for both of the platelet β3-integrins, namely αIIbβ3 and $\alpha v\beta 3$. Of major importance is the binding of vWF to the platelet receptor complex GPIb-V-IX via the A1 domain. This interaction is critically involved in the initial adhesion of platelets to damaged subendothelium under high shear (Ruggeri 1997), which occurs mostly in arteries that are mainly affected during the pathogenesis of acute coronary syndromes and stroke.

The mature vWF subunit is intracellularly assembled into dimers by two disulphide-linked C2 domains. They serve as constituents to build large multimers up to 20×10^6 in molecular weight by linkage of Cys residues in the D' and the D1 to D3 (Ruggeri 1999). The largest multimers with the greatest thrombogenic potential are stored in platelets and endothelial cells and get secreted upon cell activation or damage (Sporn et al. 1986).

Under physiological conditions, vWF does not bind or binds only with a very low affinity to its platelet receptor, GPIb-V-IX. However, when vWF is immobilized on exposed collagen of damaged subendothelial layers under flow conditions relevant for hemostasis in arterioles or thrombosis in atherosclerotic arteries, it becomes a strong adhesive substrate. Based on in vitro studies, it is believed that conformational changes in the A1 domain might cause an increase in the affinity of vWF to its interaction partners (Ruggeri 1999). Botrocetin, a viper venom protein from *Bothrops jararaca*, and the bacterial-derived antibiotic ristocetin from *Amycolatopsis lurida* (*Amycolatopsis orientalis* subsp. *lurida*) induce vWF interaction with GPIb-V-IX and both compounds bind vWF near the Cys-509/Cys-695 disulphide bond within the A1 domain (De Luca et al. 2000). Under in vivo conditions, the vWF molecule might alter conformation due to high shear forces and due to the immobilization to a surface.

The absence of vWF in humans causes severe defects in primary hemostasis and coagulation (Ewenstein 1997). Such patients have strongly reduced factor VIII levels and suffer from spontaneous bleeding. A similar phenotype is also found in vWF knock-out mice, and these animals are resistant to arterial thrombosis (Denis et al. 1998).

2.1.1
Glycoprotein Ib-V-IX

The initial tethering of platelets at sites of vascular injury requires the action of a receptor that functions irrespective of cellular activation and thereby facilitates rapid interactions that resist the shear forces acting on the cells. This process is mediated by GPIb-V-IX, a structurally unique receptor complex exclusively expressed in platelets and megakaryocytes. The interaction of GPIb-V-IX with vWF bound to subendothelium or on the surface of activated adherent platelets is essential for platelet adhesion and thrombus formation, respectively, under conditions of elevated shear (Sakariassen et al. 1986). The major importance of the receptor complex for hemostasis is exemplified by the Bernard-Soulier Syndrome, a congenital bleeding disorder which is characterized by a platelet inability to adhere to subendothelial matrices and a dramatically prolonged bleeding time (Lopez et al. 1998).

The receptor complex consists of four transmembrane subunits that belong to the leucine-rich repeat protein superfamily (Berndt et al. 2001). GPIbα (135 kDa), the dominant functional subunit of the complex, is disulphide-linked to GPIbβ (25 kDa) and non-covalently associated with GPIX (22 kDa) and GPV (88 kDa), in the ratio 2:2:2:1. While relatively little is known about the biological function of GPIX and V, much has been learned about GPIb during the past few years. The extracellular part of GPIbα contains seven tandem leucine-rich domains and carries the binding sites for all major ligands of the receptor, notably vWF, Mac-1, P-selectin, thrombin, high molecular weight kininogen, and Factor XII (Berndt et al. 2001). The cytoplasmic tail of GPIbα is 96 amino acids in length (Lopez et al. 1987), which encloses a binding site for actin-binding protein and, like GPIbβ, a recognition sequence for 14-3-3ζ (Andrews et al. 1998). A protein kinase (PK)A phosphorylation site at Ser-166 in the GPIbβ cytoplasmic tail (181 amino acids) is possibly involved in agonist-induced platelet actin polymerization (Wardell et al. 1989), whereas a putative role for the short cytoplasmic tails of GPIX and GPV has not been described yet.

The binding of GPIbα to the A1 domain of vWF is the principal interaction capable of tethering platelets to the vessel wall at high shear flow conditions. While sufficient to support platelet binding, this adhesive interaction is characterized by a rapid dissociation rate and thus cannot mediate irreversible adhesion by itself. Rather, the interaction keeps platelets in close contact with the matrix, although the cells continuously translocate in the direction of blood flow. Currently it is not clear how GPIbα-vWF interactions are regulated, but a few regions within the receptor have been proposed to be critical for this process (Marchese et al. 1995; Afshar-Kharghan and Lopez 1997; Shen et al. 2000).

The dynamics of platelet adhesion on purified immobilized vWF are principally similar to those observed in thrombus formation in vivo, in that the cells translocate before establishing firm adhesion contacts. However, on a vWF surface, platelets roll for several minutes until αIIbβ3-mediated stable adhesion is seen, which demonstrates that activation of the integrin does not efficiently occur in this isolated system. Although the importance of the GPIb-vWF interaction for platelet recruitment to the reactive surface is firmly established, it is still unclear to what extent or even whether this adhesion event contributes to platelet activation, which is a strict pre-requisite for firm adhesion and thrombus growth. Although considerable evidence has emerged during the past few years that GPIb-V-IX can elicit intracellular signals that may contribute to platelet activation, further studies will be required to identify the underlying mechanisms (Jackson et al. 2003). Furthermore, the relevance of these signals for platelet adhesion and aggregation will have to be defined, since GPIb is generally considered a very weak agonist in platelets compared to other stimuli, most notably collagen.

2.2
Collagen

In contrast to vWF, the subendothelial ECM is a highly efficient substrate to support firm platelet adhesion and thrombus formation. Among the macromolecular constituents of the ECM, fibrillar collagen type I is by far the most thrombogenic in that it not only supports platelet adhesion by direct and indirect (via bound vWF) mechanisms, but it also directly activates the cells (Baumgartner 1977). In vitro, the vast majority of platelets tethering to collagen adhere irreversibly via integrins and become highly activated, thereby providing a very reactive surface for the recruitment of additional platelets, i.e. thrombus growth. Therefore, collagens are thought to play a central role in the initiation of hemostasis and thrombosis. Besides GPIb and integrin $\alpha IIb\beta 3$, which indirectly interact with collagen via vWF, several direct collagen receptors have been identified on the platelet surface, most notably the immunoglobulin superfamily receptor, GPVI, and integrin $\alpha 2\beta 1$ (Clemetson and Clemetson 2001).

Of the more than 20 forms of collagen found in the mammalian organism, fibrillar types I and III are the major constituents of the ECM of blood vessels. In addition, the network-forming type IV collagen is the major form in the subendothelial basement membrane. Collagens consist of repeat GXY motifs where G is glycine and X and Y are frequently proline (amino acid code=P) and hydroxyproline (amino acid code=O). The major platelet-activating motif within collagen is the GPO sequence, which makes up approximately 10% of collagens I and III and is specifically recognized by the low-affinity collagen receptor, GPVI. The GXY repeat sequence forms a single left-handed helix that associates with two other chains to form a right-handed superhelix. In collagens I and III, the chains have approximately 1,000 amino acids flanked by short non-helical N- and C-terminal telopeptide extensions. The crosslinking of these monomeric collagen structures forms fibrillar collagen, the predominant structure that platelets come into contact with in the ECM.

Synthetic peptides containing five or more GPO repeats spontaneously form helical structures and act as strong platelet activators when crosslinked into large multimers, but are unable to support platelet adhesion under flow conditions, since they are not recognized by integrins (Morton et al. 1995). In contrast, the introduction of the $\alpha 2\beta 1$-reactive sequence, GFOGER, into a backbone of GPP to confer a helical sequence, results in a peptide that does not activate platelets but supports $\alpha 2\beta 1$-mediated adhesion (Knight et al. 2000). Thus, collagen contains different motifs that selectively mediate platelet activation or adhesion through distinct surface receptors.

2.2.1
Glycoprotein VI

Platelet activation by fibrillar collagen is triggered by GPVI, a platelet/megakaryocyte-specific low-affinity collagen receptor. GPVI is non-covalently complexed with the FcRγ-chain, which contains an immunoreceptor tyrosine-based activation motif (ITAM) and serves as the signal transducing subunit of the receptor complex. Signalling through the GPVI/FcRγ-chain complex is similar to that of different immune receptors, such as the T cell receptor or Fc receptors (Watson and Gibbins 1998).

GPVI is a highly glycosylated 62-kDa type I transmembrane glycoprotein belonging to the immunoglobulin superfamily and is closely related to a number of immune receptors. The receptor is composed of 339 amino acids and contains two Ig-C2-like extracellular domains formed by disulphide bonds, a mucin-like stalk, a transmembrane region and a short 51 amino acid cytoplasmic tail (Clemetson et al. 1999). GPVI harbors a positively charged arginine in its transmembrane region, which is essential for association with the FcRγ-chain. The GPVI cytosolic tail contains a proline-rich motif that binds selectively to the Src homology (SH)3 domain of the Src family tyrosine kinases, Fyn and Lyn. It is proposed that crosslinking of GPVI by collagen brings SH3-associated Fyn or Lyn to the FcRγ-chain enabling phosphorylation of the ITAM to take place and initiation of the signalling cascade. The constitutive association of Fyn and Lyn with GPVI may place the receptor in a 'ready-to-go' state, enabling rapid activation on exposure to collagen (Watson et al. 2001). The cytoplasmic part of GPVI also contains a calmodulin-binding domain. Calmodulin is constitutively associated with GPVI in platelets and undergoes delayed dissociation upon activation, although the functional significance of this is not known.

Tyrosine phosphorylation of the FcRγ-chain ITAM leads to binding and subsequent activation of the tandem SH2-domain-containing tyrosine kinase, Syk, which initiates a downstream signalling cascade that culminates in activation of a number of effector enzymes including PLCγ2, small GTPases and PI 3-kinase (see Fig. 2). The adapters linker for activation of T cells (LAT) and Src homology 2 domain-containing leukocyte protein (SLP)-76 play critical roles in this signalling cascade, which is thought to take place in cholesterol-rich membrane domains known as Gems or rafts. GPVI-mediated signalling results in strong integrin activation and release of stored mediators, which play a central role for thrombus growth (see below) (Nieswandt and Watson 2003).

Platelets in which GPVI is inhibited or absent are refractory to activation by collagen and show virtually no adhesion to the matrix protein. While the initial tethering and rolling still occurs through the interaction of GPIb with collagen-bound vWF, the transition to stable adhesion is defective due to the lack of integrin activation (Nieswandt et al. 2001b; Kato et al. 2003). Direct activation of integrins by Mn^{2+}, or the addition of agonists-which activate integrins via inside-out signalling-such as ADP can restore adhesion of such platelets. Thus, the role of GPVI is to generate intracellular signals that promote integrin activation rather than to serve as an adhesion receptor itself. Mice deficient in GPVI or the FcRγ-chain, the latter also lacking GPVI, show dramatically reduced platelet adhesion and thrombus formation on collagen in vitro and at sites of arterial injury, i.e. under high shear flow conditions in vivo (Konishi et al. 2002; Massberg et al. 2003). Thus, GPVI-mediated activation is crucial for platelet adhesion on collagen-surfaces, but the receptor is, similar to GPIb-V-IX, unable to mediate this adhesion by itself. Shear-re-

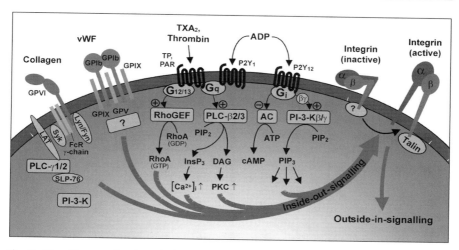

Fig. 2. Signalling mechanisms linking various platelet receptors to the inside-out activation of integrins like integrin $\alpha IIb\beta 3$ (GPIIb/IIIa) or integrin $\alpha 2\beta 1$ (GPIa/IIa). GPVI is the main receptor mediating platelet activation by collagen. Binding of collagen to GPVI results in tyrosine phosphorylation of the FcRγ chain by the kinases Lyn and Fyn, leading to recruitment of the tyrosine kinase Syk. Syk phosphorylates 'linker for activation of T cells' (*LAT*) which in turn recruits phospholipase C-γ2 (*PLCγ2*) and 'Src homology 2 domain-containing leukocyte protein of 76 kDa' (*SLP-76*) to the cell membrane. LAT also associates with phosphoinositide-3 kinase (*PI-3-K*) resulting in the formation of phosphatidylinositol(3,4,5)P$_3$. There is evidence that the vWF receptor GPIb-V-IX, after binding of ligand, also induces intracellular signalling processes. Stimulation of heptahelical G protein-coupled receptors by thromboxane A$_2$ (*TXA$_2$*), thrombin and ADP leads to the stimulation of various signalling pathways involving the G proteins G_q, $G_{i/z}$ and $G_{12/13}$. *AC*, adenylyl cyclase; *ADP*, adenosine diphosphate; *cAMP*, cyclic adenosine monophosphate; *DAG*, diacyl glycerol; *GP*, glycoprotein; *InsP$_3$*, inositol-1,4,5-trisphosphate; *PAR*, protease-activated receptor; *PI-3-Kβ/γ*, phosphoinositide-3-kinase β/γ; *PIP$_2$*, phosphatidylinositol-4,5-bisphosphate; *PIP$_3$*, phosphatidylinositol-3,4,5-trisphosphate; *PKC*, protein kinase C; *PLCβ2/3*, phospholipase C-β2/3; *RhoGEF*, Rho-specific guanine nucleotide exchange factor; *SLP*, Src homology 2 domain-containing leukocyte protein; *TP*, thromboxane A2 receptor

sistant platelet attachment on collagen requires the binding of integrins, notably $\alpha 2\beta 1$ and $\alpha IIb\beta 3$, which, respectively, directly or indirectly (via vWf) bind to the matrix protein.

2.2.2
Integrin $\alpha 2\beta 1$

Integrin $\alpha 2\beta 1$ (also known as platelet GPIa/IIa or lymphocyte VLA-2) is widely expressed in tissues and serves as a receptor for collagen (Santoro and Zutter 1995). The heterodimeric complex consists of the non-covalently associated $\alpha 2$ (GPIa, 160 kDa) and $\beta 1$ (GPIIa, 130 kDa) subunits and is present on platelets at a level of 1,000-4,000 copies/cell. Both subunits have large extracellular domains, a single transmembrane domain and short cytoplasmic tails. The extracellular part of the $\alpha 2$ subunit consists of three structural motifs. The juxtamembrane region of β-sheets is followed by the N-terminal seven tandem repeats that are folded into a seven-bladed β-propeller structure in which 3-4 Ca^{2+} coordination sites are located. Between repeats 2 and 3 of the β-propeller, the collagen binding I-domain (also termed A-domain, 200 AA) of the $\alpha 2$ subunit is inserted containing a metal coordination site with a preference for Mg^{2+}/Mn^{2+}. The Mg^{2+} ion complexed by this I-domain is critical for the interaction of $\alpha 2\beta 1$ with a glutamate residue in the collagen molecule (Emsley et al. 2000). The I-domain of the $\alpha 2$ subunit contacts a putative MIDAS (metal ion-dependent adhesion site) or I-domain-like N-terminal structure in the $\beta 1$ subunit that is located next to a plexin-semaphorin-integrin-like (PSI) domain of $\beta 1$. Above the membrane, $\beta 1$ consists of a cysteine-rich region that contains an endogenous protein disulphide isomerase activity responsible for regulating conformational changes of the integrin in response to signalling via the cytoplasmic tail.

On the platelet membrane, $\alpha 2\beta 1$ is expressed in a low-affinity conformation unable to efficiently bind to collagen. Upon cellular stimulation, the integrin undergoes a conformational change and binds to its ligand with high affinity (Jung and Moroi 2001; Nieswandt et al. 2001b). The mechanisms underlying this so-called inside-out signalling of $\alpha 2\beta 1$ are currently unclear, but they may be similar to those underlying the affinity modulation of αIIb$\beta 3$, which have been studied in more detail and are discussed below. Activated integrin $\alpha 2\beta 1$ binds preferentially to type I collagen (Xu et al. 2000) and with a lower affinity to type IV collagen (Dickeson et al 1999). The switch from the low- to the high-affinity state of $\alpha 2\beta 1$ can be induced by various agonists, most notably collagen itself (through GPVI) and ADP released from activated platelets and damaged cells at sites of injury (Jung and Moroi 2000). The role of $\alpha 2\beta 1$ in hemostasis and thrombosis has long been debated, but only the recent generation of mice lacking the receptor helped to clarify some of the issues (Nieswandt and Watson 2003). Studies with platelets from these mice demonstrated that $\alpha 2\beta 1$ contributes to shear-resistant adhesion on collagen but that it is not strictly required, since the interaction of αIIb$\beta 3$ with collagen-bound vWf is sufficient to arrest platelets on the substrate, although with reduced stability (Savage et al. 1998; Kuijpers et al. 2003). Integrin $\alpha 2\beta 1$-deficient mice display no bleeding phenotype, and the aggregation of their platelets to collagen is only slightly delayed but otherwise not affected (Holtkotter et al. 2002). Besides mediating adhesion, $\alpha 2\beta 1$ also facilitates platelet activation by stabilizing GPVI-collagen interactions, leading to sustained signalling (Kuijpers et al. 2003). Recent data demonstrate that, upon ligand binding, $\alpha 2\beta 1$ also directly contributes to intracellular signalling by regulating a similar set of intracellular signalling molecules as GPVI, including Syk, SLP-76 and phospholipase (PL) Cγ2 (Inoue et al. 2003). This 'outside-in' signalling is very similar to that mediated by ligand-occupied αIIbβ3 (see below), demonstrating that these two integrins have partially overlapping functions that may explain why α2β1 deficiency produces no major hemostatic defect.

2.2.3
Integrins α5β1/α6β1

Besides α2β1, platelets possess two more β1 integrins which serve as receptors for fibronectin (α5β1, VLA-5) and laminin (α6β1, VLA-6). Both receptors principally have the capacity to mediate firm platelet adhesion at sites of vascular injury since both ligands are highly expressed in the subendothelial ECM. Fibronectins are dimeric glycoproteins that are present in plasma (plasma fibronectin) and in tissue extracellular matrices (cellular fibronectin) (Schwarzbauer 1991). In addition to α5β1, αIIbβ3 also binds to fibronectins, and in vitro studies demonstrated that either interaction is sufficient to mediate platelet adhesion under static and flow conditions (Beumer et al. 1994). Laminins are a family of structurally related glycoproteins that are constituted by the association of three different gene products, the α, β and γ chains. These heterotrimers are tightly assembled with collagen type IV through the action of nidogen (Timpl and Brown 1994) in the basement membrane. Therefore, laminins are among the first constituents of the ECM platelets get in touch with at sites of superficial injury.

Recent studies utilizing mice with a conditional β1-integrin deficiency in their platelets demonstrated that α5β1 and α6β1 indeed significantly contribute to platelet adhesion in vivo but also that they are not essential, due to functional redundancy with αIIbβ3. Accordingly, those mice have no obvious hemostatic defect and are susceptible to arterial thrombosis (Gruner et al. 2003). Thus, α5β1 and α6β1 appear to play supportive rather than essential roles in the adhesion process, at least in those models tested so far; but further studies will be required to confirm this in different regions of the vascular system.

2.2.4
Integrin αIIbβ3 (GPIIb/IIIa)

Integrin αIIbβ3 is the most abundant cell surface glycoprotein in platelets with 50,000-80,000 copies per cell (Wagner et al. 1996). An additional pool of αIIbβ3 is present in intracellular granules and can be exposed to the surface upon platelet activation. With a few exceptions, αIIbβ3 expression is restricted to platelets and megakaryocytes. Besides its mandatory function in platelet aggregation, which is discussed below, the integrin also has a crucial role in the adhesion process as it binds to various immobilized ligands in the ECM, including vWF and fibronectin.

Integrin $\alpha IIb\beta 3$ is present on platelets as a heterodimer, which is formed during biosynthesis in megakaryocytes by non-covalent interaction between the αIIb (GPIIb) and $\beta 3$ (GPIIIa) subunits. Both subunits of $\alpha IIb\beta 3$ have large extracellular domains, a single transmembrane domain and short cytoplasmic tails of 20 and 47 amino acids in αIIb and $\beta 3$, respectively. αIIb is cleaved proteolytically in the cell into a heavy (105-kDa) and a light (25-kDa) chain, which are linked by a disulphide bond, while the $\beta 3$-subunit is a single polypeptide of about 95 kDa. The N-terminal half of the αIIb-subunit is characterized by seven homologous repeats of about 60 amino acids, which are predicted to fold in a β-propeller-like structure with 7 β-sheets arranged like blades of a propeller (Springer 1997). Four of the seven repeats have binding sites for divalent cations with homology to the EF-hand motif. The extracellular domain of $\beta 3$ contains three major domains: The N-terminal half of the protein contains an I-like domain homologous to the I-domain of other integrin α-subunits, which is predicted to have a Rossmann fold structure (Lee et al. 1995). The top surface of the I-like domain may contain a cation- and ligand-binding site called MIDAS (defined above) (Tozer et al.1996). The I-like domain of $\beta 3$ is believed to be close to the β-propeller domain of αIIb (Zang et al. 2000), and it is flanked by an N-terminal plexin/semaphorin/integrin (PSI) domain and four C-terminal epidermal growth factor (EGF)-domains. A prominent feature of $\beta 3$ is its loop-like structure produced by a disulphide bond between the N-terminal and the middle section.

Under normal conditions, $\alpha IIb\beta 3$ is expressed on the platelet surface in a low-affinity state unable to mediate platelet adhesion. One exception is immobilized fibrinogen, to which the integrin can bind even in the absence of cellular activation, but the significance of this interaction is currently not clear (Shattil 1998). In order to mediate adhesion to the ECM under flow conditions as well as aggregation, $\alpha IIb\beta 3$ has to become activated through 'inside-out' signals triggered by other receptors (see below). During the adhesion of the first platelet layer on the ECM, this signalling is predominantly provided by GPVI and, possibly, GPIb-V-IX (see above). Active $\alpha IIb\beta 3$ can bind various macromolecular glycoproteins including fibrinogen, vWF, fibronectin, fibrin or vitronectin. Most likely, all of these ligands play a role for platelet adhesion on the injured vessel wall in vivo, but it is difficult to assess their relative significance due to the complexity of the ECM. In vitro, $\alpha IIb\beta 3$ is sufficient to arrest platelets on collagen-bound vWF under high shear flow conditions, and it is believed that this interaction is also of major importance in vivo (Savage 1998). Indeed, the in vivo inhibition of $\alpha IIb\beta 3$ leads to a profound reduction of platelet adhesion at sites of vascular injury in mice, which confirms that the integrin plays a key role in this process.

From the described studies, it is now clear that platelet adhesion at sites of injury involves β1 and β3 integrins, but that none of these by themselves is essential for this process to occur (Grüner 2003). This suggests that the co-operation of multiple integrin-ligand interactions ensures a high degree of functional redundancy, enabling effective and well-controlled platelet adhesion, spreading and thrombus formation on different compositions of the subendothelial matrix. These may vary significantly between different regions in the vascular system and may also depend on the type and severity of the lesion.

3
Platelet Aggregation

Under normal conditions, platelets circulate freely in the blood and do not adhere to each other. Once platelets adhere to subendothelial matrix proteins, processes are induced which result in the accumulation of platelets into a hemostatic thrombus, a process which is based on the formation of multiple platelet-platelet-interactions (platelet aggregation). Platelet aggregation is mediated by integrin $\alpha IIb\beta 3$. On resting, non-activated platelets integrin $\alpha IIb\beta 3$ has low affinity for ligands. However, vascular injury results in the exposure and production of stimuli, which act on platelets and induce intracellular signalling processes, which influence the cytoplasmic part of $\alpha IIb\beta 3$, rapidly converting $\alpha IIb\beta 3$ in an active conformation (inside-out signalling). Active $\alpha IIb\beta 3$ binds various extracellular macromolecular ligands including fibrinogen and vWF. The dimeric structure of fibrinogen and the multimeric structure of vWF allow these ligands to crossbridge platelets and to generate a platelet aggregate. Binding of ligand to $\alpha IIb\beta 3$ induces integrin clustering and activation, which results in a series of intracellular processes (outside-in-signalling). Outside-in signalling contributes to the strengthening and stabilization of the platelet aggregate.

3.1
Integrin $\alpha IIb\beta 3$ in Aggregation

The pivotal role of $\alpha IIb\beta 3$ in mediating platelet aggregation is exemplified by platelets from patients with Glanzmann thrombasthenia, which lack the integrin and fail to aggregate in response to various physiological stimuli (Nurden 1999). Fibrinogen and vWF (Savage et al. 1998) are able to mediate platelet-platelet interaction via crossbridging. Much evidence has been provided that the top surface of the β-propeller of αIIb and the putative top surface of I-like domain, including the MIDAS motif of β3, take part in ligand binding of $\alpha IIb\beta 3$ (Plow et al. 2000). All $\alpha IIb\beta 3$ ligands contain at least one RGD sequence motif, which in most cases is critically involved in binding to

αIIbβ3 and various other integrins (Plow et al. 2000). However, the RGD motif in fibrinogen is not required for binding (Farrell and Thiagarajan 1994). Instead, the sequence KQAGDV at the C-terminus of the fibrinogen γ-chain interacts with αIIbβ3 (Holmback et al. 1996). Although the binding sites of αIIbβ3 for RGD and KQAGDV sequences are distinct, both peptide sequences show competitive binding properties (Santoro et al. 1987; Cierniewski et al. 1999; Hu et al. 1999), suggesting that both binding sites are structurally and/or functionally related. Thus, macromolecular ligands may interact with multiple sites in the extracellular domains of αIIbβ3, and interference with a single contact site may inhibit ligand receptor interaction (see also Sect. 22).

3.1.1
Inside-Out Activation

As is the case with many other integrins, αIIbβ3 is not constitutively active. In inactivated, resting platelets, αIIbβ3 is in an 'off' state in which ligand-binding affinity is low and no signalling occurs. This allows platelets to circulate freely in the blood, which contains high concentrations of fibrinogen and vWF. Only in response to appropriate stimuli produced in consequence of vascular injury does αIIbβ3 acquire an active conformation, which results in fibrinogen-/vWF-mediated platelet aggregation. The activation signals are derived from the exposed subendothelium and especially from various soluble stimuli, which act through specific receptors on the platelet surface to activate intracellular signalling cascades, which result in αIIbβ3 activation.

Different mechanisms have been postulated for the activation of integrins. Activation of ligand binding could be achieved through an increase in the affinity of individual integrins for their ligands. Some data also suggest that the clustering of integrins can increase the avidity of interaction with potential ligands without changing the affinity of integrins. In fact, both mechanisms of integrin activation may coexist (Hato et al. 1998; Shattil 1999), and recent evidence suggests that conformational activation and lateral clustering of integrins are closely linked events (Li et al. 2003). The cytoplasmic face of integrin αIIbβ3 and its potential role in mediating inside-out signalling has been intensively studied (Woodside et al. 2001). Deletion of the cytoplasmic tail as well as mutations in the membrane-proximal regions of the cytoplasmic parts results in constitutive activation of integrin αIIbβ3 (O'Toole et al. 1994; Hughes et al. 1996; Peterson et al. 1998). Recent analysis of the conformational changes that underlie αIIbβ3 activation (Takagi et al. 2002; Vinogradova et al. 2002; Garcia-Alvarez et al. 2003) has resulted in a new model describing integrin activation (Hynes 2002; Liddington and Ginsberg 2002). The two cytoplasmic domains of αIIbβ3 are closely associated when the integrin is in a low-affinity state. They may be held together by unknown cytosolic proteins. In the inactive state, the extracellular portion of the integrin heterodimer is bent over, unable to bind ligand. Inside-out signals are believed to activate $\alpha IIb\beta 3$ by breaking the interaction between the cytoplasmic and transmembrane regions of the two subunits. This then results in straightening of the extracellular portion of the heterodimer and activation of the ligand-binding domain (see Fig. 2). This model suggests that the regulation of the association/dissociation of the cytoplasmic domains of integrins, presumably by interacting cytoplasmic proteins, controls the activation state of the integrin. Several proteins have been shown to interact with the cytoplasmic tails of $\alpha IIb\beta 3$ (Shattil 1999). Especially the 250-kDa protein talin (Critchley 2000) has attracted interest. Talin has been shown to play an important role in integrin clustering into focal adhesions by linking $\alpha IIb\beta 3$ to actin (Priddle et al. 1998), and binding of talin to the integrin β tail has recently been shown to be the final step resulting in integrin activation (Tadokoro et al. 2003). Through its head domain, talin binds to the short cytoplasmic domain of integrin β-subunits (Liu et al. 2000). The expression of the head domain of talin alone is sufficient to activate integrins (Calderwood et al. 1999, 2002). In intact talin, the integrin-binding domain appears to be masked and can be uncovered by calpain-mediated proteolytic cleavage or binding to phosphoinositides (Yan et al. 2001; Martel et al. 2001). Whether any of these or other mechanisms, which allow talin to bind to and to activate integrin $\alpha IIb\beta 3$, are involved in inside-out signalling in intact platelets is unclear (Calderwood and Ginsberg 2003).

The intracellular signals directly regulating the proteins involved in integrin $\alpha IIb\beta 3$ activation have not been clearly identified yet. These intracellular signals are induced by various platelet stimuli like collagen, vWF as well as thrombin, ADP or TXA_2. Several of the initial upstream signalling events induced by these platelets stimuli have been characterized in recent years.

During the initial phase of platelet aggregation, collagen and, especially under conditions of high flow rate, vWF play central roles in the induction of signalling pathways leading to $\alpha IIb\beta 3$ activation. Stimulation of inside-out signalling of platelets by collagen is mediated by GPVI (see above), and the GPIb-IX-V complex contributes to inside-out activation of $\alpha IIb\beta 3$ as a receptor for vWF. While these processes are intimately linked to platelet adhesion, subsequent growth of the platelet plug requires the recruitment of additional platelets from the circulation via $\alpha IIb\beta 3$-mediated aggregation. These processes are regulated by diffusible ligands such as thrombin, which can be produced on the surface of activated platelets, as well as stimuli secreted or released from activated platelets like ADP/ATP or TXA_2. Most of these stimuli, which induce platelet aggregation into a platelet plug, act through G protein-coupled receptors. The effects of ADP/ATP are mediated by three receptors (Gachet 2001; Kunapuli 2003). The $P2X_1$ receptor is a ligand-operated cation channel, which mediates ATP-induced Ca^{2+} influx. However, its role in inside-out signalling to $\alpha IIb\beta 3$ is not clear. ADP activates two G protein-coupled receptors, P2Y$_1$, which couples to G$_{q/11}$ and P2Y$_{12}$, which couples to G$_i$. TXA$_2$ receptors couple to G$_q$ and G$_{12}$/G$_{13}$. Finally, thrombin-induced inside-out activation is mediated primarily by G protein-coupled protease-activated receptors (PARs) (Coughlin 1999). PAR1 and PAR4 are expressed in human platelets, while mouse platelets express PAR3 and PAR4. PAR receptors are coupled to G$_{q/11}$, G$_{12}$/G$_{13}$ and, in some cases, to G$_i$. While the exact signalling mechanisms which link these stimulatory receptors to the cytoplasmic domains of αIIbβ3 are incompletely understood, the role of different G protein subfamilies and the role of some of their immediate effectors have been described in recent years. There is good evidence indicating that activation of phospholipase Cβ through G$_q$, which results in the formation of IP$_3$ and diacyl glycerol, plays an essential role in mediating αIIbβ3 activation. IP$_3$-mediated increase in cytosolic free Ca^{2+} appears to be required for integrin activation, but Ca^{2+} alone is obviously not sufficient to induce inside-out signalling (Shattil and Brass 1987). Activation of PKC by diacyl glycerol or phorbol esters leads to αIIbβ3 activation (Shattil and Brass 1987). However, conflicting data exist as to the sensitivity of agonist-induced aggregation to inhibition of PKC isoforms (Toullec et al. 1991; Shattil et al. 1992; Walker and Watson 1993; Pulcinelli et al. 1995). The requirement of G$_q$-mediated signalling for agonist-induced αIIbβ3 activation has been also demonstrated by the phenotype of Gα_q-deficient platelets, which fail to aggregate and to secrete in response to thrombin, ADP and the TXA$_2$-mimetic U46619 due to a lack of agonist-induced phospholipase C activation (Offermanns et al. 1997). There is, however, clear evidence for additional G$_q$-independent signalling processes, which are involved in αIIbβ3 activation. Platelets deficient in P2Y$_1$ or in which P2Y$_1$ was blocked pharmacologically do not aggregate in response to low and intermediate concentrations of ADP (Hechler et al. 1998; Jin and Kunapuli 1998; Savi et al. 1998; Fabre et al. 1999; Leon et al. 1999). Aggregation could be restored by serotonin, which induces G$_q$-mediated phospholipase C activation but is alone not able to induce platelet aggregation. Platelet activation by ADP obviously requires an additional signal, which appears to be mediated by the G$_i$-coupled P2Y$_{12}$ receptor. Platelets lacking P2Y$_{12}$ or in which P2Y$_{12}$ was blocked did not aggregate in response to ADP unless the G$_i$-mediated pathway was activated by alternative mechanisms (Savi et al. 1998; Foster et al. 2001). Thus, G$_q$ and G$_i$ synergize to induced platelet activation. It is currently not clear how G$_i$ contributes to integrin αIIbβ3 activation in platelets. The decrease in cyclic adenosine monophosphate (cAMP) levels by G$_i$-mediated inhibition of adenylyl cyclase is obviously not involved, since an adenylyl cyclase inhibitor failed to induce aggregation in platelets in which the G$_q$-mediated pathway was selectively activated (Savi et al. 1996; Pulcinelli et al. 1998; Daniel et al. 1999). Alternatively, $\beta\gamma$-subunits released from activated G$_i$ may contribute to αIIbβ3 activation and aggregation. A candidate effector which mediates $\beta\gamma$ effects is phosphatidylinositol-3 kinase (Vanhaesebroeck et al.

2001; Wyman et al. 2003). The $\beta\gamma$-complex activates the p110β and p110γ isoform of PI-3 kinase (Vanhaesebroeck et al. 2001), and inhibition of PI-3 kinase has been shown to affect agonist-induced platelet aggregation (Zhang et al. 1996).

Recent data also indicate that some level of αIIbβ3 activation can be induced in the absence of G_q-mediated signalling by activating G_i- and G_{12}/G_{13}-dependent mechanisms (Dorsam et al. 2002; Nieswandt et al. 2002). Under in vivo conditions the extension of the platelet plug through platelet aggregation has to occur very rapidly, since the potential contact time for platelets under high shear rate conditions is very short, and it has to occur with sufficient efficacy, since soluble mediators are rapidly washed away. In order to rapidly activate platelet aggregation with high efficiency, all three signalling pathways mediated by the heterotrimeric G proteins G_q, G_{12}/G_{13} as well as G_i need to be activated. In the absence of one of the three pathways, activation of αIIbβ3 still occurs, although with lower efficacy. Recent studies showed that G_{13} deficiency in platelets leads to impaired thrombus formation under flow conditions in vitro and defective hemostasis and arterial thrombosis in vivo, underscoring the important role of this signalling pathway for platelet function (Moers et al. 2003).

Signalling pathways, leading to αIIbβ3 activation, have to be balanced by pathways that inhibit αIIbβ3, in order to counteract uncontrolled platelet aggregation, which would be deleterious under inappropriate conditions. These pathways are activated by various mediators, especially those produced by the intact endothelium. Endothelial cells produce prostacyclin (PGI$_2$) and NO, which raise platelet cAMP and cyclic guanosine monophosphate (cGMP) levels, respectively. In general, the effects of the cyclic nucleotides cAMP and cGMP are thought to be mediated by PKA and cGMP-dependent kinase (PKG). PKA and PKG are likely to exert their inhibitory effects on αIIbβ3 at several levels of the inside-out activation pathway (Schwarz et al. 2001). The phosphorylation of the 50-kDa protein vasodilator-stimulated phosphoprotein (VASP) on specific serine residues by PKA or PKG has been shown to correlate with the inhibition of αIIbβ3 activation. However, mice deficient in VASP show only small defects in cAMP-/cGMP-mediated inhibition of platelet aggregation (Aszodi et al. 1999; Hauser et al. 1999), indicating that VASP, which is localized to focal adhesions, is only one of probably many mediators of the effect of cyclic nucleotides. Mice lacking the PKG I-isoform show defective cGMP effects on platelets in vitro as well as an increased in vivo platelet adhesion and aggregation (Massberg et al. 1999).

3.1.2
Outside-In Activation

Once fibrinogen and vWF have bound to $\alpha IIb\beta 3$ to mediate platelet aggregation, integrins cluster and their cytoplasmic domains undergo a conformational change. These processes are believed to affect the interaction of the cytoplasmic domains with intracellular proteins, which results in the induction of intracellular signalling events, summarized as 'outside-in signalling'. These signalling processes eventually contribute to firm adhesion of platelets and stabilization of platelet aggregates, and they promote further secretion of platelet granules. Multiple intracellular proteins have been shown to be regulated by outside-in signalling of $\alpha IIb\beta 3$ (Shattil et al. 1999). Clustering of integrin $\alpha IIb\beta 3$ rapidly induces activation of non-receptor tyrosine kinases like Syk and Src. Activated Syk can phosphorylate and activate Vav, a guanine-nucleotide exchange factor for Rac (Miranti et al. 1998). Outside-in signalling induces cytoskeletal reorganization and the formation of focal adhesion-like structures. This is associated with the recruitment of multiple proteins into a complex attached to the cytoplasmic part of the integrin (Hynes 2002; Brakebusch and Fässler 2003; DeMali et al. 2003).

Part of the signalling events induced by $\alpha IIb\beta 3$ clustering and activation require the phosphorylation of Tyr747 and Tyr759 in the cytoplasmic domain of $\beta 3$ by non-receptor tyrosine kinases (Phillips et al. 2001). This generates docking sites for proteins like Shc or myosin. The physiological significance of $\beta 3$ phosphorylation has been demonstrated in mice in which Tyr747 and Tyr759 were mutated to phenylalanine. Platelets of these animals have a reduced ability to stably aggregate, which results in an instability of the hemostatic plug in vivo (Law et al. 1999).

3.2
Stabilization of Platelet Aggregates

Stabilization of the platelet aggregate is important for the perpetuation of the platelet hemostatic plug as well as for pathological processes underlying arterial thrombosis under high shear conditions. There is growing evidence that in addition to $\alpha IIb\beta 3$ outside-in signalling, other mechanisms contribute to the stabilization of platelet-platelet interactions once aggregation has occurred.

CD40 ligand is expressed on the surface of activated platelets where it binds to $\alpha IIb\beta 3$ via its KGD integrin recognition sequence. In mice lacking CD40 ligand, stability of arterial thrombi is reduced and arterial occlusion is delayed (Andre et al. 2002). Growth arrest-specific gene 6 (Gas6) is a protein related to protein S, which is secreted from activated platelets (Ishimoto et al. 2000). It is a ligand for the receptor tyrosine kinase Axl, which is expressed on platelets. Platelets from Gas6-deficient mice disaggregate after an initially normal aggregation response, and Gas6-deficient mice are resistant to thrombosis (Angelillo-Scherrer et al. 2001), suggesting that Gas6 may promote platelet plug formation and stabilization by binding to Axl or negatively charged phospholipids on platelet surfaces. Recently, the Eph-family of receptor tyrosine kinases has also been involved in perpetuating platelet aggregation. Platelets express EphA4 and EphB1 as well as the transmembrane protein ephrin B1, which is a ligand for Ephs of the B-family (Prevost et al. 2002). When Eph/ephrin interaction is disrupted on activated platelets, platelets disaggregate (Prevost et al. 2002), indicating that this system contributes to stabilization of the platelet plug (Prevost et al. 2003). P-selectin, which is exposed on the surface of activated platelets, is involved in platelet-vessel wall and platelet-leukocyte interaction. Recent evidence suggests that P-selectin might also play a role in stabilizing platelet aggregates. P-selectin-deficient mice formed less compact platelet layers on injured arterial surfaces in vivo (Smith et al. 2001). Various data indicate that a ligand different from P-selectin glycoprotein ligand (PSGL)-1 is involved in this P-selectin function (Merten and Thiagarajan 2000).

4
Platelet-Endothelial Cell Interaction

Under normal conditions, circulating platelets do not attach to endothelial cells whose luminal surface is highly anti-thrombogenic. In certain disease states, however, endothelial dysfunction can create the conditions for platelet activation and adhesion even without the exposure of the ECM, and it has been recognized that this interaction is a central step in the pathogenesis of inflammatory vascular diseases, such as reperfusion or atherosclerosis (Gawaz et al. 1997; Massberg et al. 2002; Ruggeri 2003). Currently, the mechanisms by which platelets adhere to endothelium are not fully understood. Endothelial cell membranes contain a variety of adhesion receptors, including integrins, Ig receptors, selectins and adhesive ligands, which are also present on platelet membranes. The expression levels of these receptors/ligands are to a large extent determined by the activation state of the cells, with many being upregulated upon stimulation. For example, in the early phase of atherogenesis, enhanced release of vWF from endothelial cells in response to inflammatory stimuli can lead to local recruitment of platelets (Theilmeier et al. 2002) a process that is entirely GPIb dependent (Massberg et al. 2002). In addition to vWF, GPIb also binds to P-selectin expressed on activated endothelium, and this interaction appears to play a role in the initial 'rolling' of platelets, a transient interaction similar to that typically described for leukocytes. This may at least partly explain why deficiency in vWF or P-selectin, as well as a prolonged inhibition of GPIb, is protective in experimental atherogenesis (Dong et al. 2000; Methia et al. 2001; Massberg et al. 2002). Firm platelet adhesion on activated/dysfunctional endothelium has been shown to be dependent on integrin $\alpha IIb\beta 3$, but it is not clear to which ligands the receptor binds under these conditions. Besides vWF, soluble fibrinogen is the strongest candidate, since it can be bound by endothelial cells either through intercellular cell adhesion molecule (ICAM)-1 or integrin $\alpha v\beta 3$, both of which are strongly upregulated by endothelial agonists (Ruggeri 2003). Activated platelets have been shown to alter the chemotactic and adhesive properties of the endothelium by expression of two inflammatory cytokines, CD40 ligand (CD40L) and interleukin (IL)-1 β, thereby facilitating endothelial activation and recruitment of additional platelets and subsequently immune cells.

While it is clear that platelet attachment to endothelium involves GPIb-V-IX mediating the initial contact and $\alpha IIb\beta 3$ mediating firm adhesion, it is currently unclear which structure(s) on dysfunctional/activated endothelium are responsible for platelet activation and which platelet receptors are involved.

5
Pharmacological Strategies to Interfere with Platelet Adhesive Functions

For a long time, platelets have been regarded as mere bystanders in hemostasis and thrombosis. Only during the last 25 years has their central role in arterial thrombosis and, more recently, also in inflammatory processes associated with atherosclerotic processes been recognized. The recognition of the importance of platelets in cardiovascular diseases was facilitated not the least by the understanding that the beneficial anti-thrombotic effects of aspirin (see below) are based on its ability to interfere with the activation of platelets. Adhesive processes between platelets and the altered vessel wall as well as between platelets themselves are at the center of the pathological events which underlie thrombus formation. Thus, inhibition of platelet adhesion and aggregation has become one of the major strategies for the development of drugs used to prevent and/or treat cardiovascular diseases.

5.1
Inhibition of Platelet Adhesion

During the past decades drugs have been developed and are in clinical use that inhibit platelet aggregation (see below). However, the development of compounds that interfere with the first step of hemostasis, i.e. platelet adhesion to the injured vessel wall, has lagged behind and has only received more attention during the past few years.

Inhibition of GPIb-V-IX receptor-mediated platelet adhesion to subendothelial collagen might be a promising approach to improve strategies in the treatment of cardiovascular diseases, as the receptor complex is exclusively expressed on platelets, and is essential for the formation of thrombi in arteries. Accordingly, Fab-fragments of the monoclonal anti-human GPIb antibody 6B4 had a protective effect in an in vivo model of arterial thrombosis in baboons without affecting platelet count and bleeding time in these animals (Wu et al. 2002). Similarly, VCL, which is a recombinant fragment of vWF (Leu504-Lys728) carrying the GPIb-V-IX-binding motif was proved to be beneficial in several models of thrombosis (McGhie et al. 1994; Yao et al. 1994; Azzam et al. 1995). However, a significant decrease in platelet number and a moderately prolonged bleeding time due to VCL treatment was observed in some of these studies. The rapid clearance of VCL (22-29 min) additionally limits its use as a therapeutical drug. Other antagonists of GPIb, most prominently snake venom proteins such as crotalin or agkistin display a clear antithrombotic effect, but cause prolonged bleeding and thrombocytopenia too (Chang et al. 1998; Yeh et al. 2001).

Inhibitors of vWF or collagen-vWF interaction appear to be equally promising for the development of antithrombotic compounds. For example, the humanized monoclonal anti-vWF antibody AJW200, which is known to bind vWF A1 domain, clearly prevented arterial occlusion in monkeys and dogs and had only minor effects on bleeding time and platelet count (Kageyama et al. 2002). In addition, coagulation parameters were unaffected. Similar results were obtained in baboons with the monoclonal anti-vWF 82D6A3, which is directed against the collagen type I and III vWF-binding domain A3 (Wu et al. 2002). Of particular interest is the collagen-binding protein saratin, which is a recombinant protein isolated from *Hirudo medicinalis*. Topically administered at the luminal side of the carotid artery, platelet adhesion to the injured vessel was massively reduced as determined by electron microscopy (Cruz et al. 2001). By binding collagen, saratin does not alter platelet count or bleeding time, suggesting this mechanism to be potentially useful as an antithrombotic compound.

While an antithrombotic benefit by pharmacological inhibition of the collagen receptor integrin $\alpha 2\beta 1$ lacks clear in vivo evidence, present findings strongly suggest the other platelet collagen receptor, GPVI, as a promising target. Injection of the monoclonal anti-GPVI antibody JAQ1 in mice resulted in a profound protection from thrombosis in several in vivo models (Nieswandt et al. 2001a; Massberg et al. 2003). This treatment caused GPVI depletion from the platelet surface, resulting in a knock-out-like phenotype of these mice (Nieswandt et al. 2001a). Platelets of these mice do not adhere to collagen, demonstrating an essential role for GPVI in the initiation of adhesion in vitro and in vivo. Interestingly, these mice display only moderately prolonged bleeding times and normal platelet numbers. Thus, it appears that other receptors/signalling pathways can substitute for GPVI during normal hemostasis, but the exact mechanisms are, at present, not clear.

5.2
Inhibition of Integrin $\alpha IIb\beta 3$ (GPIIb/IIIa) Activation

The current most widely used anti-platelet drugs do not interfere with the function of adhesive proteins directly but block processes which mediate the inside-out activation of integrin $\alpha IIb\beta 3$.

5.2.1
Inhibition of TXA$_2$ Production or Action

Acetylsalicylic acid (aspirin) is the clinically best established anti-platelet drug, which because of its relative safety and low costs serves as a standard for the development of new anti-platelet agents. Although aspirin has well proven clinical effects (see below), it is a weak inhibitor of platelets, since it blocks only TXA$_2$-dependent platelet activation, while other mechanisms remain unaffected. Aspirin interferes with the biosynthesis of prostanoids by irreversible acetylation of a serine residue of cyclooxygenase (COX)-1, which prevents the enzyme from binding its substrate arachidonic acid. In platelets, the product of COX-1, PGH$_2$, is rapidly converted into TXA$_2$ by a specific TXA$_2$ synthase. Thus, aspirin prevents the formation of TXA$_2$ in platelets, which is produced after platelet activation and functions as an amplifier in platelet plug formation by contributing to $\alpha IIb\beta 3$ inside-out activation (see Sect. 13). Although COX-1 is a widely expressed enzyme, platelets are particularly sensitive to low doses of aspirin. Since platelets, as anuclear cells, lack the machinery to resynthesize COX-1, the effects of aspirin last for the lifetime of platelets, approximately 10 days. Aspirin is rapidly deacetylated by the liver to salicylic acid, which cannot block COX-1 irreversibly. However, platelet COX-1 can be irreversibly inhibited by orally administered aspirin in the portal circulation before it enters the systemic circulation, thereby providing an additional mechanism which allows low doses of aspirin to act in a somewhat platelet-specific manner. A complete inhibition of platelet COX-1 can be obtained after a single dose of 160 mg of aspirin, and the same effect can be achieved with daily doses of 30-50 mg given for 7-10 days (Patrono 1994).

Aspirin has been shown to be clinically effective in the treatment of acute cardiovascular diseases, which are characterized by platelet activation and intravascular thrombosis, like unstable angina or acute myocardial infarction (Foster et al. 1992). A decrease in mortality could be demonstrated in patients who received aspirin after acute myocardial infarction (ISIS-2 1988) as well as in patients with unstable angina (Antiplatelet Trialists' Collaboration 1994). There is also clear evidence that aspirin is beneficial in the acute treatment of ischemic stroke (CAST Collaborative Group 1997; IST Collaborative Group 1997). Under these acute conditions, aspirin is given at a dose of 160-325 mg followed by 75-160 mg daily. Patients with a history of acute cardiovascular diseases or acute cerebrovascular accidents clearly benefit from long-term treatment with low-dose aspirin (75-160 mg/day), which serves as a secondary prevention (Antiplatelet Trialists' Collaboration 1994). However, there is little support for the routine use of aspirin for primary prevention in patients with a low risk for cardiovascular diseases. Nevertheless, a subgroup of patients without prior ischemic events but with an increased risk may benefit from primary prevention with low-dose aspirin (Platelet Trialists' Collaboration 2002).

In order to selectively inhibit TXA_2-mediated platelet activation, TXA_2 synthase inhibitors as well as TXA_2 receptor antagonists have been developed. However, clinical studies have been rather disappointing (Serroys et al. 1991; RAPT Investigators 1994).

5.2.2
Inhibition of ADP-Dependent Platelet Activation

Other than TXA_2, ADP is an important mediator which also augments platelet activation by a positive feedback mechanism. The thienopyridines ticlopidine and clopidogrel inhibit ADP-induced platelet activation by blocking the effect of ADP on the $P2Y_{12}$ receptor (Foster et al. 2001; Hollopeter et al. 2001). This particular ADP receptor subtype mediates the amplification of platelet activation by ADP released from activated platelets (Gachet 2001; Kunapuli et al. 2002). Both thienopyridine derivatives function through active metabolites produced mainly in the liver, which irreversibly bind to $P2Y_{12}$. The requirement for metabolism of the thienopyridines results in a delayed onset of the effect. Like aspirin, the inhibitory effects of thienopyridines are irreversible and last for the life span of platelets, since their active metabolites covalently modify $P2Y_{12}$ (Quinn and Fitzgerald 1999). Ticlopidine and clopidogrel appear similar in efficacy. However, a higher rate of side-effects-especially neutropenia-of ticlopidine has resulted in its replacement by clopidogrel in clinical practice. A large study on the secondary prevention of cardiovascular diseases, the CAPRIE trial, has suggested a marginal advantage for clopidogrel over aspirin in patients with a history of cerebrovascular accidents (The CAPRIE Steering Committee 1996). Since aspirin and clopidogrel interfere with different amplification mechanisms during platelet activation by inhibiting the effect of TXA_2 and ADP, respectively, coapplication of both drugs may have additive or even more than additive effects. Several trials are under way to test this rationale. Recent data indicate that clopidogrel given in addition to aspirin before and after percutaneous coronary interventions reduces the cardiovascular risk in patients with acute coronary syndrome (Mehta et al. 2001; Yussuf et al. 2001; Steinhubl et al. 2002).

The slow onset of action and the irreversible nature of the effect of clopidogrel are a problem for its use in acute coronary syndromes. Therefore, rapid onset, reversible ADP receptor antagonists are under development. The intravenous compound AR-C69931MX is a specific $P2Y_{12}$ receptor antagonist and effectively inhibits platelet function in vivo (Storey et al. 2002; Wang et al. 2003). There are also indications that the blockade of two ADP receptors, $P2Y_1$ and $P2Y_{12}$, may be advantages to single blockade of $P2Y_{12}$ (Turner et al. 2001). ADP can be removed from the vicinity of platelets by the activity of CD39, an ecto-ADTPase expressed on endothelial cells (Marcus et al. 1997). Recombinant CD39 has been shown to decrease the infarct volume in an animal stroke model (Pinsky et al. 2002). Thus, there is still room for improvements of antiplatelet drugs based on ADP antagonism.

5.3
Glycoprotein IIb/IIIa Antagonists

After the discovery of the importance of platelets in thromboembolic diseases and the clinical evidence for a benefit in treating patients with antiplatelet drugs like aspirin, the search for more specific and efficacious drugs intensified. Although various stimuli like TXA_2, thrombin, ADP or collagen induce platelet aggregation and plug formation, the common final pathway which mediates thrombus formation is the $\alpha IIb\beta 3$ crosslinking of platelets by fibrinogen and vWF. $\alpha IIb\beta 3$ became a new target for antithrombotic therapy when blockade of its interaction with macromolecular proteins was shown to inhibit platelet aggregation (Coller et al. 1983). Currently, three $\alpha IIb\beta 3$ antagonists have been approved for clinical use.

Abciximab is a chimeric (murine/human) F_{ab} fragment derived from a mouse monoclonal antibody which binds with high affinity (K_D: 5 nm) to the $\beta 3$ subunit of integrin $\alpha IIb\beta 3$ and $\alpha V\beta 3$ (Jordan et al. 1996). A crossreactivity with $\alpha M\beta 2$ (CD11/CD18, MAC-1) has also been described (Coller 1999). After intravenous injection the free plasma levels of abciximab decrease rapidly with an initial half-life of 30 min. However, the platelet-bound antibody can be exchanged between platelets, and abciximab is detected for more than two weeks after the initial infusion. When a standard dose (0.25 mg/kg bolus, followed by 0.125 µg/kg/min for 12 h) is given, more than 80% of $\alpha IIb\beta 3$ occupancy is maintained. After infusion is stopped, platelet function is effectively inhibited for 12-26 h.

Barbourin, a snake venom, specifically inhibits $\alpha IIb\beta 3$ (Scarborough et al. 1991). This specificity for $\alpha IIb\beta 3$ is due to a single lysine-to-arginine substitution in the RGD sequence, resulting in the amino acid sequence KGD. Based on this observation, a KGD containing cyclic heptapeptide called eptifibatide was developed (Scarborough 1999). Eptifibatide binds selectively to $\alpha IIb\beta 3$ with a dissociation constant of 120 nM. It has a relatively short half-life, which results in normalization of platelet function within a few hours after the end of infusion. Starting from a RGD containing disintegrin, a nonpeptide $\alpha IIb\beta 3$ antagonist, tirofiban, was generated which binds to $\alpha IIb\beta 3$

with high selectivity and affinity (K_D 15 nM) (Bednar et al. 1995). Similar to eptifibatide, tirofiban has a short half-life (1.5 h).

The currently available $\alpha IIb\beta 3$ antagonists have to be given parenterally, which restricts their use to defined clinical situations. Multiple studies have proved the beneficial effect of $\alpha IIb\beta 3$ antagonists during coronary interventions, such as percutaneous transluminal coronary angioplasty (PTCA) or coronary stenting (Bhatt and Topol 2000). There are some indications that $\alpha IIb\beta 3$ blockade with abciximab may be advantageous compared with tirofiban (Topol et al. 2001). Integrin $\alpha IIb\beta 3$ blockade has also been shown to be of benefit in acute coronary syndromes, although the risk reduction is less pronounced than during coronary intervention. Various trials are currently going on to resolve the question of potential differences in the clinical effects of individual $\alpha IIb\beta 3$ antagonists.

Since intravenous $\alpha IIb\beta 3$ antagonists showed high clinical efficacy with a reasonable risk of side-effects, orally available blockers have been developed recently. However, the first clinical phase III trials were rather disappointing (O'Neill et al. 2000; Cannon et al. 2000; The SYMPHONY Investigators 2000). None of the trials showed a clear benefit for oral $\alpha IIb\beta 3$ antagonists in the prevention of cardiovascular events after an episode of acute coronary syndrome or after undergoing percutaneous coronary interventions, and the risk of major bleeding appeared to be increased.

In contrast to the inhibition of ADP- or TXA_2-mediated platelet activation, blockade of $\alpha IIb\beta 3$ can completely inhibit platelet aggregation. This relatively high efficacy together with the steep dose-response relationship of $\alpha IIb\beta 3$ antagonists results in a comparably high risk for bleeding complications when $\alpha IIb\beta 3$ antagonists are clinically applied. This risk can be well controlled under acute clinical conditions, such as coronary interventions or in acute coronary syndromes, where intravenous $\alpha IIb\beta 3$ antagonists have been demonstrated to improve the long-term prognosis of patients. However, currently it appears rather unlikely that $\alpha IIb\beta 3$ antagonists are of benefit beyond the acute period.

6
Conclusions

Platelets are a fascinating model for cell adhesion and its regulation under physiological and pathological conditions. Under normal conditions platelets freely circulate in the blood. However, at places of vascular injury platelets rapidly adhere and aggregate in order to protect the organism from blood loss. The physiological role of platelets is based on their ability to become adhesive under defined conditions. Platelet activation and adhesion is a multi-step process which involves many different stimuli and adhesive molecules. During the past decade, many of the stimuli and their mechanisms of action as well as the nature of the receptors which mediate platelet adhesion have been identified and characterized. It has also become clear that most of the activation and adhesion processes required for the hemostatic function of platelets are also involved in inappropriate platelet activation and platelet plug formation, which leads to pathological clogging of arterial blood vessels. Understanding the molecular mechanisms of the regulated adhesiveness of platelets is the key to the development of new pharmacological approaches to prevent or/and treat diseases caused by platelet-dependent thrombosis.

References

Afshar-Kharghan V, Lopez JA (1997) Bernard-Soulier syndrome caused by a dinucleotide deletion and reading frameshift in the region encoding the glycoprotein Ib alpha transmembrane domain. Blood 90:2634-2643

Andre P, Prasad KS, Denis CV, He M, Papalia JM, Hynes RO, Phillips DR, Wagner DD (2002) CD40L stabilizes arterial thrombi by a beta3 integrin-dependent mechanism. Nat Med 8:247-252

Andrews RK, Harris SJ, McNally T, Berndt MC (1998) Binding of purified 14-3-3 zeta signaling protein to discrete amino acid sequences within the cytoplasmic domain of the platelet membrane glycoprotein Ib-IX-V complex. Biochemistry 37:638-647

Angelillo-Scherrer A, de Frutos P, Aparicio C, Melis E, Savi P, Lupu F, Arnout J, Dewerchin M, Hoylaerts M, Herbert J, Collen D, Dahlback B, Carmeliet P (2001) Deficiency or inhibition of Gas6 causes platelet dysfunction and protects mice against thrombosis. Nat Med 7:215-221

Aszodi A, Pfeifer A, Ahmad M, Glauner M, Zhou XH, Ny L, Andersson KE, Kehrel B, Offermanns S, Fassler R (1999) The vasodilator-stimulated phosphoprotein (VASP) is involved in cGMP- and cAMP-mediated inhibition of agonist-induced platelet aggregation, but is dispensable for smooth muscle function. EMBO J 18:37-48

Azzam K, Garfinkel LI, Bal dit Sollier C, Cisse Thiam M, Drouet L (1995) Antithrombotic effect of a recombinant von Willebrand factor, VCL, on nitrogen laser-induced thrombus formation in guinea pig mesenteric arteries. Thromb Haemost 73:318-323

Baumgartner HR (1977) Platelet interaction with collagen fibrils in flowing blood. I. Reaction of human platelets with alpha chymotrypsin-digested subendothelium. Thromb Haemost 37:1-16

Bednar RA, Bednar B, Gaul SL, Chang CT, Hamill T, Egbertson MS, Halxzenko W, Hartrnan GD, Schafer JA, Gould RJ (1996) Binding of the fibrinogen receptor antagonist MK-O383 to purified GP IIb/IIIa and to platelets. FASEB J 9:A56

Berndt MC, Shen Y, Dopheide SM, Gardiner EE, Andrews RK (2001) The vascular biology of the glycoprotein Ib-IX-V complex. Thromb Haemost 86:178-188

Beumer S, IJsseldijk MJ, de Groot PG, Sixma JJ (1994) Platelet adhesion to fibronectin in flow: dependence on surface concentration and shear rate, role of platelet membrane glycoproteins GP IIb/IIIa and VLA-5, and inhibition by heparin. Blood 84:3724-3733

Bhatt DL, Topol EJ (2000) Current role of platelet glycoprotein IIb/IIIa inhibitors in acute coronary syndromes. JAMA 284:1549-1558

Brakebusch C, Fassler R (2003) The integrin-actin connection, an eternal love affair. EMBO J 22:2324-2333

Calderwood DA, Ginsberg MH (2003) Talin forges the links between integrins and actin. Nat Cell Biol 5:694-697

Calderwood DA, Zent R, Grant R, Rees DJ, Hynes RO, Ginsberg MH (1999) The Talin head domain binds to integrin beta subunit cytoplasmic tails and regulates integrin activation. J Biol Chem 274:28071-28074

Calderwood DA, Yan B, de Pereda JM, Alvarez BG, Fujioka Y, Liddington RC, Ginsberg MH (2002) The phosphotyrosine binding-like domain of talin activates integrins. J Biol Chem 277:21749-21758

Cannon CP, McCabe CH, Wilcox RG, Langer A, Caspi A, Berink P, Lopez-Sendon J, Toman J, Charlesworth A, Anders RJ, Alexander JC, Skene A, Braunwald E (2000) Oral glycoprotein IIb/IIIa inhibition with orbofiban in patients with unstable coronary syndromes (OPUS-TIMI 16) trial. Circulation 102:149-156

CAPRIE Steering Committee (1996) A randomised, blinded, trial of clopidogrel versus aspirin in patients at risk of ischaemic events (CAPRIE). Lancet 348:1329-1339

Chang MC, Lin HK, Peng HC, Huang TF (1998) Antithrombotic effect of crotalin, a platelet membrane glycoprotein Ib antagonist from venom of Crotalus atrox. Blood 91:1582-1589

Cierniewski CS, Byzova T, Papierak M, Haas TA, Niewiarowska J, Zhang L, Cieslak M, Plow EF (1999) Peptide ligands can bind to distinct sites in integrin alphaIIbbeta3 and elicit different functional responses. J Biol Chem 274:16923-16932

Clemetson JM, Polgar J, Magnenat E, Wells TN, Clemetson KJ (1999) The platelet collagen receptor glycoprotein VI is a member of the immunoglobulin superfamily closely related to FcalphaR and the natural killer receptors. J Biol Chem 274:29019-29024

Clemetson KJ, Clemetson JM (2001) Platelet collagen receptors. Thromb Haemost 86:189-197

Coller BS (1999) Potential non-glycoprotein IIb/IIIa effects of abciximab. Am Heart J 138:1-5

Coller BS, Peerschke EI, Scudder LE, Sullivan CA (1983) A murine monoclonal antibody that completely blocks the binding of fibrinogen to platelets produces a thrombasthenic-like state in normal platelets and binds to glycoproteins IIb and/or IIIa. J Clin Invest 72:325-338

Coughlin SR (1999) Protease-activated receptors and platelet function. Thromb Haemost 82:353-356

Critchley DR (2000) Focal adhesions-the cytoskeletal connection. Curr Opin Cell Biol 12:133-139

Cruz CP, Eidt J, Drouilhet J, Brown AT, Wang Y, Barnes CS, Moursi MM (2001) Saratin, an inhibitor of von Willebrand factor-dependent platelet adhesion, decreases platelet aggregation and intimal hyperplasia in a rat carotid endarterectomy model. J Vasc Surg 34:724-729

Daniel JL, Dangelmaier C, Jin J, Kim YB, Kunapuli SP (1999) Role of intracellular signaling events in ADP-induced platelet aggregation. Thromb Haemost 82:1322-1326

De Luca M, Facey DA, Favaloro EJ, Hertzberg MS, Whisstock JC, McNally T, Andrews RK, Berndt MC (2000) Structure and function of the von Willebrand factor A1 domain: analysis with monoclonal antibodies reveals distinct binding sites involved in recognition of the platelet membrane glycoprotein Ib-IX-V complex and ristocetin-dependent activation. Blood 95:164-172

DeMali KA, Wennerberg K, Burridge K (2003) Integrin signaling to the actin cytoskeleton. Curr Opin Cell Biol 15:572-782

Denis C, Methia N, Frenette PS, Rayburn H, Ullman-Cullere M, Hynes RO, Wagner DD (1998) A mouse model of severe von Willebrand disease: defects in hemostasis and thrombosis. Proc Natl Acad Sci U S A 95:9524-9529

Dong ZM, Brown AA, Wagner DD (2000) Prominent role of P-selectin in the development of advanced atherosclerosis in ApoE-deficient mice. Circulation 101:2290-2295

Dorsam RT, Kim S, Jin J, Kunapuli SP (2002) Coordinated signaling through both G12/13 and G(i) pathways is sufficient to activate GPIIb/IIIa in human platelets. J Biol Chem 277:47588-47595

Emsley J, Knight CG, Farndale RW, Barnes MJ, Liddington RC (2000) Structural basis of collagen recognition by integrin alpha2beta1. Cell 101:7-56

Ewenstein BM (1997) Von Willebrand's disease. Annu Rev Med 48:25-542

Fabre JE, Nguyen M, Latour A, Keifer JA, Audoly LP, Coffman TM, Koller BH (1999) Decreased platelet aggregation, increased bleeding time and resistance to thromboembolism in P2Y1-deficient mice. Nat Med 5:1199-1202

Farrell DH, Thiagarajan P (1994) Binding of recombinant fibrinogen mutants to platelets. J Biol Chem 269:226-231

Foster CJ, Prosser DM, Agans JM, Zhai Y, Smith MD, Lachowicz JE, Zhang FL, Gustafson E, Monsma FJ Jr, Wiekowski MT, Abbondanzo SJ, Cook DN, Bayne ML, Lira SA, Chintala MS (2001) Molecular identification and characterization of the platelet ADP receptor targeted by thienopyridine antithrombotic drugs. J Clin Invest 107:1591-1598

Fuster V, Badimon L, Badimon JJ, Chesebro JH (1992) The pathogenesis of coronary artery disease and the acute coronary syndromes (1). N Engl J Med 326:242-250

Gachet C (2001) ADP receptors of platelets and their inhibition. Thromb Haemost 86:222-232

Garcia-Alvarez B, de Pereda JM, Calderwood DA, Ulmer TS, Critchley D, Campbell ID, Ginsberg MH, Liddington RC (2003) Structural determinants of integrin recognition by talin. Mol Cell 11:49-58

Gawaz M, Neumann FJ, Dickfeld T, Reininger A, Adelsberger H, Gebhardt A, Schomig A (1997) Vitronectin receptor (alpha(v)beta3) mediates platelet adhesion to the luminal aspect of endothelial cells: implications for reperfusion in acute myocardial infarction. Circulation 96:809-1818

Gruner S, Prostredna M, Schulte V, Krieg T, Eckes B, Brakebusch C, Nieswandt B (2003) Multiple integrin-ligand interactions synergize in shear-resistant platelet adhesion at sites of arterial injury in vivo. Blood 102:4021-4027

Hato T, Pampori N, Shattil SJ (1998) Complementary roles for receptor clustering and conformational change in the adhesive and signaling functions of integrin alphaIIb beta3. J Cell Biol 141:1685-1695

Hauser W, Knobeloch KP, Eigenthaler M, Gambaryan S, Krenn V, Geiger J, Glazova M, Rohde E, Horak I, Walter U, Zimmer M (1999) Megakaryocyte hyperplasia and enhanced agonist-induced platelet activation in vasodilator-stimulated phosphoprotein knockout mice. Proc Natl Acad Sci U S A 96:8120-8125

Hechler B, Leon C, Vial C, Vigne P, Frelin C, Cazenave JP, Gachet C (1998) The P2Y1 receptor is necessary for adenosine 5-diphosphate-induced platelet aggregation. Blood 92:152-159

Hollopeter G, Jantzen HM, Vincent D, Li G, England L, Ramakrishnan V, Yang RB, Nurden P, Nurden A, Julius D, Conley PB (2001) Identification of the platelet ADP receptor targeted by antithrombotic drugs. Nature 409:202-207

Holmback K, Danton MJ, Suh TT, Daugherty CC, Degen JL (1996) Impaired platelet aggregation and sustained bleeding in mice lacking the fibrinogen motif bound by integrin alpha IIb beta 3. EMBO J 15:5760-5771

Holtkotter O, Nieswandt B, Smyth N, Muller W, Hafner M, Schulte V, Krieg T, Eckes B (2002) Integrin alpha 2-deficient mice develop normally, are fertile, but display partially defective platelet interaction with collagen. J Biol Chem 277:10789-10794

Hoylaerts MF, Yamamoto H, Nuyts K, Vreys I, Deckmyn H, Vermylen J (1997) Von Willebrand factor binds to native collagen VI primarily via its A1 domain. Biochem J 324:185-191

Hu DD, White CA, Panzer-Knodle S, Page JD, Nicholson N, Smith JW (1999) A new model of dual interacting ligand binding sites on integrin alphaIIbbeta3. J Biol Chem 274:4633-4639

Hughes PE, Diaz-Gonzalez F, Leong L, Wu C, McDonald JA, Shattil SJ, Ginsberg MH (1996) Breaking the integrin hinge. A defined structural constraint regulates integrin signaling. J Biol Chem 271:6571-6574

Hynes RO (2002) Integrins: bidirectional, allosteric signaling machines. Cell 110:673-687

Inoue O, Suzuki-Inoue K, Dean WL, Frampton J, Watson SP (2003) Integrin alpha2beta1 mediates outside-in regulation of platelet spreading on collagen through activation of Src kinases and PLCgamma2. J Cell Biol 160:769-780

Ishimoto Y, Nakano T (2000) Release of a product of growth arrest-specific gene 6 from rat platelets. FEBS Lett 466:197-199

ISIS-2 (Second International Study of Infarct Survival) Collaborative Group (1988) Randomised trial of intravenous streptokinase, oral aspirin, both, or neither among 17,187 cases of suspected acute myocardial infarction: ISIS-2. Lancet 2:349-360

Jackson SP, Nesbitt WS Kulkarni S (2003) Signaling events underlying thrombus formation. J Thromb Haemost 1:1602-1612

Jin J, Kunapuli SP (1998) Coactivation of two different G protein-coupled receptors is essential for ADP-induced platelet aggregation. Proc Natl Acad Sci U S A 95:8070-8074

Jordan RE, Wagner CL, Mascelli MA, Treacy G, Nedelman MA, Woody JN, Weisman HF, Coller BS (1996) Preclinical development of c7E3 Fab: a mouse/human chimeric monoclonal antibody fragment that inhibits platelet function by blockade of GP IIb/IIIa receptors with observations on the immunogenicity of c7E3 Fab in humans. In: Horton MA (ed) Adhesion receptors as therapeutic targets. CRC Press, Boca Raton, pp 281-305

Jung SM, Moroi M (2000) Signal-transducing mechanisms involved in activation of the platelet collagen receptor integrin alpha(2)beta(1). J Biol Chem 275:8016-8026

Jung SM, Moroi M (2001) Platelet collagen receptor integrin alpha2beta1 activation involves differential participation of ADP-receptor subtypes P2Y1 and P2Y12 but not intracellular calcium change. Eur J Biochem 268:3513-3522

Kageyama S, Matsushita J, Yamamoto H (2002) Effect of a humanized monoclonal antibody to von Willebrand factor in a canine model of coronary arterial thrombosis. Eur J Pharmacol 443:143-149

Kageyama S, Yamamoto H, Nakazawa H, Matsushita J, Kouyama T, Gonsho A, Ikeda Y, Yoshimoto R (2002) Pharmacokinetics and pharmacodynamics of AJW200, a humanized monoclonal antibody to von Willebrand factor, in monkeys. Arterioscler Thromb Vasc Biol 22:187-192

Kato K, Kanaji T, Russell S, Kunicki TJ, Furihata K, Kanaji S, Marchese P, Reininger A, Ruggeri ZM, Ware J (2003) The contribution of glycoprotein VI to stable platelet adhesion and thrombus formation illustrated by targeted gene deletion. Blood 102:1701-1707

Knight CG, Morton LF, Peachey AR, Tuckwell DS, Farndale RW, Barnes MJ (2000) The collagen-binding A-domains of integrins alpha(1)beta(1) and alpha(2)beta(1) recognize the same specific amino acid sequence, GFOGER, in native (triple-helical) collagens. J Biol Chem 275:35-40

Konishi H, Katoh Y, Takaya N, Kashiwakura Y, Itoh S, Ra C, Daida H (2002) Platelets activated by collagen through immunoreceptor tyrosine-based activation motif play pivotal role in initiation and generation of neointimal hyperplasia after vascular injury. Circulation 105:912-916

Kuijpers MJ, Schulte V, Bergmeier W, Lindhout T, Brakebusch C, Offermanns S, Fassler R, Heemskerk JW, Nieswandt B (2003) Complementary roles of platelet glycoprotein VI and integrin alpha2beta1 in collagen-induced thrombus formation in flowing whole blood ex vivo. FASEB J 17:685-687

Kunapuli SP, Dorsam RT, Kim S, Quinton TM (2003) Platelet purinergic receptors. Curr Opin Pharmacol 3:175-180

Law DA, DeGuzman FR, Heiser P, Ministri-Madrid K, Killeen N, Phillips DR (1999) Integrin cytoplasmic tyrosine motif is required for outside-in alphaIIbbeta3 signalling and platelet function. Nature 401:808-811

Lee JO, Rieu P, Arnaout MA, Liddington R (1995) Crystal structure of the A domain from the alpha subunit of integrin CR3 (CD11b/CD18). Cell 80:631-638

Leon C, Hechler B, Freund M, Eckly A, Vial C, Ohlmann P, Dierich A, LeMeur M, Cazenave JP, Gachet C (1999) Defective platelet aggregation and increased resistance to thrombosis in purinergic P2Y(1) receptor-null mice. J Clin Invest 104:1731-1737

Li R, Mitra N, Gratkowski H, Vilaire G, Litvinov R, Nagasami C, Weisel JW, Lear JD, DeGrado WF, Bennett JS (2003) Activation of integrin alphaIIbbeta3 by modulation of transmembrane helix associations. Science 300:795-798

Liddington RC, Ginsberg MH (2002) Integrin activation takes shape. J Cell Biol 158:833-839

Liu S, Calderwood DA, Ginsberg MH (2000) Integrin cytoplasmic domain-binding proteins. J Cell Sci 113:3563-3571

Lopez JA, Chung DW, Fujikawa K, Hagen FS, Papayannopoulou T, Roth GJ (1987) Cloning of the alpha chain of human platelet glycoprotein Ib: a transmembrane protein with homology to leucine-rich alpha 2- glycoprotein. Proc Natl Acad Sci U S A 84:5615-5619

Lopez JA, Andrews RK, Afshar-Kharghan V, Berndt MC (1998) Bernard-Soulier syndrome. Blood 91:4397-4418

Marchese P, Murata M, Mazzucato M, Pradella P, De Marco L, Ware J, Ruggeri ZM (1995) Identification of three tyrosine residues of glycoprotein Ib alpha with distinct roles in von Willebrand factor and alpha-thrombin binding. J Biol Chem 270:9571-9578

Marcus AJ, Broekman MJ, Drosopoulos JH, Islam N, Alyonycheva TN, Safier LB, Hajjar KA, Posnett DN, Schoenborn MA, Schooley KA, Gayle RB, Maliszewski CR (1997) The endothelial cell ecto-ADPase responsible for inhibition of platelet function is CD39. J Clin Invest 99:1351-1360

Martel V, Racaud-Sultan C, Dupe S, Marie C, Paulhe F, Galmiche A, Block MR, Albiges-Rizo C (2001) Conformation, localization and integrin binding of talin depend on its interaction with phosphoinositides. J Biol Chem 276:21217-21227

Massberg S, Sausbier M, Klatt P, Bauer M, Pfeifer A, Siess W, Fassler R, Ruth P, Krombach F, Hofmann F (1999) Increased adhesion and aggregation of platelets lacking cyclic guanosine 3',5'-monophosphate kinase I. J Exp Med 189:1255-1264

Massberg S, Brand K, Gruner S, Page S, Muller E, Muller I, Bergmeier W, Richter T, Lorenz M, Konrad I, Nieswandt B, Gawaz M (2002) A critical role of platelet adhesion in the initiation of atherosclerotic lesion formation. J Exp Med 196:887-896

Massberg S, Gawaz M, Gruner S, Schulte V, Konrad I, Zohlnhofer D, Heinzmann U, Nieswandt B (2003) A crucial role of glycoprotein VI for platelet recruitment to the injured arterial wall in vivo. J Exp Med 197:41-49

McGhie AI, McNatt J, Ezov N, Cui K, Mower LK, Hagay Y, Buja LM, Garfinkel LI, Gorecki M, Willerson JT (1994) Abolition of cyclic flow variations in stenosed, endothelium-injured coronary arteries in nonhuman primates with a peptide fragment (VCL) derived from human plasma von Willebrand factor-glycoprotein Ib binding domain. Circulation 90:2976-2981

Mehta SR, Yusuf S, Peters RJ, Bertrand ME, Lewis BS, Natarajan MK, Malmberg K, Rupprecht H, Zhao F, Chrolavicius S, Copland I, Fox KA; Clopidogrel in Unstable angina to prevent Recurrent Events trial (CURE) Investigators (2001) Effects of pretreatment with clopidogrel and aspirin followed by long-term therapy in patients undergoing percutaneous coronary intervention: the PCI-CURE study. Lancet 358:527-533

Merten M, Thiagarajan P (2000) P-selectin expression on platelets determines size and stability of platelet aggregates. Circulation 102:1931-1936

Methia N, Andre P, Denis CV, Economopoulos M, Wagner DD (2001) Localized reduction of atherosclerosis in von Willebrand factor-deficient mice. Blood 98:1424-1428

Miranti CK, Leng L, Maschberger P, Brugge JS, Shattil SJ (1998) Identification of a novel integrin signaling pathway involving the kinase Syk and the guanine nucleotide exchange factor Vav1. Curr Biol 8:1289-1299

Moers A, Nieswandt B, Massberg S, Wettschureck N, Grüner S, Konrad I, Aktas B, Schulte V, Gratacap MP, Simon MI, Gawaz M, Offermanns S (2003) G13 is an essential mediator of platelet activation in haemostasis and thrombosis. Nat Med 9:1418-1422

Morton LF, Hargreaves PG, Farndale RW, Young RD, Barnes MJ (1995) Integrin alpha 2 beta 1-independent activation of platelets by simple collagen-like peptides: collagen tertiary (triple-helical) and quaternary (polymeric) structures are sufficient alone for alpha 2 beta 1-independent platelet reactivity. Biochem J 306:337-344

Nieswandt B, Watson SP (2003) Platelet-collagen interaction: is GPVI the central receptor? Blood 102:449-461

Nieswandt B, Schulte V, Bergmeier W, Mokhtari-Nejad R, Rackebrandt K, Cazenave JP, Ohlmann P, Gachet C, Zirngibl H (2001a) Long-term antithrombotic protection by in vivo depletion of platelet glycoprotein VI in mice. J Exp Med 193:459-470

Nieswandt B, Brakebusch C, Bergmeier W, Schulte V, Bouvard D, Mokhtari-Nejad R, Lindhout T, Heemskerk JW, Zirngibl H, Fassler R (2001b) Glycoprotein VI but not alpha2beta1 integrin is essential for platelet interaction with collagen. EMBO J 20:2120-2130

Nieswandt B, Schulte V, Zywietz A, Gratacap MP, Offermanns S (2002) Costimulation of Gi- and G12/G13-mediated signaling pathways induces integrin alpha IIbbeta 3 activation in platelets. J Biol Chem 277:39493-39498

Nurden AT (1999) Inherited abnormalities of platelets. Thromb Haemost 82:468-480

O'Neill WW, Serruys P, Knudtson M, van Es GA, Timmis GC, van der Zwaan C, Kleiman J, Gong J, Roecker EB, Dreiling R, Alexander J, Anders R (2000) Long-term treatment with a platelet glycoprotein-receptor antagonist after percutaneous coronary revascularization. EXCITE Trial Investigators. Evaluation of Oral Xemilofiban in Controlling Thrombotic Events. N Engl J Med 342:1316-13240

O'Toole TE, Katagiri Y, Faull RJ, Peter K, Tamura R, Quaranta V, Loftus JC, Shattil SJ, Ginsberg MH (1994) Integrin cytoplasmic domains mediate inside-out signal transduction. J Cell Biol 124:1047-1059

Offermanns S, Laugwitz KL, Spicher K, Schultz G (1994) G proteins of the G_{12} family are activated via thromboxane A_2 and thrombin receptors in human platelets. Proc Natl Acad Sci U S A 91:504-508

Offermanns S, Toombs CF, Hu YH, Simon MI (1997) Defective platelet activation in G alpha(q)-deficient mice. Nature 389:183-186

Parise LV (1999) Integrin alpha(IIb)beta(3) signaling in platelet adhesion and aggregation. Curr Opin Cell Biol 11:597-601

Patrono C (1994) Aspirin as an antiplatelet drug. N Engl J Med 330:1287-1294

Peterson JA, Visentin GP, Newman PJ, Aster RH (1998) A recombinant soluble form of the integrin alpha IIb beta 3 (GPIIb-IIIa) assumes an active, ligand-binding conformation and is recognized by GPIIb-IIIa-specific monoclonal, allo-, auto-, and drug-dependent platelet antibodies. Blood 92:2053-2063

Phillips DR, Prasad KS, Manganello J, Bao M, Nannizzi-Alaimo L (2001) Integrin tyrosine phosphorylation in platelet signaling. Curr Opin Cell Biol 13:546-554

Pinsky DJ, Broekman MJ, Peschon JJ, Stocking KL, Fujita T, Ramasamy R, Connolly ES Jr, Huang J, Kiss S, Zhang Y, Choudhri TF, McTaggart RA, Liao H, Drosopoulos JH, Price VL, Marcus AJ, Maliszewski CR (2002) Elucidation of the thromboregulatory role of CD39/ectoapyrase in the ischemic brain. J Clin Invest 109:1031-1040

Plow EF, Haas TA, Zhang L, Loftus J, Smith JW (2000) Ligand binding to integrins. J Biol Chem 275:21785-21788

Prevost N, Woulfe D, Tanaka T, Brass LF (2002) Interactions between Eph kinases and ephrins provide a mechanism to support platelet aggregation once cell-to-cell contact has occurred. Proc Natl Acad Sci U S A 99:9219-9224

Prevost N, Woulfe D, Tognolini M, Brass LF (2003) Contact-dependent signaling during the late events of platelet activation. J Thromb Haemost 1:1613-1627

Priddle H, Hemmings L, Monkley S, Woods A, Patel B, Sutton D, Dunn GA, Zicha D, Critchley DR (1998) Disruption of the talin gene compromises focal adhesion assembly in undifferentiated but not differentiated embryonic stem cells. J Cell Biol 142:1121-1133

Pulcinelli FM, Ashby B, Gazzaniga PP, Daniel JL (1995) Protein kinase C activation is not a key step in ADP-mediated exposure of fibrinogen receptors on human platelets. FEBS Lett 364:87-90

Pulcinelli FM, Pesciotti M, Pignatelli P, Riondino S, Gazzaniga PP (1998) Concomitant activation of Gi and Gq protein-coupled receptors does not require an increase in cytosolic calcium for platelet aggregation. FEBS Lett 435:5-8

Quinn MJ, Fitzgerald DJ (1999) Ticlopidine and clopidogrel. Circulation 100:1667-1672

Ruggeri ZM (1997) Von Willebrand factor [published erratum appears in J Clin Invest 1998 Feb 15;101(4):919]. J Clin Invest 100:41-46

Ruggeri ZM (1999) Structure and function of von Willebrand factor. Thromb Haemost 82:576-584

Ruggeri ZM (2003) Von Willebrand factor, platelets and endothelial cell interactions. J Thromb Haemost 1:1335-1342

Sakariassen KS, Nievelstein PF, Coller BS, Sixma JJ (1986) The role of platelet membrane glycoproteins Ib and IIb-IIIa in platelet adherence to human artery subendothelium. Br J Haematol 63:681-691

Santoro SA, Lawing WJ Jr (1987) Competition for related but nonidentical binding sites on the glycoprotein IIb-IIIa complex by peptides derived from platelet adhesive proteins. Cell 48:867-873

Santoro SA, Zutter MM (1995) The alpha 2 beta 1 integrin: a collagen receptor on platelets and other cells. Thromb Haemost 74:813-821

Savage B, Almus-Jacobs F, Ruggeri ZM (1998) Specific synergy of multiple substrate-receptor interactions in platelet thrombus formation under flow. Cell 94:657-666

Savage B, Almus-Jacobs F, Ruggeri ZM (1998) Specific synergy of multiple substrate-receptor interactions in platelet thrombus formation under flow. Cell 94:657-666

Savi P, Beauverger P, Labouret C, Delfaud M, Salel V, Kaghad M, Herbert JM (1998) Role of P2Y1 purinoceptor in ADP-induced platelet activation. FEBS Lett 422:291-295

Scarborough RM (1999) Development of eptifibatide. Am Heart J 138:1093-1094

Scarborough RM, Naughton MA, Teng W, Rose JW, Phillips DR, Nannizzi L, Arfsten A, Campbell AM, Charo IF (1993) Design of potent and specific integrin antagonists. Peptide antagonists with high specificity for glycoprotein IIb-IIIa. J Biol Chem 268:1066-1073

Schwarz UR, Walter U, Eigenthaler M (2001) Taming platelets with cyclic nucleotides. Biochem Pharmacol 62:1153-1161

Schwarzbauer JE (1991) Fibronectin: from gene to protein. Curr Opin Cell Biol 3:786-791

Shattil SJ (1999) Signaling through platelet integrin alpha IIb beta 3: inside-out, outside-in, and sideways. Thromb Haemost 82:318-325

Shattil SJ, Brass LF (1987) Induction of the fibrinogen receptor on human platelets by intracellular mediators. J Biol Chem 262:992-1000

Shattil SJ, Cunningham M, Wiedmer T, Zhao J, Sims PJ, Brass LF (1992) Regulation of glycoprotein IIb-IIIa receptor function studied with platelets permeabilized by the pore-forming complement proteins C5b-9. J Biol Chem 267:18424-18431

Shattil SJ, Kashiwagi H, Pampori N (1998) Integrin signaling: the platelet paradigm. Blood 91:2645-2657

Shelton-Inloes BB, Titani K, Sadler JE (1986) cDNA sequences for human von Willebrand factor reveal five types of repeated domains and five possible protein sequence polymorphisms. Biochemistry 25:3164-3171

Shen,Y, Romo GM, Dong JF, Schade A, McIntire LV, Kenny D, Whisstock JC, Berndt MC, Lopez JA, Andrews RK (2000) Requirement of leucine-rich repeats of glycoprotein (GP) Ibalpha for shear-dependent and static binding of von Willebrand factor to the platelet membrane GP Ib-IX-V complex. Blood 95:903-910

Smyth SS, Reis ED, Zhang W, Fallon JT, Gordon RE, Coller BS (2001) Beta(3)-integrin-deficient mice but not P-selectin-deficient mice develop intimal hyperplasia after vascular injury: correlation with leukocyte recruitment to adherent platelets 1 hour after injury. Circulation 103:2501-2507

Sporn LA, Marder VJ, Wagner DD (1986) Inducible secretion of large, biologically potent von Willebrand factor multimers. Cell 46:185-190

Springer TA (1997) Folding of the N-terminal, ligand-binding region of integrin alpha-subunits into a beta-propeller domain. Proc Natl Acad Sci U S A 94:65-72

Steinhubl SR, Berger PB, Mann JT 3rd, Fry ET, DeLago A, Wilmer C, Topol EJ (2002) CREDO investigators. Clopidogrel for the reduction of events during observation. Early and sustained dual oral antiplatelet therapy following percutaneous coronary intervention: a randomized controlled trial. JAMA 288:2411-2420

Storey RF, Judge HM, Wilcox RG, Heptinstall S (2002) Inhibition of ADP-induced P-selectin expression and platelet-leukocyte conjugate formation by clopidogrel and the

P2Y12 receptor antagonist AR-C69931MX but not aspirin. Thromb Haemost 88:488-494

Tadokoro S, Shattil SJ, Eto K, Tai V, Liddington RC, De Pereda JM, Ginsberg MH, Calderwood DA (2003) Talin binding to integrin beta tails: a final common step in integrin activation. Science 302:103-106

Takagi J, Petre BM, Walz T, Springer TA (2002) Global conformational rearrangements in integrin extracellular domains in outside-in and inside-out signaling. Cell 110:599-611

The SYMPHONY Investigators (2000) Sibrafiban versus Aspirin to Yield Maximum Protection from Ischemic Heart Events Post-acute Coronary Syndromes. Comparison of sibrafiban with aspirin for prevention of cardiovascular events after acute coronary syndromes: a randomised trial. Lancet 355:337-345

Theilmeier G, Michiels C, Spaepen E, Vreys I, Collen D, Vermylen J, Hoylaerts MF (2002) Endothelial von Willebrand factor recruits platelets to atherosclerosis-prone sites in response to hypercholesterolemia. Blood 4486-4493

Timpl R, Brown JC (1994) The laminins. Matrix Biol 14:275-281

Topol EJ, Moliterno DJ, Herrmann HC, Powers ER, Grines CL, Cohen DJ, Cohen EA, Bertrand M, Neumann FJ, Stone GW, DiBattiste PM, Demopoulos L (2001) TARGET investigators. Do tirofiban and Reopro give similar efficacy trial. Comparison of two platelet glycoprotein IIb/IIIa inhibitors, tirofiban and abciximab, for the prevention of ischemic events with percutaneous coronary revascularization. N Engl J Med 344:1888-1894

Toullec D, Pianetti P, Coste H, Bellevergue P, Grand-Perret T, Ajakane M, Baudet V, Boissin P, Boursier E, Loriolle F, et al (1991) The bisindolylmaleimide GF 109203X is a potent and selective inhibitor of protein kinase C. J Biol Chem 266:15771-15781

Tozer EC, Liddington RC, Sutcliffe MJ, Smeeton AH, Loftus JC (1996) Ligand binding to integrin alphaIIbbeta3 is dependent on a MIDAS-like domain in the beta3 subunit. J Biol Chem 271:21978-21984

Turner NA, Moake JL, McIntire LV (2001) Blockade of adenosine diphosphate receptors P2Y(12) and P2Y(1) is required to inhibit platelet aggregation in whole blood under flow. Blood 98:3340-3345

Vinogradova O, Velyvis A, Velyviene A, Hu B, Haas T, Plow E, Qin J (2002) A structural mechanism of integrin alpha(IIb)beta(3) "inside-out" activation as regulated by its cytoplasmic face. Cell 110:587-597

Wagner CL, Mascelli MA, Neblock DS, Weisman HF, Coller BS, Jordan RE (1996) Analysis of GPIIb/IIIa receptor number by quantification of 7E3 binding to human platelets. Blood 88:907-914

Walker TR, Watson SP (1993) Synergy between Ca^{2+} and protein kinase C is the major factor in determining the level of secretion from human platelets. Biochem J 289:277-282

Wang K, Zhou X, Zhou Z, Tarakji K, Carneiro M, Penn MS, Murray D, Klein A, Humphries RG, Turner J, Thomas JD, Topol EJ, Lincoff AM (2003) Blockade of the platelet P2Y12 receptor by AR-C69931MX sustains coronary artery recanalization and improves the myocardial tissue perfusion in a canine thrombosis model. Arterioscler Thromb Vasc Biol 23:357-362

Wardell MR, Reynolds CC, Berndt MC, Wallace RW, Fox JE (1989) Platelet glycoprotein Ib beta is phosphorylated on serine 166 by cyclic AMP-dependent protein kinase. J Biol Chem 264:15656-15661

Watson SP, Gibbins J (1998) Collagen receptor signalling in platelets: extending the role of the ITAM. Immunol Today 19:260-264

Watson SP, Asazuma N, Atkinson B, Berlanga O, Best D, Bobe R, Jarvis G, Marshall S, Snell D, Stafford M, Tulasne D, Wilde J, Wonerow P, Frampton J (2001) The role of ITAM- and ITIM-coupled receptors in platelet activation by collagen. Thromb Haemost 86:276-288

Woodside DG, Liu S, Ginsberg MH (2001) Integrin activation. Thromb Haemost 86:316-323

Wu D, Meiring M, Kotze HF, Deckmyn H, Cauwenberghs N (2002) Inhibition of platelet glycoprotein Ib, glycoprotein IIb/IIIa, or both by monoclonal antibodies prevents arterial thrombosis in baboons. Arterioscler Thromb Vasc Biol 22:323-328

Wymann MP, Zvelebil M, Laffargue M (2003) Phosphoinositide 3-kinase signalling-which way to target? Trends Pharmacol Sci 24:366-376

Xu Y, Gurusiddappa S, Rich RL, Owens RT, Keene DR, Mayne R, Hook A, Hook M (2000) Multiple binding sites in collagen type I for the integrins alpha1beta1 and alpha2beta1. J Biol Chem 275:38981-38989

Yan B, Calderwood DA, Yaspan B, Ginsberg MH (2001) Calpain cleavage promotes talin binding to the beta 3 integrin cytoplasmic domain. J Biol Chem 276:28164-28170

Yao SK, Ober JC, Garfinkel LI, Hagay Y, Ezov N, Ferguson JJ, Anderson HV, Panet A, Gorecki M, Buja LM (1994) Blockade of platelet membrane glycoprotein Ib receptors delays intracoronary thrombogenesis, enhances thrombolysis, and delays coronary artery reocclusion in dogs. Circulation 89:2822-2828

Yeh CH, Chang MC, Peng HC, Huang TF (2001) Pharmacological characterization and antithrombotic effect of agkistin, a platelet glycoprotein Ib antagonist. Br J Pharmacol 132:843-850

Yusuf S, Zhao F, Mehta SR, Chrolavicius S, Tognoni G, Fox KK, Clopidogrel in Unstable Angina to Prevent Recurrent Events Trial Investigators (2001) Effects of clopidogrel in addition to aspirin in patients with acute coronary syndromes without ST-segment elevation. N Engl J Med 345:494-502

Zang Q, Lu C, Huang C, Takagi J, Springer TA (2000) The top of the inserted-like domain of the integrin lymphocyte function-associated antigen-1 beta subunit contacts the alpha subunit beta -propeller domain near beta-sheet 3. J Biol Chem 275:22202-22212

Zhang J, Zhang J, Shattil SJ, Cunningham MC, Rittenhouse SE (1996) Phosphoinositide 3-kinase gamma and p85/phosphoinositide 3-kinase in platelets. Relative activation by thrombin receptor or beta-phorbol myristate acetate and roles in promoting the ligand-binding function of alphaIIbbeta3 integrin. J Biol Chem 271:6265-6272

Subject Index

abciximab 460
acetylcholine receptor 231
acetylsalicylic acid 458
acid sphingomyelinase 315
actin 42, 107, 200, 302
– actin-binding domain (ABD) 153
– cytoskeleton 266, 350, 355, 359
– filament 348
α-actinin 41
activation 201
actodermal dysplasia 162
acute
– cellular response 417
– coronary syndrome 440, 459
– myocardial infarction 458
– vascular response 417
(ADAM)10 390
addressin 415
adenocarcinoma 306
adenomatous polyposis coli (APC) 72, 107
adenosine diphosphate (ADP) 440, 444, 446, 451, 452, 459
– receptor 460
adenovirus 365
– receptor 364
adherens junction 23, 24, 40, 105, 106, 345
adhesion 456
afadin 347, 348, 353
– l-afadin 37, 41, 42
aggregation 448, 456
agkistin 457
agrin 231

allantois 223
alpha helical rod domain 152
Alzheimer's disease 122
angiogenesis 225, 319
annexin II 303
antiplatelet
– drug 437
– trialists' collaboration 458
anti-thrombotic effect 456
anti-vWF antibody AJW200 457
APC 107
– tumor suppressor 108
apoptosis 223
arachidonic 458
AR-C69931MX 460
armadillo repeat 105, 111
– domain 146
arrhythmogenic right ventricular cardiomyopathy (ARVC) 162, 163
arterial thrombosis 439, 441, 447, 456
Asef 111
aspirin 456, 458, 459, 460
atherosclerosis 455
ATP 451
axin/conductin 109
axon 230
– guidance 60, 375
axonin-1 378, 381
– TAG-1 376, 390

B lymphocytes 311
barbourin 460
basement membrane (BM) 219
Bernard–Soulier syndrome 441

beta-propeller-like structure 446, 448
blastocyst 222
bone 227
botrocetin 441
BP180 250, 263
BP230 254, 258, 263
BPAG1 160
B-Raf 202
branching morphogenesis 64
breast carcinoma 81
bullous
– impetigo 156
– pemphigoid 258

Ca^{++}-binding 6
cadherin 3, 29, 55, 296, 353, 385
– 6B 59
– 7 59
– 11 57
– C 59
– classical cadherin 25
– DE-cadherin 60
– repeat (EC1-EC4) 141
– T 84
Caenorhabditis elegans 362
Cajal Retzius cell 232
calmodulin 300
calpain 199
calponin 256
cAMP 452, 453
cancer 111
CAPRIE trial 459
carcinoembryonic antigen-related cellular adhesion molecule, see CEACAM
carcinoma 70, 73
– β-cell carcinoma 70
cardiovascular disease 456, 458
casein kinase II (CKII) 26, 32
Caspr 385
catenin 71, 72, 105, 353
– α 16, 37, 39, 40, 43, 74, 105, 123
– – target gene 114
– β 25, 27, 28, 30, 33, 37, 72, 74, 105, 111–113, 116
– γ 105, 124
caveolin 424
– 1 201
CD31 411, 426
CD39 460
CD40 ligand 454, 456

CD44 428
CD87 424
CD98 424
CD99 413, 428
CDC42 126, 202
CEA, pregnancy 295
CEACAM 284
– CEACAM1 protein 287
– expression pattern 291
cell
– adhesion molecule (CAM) 55, 299, 406, 456
– cell–cell adhesion 3, 105
– cycle 209
– migration 115, 382
– motility 59
– polarity 57, 361, 363
– signaling 359
– surface proteoglycan 83
– trafficking 414
cerebellum 382, 384
cerebral cortex 232
cerebrovascular accident 459
cGMP 453
– cGMP-dependent kinase (PKG) 453
chemotactic peptide 309
cholangiocarcinoma 81
chorion 223
chromosome 16q22.1 75
circula dichroism spectroscopy 6
cis
– dimer 4, 32
– interaction 375, 387-389
c-Jun N-terminal kinase (JNK) 361
claudin 352
– family 411
CLEVER-1 416
clopidogrel 459
clustering 450, 454
cochlea 227
cofilin 203
collagen 437, 439, 442, 451, 457
– type I 440, 442, 446
– type IV 219, 446, 447
– type VI 440
– type XVII 250
colon 294
colorectal cancer 112
commissural axon 377, 389, 391
conductin/axin-2 120

conformational change 381, 389, 391
congenital bleeding disorder 441
convergent extension 59
corneodesmosin 155
coronary stenting 461
coxsackievirus 364
crotalin 457
C-terminal-binding system 79
cyclic
– adenosine monophosphate
  (cAMP) 288, 452, 453
– nucleotide 453
cyclin D1 208
cyclooxygenase (COX) 79, 458
cytochalasin D 304
cytokeratin 252
cytoplasmic domain 200
cytoskeleton 107, 206, 207, 388
cytosolic truncation 82

Darier's disease 171
Dcc (deleted in colorectal cancer) 380, 383, 391
deafness 227
DE-cadherin 60
dedifferentiation 114
delayed-type hypersensitivity 312
desmin 264
desmocalmin 155
desmocollin (Dsc) 26, 36, 85, 141, 172
desmoglein (Dsg) 26, 36, 85, 141, 172
desmoplakin (DP) 44, 139, 150, 151, 172, 173, 259
– I 151, 160
– II 151, 160
– model 169
– particle 172
desmosomal
– cadherin 35, 46, 141
– – model 166
– precursor 171
desmosome 23, 24, 105, 138, 345, 349
– assembly 147, 171–174
– dissolution 174
– precursor 172
– protein 174
– – modification by kinase
– – modification by phosphatase 174
diacyl glycerol 452
diapedesis 410

differential adhesion 55
dilated left ventricular cardiomyopathy 163
disintegrin 460
dorsal root entry zone 381
DPCT 44
Drosophila 362
DTD 143
dynactin complex 107
dynein complex 107
dysplasia 352
dystonin 256
dystrophin 229, 256
– glycoprotein complex 229

EA 141, 142
EC5 142
E-cadherin 6, 25, 30, 56, 70, 71, 107, 345, 348
– as tumor suppressor 71
– downregulation 77
– gene (CDH1) 75, 78
– upregulation 85, 86
ectodysplasin A 250
ectoplasmic specialization 357
EF-hand motif 448
EGF 117
embryoid body 222
embryonic stem (ES) cell 222
EMT 109
endocytosis 307
endoderm 222, 223
endothelial
– cell 455
– – junction 412
– dysfunction 455
endothelium 319, 456
enterocytes 295
envoplakin 150, 154, 160, 254, 260
Eph
– A4 455
– B1 455
– family 455
– interaction 455
epicardium 223
epidermal growth factor (EGF)
– domain 448
– receptor 72
epidermolysis bullosa 264
epiplakin 254, 261

epithelial-mesenchymal transition 80
eptifibatide 460
ErbB2 79
E-selectin 74, 320
– ligand-1 419
estrogen receptor 80
exfoliative toxin
– A (ETA) 160
– B (ETB) 160
– D (ETD) 160
external granule cell layer 384
extracellular
– anchor (EA) 141, 142
– domain 3
– matrix 117, 217, 437–439, 443, 447, 448, 455
– signal-regulated kinase (ERK) 361

F11/contactin 381
F11/F3/contactin 385, 390
F-actin 43
factor
– VIII 441
– XII 442
FAK 198
FAP 111, 121
fat 57
FcRγ-chain 443
fibrillar actin 303
fibrin 448
fibrinogen 448, 449, 454, 460
fibroblast growth factor (bFGF) 62
fibronectin 223, 248, 314, 439, 447, 448
filopodia 203, 350, 360
fimbriae type 1 313
firm adhesion 410
flamingo (fmi) 57
floor plate 377, 389, 391
focal adhesion 424
Fyn 388, 444
G protein 248
– coupled receptor 209
– subfamily 452
G_{12} 452
G_{13} 452
galectin-3 310
gastric cancer 77
gastrulation 110
generalized atrophic benign epidermolysis bullosa (GABEB) 251

G_i 452
Glanzmann thrombasthenia 449
globular actin 303
glomerulus 225
glycine-rich desmoglein terminal domain 143
glycine-serine-arginine (GSR) domain 153
glycogen synthase kinase 3β (GSK-3β) 26
glycoprotein (GP)
– Ia 445
– Ib 439, 443, 455
– – IX-V complex 451
– – β 441
– – V-IX 437, 440, 441, 444, 448, 456, 457
– IIa 445
– IIb 448
– – antagonist 437, 460
– IIIa 448
– – antagonist 437, 460
– IX 441
– V 441
– VI 437, 443
– VI 440, 443, 444, 447, 448, 451, 457
– pregnancy-specific 285
glycosylphosphatidylinositol 288, 289, 388
gonococci 314
GPI-linked IgSF 388
GPO
– repeat 443
– sequence 443
$G_{q/11}$ 452
G_q-mediated signaling 452
granule 410, 420
– α 410, 420
– cell 382
Grb-2 adaptor protein 201
groucho 109
growth
– arrest-specific gene 6 (Gas6) 454
– cone 376
– – morphology 389
GSK-3β 26, 32, 109
GTPase Rac 63
guanosine triphosphate 345
guidepost cell 377

Haemophilus influenzae   316
Hailey–Hailey's disease   171
hair cell   227
haptotactic gradient   412
HAV   8
heart   223
helix exchange   39
hemidesmosome   226, 245
- assembly   266
hemostasis   441, 443, 446, 456
hepatocyte growth factor (HGF)   249
herpes simplex virus   346
herpetiformis pemphigus (HP)   156
hetero-*trans*-dimer   349
HGF   117
high endothelial venule (HEV)   413
high mannose   289, 313
high molecular weight kininogen   442
high shear
- condition   454
- flow   444
hippocampus   233, 353
Hirudo medicinalis   457
hmr-1   56
HNPCC   119
homeostasis   175
homophilic
- adhesion   4
- interaction   107
homo-*trans*-dimer   349
human cancer   121
hydrophobic pocket   11
hypermethylation   78
hypotrichosis   155

IGF
- I   117
- II   117
I-like domain   448
immunoglobulin (Ig)   284
- superfamily   284, 424, 444
immunohistochemistry   77
implantation   222
inflammatory
- disease   456
- vascular disease   455
inner cell mass   222
inner ear   227
inside-out
- activation   450

- signalling   422, 446, 449, 450
insulin receptor   307
integrin   198, 217, 246, 413, 421, 450
- $\alpha 6\beta 4$ 246
- $\alpha 2\beta 1$ 437, 440, 443, 445, 446, 457
- $\alpha 4$ 410
- $\alpha 5\beta 1$ 437, 440, 447
- $\alpha 6\beta 1$ 437, 440, 447
- $\alpha IIb\beta 3$ 440, 442, 443, 445, 447, 450, 454, 456, 458
- - oral antagonist   461
- - activation   450
- - cross-linking   460
- - outside-in signaling   454
- $\alpha v\beta 3$ 413, 423, 440, 456
integrin-associated protein (IAP)   233, 412, 423
integrin-linked kinase (ILK)   79, 118
intercellular adhesion molecule (ICAM)   364, 410, 425
- 1   425
- 2   426
- 3   426
interferon   288
interleukin (IL)   80, 288
- 1 β 456
intermediate filament (IF)   44, 226, 252, 255
internal granule cell layer   384
intracellular
- anchor (IA)   142, 143
- cadherin-type sequence (ICS)   143
- proline-rich linker region (IPL)   143
- signaling pathway   387
intravascular thrombosis   458
invasion   113
invasive front   113
IP_3 452
IQGAP   107
ischemic stroke   439
ITAM   443, 444
JNK   202
junction assembly   175
junctional adhesion molecule (JAM)   345, 411, 427
junctional epidermolysis bullosa (JEB)   247

keratin   139, 252, 264
keratinocyte   226, 253, 256, 264

keratocalmin 155
kidney 224

labeled pathway hypothesis 377
lamellipodia 203, 350, 360
laminin 219, 439, 447
– 5 245, 246, 251
lateral dimer 32
LEF1 118
left-right asymmetry 58
leukocyte 455
ligand binding 201
LIM-kinase 203
linker for activation of T cell (LAT) 444
lithium 123
long-range guidance cues 378
long-term potentiation (LTP) 233, 387
low-dose aspirin 459
lymphatic endothelium 415
lymphocyte VLA-2 445
Lyn 444
LYVE-1 416

M domain 40, 42
Mac-1 424, 441
Madin-Darby canine kidney 349
MAP kinase 199
matrix protein 444
mechanical force 217
mechanosensor 227
megakaryocyte 438, 447
memory and learning 387
mesoderm 223
Met receptor 249
metal ion-dependent adhesion site (MIDAS) 422, 446, 448, 449
metanephric mesenchyme 224
metastasis 113, 118
metazoan 218
microsatellite instability 119
microtubule
– microtubule-actin cross-linking factor (MACF) 254, 260
– network 107
migration 217
MLC kinase 209
MN-cadherin 59
morphogenesis 53
motility 115
motoneuron 231

mucin-like glycoprotein 83
mucosal addressin cell adhesion molecule 1 426
mucosal endothelial addressin (Mad)CAM-1 410
multistep adhesion cascade 408
murine hepatitis virus 317
muscle 228
– fiber 228
muscular dystrophy 229
myelination 230
myoblast 228
– fusion 228
myocardial infarction 439
myogenesis 296
myosin 454
myotendinous junction 229
myotube 229

natural killer cells 312
naxos disease 162
N-cadherin 6, 56
NCAM 377, 386
nectin 346
– synapsis 352
Neisseria 312, 314, 316
nephron 225
nephronectin 224
netrin 380, 383
– 1 377
neuregulin 233
neurexin 385
neurodegenerative disease 122
neurofascin 385
neuroligin 385
neuromuscular junction 230, 385, 386
neuron-glia CAM/L1 376, 377, 382
neutrophilic granulocytes 309
nidogen/entactin 219
NO 453
nociceptive fiber 382
node of Ranvier 385
NrCAM 378, 382
NSAID 120
nucleus 207

occludin 352, 411
odontogenesis 296
oligodendrocyte 233

oligosaccharide 310
opacity protein 314
organogenesis 224
ossification 227
osteoclast 227
osteopontin 224
osteosarcoma 84
outside-in
– activation 454
– signalling 422, 447, 449

p0071 (plakophillin 4) 149
p120
– catenin 105, 126
– ctn 30, 33
p130Cas 198
p150,95 (CD11c/CD18) 424
p16 119
P2X$_1$ receptor 451
P2Y$_1$ 452, 460
P2Y$_{12}$ 452, 460
– receptor 459
p38
– kinase 202
– MAP kinase 361
p53 75
PAK 203
parachute (pac) 57
paraneoplastic pemphigus (PNP) 156, 160
paraxial protocadherin (PAPC) 60
pathogen 219
paxillin 198, 302
P-cadherin 7, 84
pemphigus
– foliaceus (PF) 156, 159
– vulgaris (PV) 156, 159
percutaneous
– coronary intervention 459
– transluminal coronary angioplasty (PTCA) 461
peripheral nerve 230
periplakin 150, 154, 160, 254, 261
perlecan 219
pervanadate 302
PEST motif 26
PGH2 458
phorbol ester 452
phosphoinositide-3-OH-kinase 249, 360, 452

phospholipase (PL)
– C 452
– Cβ 452
– Cγ2 447
phosphorylation 32, 34
– of the desmoplakin 175
– of the plakoglobin 174
PI 3-kinase 444
pial surface 232
pinin 155
placentation 223
plakin 44, 46, 150, 253
– domain 152
– repeat domain (PRD) 152
plakoglobin 28, 35, 36, 74, 82, 124, 146, 162, 168, 173, 259
– associate 172
plakophilin 30, 46, 148
– 1 259
planar polarity 57
platelet 437
– activation 437, 442
– adhesion 437, 439, 449, 456
– aggregation 437, 439, 449, 450, 452–454, 456
– platelet/endothelial CAM (PECAM)-1 (CD31) 408
– platelet-derived growth factor 233
– platelet–platelet-interaction 449
– plug 439
PLCβ 209
plectin 150, 153, 160, 254, 255, 262
plexin/semaphorin/integrin (PSI) domain 446, 448
podocyte 225
podoplanin 416
poliovirus 347
polycystin-1 155
ponsin 355
postsynapsis 231
– density 346
presynapsis 231
primary
– adhesion 409
– hemostasis 437, 439, 441
proliferation 217
proprioceptive 382
prostacyclin (PGI2) 453
protease 219
protease-activated receptor (PAR) 452

protein kinase (PK) 72, 79
– A   199, 442
– C   452
– G   453
– P
– – 1   148
– – 2   148
– – 3   149
protein S   454
proteolysis   115
P-selectin   441, 455
– glycoprotein ligand (PSGL)-1   419, 455
pseudogene   285
pseudorabies virus   364
PSI   446
PTB domain   200

Rac   126, 202, 360, 454
radial glial   232
Raf-1   202
Ras   201
R-cadherin   61
receptor tyrosine   205
–   kinase Axl   454
redifferentiation   118
reelin   232
repeat domain (EC1-EC4)   142
repeated unit domain (RUD)   143
resolution phase   418
retinoblastoma protein   306
RGD sequence motif   449
Rho   202, 360
– A   126
– GDI   203
– guanosine triphosphatase   304
– Rho/rac-family of small GTPase   106
ristocetin   441
Robo (Roundabout)   379, 380, 391
rolling   409, 421
Salmonella typhimurium   313
saratin   457
scaffold   199
Schwann cells   230
secondary
– adhesion   410
– lymph node   413
segregation   59
selectin   410, 4118
sensory
– afferent   381

– axon pathfinding   382
serine/threonine kinase   79
Sertoli cell   357
severe combined immunodeficiency (SCID)   320
Shc   454
shear stress   204
short-range guidance cues   378
sialyl-Lewis X   419
signal transduction   285, 301, 305, 322
signaling   375, 391
silencing   391
skeleton   227
skin   226
– fragility   162
Slit   379, 380, 391
SLP-76   447
small cell carcinoma   306
small GTPase   444
snake venom protein   457
sperm   218
squamous cell carcinoma   306
Src   201, 454
– homology   444
staphylococcal scalded skin syndrome (SSSS)   156, 160
stem-cell compartment   110
stereocilia   227
stress fiber   203
striate palmoplantar keratoderma (SPPK)   161, 163
stroke   440
subendothelial matrix   440, 449
subendothelium   440, 450
support cell   227
surface force apparatus   14
survival   61, 217
Syg-1   386
Syk   444, 447, 454
synapsis   233, 385
– contact   230
– formation   385
– plasticity   387
SynCAM   386
syncytiotrophoblast   293

T cell homing   414
T lymphocytes   311
TAG-1   384
– axonin-1   388